Archaeological Chemistry

ACS SYMPOSIUM SERIES **968**

Archaeological Chemistry

Analytical Techniques and Archaeological Interpretation

Michael D. Glascock, Editor
University of Missouri

Robert J. Speakman, Editor
Smithsonian Institution

Rachel S. Popelka-Filcoff, Editor
University of Missouri–Columbia

**Sponsored by the
ACS Divisions of Nuclear Chemistry and Technology
and the History of Chemistry**

American Chemical Society, Washington, DC

Library of Congress Cataloging-in-Publication Data

Archaeological chemistry : analytical techniques and archaeological interpretation / Michael D. Glascock, editor , Robert J. Speakman, editor, Rachel S. Popelka-Filcoff, editor.

p. cm.—(ACS symposium series ; 968)

"Sponsored by the ACS Divisions of Nuclear Chemistry and Technology and the History of Chemistry."

"The symposium upon which this book is based was held at the 231st National Meeting of the American Chemical Society, March 26–27, 2006 in Atlanta, Georgia".—Pref.

Includes bibliographical references and index.

978–0–8412–7413–6 (alk. paper)

1. Archaeological chemistry—Congresses. 2. Archaeology—Methodology—Congresses. III. Antiqities—Analysis—Congresses.

I. Glascock, Michael D. II. Speakman, Robert J., 1970- III. Popelka-Filcoff, Rachel S., 1977- IV. American Chemical Society. Meeting (231st : 2006 : Atlanta, Ga.)

CC79.C5A726 2007
930.1—dc22

2007060680

The paper used in this publication meets the minimum requirements of American National Standard for Information Sciences—Permanence of Paper for Printed Library Materials, ANSI Z39.48–1984.

Distributed by Oxford University Press

PRINTED IN THE UNITED STATES OF AMERICA

Foreword

The ACS Symposium Series was first published in 1974 to provide a mechanism for publishing symposia quickly in book form. The purpose of the series is to publish timely, comprehensive books developed from ACS sponsored symposia based on current scientific research. Occasionally, books are developed from symposia sponsored by other organizations when the topic is of keen interest to the chemistry audience.

Before agreeing to publish a book, the proposed table of contents is reviewed for appropriate and comprehensive coverage and for interest to the audience. Some papers may be excluded to better focus the book; others may be added to provide comprehensiveness. When appropriate, overview or introductory chapters are added. Drafts of chapters are peer-reviewed prior to final acceptance or rejection, and manuscripts are prepared in camera-ready format.

As a rule, only original research papers and original review papers are included in the volumes. Verbatim reproductions of previously published papers are not accepted.

ACS Books Department

Contents

Preface..xi

1. **Expanding the Range of Electron Spin Resonance Dating....................1**
   Anne R. Skinner, Bonnie A. B. Blackwell, Maysun M. Hasan, and Joel I. B. Blickstein

2. **Toward the Classification of Colorants in Archaeological Textiles of Eastern North America..15**
   Christel M. Baldia and Kathryn A. Jakes

3. **Infrared Examination of Fiber and Particulate Residues from Archaeological Textiles..44**
   Kathryn A. Jakes, Christel M. Baldia, and Amanda J. Thompson

4. **Extraction and Analysis of DNA from Archaeological Specimens......78**
   Brian M. Kemp, Cara Monroe, and David Glenn Smith

5. **Using Archaeological Chemistry to Investigate the Geographic Origins of Trophy Heads in the Central Andes: Strontium Isotope Analysis at the Wari Site of Conchopata.................................99**
   Kelly J. Knudson and Tiffiny A. Tung

6. **Interpreting Stable Isotopic Analyses: Case Studies on Sardinian Prehistory..114**
   Luca Lai, Robert H. Tykot, Jessica F. Beckett, Rosalba Floris, Ornella Fonzo, Elena Usai, Maria Rosaria Manunza, Ethan Goddard, and David Hollander

7. **Bitumen in Neolithic Iran: Biomolecular and Isotopic Evidence.......137**
   Michael W. Gregg, Rhea Brettell, and Benjamin Stern

8. **Surface Analysis of a Black Deposit from Little Lost River Cave, Idaho....152**
Reshmi Perumplavil and Ruth Ann Armitage

9. **Shell Bead Sourcing: A Comparison of Two Techniques on *Olivella biplicata* Shells and Beads from Western North America....167**
Jelmer W. Eerkens, Jeffrey S. Rosenthal, Howard J. Spero, Ryoji Shiraki, and Gregory S. Herbert

10. **Archaeological Soils and Sediments: Application of Microfocus Synchrotron X-ray Scattering, Diffraction, and Fluorescence Analyses in Thin-Section....194**
W. Paul Adderley, Ian A. Simpson, Raymond Barrett, and Timothy J. Wess

11. **Quantitative Modeling of Soil Chemical Data from Inductively Coupled Plasma–Optical Emission Spectroscopy Reveals Evidence for Cooking and Eating in Ancient Mesoamerican Plazas....210**
E. Christian Wells, Claire Novotny, and James R. Hawken

12. **Chemical Composition of Song Dynasty, Chinese, Copper-Based Coins via Energy Dispersive X-ray Fluorescence....231**
Jessica Misner, Jeffe Boats, and Mark A. Benvenuto

13. **Elemental Compositions of Herodian Prutah, Copper Coins—of the Biblical "Widow's Mites" Series—via Energy Dispersive X-ray Fluorescence....246**
Meghann Mouyianis, Jeffe Boats, and Mark A. Benvenuto

14. **Chemical Composition of the Isfiya and Qumran Coin Hoards....258**
Michael Notis, Aaron Shugar, Danny Herman, and Donald T. Ariel

15. **Selected Applications of Laser Ablation Inductively Coupled Plasma–Mass Spectrometry to Archaeological Research....275**
Robert J. Speakman, Michael D. Glascock, Robert H. Tykot, Christophe Descantes, Jennifer J. Thatcher, Craig E. Skinner, and Kyra M. Lienhop

16. **Evaluating the Precision Requirements for Isotope Ratio Determination of Archaeological Materials Using Laser Ablation–Time-of-Flight–Inductively Coupled Plasma–Mass Spectrometry Increasing Ratio Precision**.......297
John V. Dudgeon, Hector Neff, Andrew "Flynn" Saint, and William Balsanek

17. **Lead Isotope Analysis of Roman Carthage Curse Tablets**.......311
Sheldon Skaggs

18. **Laser Ablation–Inductively Coupled Plasma–Mass Spectrometry Analysis of Ancient Copper Alloy Artifacts**.......336
Laure Dussubieux

19. **Laser Ablation–Inductively Coupled Plasma–Mass Spectrometry Analysis Applied to the Characterization of Peruvian Wari Ceramics**.......349
Laure Dussubieux, Mark Golitko, Patrick Ryan Williams, and Robert J. Speakman

20. **Characterization of Building Materials from the Brick Chapel at Historic St. Mary's City**.......364
Ruth Ann Armitage, Leah Minc, Silas Hurry, and Melissa Doolin

21. **Characterization of 15th–16th Century Majolica Pottery Found on the Canary Islands**.......376
Javier Garcia Iñañez, Jaume Buxeda i Garrigós, Robert J. Speakman, Michael D. Glascock, and Elena Sosa Suárez

22. **Intraregional Provenancing of Philistine Pottery from Israel**.......399
David Ben-Shlomo

23. **The Technology of Mesopotamian Ceramic Glazes**.......423
David V. Hill, Robert J. Speakman, Michael D. Glascock, and Hector Neff

24. **Analysis of Historic Latter-day Saint Pottery Glazes by Laser Ablation–Inductively Coupled Plasma–Mass Spectrometry**.......447
Nicole C. Little, Timothy J. Scarlett, Robert J. Speakman, Michael D. Glascock, and Christopher W. Merritt

25. **Fingerprinting Specular Hematite from Mines in Botswana, Southern Africa** 460
Adam V. Kiehn, George A. Brook, Michael D. Glascock, Jonathan Z. Dake, Lawrence H. Robbins, Alec C. Campbell, and Michael L. Murphy

26. **Instrumental Neutron Activation Analysis of Ochre Artifacts from Jiskairumoko, Peru** 480
Rachel S. Popelka-Filcoff, Nathan Craig, Michael D. Glascock, J. David Robertson, Mark Aldenderfer, and Robert J. Speakman

27. **Feasibility of Field-Portable XRF to Identify Obsidian Sources in Central Petén, Guatemala** 506
Leslie G. Cecil, Matthew D. Moriarty, Robert J. Speakman, and Michael D. Glascock

28. **Sources of Archaeological Obsidian in Peru: Descriptions and Geochemistry** 522
Michael D. Glascock, Robert J. Speakman, and Richard L. Burger

Indexes

Author Index 555

Subject Index 557

Preface

Archaeological chemistry is an interdisciplinary field of archaeological research in which techniques and approaches from the chemical, biological, physical, geological, and statistical sciences are employed to extract more information from the material record of past human activity. The range of research conducted is so broad that any attempt to describe the field in a comprehensive summary is difficult. Major developments within archaeological chemistry have relied on the development and improvement of new techniques and procedures in conjunction with their application to significant archaeological problems.

The symposium upon which this book is based was held at the 231st National Meeting of the American Chemical Society, March 26–27, 2006 in Atlanta, Georgia. The chapters included in this volume were selected from the oral presentations at the symposium to demonstrate the interdisciplinary nature of archaeological chemistry. The editors felt that it was important for the papers in this volume to describe an archaeological problem, to explain the analytical techniques and procedures used to investigate the problem, and most importantly to present an interpretation of their results for appreciation by archaeologists, chemists, and others.

Michael D. Glascock
Research Reactor Center
University of Missouri
Columbia, MO 65211

Robert J. Speakman
Smithsonian Institution
Washington, DC 20560

Rachel S. Popelka-Filcoff
National Institute of Standards and Technology
Gaithersburg, MD 20899

Chapter 1

Expanding the Range of Electron Spin Resonance Dating

Anne R. Skinner[1], Bonnie A. B. Blackwell[1,2], Maysun M. Hasan[2], and Joel I. B. Blickstein[2]

[1]Department of Chemistry, Williams College, Williamstown, MA 01267
[2]RFK Science Institute, 75–40 Parsons Boulevard, Flushing, NY 11366

Archaeologists often turn to physical science for help in determining ages of sites and/or materials. Electron spin resonance (ESR) has proven a valuable tool for dating eutherian teeth from archaeological, paleoanthropological, and paleontological sites ranging in age from 10 ka to 5 Ma. Because tooth enamel is the body's hardest tissue, teeth are frequent components of fossil assemblages. The ESR signal stability in enamel exceeds 100 Ma. In eutherian teeth, the ESR signal does not depend on species. We have tested metatherian, reptilian and chondrichthyian teeth to see whether this method can be extended to additional environments and time periods. In diprotodontid marsupial teeth, the ESR signal appears identical to that in eutherian enamel. The crocodylian teeth tested here contained significant iron concentrations, which interfered with the ESR signal. Their fossilization in fluvial sediment may have caused the iron contamination. Reliability in other reptile teeth remains to be tested. Although shark teeth contain enameloid rather than enamel, the ESR signal morphology differs only slightly. However, the signal stability is reduced and calculated ages are much younger than expected.

Archaeological and paleontological applications of chemistry predominantly take two forms. One is the direct determination of some archaeologically significant factor, perhaps the age, composition, or provenance of an artifact. The other is a methodological development, either a new technique or an improvement to an existing one, that enables scientists to analyze new objects. Electron spin resonance (ESR) has been very successful in dating teeth as young as a few thousand years and as old as several million years. Ages can, therefore, be cross-calibrated both with ^{14}C and with $^{39}Ar/^{40}Ar$ results. Most archaeological applications have used enamel from teeth of large placental mammals such as bovids and equids, in part because teeth, being the hardest tissues in the body, are often preserved (*1*). There are many environments and time periods where placental mammals are rare or non-existent. To be able to date teeth from a wider range of species would expand the physical and temporal range of this method. Therefore we have examined three other taxa: marsupials, crocodiles and sharks.

Dating methods fall into three main groups. Radiometric methods measure the present radioisotope content of the sample and, by knowing the initial concentration and the decay rate, calculate an age that is largely independent of environment. Those most often part of archaeological and paleoanthropological investigations are ^{14}C and $^{39}Ar/^{40}Ar$. These radiometric methods (like all methods) have limitations. Dating with ^{14}C requires some calibration because the initial concentration has varied; $^{39}Ar/^{40}Ar$ is useful only for volcanic rocks. Other chronometric methods including, amino acid racemization, uranium series and ESR require knowledge of the environment for accurate results, but can yield a numerical age. Relative methods such as stratigraphy simply enable one to tell which of two objects is older.

ESR Dating

As one of the 'trapped charge' methods, ESR dates materials by measuring damage caused through environmental radiation, rather than by measuring radioisotope content. The general principles of the method have been covered in numerous reviews (*2–4*).

ESR measures the number of unpaired electrons created when radiation cleaves a stable chemical bond (or knocks an electron out of a lone pair). To be used in dating, the ESR signal must grow in a reproducible fashion with radiation dose, whether natural or artificial. Most trapped charge methods, such as TL and OSL, as well as ESR, study a signal response as a function of this dose, creating a growth curve (Figure 1). TL, however, cannot study teeth, because the heating needed to produce luminescence chars the teeth. Saturation limits ESR dating of teeth; samples older than 5–8 million are saturated or so close to saturation that the difference cannot be detected with current technology.

In enamel the ESR signal arises from hydroxyapatite (HAP). The shape of the signal (Figure 2) reflects the existence of both $g_{\perp}$ and $g_{\parallel}$ components. Teeth have several advantages as dating samples. The signal is extremely stable, with a mean lifetime exceeding 10^{10} years (*5*). The signal is also unaffected by light, which means samples can be collected and prepared in ambient light. Museum samples can be dated without fear that the sample has been affected by storage. Another advantage is that measuring the signal does not destroy the signal, allowing samples to be remeasured repeatedly, perhaps using different measurement parameters.

For ESR in teeth, enamel is separated from dentine and then powdered. Aliquots are irradiated to obtain a growth curve. The accumulated dose (AD) is then converted to an age by factoring in the external and internal dose rates. Standard parameters, used for all dates in all tables here, assume sedimentary water concentrations of 10 ± 5%, no radon loss, an initial U activity ratio, $(^{234}U/^{238}U)_0$=1.20 ±0.20, and an α-efficiency factor, k_α = 0.15 ± 0.02. While the external sample dose and our artificial doses are supplied by γ-radiation, uranium within the enamel delivers primarily α particles. All ESR dating, therefore, requires knowing the relative efficiency of these two types in inducing the measured signal. ESR ages generally show a precision of about ± 5%.

There are two caveats with this method. First, the dose to the sample derives in part from the environment around the sample during the entire deposition period. Numerous environmental factors must be measured or modeled in order to obtain a result that is both accurate and precise. With increasing time, our ability to do so becomes more problematic. This problem is, of course, common to all materials and all trapped charge methods. Second, teeth (and some other materials, e.g., bones and shells) are open systems with respect to uranium uptake, which means the internal dose rate changes with time. Traditionally researchers have simply assumed one of three models: early uptake (EU; all uranium enters the tooth shortly after deposition), linear uptake (LU; continuous uptake), and recent uptake (RU; most of the modern concentration of uranium entered the tooth, and in particular the enamel, shortly before excavation). It is now possible to use coupled $^{230}Th/^{234}U$ dating to obtain a more accurate model. While U-series ages such as those from $^{230}Th/^{234}U$ disequilibrium are even more problematic for open systems, by iterating the two results a reliable age can be obtained (*6*). Where the sample age exceeds the $^{230}Th/^{234}U$ limit (~500 ka), or if the tooth does not contain enough uranium for a precise $^{230}Th/^{234}U$ age determination, modeled ages are still reported.

Marsupial Teeth

The Lake Eyre Basin in Australia contains the world's largest ephemeral lake. During the Quaternary the climate fluctuated between relatively arid and

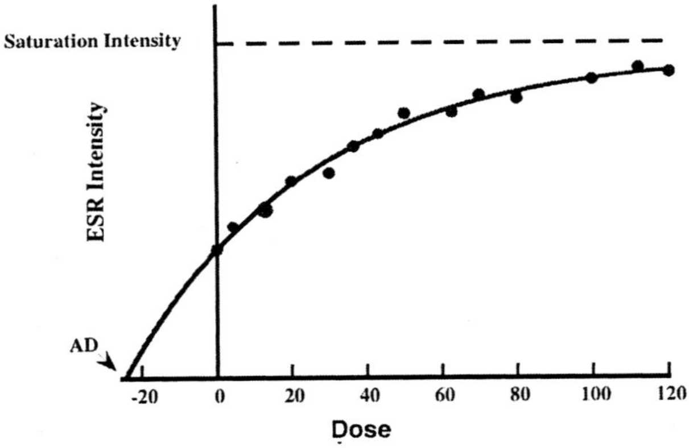

Figure 1. Typical ESR growth curve. Aliquots of enamel or other sample material are irradiated artificially and the increase in intensity measured. A minimum of 10–12 points is required to provide statistical accuracy. The extrapolation of the growth curve to the x-axis gives the accumulated dose, AD.

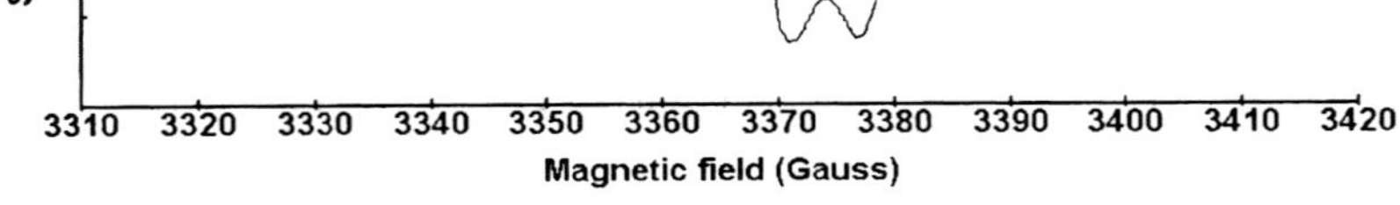

Figure 2. Typical HAP ESR spectrum

relatively humid episodes. The wet phases correlate to Northern Hemisphere glaciations. Faunal correlations appeared to place the Katipiri Formation, the younger of the two fluvially deposited Quaternary formations in the basin, in Oxygen Isotope Stage (OIS) 5. Faunal correlations at best provide only relative ages, especially when temporally mixed assemblages are detected. While fauna adapted to arid conditions can coexist with those adapted to wetter environments, any fossil assemblage in a fluvial deposition could be reworked from older deposits and, therefore, not provide accurate correlations.

Studying the ages of these fauna would clarify temporal variability in the Katipiri sands and help understand paleoenvironments during this period. In addition, at some point during the Quaternary megafauna including *Diprotodon*, *Genyornis* and other giant marsupials became extinct. Miller *et al.* (*7*) suggested that this extinction might be related to the arrival of hominins. Alternatively extinction might have resulted from environmental changes. Chronometric dates for the fauna would provide some answers to this problem.

Several diprotodontid teeth from the Lake Eyre Basin were prepared by standard methods. While collected from seven different locations, all were expected to be the same age. Diprotodontid canines and incisors have a single enamel layer around a dentine core. The signal from diprotodontid enamel does not differ significantly from that in eutherian mammals. The results, however, showed that deposition in the basin was quite complex (Table I). KT5, with an LU age of 46.3 ± 2.2 ka, is probably *in situ*. The wide age range for the other teeth clearly indicates they have been reworked from older formations. Therefore, KT5 provides a maximum age for the Katipiri Formation deposition and dates it to OIS 3 rather than OIS 5. It also shows that diprotodonids, at least, were not extinct before that time. Hominins are believed to have arrived in Australia between 50 ka and 60 ka (*8*), and at least some megafauna survived for more than 15 ka after this appearance in southeastern Australia (*9*). ESR dating shows their survival in southwestern Australia as well. *Diprotodon*, about the size of a bear, should have been an excellent food source. Although hominins could well have played a role in the extinction, the pace of their influence may have been slow in comparison to similar events in other parts of the world.

Crocodylian Teeth

Crocodylians, which include crocodiles, gavials, alligators, and their immediate relatives, are poikilothermic quadrupedal reptiles that first evolved in the late Triassic. Today, crocodylians inhabit the subtropical to tropical zones on all continents. Their wide geographic distribution and evolutionary stability make them valuable dating samples (*10*).

Three crocodylian teeth plus one elephant tooth from the Siwalik Group of India were dated by standard ESR methods. From oldest to youngest, the Upper

Table I. Mean ESR Ages for Teeth from the Katipiri Formation

Sample (subsamples)		*Accumulated Dose, AD (Gray)*	*Weighted Mean Standard ESR Ages* EU (ka)	LU (ka)	RU (ka)
KT5		36.61	41.6	46.3	48.8
(4)	±	0.76	1.8	2.2	2.5
KT4		221.1	51.7	85.2	173.6
(4)	±	6.7	2.9	4.5	8.8
KT7		529.1	125.0	210.8	376.7
(6)	±	8.1	4.4	7.1	12.9
KT6		551.4	155.6	254.6	597.0
(3)	±	12.1	6.7	10.6	30.2
KT3		1716.	320.7	562.8	1411.
(4)	±	26.	15.6	25.0	58.
KT8		1793.	369.1	636.5	1516.
(2)	±	117.	41.4	66.7	147.
KT1		2810.	559.6	973.3	2439.
(4)	±	68.	29.2	46.7	113.
KT2		2280.	767.3	1196.	2337.
(4)	±	42.	33.1	41.	90.
KT10		4419.	876.2	1316.	3601.
(1)	±	216.	77.6	115.	282.
KT9		4943.	1671.	2657.	5096.
(2)	±	318.	137.	211.	437.

Abbreviations:
EU = assuming early U uptake
LU = assuming linear (continuous) U uptake
RU = assuming recent U uptake

Siwalik Group contains three main formations, Tatrot, Pinjore, and the Boulder Conglomerate. The teeth in this study came from the Pinjore Formation. From its magnetostratigraphy, the Pinjore Formation ranges from 2.48 to 0.63 Ma (*11*). Based on their faunal associations, the teeth probably date from 1.5 to 2.5 Ma, but the lower units in the formation have not been dated chronometrically.

The Siwalik Group has long been famous for its abundant vertebrate fossils, among which are some early hominid ancestors. Its primates include *Ramapithecus* and *Sivapithecus* from the older beds, and younger specimens, *cf. Homo erectus*, from the Upper Siwalik Group, including the Pinjore Formation (*12*). Dates would therefore help establish the arrival time of hominins in the Indian subcontinent.

Preparing these teeth was extremely difficult. Crocodylian teeth are small, and dark mineralization made it difficult to detect the boundary between dentine and enamel in order to separate the two tissues. To ensure an accurate ESR age, all dentine must be removed from the enamel, not because dentine affects the ESR signal, but because the uranium concentration in dentine is often as much as an order of magnitude greater than in enamel. Contamination of enamel by dentine, therefore, affects the calculation of the internal dose. Once prepared, the ESR spectra showed that the fossil enamel contained significant amounts of Fe^{3+}, interfering with–almost concealing in some cases–the HAP peak. The presence of Fe^{3+} causes a sloping baseline, with the slope directly proportional to the iron(III) concentration (Figure 3A). Dating requires knowing the intensity of the peak, and unfortunately could not be calculated for most crocodylian subsamples.

Signal subtraction demonstrates that an HAP dating peak exists (Figure 3B and 3C). The signal shape looks different, reflecting a decrease in resolution due to the subtraction process. Nonetheless, the signal grows with radiation in a manner similar to other enamel samples (Figure 4).

Despite these problems, growth curves could be constructed for some samples, and ages calculated (Table II). The ages of these teeth appear to be too young, unless the RU, or recent uptake, model is invoked. This model has been shown to be more probable for teeth >1Ma in age (*13*). This appears counterintuitive, but is related to changes in enamel crystallinity and porosity that allow more uranium to penetrate an older tooth than a younger one. Invoking the RU model also yields an age for the elephant tooth more in agreement with expectations (Table II). Even so, the crocodile teeth appear younger than the elephant tooth, perhaps because the Fe^{3+} interference has led to inaccurate accumulated dose estimates or possibly due to reworking of the crocodile teeth. Determining signal stability awaits the discovery of some crocodile teeth without contamination. A single aliquot was annealed for 9 hours at 200°C without significantly reducing the apparent peak intensity, suggesting that the signal stability resembles that in mammalian enamel.

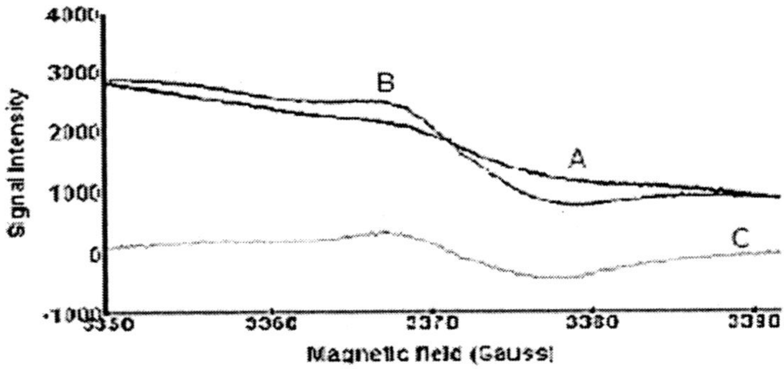

Figure 3. Crocodylian ESR spectra. A: Spectrum of natural sample. B: Spectrum after 30 kGy artificial irradiation. C: Result of subtracting A from B. Note that the subtracted spectrum has only one minimum, compared to two in Figure 2. The overall peak width is the same in both this spectrum and the typical mammalian example.

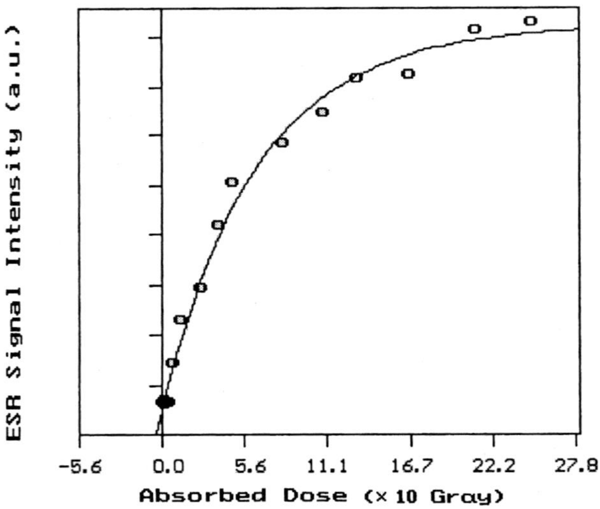

Figure 4. Growth curve of subtracted crocodile spectra. The shape is normal for ESR growth curves. That the x-axis intercept (AD) is not zero is due to imprecision in reading the spectra.

There is another potential explanation for the young ages. Oduwole and Sales (*14*) previously observed in bones that Fe^{3+} interferes with the HAP signal. They proposed several possible explanations, ranging from the possibility that the dating signal, always weak in bones, simply could not be seen, to the more intriguing possibility that Fe^{3+} reacts with electrons that otherwise would have been trapped in the HAP, reducing Fe^{3+} to Fe^{2+} and also, of course, reducing the

Table II. Preliminary ESR Ages for Teeth from Devni-Khadri, India.

Sample (subsamples)		*Accumulated Dose, AD (Gray)*	*Mean Standard ESR Ages*[a] *EU (ka)*	*LU (ka)*	*RU*[b] *(Ma)*	*RU*[c] *(Ma)*
a. Crocodylian teeth						
FT46		5098.	129.	237.	0.79	1.30
(2)	±	334.	10.	18.	0.06	0.09
FT47		6052.	171.	305.	0.85	1.17
(2)	±	438.	15.	26.	0.07	0.09
FT48		4394.	173.	307.	0.80	1.06
(3)	±	145.	9.	16.	0.03	0.04
Crocodylian mean			162.	290.	0.81	1.13
			6.	11.	0.03	0.04
			3.8%	3.8.%	3.4%	3.5%
b. Elephantid tooth						
FT38		3693.	336.	639.	2.59	-
(8)	±	81.	15.	27.	0.09	-
		2.2%	4.4%	4.3%	3.4%	-

[a] Abbreviations:
EU = assuming early U uptake
LU =assuming linear (continuous) U uptake
RU =assuming recent U uptake

[b] Calculated using U uptake parameter, $p = 10$.

[c] Calculated using U uptake parameter, $p = 20$

dating peak intensity. The process of creating the growth curve for one subsample demonstrated some direct evidence for the latter. The slope of the Fe^{3+} baseline decreased exponentially with increasing applied radiation, suggesting that the Fe^{3+} concentration was decreasing (Figure 5). Not all samples in this study showed this phenomenon, however. The explanation may well lie in the same factor that generally accounts for the signal stability in enamel – the large crystal size. Radicals formed in the interior of such a crystal would normally be sheltered from destruction by reaction with water, or oxygen. If, in these teeth, the Fe^{3+} was incorporated into the enamel at the time of crystallization, due to iron in ground water, it could easily react with other radicals formed by radiation. Other samples, where the Fe^{3+} was perhaps adsorbed on the surface of the crystal, would not show this effect, explaining why it was not universally observed. Studies presently underway into the uptake of Fe^{3+} into crocodylian and other teeth may differentiate these possibilities or even suggest other explanations.

Shark Teeth

Clearly environments hospitable to sharks are not likely to be occupied by humans. Dating shark teeth will thus not provide information on hominins directly, but when combined with other studies the dates can reveal paleoenvironmental information, which could include factors affecting hominid behavior. Shark teeth are sometimes of archaeological interest. People have been known to use shark teeth as decorative items. In principle, if the teeth were mined from a deposit rather than collected from the shore, the date of the teeth might indicate provenance, and hence elucidate such evidence of human mobility as trade patterns. This experiment, however, simply assesses whether shark teeth might yield a signal and, therefore, a reasonable age. As with the marsupials, in looking at a population the youngest tooth would provide a maximum age for the formation, with older samples indicating the extent of reworking. Given that shark teeth are easy to prepare, analyzing a population would not be excessively challenging. The shark teeth in this study came from a museum collection and were catalogued as Pleistocene or Pliocene. The structure of these teeth differs from the other taxa in this chapter in that the material around the dentine core is enameloid, not enamel.

The ESR signal in shark teeth (Figure 6) also differs from that in mammalian enamel. The minor peaks in Figure 6 can be attributed to organic matter. Mammalian enamel contains less than 2% organic material; dentine contains around 20% and its ESR spectrum shows some of the same organic structure peaks. The HAP signal in Figure 6 also exhibits different contributions of $g_{\parallel}$ and $g_{\perp}$ components when compared to Figure 2.

The calculated ages for shark teeth are younger than predicted from their provenance (Table III). The signal might have faded due to reduced signal stability, suggested by annealing experiments. Signal stability is expected to

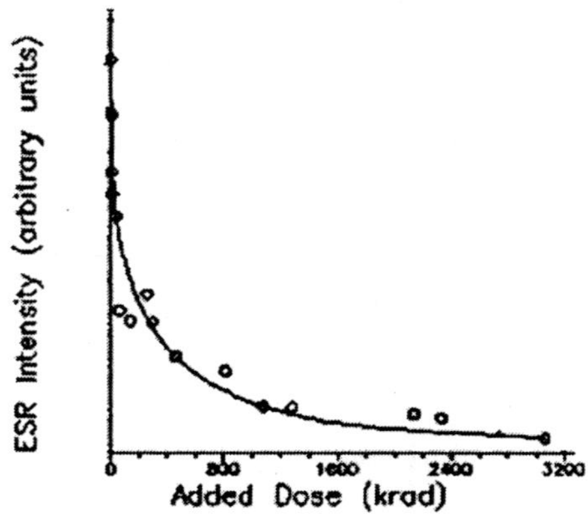

Figure 5. Effect of artificial irradiation on ESR intensity for FT48en3. The intensity appears to decrease exponentially with added dose.

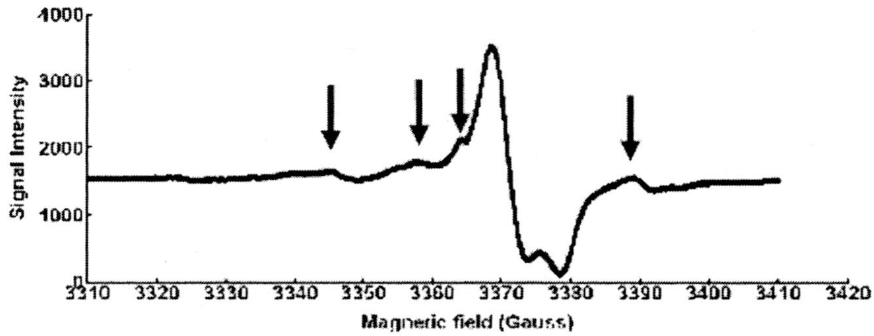

Figure 6. ESR Spectrum of shark enameloid. Arrows indicate position of organic radical peaks.

Table III. Mean ESR Ages for Shark Teeth

Sample (subsamples)		*Accumulated Dose, AD (Gray)*	*Weighted Mean Standard ESR Ages*[a] *EU (ka)*	*LU (ka)*	*RU (Ma)*
RST1A		2980.9	265.2	435.1	1.7
(2)	±	314.2	25.1	47.2	0.08
		10.5%	9.5%	11.4%	4.96%
RST3		1661.3	277.4	506.4	1.61
(2)	±	129.1	27.8	51.0	0.17
		7.8%	10.0%	10.1%	10.28%

[a]Assuming: $\kappa_\alpha = 0.15$ (as is true for enamel), and enameloid density is the same as that of enamel

decrease with crystal size. Both enameloid and dentine have much smaller HAP crystals than does enamel. SEM photos illustrate clearly the order in enamel compared to the disorder and small crystal size in enameloid (Figures 7 and 8). Alternatively, radiation may affect enameloid and enamel in different ways. Consider, specifically, the α-efficiency factor, κ_α. As noted earlier, in enamel it has been measured as 0.15 ± 0.02. Experiments in our laboratory suggest this factor is substantially less in dentine, although a precise value has not yet been determined. If the same were true for enameloid, the calculated ages would be considerably greater.

Conclusions

Methodological developments are essential to improving the ability of physical science to solve archaeological problems. In this case, we have moved ESR dating forward slightly. The marsupial results open up archaeological possibilities on the Australian continent, for example dating aboriginal settlements beyond the ^{14}C limit. The Fe^{3+} in crocodylian teeth requires further study. Modern crocodylians do not have this impurity. Investigating Fe^{3+} scavenging in these teeth can illuminate the mechanism of radical formation and decay in teeth. Also, of course, clearly not all crocodylian teeth will be contaminated with iron, so their dating potential merely awaits discovery of

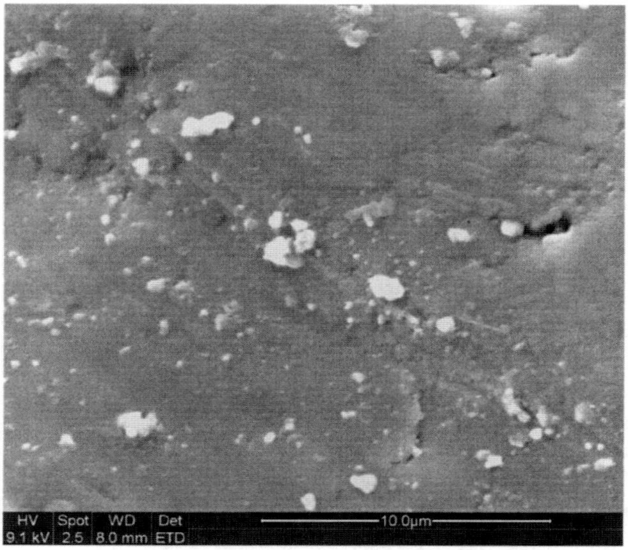

Figure 7. SEM micrograph of mammalian enamel. Note smooth surface.

Figure 8. SEM micrograph of enameloid at the same scale. Note small crystals and random orientations.

appropriate samples. While shark teeth are not suitable dating material, at least with current technology, the SEM results reinforce other observations, such as the weak and unstable signals in dentine and bone, both of which, like enameloid, contain only very small HAP crystals.

References

1. Falguères, C. *Quatern. Sci. Rev.* **2003**, *22*, 1345–1351.
2. Blackwell, B. A. B. In *Tracking Environmental Change Using Lake Sediments: Basin Analysis, Coring, and Chronological Techniques*; Last, W. M.; Smol, J. P., Eds.; Kleuwer: Dordecht, 2001; Vol. 1, pp 283–368.
3. Grün, R. *Quatern. Int.* **1989**, *1*, 65–109.
4. Rink, W. J. *Radiat. Meas.* **1997**, *27*, 975–1025.
5. Skinner, A. R.; Blackwell, B. A. B.; Chasteen, D. E.; Shao, J. M.; Min, S. S. *Appl. Rad. Isot.* **2000**, *52*, 1337–1344.
6. Grün, R. *Ancient TL* **2000**, *18*, 1–4.
7. Miller, G. H.; Magee, J. W.; Johnson, B. J.; Fogel, M. L.; Spooner, N. A.; McCulloch, M. T.; Ayliffe, L.K. *Science (Washington, DC)* **1999**, *283*, 205–208.
8. Bowler, J. M.; Johnston, H.; Olley, J. M.; Prescott, J. R.; Roberts, R. G.; Shawcross, W.; Spooner, N. A. *Nature (London)* **2003**, *421*, 837–840.
9. Trueman, C. N. G.; Field, J. H.; Dortch, J.; Charles, B.; Wroe, S. *Proc. Natl. Acad. Sci.* **2005**, *102*, 8381–8385.
10. Naish, D. *Geol. Today* **2001**, *17*, 71–77.
11. Nanda, A. C. *J. Asian Earth Sci.* **2002**, *21*, 47–58.
12. Patnaik, R. *pers. comm.,* **2005**.
13. Skinner, A. R.; Chasteen, D. E.; Shao, J. M.; Goodfriend, G. A; Blackwell, B. A. B. *Quatern. Int.* **2005**, *135*, 13–20.
14. Oduwole, A. D.; Sales, K. D. *Nucl. Tracks* **1991**, *13*, 213–221.

Chapter 2

Toward the Classification of Colorants in Archaeological Textiles of Eastern North America

Christel M. Baldia[1] and Kathryn A. Jakes[2]

[1]Department of Human Development and Environmental Studies, Indiana University of Pennsylvania, Indiana, PA 15701
[2]Department of Consumer Sciences, The Ohio State University, Columbus, OH 43210

Due to degradation in archaeological contexts of eastern North America, the few textiles that have been recovered are discolored brown or black; some are green due to mineralization. The original coloration goes unrecognized. A protocol was developed to classify colorants used in these materials that is minimally destructive, and includes forensic photography, optical microscopy of strategically sampled fibers, scanning electron microscopy and energy dispersive spectrometry. Because ultraviolet reflected/fluorescent and infrared photography capture differences in chemical composition that cannot necessarily be seen by the unaided eye, purposive sampling is facilitated. Observed under the optical microscope, thoroughly saturated color in the fibers is indicative of dyeing in comparison to surface deposited pigment. Elemental composition of fibers and surface particulate can be correlated to potential mineral or organic colorant sources, as well as reflects fiber mineralization.

Although many ethnographic accounts *(1, 2)* and the observations of early travelers (*3*) describe color as having been used abundantly, textiles recovered from eastern North American sites do not necessarily indicate colorant use. While a few archaeological textiles show coloration, including some fragments from the Hopewellian Seip Mound site in Ohio, in most burial contexts textiles typically degrade leaving them brown or black in appearance. Even though these latter textiles do not appear to be colored by dyes or pigments, it cannot be assumed that they were not decorated with some color in the past.

North American archaeological sites reflect the use of colorants, exemplified by the abundance of the inorganic mineral pigment red ochre in domestic and mortuary contexts in sites from the Clovis through the Mississippian periods (10000 B.P.–1400 A.D.). Typically the presence of these accumulations of pigment minerals is interpreted as material used in ceramics, or for body painting. Color on textiles is not usually considered.

Coloration of fabrics can be achieved by the use of a variety of organic or inorganic materials, which require significantly different techniques of application, but also leaving specific chemical signatures behind. This paper describes a protocol that was developed as a starting point to classify colorants used to decorate textiles in eastern North America. Prototypes that simulated expected properties of colored textiles from the Hopewell period were prepared and evaluated using forensic photography, optical microscopy (OM), scanning electron microscopy (SEM) and energy dispersive spectrometry (EDS), inductively coupled plasma-mass spectrometry (ICP-MS) and ICP-optical emission spectroscopy (ICP-OES). After the protocol was tested on the prototypes, it was then applied to artifacts from Seip Mound. The sequence of methods was designed to glean the most information from the textile prior to sampling, to facilitate selective sampling as informed by the visible evidence, and to maximize data with the least amount of sampled materials.

Literature Review

Eastern North American Textile Research with Respect to Coloration

The Hopewell Mound Group, located 17 miles southwest of Chillicothe, Ohio, yielded the largest number of textiles recovered from any Hopewellian burial mound in Ohio, comprising about 50% of the total textiles recovered from 10 sites (*4*). Mills (*5)* reported finding great quantities of charred cloth," ...and "... a portion of fabric or burial shroud bearing a design in color..." (6). The fabric structure of Hopewell textiles were described and "...the designs are not the result of weaving colored threads into the fabric but were affected by means of staining or dyeing with mineral pigments, possibly by the use of stamps...

(7)". Willoughby also recognized that exposure and the handling of the specimen "practically destroyed both design and coloring… [that had been] a dark maroon, the designs…clear yellow outlined in black…" *(8)*.

Church *(9, 10)* proposed that the decorations on the "painted" Seip textiles and their associations with the burials were meant to communicate social identities or roles. Song (*11, 12*) found that specific structures were correlated to specific fiber contents. However, the study of the colorants was not pursued.

The use of color by prehistoric peoples was also addressed in Willoughby's discussion of textiles found at the Mississippian temple mound sites of Etowah, Georgia and Spiro, Oklahoma (*8*), which also yielded colored textiles. Kuttruff proposed that the complexity of these textiles including the colored patterning and design motifs is indicative of sophisticated artisanship and status differences that were expressed through clothing (*13*), but her study did not involve the identification of colorant types. Saltzman (*14*) identified madder as the dye plant that had been used to color some of the Spiro textiles.

Martoglio et al. (*15, 16*) and Jakes et al. (*17*) used microspectroscopy to find evidence of dyes, but the results were inconclusive due to the small sample size, and the lack of a comparative dye plant collection and associated spectra. Sibley and Jakes (*18*) researched the colored textiles of Etowah burial 57. Since it contained no iron, the iron-containing minerals that are thought of as ochre[1] could not have been the source of the colorant. Furthermore, the researchers found organic bands in infrared spectra of these fibers that revealed not only the proteinaceous nature of the mineralized fibers, but also other bands that were attributed to dye.

Colorants

The study of colorants that can be applied to textiles is inherently complex because it cuts across several disciplines and traditions: geology, botany, dye chemistry, material analysis, textile science, art, history, and prehistory. When dealing with archaeological specimens, the process is even more complicated because materials are likely to have been altered due to their environment during formation, subsequent degradation, and handling throughout excavation and in curation. Furthermore, dyes and pigments dictate the methods by which they can be applied to produce the desired colors. All colorants produce color as a result of their chemical structure, i.e., by the presence of conjugated double bonds and associated functional groups. These systems are called chromophores for their ability to absorb visible light and to reflect some of that energy, which is then perceived by humans as color.

---

[1] The term ochre is problematic. Please see *Known Inorganic Mineral Pigments*

Pigment or Dye, Organic or Inorganic?

Color for dress or body adornment purposes can be achieved by the use of either pigments or dyestuff. Physically, dyes and pigments differ in molecular size, pigments being much larger than dye molecules. Furthermore, natural pigments generally are inorganic, and they adhere to the surface of the fiber aided by a binder or glue. Historically, media such as egg white, water, clays, oils and resins have been used as pigment binders.

By contrast, a dye molecule interacts chemically with the substrate at a molecular level. Chemical bonds or associations are formed, thus best results are achieved when the two components are compatible. If the substrate/dye combination is ideal, a more durable, fast and level color is created. Dyes can be applied at any time during the textile production process, while the pigments generally are applied to a finished textile. Pigments can also rub off more easily, because they are not chemically bound to the substrate.

An organic dye that has entered fibers at a molecular level can form a much larger complex molecule, sometimes called a lake, in the presence of a metal ion and a chelating agent. In this process called mordanting, this complex acts as chelate between dye and fibers creating much more long lasting colors because the now larger molecule is confined to the inside of the fibers. Sometimes this lake is also labeled as pigment even though the source of color is organic.

This label as organic pigment describes the large molecule it forms but it is also potentially confusing, because the colorant is located within the fiber. According to Berns (*19*), the "greatest validity" to differentiate between pigments and dyes is the presence or absence of the binder, which makes the pigment adhere to the fiber surface.

Pigments from inorganic (mineral) sources have been found in many North American archaeological sites, suggesting plentiful use. For instance, Brown (*20*) lists a variety of minerals that were associated with Spiro as follows: ochre as source for red and yellow, glauconite for green, manganese ore for black and purple, and galena and kaolin as a source for gray and white. The Haley site yielded limonite which is a source for yellow and cinnabar which is a source for green (*21*).

However, some of the colors found in archaeological remains such as textiles cannot totally be explained by the use of the inorganic pigments alone. If pigments were used, elemental analyses should disclose the presence of metal ions.

The colors in the Spiro textiles include a rose red, pink, yellow, grey, tan, and brown-black. King and Gardner (*22*) claim that the textiles were colored by some sort of resist dye methodology. Because pigments must have a binder to "paint" or coat fiber surfaces they could not have been the source for these colors if a resist technique was in fact the method of Spiro textile coloration.

Furthermore, the types of yarn structures that Willoughby (*8*) described in the Seip textiles as "vegetal core with red-stained rabbit hair wrapped around it" can only be achieved in coloring the fibers before spinning. Therefore, the prehistoric artisans had to have specific knowledge about dyes and how to apply them.

While the identification of inorganic pigments is relatively straightforward and their sources might be recognized through comparison of trace elemental composition, the organic remains from dyes are much more difficult to detect and interpret in an archaeological assemblage since organic materials degrade readily. Some evidence might be gained by looking for plant evidence of dyeing (e.g., *23, 24*); however, since plants that were probably the most common source for dyes in the past also could have had other uses such as food or medicine, the presence of plant remains in archaeological contexts does not necessarily support their use as dyes. For example, walnuts are food but their hulls yield a black or brown colorant (*25–27*). Sumac and bloodroot are identified as medicine, but the authors also suggested their use as dyestuffs. Ethnobotanists have labeled seed assemblages as evidence of prehistoric diet, but some might also be related to coloration (*28*). These multipurpose plants make interpretation difficult and justify further material analysis of the archaeological textile remains themselves rather than focusing on the correlates.

Known Inorganic Mineral Pigments

The most prominent pigments that were used prehistorically are the iron-oxide pigments, often referred to as "ochre". While the term is used widely, it is also very problematic. In the strictest sense, it only denotes the color it produces, even though it is often used interchangeably with hematite (Fe_2O_3). In fact, the term "ochre" does not describe any elemental or mineralogical composition, which can vary greatly; and in some cases, the ochre-colored material that is described does not contain measurable amounts of iron.

Iron oxides and hydroxides occur in a range of colors and crystal structures. Some of these exist as polymorphs, i.e., the same chemical composition has a different crystal structure and a different name. All but magnetite are anisotropic, and impurities such as manganese, clays, and organics enhance the variety of colors of natural iron-oxide pigments. Glauconite for green, manganese ore for black and purple, galena and kaolin as a source for gray and white are included, as are several others that were known to have been used historically (*31*).

The mineral names, a general elemental composition, and the colors that they produce are summarized in Table I (*29,30*). Minerals may vary greatly in trace element ratios, and these can be used as a fingerprint to identify mineral source locations, which was not considered here.

Table I. Mineral Content of Inorganic Pigments

Mineral	*Color*	*Chemical Composition*	*Possible Traces of*	*Notes*	*Occurrence*
Hematite	dull to bright red	Fe_2O_3	Al, Ti, Cr, V		Great Lakes, southern Appalachia
Magnetite	black with metallic luster	Fe_3O_4	Ti, Mg, Ni, Mn, Al, V, Cr	Occurs in context with limestone, hematite, pyrite & other sulfides	New York, Pennsylvania and Arkansas
Goethite	yellowish to reddish brown	α FeOOH x H_2O	Mn		Great Lakes area
Lepdocrocite	yellow, red, reddish-brown	γ FeOOH x H_2O		polymorph of goethite	Great Lakes area.
Galena	bluish gray	PbS	Ag, Bi, Sb, Hg, Cu		New York, Missouri, Kansas, Oklahoma, Colorado, and Idaho and Arkansas.
Kaolin	white, yellow, green, blue	$AL_2Si_2O_5(OH)_4$		Type of clay	Georgia, South Carolina, North Carolina, Arkansas, New Mexico and New Jersey
Cinnabar	scarlet or dark red	HgS			Arkansas, Texas, Nevada, California

Table I. *Continued.*

Mineral	*Color*	*Chemical Composition*	*Possible Traces of*	*Notes*	*Occurrence*
Glauconite	yellow-green, green or blue-green	(K, Na, Fe^{2+}) 0.33 (Fe^{3+} , Al) 1.67 [(Si, Al) $4O_{10}$]$(OH)_2$		Silicate of iron & potassium, occurs with clay	Dolomite environment of Missouri
Manganese	black or purple	MnO (OH)			
Limonite	yellow, orange, reddish brown or brownish black	Iron oxides		term for mixture of different hydrated iron oxides, mostly Goethite	
Gypsum	white or grey	$CaSO_4 \cdot 2 H_2O$	Ba, Sr, Mg	Also called hydrated gypsum, lime, gips, selenite	New York, Maryland, Oklahoma

Table II. Chromophores and Potential Native North American Plant Sources

Chromophore	*Possible Color*	*Chemical Composition*	*Known Example or Constituent*	*Examples of Potential Plant Sources*	*Notes*
Carotenoid	Yellow to violet	Dehydrated isoprene	Carotene	Pigweed (*Amaranthus* sp., Amaranthaceae) Sorrel (*Rumex* sp.) Little hogweed (*Portulaca oleracea* L.)	
Quinone	Colors depend on mordant and pH	1. benzoquinone 2. naphthaquinone 3. anthraquinone	Carthamine Juglone Alizarin Purpurin	Walnut or butternut (*Juglans* sp.) Rubiaceae family Sorrel (*Rumex* sp.) Poligonaceae family	Forms lake with mordant For 3. pH >7 → violet-blue; pH < 7→ yellow-red; AL=rose-red; Ca bluish-red; Zn=red-violet; Fe=black-violet; Cr=red brown
Flavonoid	Yellows	Structures containing Phenols, names depending OH or OCH position	Quercetin	Black oak (*Quercus velutina* Lam.) Onions (*Allium* sp.) Red grapes (*Vitis* sp.)	Requires mordant. Metal ions in a mordant form insoluble lake with OH and OCH groups.
Anthocyanine	Red or blue depending on pH	Reduced flavonoid	Sambucin	Elder (*Sambucus* sp.) Blackberries, dewberries (*Rubus* sp. L.) Grapes (*Vitis* sp.) Cherries (*Prunus* sp. L.)	

Betalain	Color depends on pH	Reduced flavonoid	Betanin	Pokeweed (*Phytolacca Americana* L.) Lambsquarters (*Chenopodium americana* L.)	Named after beets. Occur in many flowers and fruits
Basic dyes		Contain cations of quaternary NH group forming watersoluble salts with organic or inorganic anions	Berberine	Berberidaceae family such as Barberry (*Berberis vulgaris* L.); Oregon grape (*Mahonia aquifolium* (Pursh) Nutt.); goldenseal (*Hydrastis Canadensis* L.); bloodroot (*Sanguinaria Canadensis* L.)	Sometimes labeled as alkaloids. These however only contain secondary or tertiary NH groups
Tannins		1. hydrolysable tannins 2. condensed tannins		Sumac (*Rhus* Sp.); oak (*Quercus* sp.); chestnut (*Castanea* sp.); alder (*alnus* sp.) Pineaceae family such as pine, spruce, hemlock	Also called: 1. gallotannins because they from esters with gallic acid 2. hydrated flavonols or proanthocyanidins Historically labeled as tanning agents
Indigoid	Blues		Indigo	Coastal indigo (*Indigofera miniata* Ortega); Carolina indigo (*Indigofera caroliniana* P. Mill.); anil de pasto (*Indigofera suffruticosa* P. Mill.)	

Known Dyestuff and Possible Plant Sources

Plants, fungi and animals are the sources of organic dyes. Any source can contain several different chromophores with each producing different colors. On the other hand, a single chromophore can be extracted from several different sources. Therefore, to identify the source of a colorant, the relative composition of several chromophores must be identified (*32, 33*). Dyes are often classified into categories according to the chemistry of their attachment to the textile, i.e., direct, vat or metal complex dyes, acid, base, fiber reactive. They are also classified by chromophore, i.e., carotenoid, quinone, and others. Table II summarizes those chromophores that are found in some North American plants that are known to have been used by native peoples. The information is adapted from Schweppe (*34*) and supplemented with ethnobotanical data from Moerman (*26, 27*), and phytochemical data from Duke (*35*). Additional data concerning plant locales was gathered from the plants database of the United States Department of Agriculture (*36*).

Photography

The wavelengths of the electromagnetic spectrum from 400–700 nm are visible to the human eye. They range from blue light at 400 nm through red at 700 nm. The ultraviolet (UV) region is comprised of wavelengths less than 400 nm and extends into the x-ray region, while the infrared spectrum is composed of frequencies of more than 700 nm. Both of these spectra are not visible to the human eye, but their use in photography enables the camera to capture these otherwise non-visible regions and they are used extensively as tools to recover more information about the subject.

In UV reflected photography, a light source emitting ultraviolet light is directed at the subject which then reflects this radiation into the camera. Visible radiation from the room or reflected from the subject will be absorbed by an ultraviolet transmission filter over the camera lens thereby preventing any visible light from reaching the film.

In UV fluorescence photography, the fluorescence of a substance excited by UV illumination is captured. The source of ultraviolet radiation filtered with an ultraviolet transmission filter, or excitation filter, is aimed at the subject in a completely darkened room. The subject reflects the ultraviolet light, but can also emit a visible fluorescence. The ultraviolet light is then prevented from reaching the film by a barrier filter that only allows visible light to be transmitted to the film.

Materials such as pigments, iron stains and paint found on a textile may fluoresce. The presence of certain dyes can be determined based on their fluorescence (*37*). For instance, "one of the oldest organic pigments to be found is the natural madder lake...of alizarin...but also containing purpurin (1,2,4–trihydroxyanthraquinone), [which] fluoresces in a bright yellow-red...[which is] activated by energy of 400 nm...[and] weaker fluorescence at 345 nm (38). Berberine is another example of a natural dye that fluoresces (*39*).

In infrared photography, the subject is exposed to light that contains the infrared spectrum such as a tungsten light or photoflood lamp. The visible part of the spectrum is filtered out by using an IR transmission filter, which transmits a specific energy range from the IR region. An IR sensitive film has to be used to record images.

In conservation, paintings are often examined with IR illumination because pigments and varnishes vary in their reflectance and absorption in that portion of the electromagnetic spectrum. The wavelength is longer than visible light; therefore, the small particles in pigments and varnishes scatter the light less efficiently than visible light. IR can be used to "reveal underpaint, underdrawing in carbon black, retouching, and changes in composition as well as distinguish and penetrate beneath discoloration or soiling". The different reflectances of vegetable dyes can be used when examining textiles for evidence of restoration for instance (*37*).

Microscopy

Techniques of optical microscopy (OM) are well known and often used for the examination of fibers and yarns from archaeological textiles. Many texts provide the fundamentals of the technique (e.g. *40–43*). Some manuscripts describe the methods that may be employed in the study of archaeological materials in particular (*44, 45*), while others report the results of optical microscopic examination in identification and characterization of archaeological fibers (e.g., *12, 46*).

Similarly, SEM is a well known method of examining and imaging fibers and particles (*47*). Many papers report the use of this technique in the study of archaeological materials (e.g., *48–50)*.

SEM is particularly useful when integrated with an energy dispersive spectrometer (EDS), thereby allowing the determination of elemental composition of the materials that are also being observed and micrographed. Elemental composition of fibers and deposits has been studied in textiles from Etowah (*51*). The elemental composition reflects their burial environment in association with copper as well as their constituent plant fibers. Rowe (*52*) applied this technique successfully to pigments used in rock art, and it has been used in the study of archaeological fibers (*11, 53–55*).

Detection of Inorganic Constituents

Inorganic pigments and lakes (organic dyes bonded to an inorganic support) can be recognized by the ratio of elements in their composition, making elemental analysis an important tool in their identification. EDS may facilitate an initial qualitative analysis, but quantitative analysis and the detection of trace elements are needed to identify the inorganic colorant components. Due to sample size restrictions, the methods that can be employed are limited. The techniques of inductively-coupled plasma mass spectrometry (ICP-MS), ICP-optical emission spectroscopy (ICP-OES), and laser ablation ICP-MS are described in the literature (*56*).

ICP spectroscopy has been applied in quantitative elemental analysis in forensic examinations (*57*), and for such issues as the determination of source provenance based on these data for materials such as ochre (*58*) or other pigments. Speakman et al. (*59*) report on the characterization of archaeological materials with LA-ICP-MS, while others analyzed pigments successfully on pottery from the American Southwest (*60*), including the Mesa Verde region (*61*) among others.

Detection of Organic Constituents

There is a vast body of literature focused on the identification of dyes. Verhecken (*62*) reviews the types of methods used in the past for colorant identification. He classifies these as: 1) chemical methods which might yield colorimetric results, 2) chromatography, which separates components of differing chemical composition, and which requires some subsequent method of detection of the separated chromophores, and 3) spectroscopy, including visible, UV-visible, fluorescence, infrared, Raman, mass, and nuclear magnetic resonance techniques. No single technique is outstanding in its performance, each has advantages and disadvantages. In fact, Wouters (*63*) states the need for further development of colorant identification procedures that address issues raised by currently used methods. Since the concentration of dye on a fiber is relatively small, colorant identification requires the identification of a very small quantity on a material whose sample size is very limited. In addition, some methods employed require dissolution of the material, which breaks dye-mordant complexes and thus might cause some alteration that would affect identification. Finally, dyes can be complex mixtures of multiple colorants, thus the determination of relative ratios of these components is needed in order to distinguish their sources.

Most research has focused on the identification of typical colorants used in Europe and the Middle East, with some work on those used in the Far East, and in South America. Very little is known concerning North American colorants and their use and much more needs to be done in the field of colorant identification. Reliable plant and animal sources that will serve as comparative materials are needed for accurate determination of the unknown colorants of North America. Therefore, the protocol developed herein does not indicate one single particular analytical method for the identification of organic colorants. The protocol provides guidance for the sample taking once the methods can be developed and tested.

Methods Used to Develop the Protocol

Replicated Materials

To prepare a painted fabric for comparative purposes, small bands of ferrous oxide, an alizarin/purpurin mixture, ground up bloodroot *(Sanguinaria canadensis* L.) and bedstraw (*Galium verum* L.) were painted on linen fabric (Testfabrics L-57). As carriers or binder for the pigment colorants to be painted on the fibrous materials, beef fat was used as a substitute for bear grease and egg white was used as source for albumin. The various combinations of these materials were mixed and painted on areas of the test fabric as outlined in Table III. The painted test fabric was left to dry for 24 hours.

Commercially sold rabbit hair (Joseph Galler Inc.) and milkweed fibers that had been collected by the researcher in 2004 were each colored with aqueous solutions of lab grade hematite (Fe_2O_3) as a substitute for ochre and copper sulfate. Additionally, rabbit hair from a breeder (Jennings, T.), commercially produced rabbit yarn, and milkweed fibers were also colored in an aqueous bloodroot dye bath that did not contain any dyeing aids.

Analysis

To determine the best photographic method for the study of prehistoric textiles, experimentation with different light sources was necessary. A single lens reflex (SLR) film camera (Canon AE1) with an option for manual adjustments, and different macro and zoom lenses and filters were used for all photography. All pictures were taken with the samples placed on flocking paper (Edmund Scientific), which creates an optically inert background and therefore does not interfere with the optical properties of the textile to be examined. Details of the forensic photography methods are described in Baldia and Jakes (*64, 65*).

For optical microscopic examination, a Zeiss Axioplan microscope and brightfield (BF), darkfield (DF), polarized light (P) and differential interference

Table III. Colorant Binder Combinations and Fluorescing Results

	#	*Colorant*	*Binder*	*Aid*	*Fluorescing*
Painted Fabrics	1	Fe_2O_3	Beef Fat	None	None
	2	Fe_2O_3	Albumin	None	None
	3	Alizarin/purpurin	Beef Fat	None	None
	4	Alizarin/purpurin	Albumin	None	None
	5	Alizarin/purpurin	Beef Fat	Tannic Acid	None
	6	Alizarin/purpurin	Albumin	Tannic Acid	None
	7	Dried, ground bloodroot	None	None	None
	8	Dried, ground bloodroot	Albumin	None	Pale yellow
	9	Dried, ground bloodroot	Beef Fat	None	None
	10	Bedstraw	Beef Fat	None	None
	11	Bedstraw	Albumin	None	Orange or pale salmon
	12	Albumin only	None	None	Bluish white
Colored Fibers	13	Bloodroot dyed milkweed fibers	None	None	Pale yellow
	14	Bloodroot dyed rabbit hair yarn	None	None	Bright pale yellow
	15	Undyed rabbit hair yarn	None	None	Bright bluish-white
	16	Fe_2O_3 dyed milkweed fibers	None	None	None
	17	Fe_2O_3 dyed rabbit hair	None	None	None
	18	Bloodroot dyed rabbit hair	None	None	Pale yellow, weak
	19	Undyed rabbit hair	None	None	None
	20	Copper sulfate colored rabbit hair	None	None	None
	21	Copper sulfate colored milkweed	None	None	None

contrast (DIC) techniques were used to examine the morphology of the fibers and particulates on the fiber surfaces. Nominal 100 X and 400 X magnifications were used. For the image capture and analysis a ProgRes 3008 Digital Camera, Adobe Photoshop version 5.0 and Zeiss AxioVision version 3.1 software were used.

A JEOL JSM-820 scanning electron microscope with an Oxford eXL energy dispersive X-ray analyzer were employed in examining the fibers at magnifications higher than the light microscope. Samples were mounted on aluminum planchettes covered with double sided carbon tape (SPI) and carbon coated with a Denton Vacuum Desk II. Although images were collected, the objective was not to identify the fibers themselves beyond classifying them as hair or bast fiber. Rather, the aim was to capture images and to magnify the fibers enough that adhering deposits could be separated from the fibers and an EDS analysis could be performed on the deposits and fibers separately.

Elemental analysis by mass spectrometry was performed using a Perkin-Elmer Sciex ELAN 6000 Inductively Coupled Plasma—Mass Spectrometer (ICP-MS) and a ThermoFinnigan Element 2 Inductively Coupled Plasma Sector Field Mass Spectrometer. The iron oxide colored fibers (Table III, #16, #17) were weighed into the following aliquots: (1) 1 mg rabbit hair (RH), (2) 5 mg RH, (3) 1 mg of milkweed (MW), (4) 5 mg MW and (5) a mixture of 2.5 mg RH and MW each. These samples were combined with concentrated nitric acid. Warming the mixture to just below boiling temperature for 5 to 7 minutes did not result in total digestion. Hence, the incubation time in the water bath was increased to 30 minutes. Afterward, these samples were analyzed in the spectrometer.

Results of Testing the Simulations and the Sequence of Methods

Under ultraviolet light milkweed and rabbit hair fibers that had been dyed with bloodroot fluoresced in a pale yellow/orange and a bright yellow/orange respectively. Because the commercially obtained undyed rabbit hair yarn fluoresced intensely, untreated rabbit hair obtained directly from a breeder was used for testing and to use as control (Table III).

SEM-EDS was not performed with the replicated samples before using it on the actual artifacts, because its frequent application to analyze historic and archaeological fibers has verified its usefulness.

Because the objective of the protocol was not only to get a quantitative elemental analysis, but also to keep the sample size as small as possible, EDS must be considered as preliminary test for elemental analysis by ICP-MS or ICP-OES, thus allowing for the chemical composition of the material to dictate what must be done next.

To keep the sample size as small as possible and to measure Fe, ICP-MS was evaluated before ICP-OES, which requires higher concentrations or a much larger sample size. However, the quadrupole MS did not have sufficient resolution to measure the main isotope of iron (^{56}Fe) in the small concentrations able to be achieved. Therefore, the use of a high resolution spectrometer was found to be necessary if the identification of iron containing minerals is desired.

While using the Element 2 spectrometer, the dilute sample clogged the tubing to the instrument; the samples were not digested well enough to pass through the tubing despite extending the digestion time and the apparent visual clarity of the solution. Therefore, the digestion must be adapted in future trials.

The results of testing the methods and the sequence in which the steps of the protocol must be applied are summarized in Figure 1. Through extensive visual examination and forensic photography of archaeological textiles, regions of particular interest may be determined, even when these are not visible to the naked eye. Informed by these photographs, samples may be selected which are more likely to contain important information about the chemistry of the fabric. That is, if there are variations in the chemistry due to coloration for example, forensic photography aids in displaying that variation and representative samples can be chosen. Sequencing of subsequent examinations maximizes the data obtained from a single small sample.

Once a sample is taken, further study under low magnification macroscopy can reveal additional features including color or structure variation. This information aids in the selection of sub-samples from which OM fibers and particulate can be identified. Furthermore, the general type of colorant can be determined by its location, thoroughly saturated color in the fibers is indicative of dyeing in comparison to surface deposited pigment particles. A single small fiber which has been chosen in an informed manner can yield not only morphological information through microscopic study, but also elemental composition through EDS. The resulting EDS data of the fibers and of particulate adhering to the fibers provide information about the ratio of inorganic elements. These data in turn lead to determination of subsequent analyses for organic or inorganic components, because informed choices are possible. For example, the solvents needed to digest sampled material and the controls for ICP-MS can be determined without wasting material, or a high carbon content can suggest the use of methods for detection of organic components such as chromatography or infrared and Raman molecular spectroscopy.

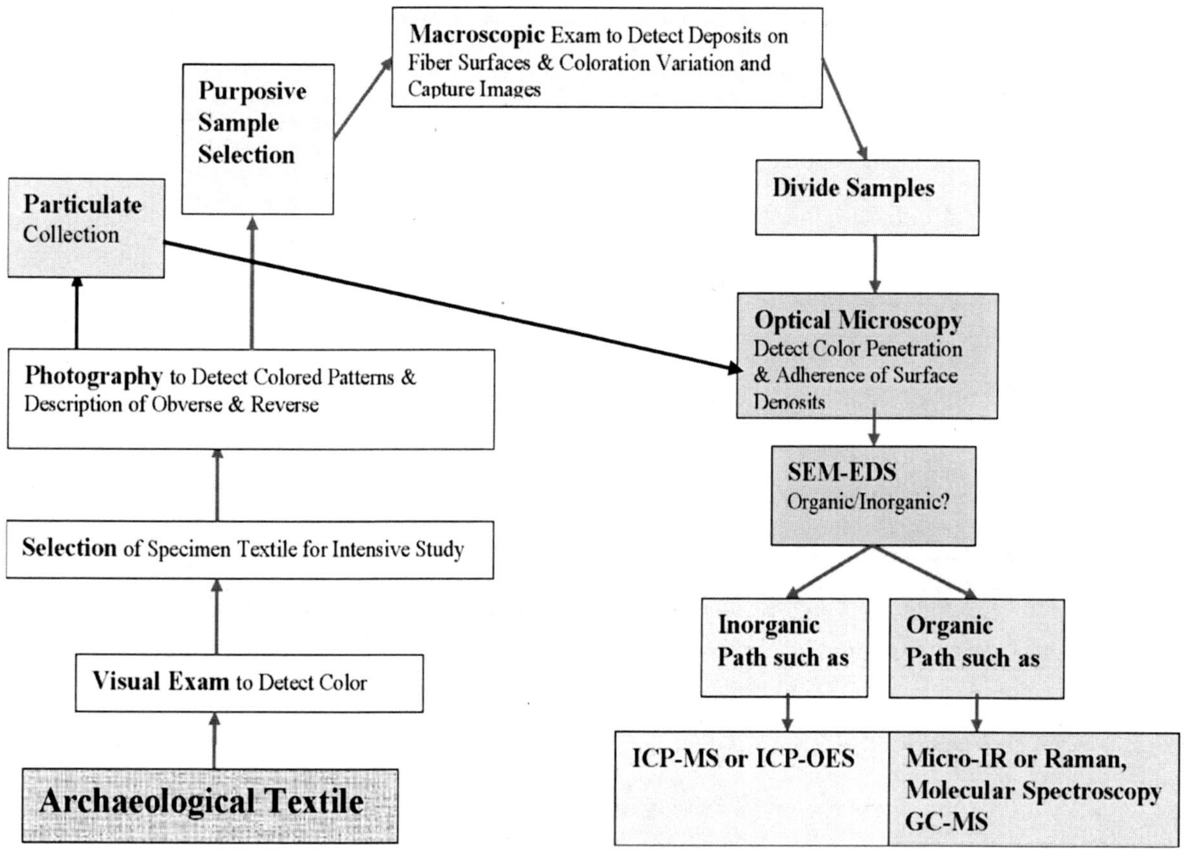

Figure 1. Ideal Protocol Overview and Sequence.

Table IV. Seip Textiles Selected for Further Analysis

Textile # (Song)	*Textile # (Thompson)*	*Apparent Color of the Fragments*	*Charred Yes/no*	*Samples Taken from colored in:*
4	0701'-'231	Yellow/Brown	No	Yellow/Brown/Green
23	1500'-'210	Turquoise/White	No	Shades of Turquoise/ off- white/white
36	2201'-'100 ; 2202'-'100	Different shades of black	Yes	Orange/red, black
37	2101'-'110 ; 2102'-'110; 2103'-'110	Different shades of black	Yes	Blue

Results of the Application of the Protocol

Selection of Samples Suited for Analyses

Permission to take photographs and remove samples from Hopewell textile fragments was requested from the appropriate entities at the Ohio Historical Society (OHS). There are a total of 226 Seip textiles curated therein. Only those textiles that did not adhere to copper or were otherwise not directly associated with copper were examined for evidence of coloration. These were divided into three groups based on patterns in: (1) yellow/brown, (2) white/turquoise, and (3) black on black colors. The latter materials are charred textiles and the patterns in black may indicate previous patterns of coloration. Eleven textiles representing each of these groups (four yellow/brown, two white/turquoise and four charred) were selected for photography and sampling. Hereinafter this group will be labeled the "selected" textiles. In addition, one textile that upon visual examination did not show any apparent coloration was chosen as a "non-colored control" (Table IV).

Due to the constraints of the glass plate housing, initial examination was limited to the obverse side of each textile. Later, the obverse and the reverse side of each of the selected textiles were examined more thoroughly under fluorescent white light to determine if coloration was applied to each side. Each glass case was opened and the textiles were carefully transferred to flocking paper. The obverse side of the different fragments that are housed together were examined, described and photographed. Although fabric structure was noted, it was not considered further since Thompson (*66*) had already evaluated the structures of the Seip textiles. To be able to examine the reverse side of the textiles, they had to be turned over. Instead of doing this one fragment at a time and compromising the integrity of the fragile pieces, another sheet of flocking paper was placed on top of the textile assemblage. Bottom and top flocking paper were held together firmly and flipped over in one steady motion to display the reverse side of the textile fragments arranged as they were placed in the housing. This reverse side was also observed, described and photographed.

The obverse side of all of the selected textiles was photographed in UV, photoflood (3400 K), and simulated daylight (Xenon 6420 K) conditions. The reverse side of all textiles was photographed with the daylight and the photoflood lamp, but to limit the textiles' exposure to UV, the reverse side of only certain textiles was photographed in UV. When the results of the previous photography of the textiles' obverse side showed patterns of coloration, then it was decided that further UV exposure was justified.

Acquisition of Samples

With aging, textiles can become very brittle and fall apart into a dust-like material containing fiber fragments. At the time of moving the selected textiles from their housing onto the flocking paper, the particulate material that had accumulated within the glass housing cases was collected. When possible, the dust under differently colored sections of the individual textiles was collected separately even though contamination from one color into the next could not be totally eliminated. This material was examined using optical microscopy (OM) as described in detail later.

The patterns detected by the photography, particularly UV, and the apparent degree of contamination by other materials, such as soil, were taken into consideration when the sampling sites on the textile fragments were selected. Due to the great fragility of the textiles, and their limited quantity, not all eleven of the selected textiles were sampled for SEM-EDS analysis. Samples were taken that were representative of the three textile groups that had been identified during the initial examination of the Seip textiles. Therefore, samples were taken from textile #4 representing the yellow/brown group and textile # 23 representing the turquoise on white group. The charred textiles were represented by textile # 36, which showed orange/red and some blue coloration in UV light, and textile # 37, which had enough blue coloration to justify sampling. Textile #36 did not have enough blue area for sampling without permanently altering the appearance of this textile (Table IV).

The yarn samples that were taken directly from the textiles were chosen from the apparently colored and apparently non-colored sections of the selected

Seip Mound textile fragments. Great care was taken to remove the smallest sample possible, which also contained as much information as possible. Often that meant that the yarn samples displayed several colors on the single pieces that were taken. These yarns, or sometimes just fibers, were not any smaller than 3 mm but did not exceed 8 mm. The sample taking was conducted in such a way that minimal damage occurred, and the removal did not leave a noticeable void in the textile fragment. Surgical scissors (SPI Supplies) and very fine tipped tweezers were used to cut the material. The selected yarn samples were placed on an aluminum planchette covered with double sided carbon tape (8 X 20 mm, # 5072, SPI Supplies). The areas where these samples were removed from the textiles and the lay-out of the samples on the planchettes were carefully drawn and documented on photographs and these were placed with the Ohio Historical Society as reference for future research. Each yarn specimen was observed with a Bausch and Lomb macroscope under 3.18 X to 21.8 X magnification to determine if any encrustations or minerals were attached to the outside of the fibers. Then the samples were photographed at 3.18 X, and divided further for the other examinations.

While viewing the yarn samples under magnification, individual fibers were picked from the collected samples using fine tweezers and carefully placed onto microscope slides for further examination.

Microanalysis

In analyzing the fibers at a microscopic level, the general fiber classes were identified (hair or bast). Furthermore, adherence of particles to the fibers, be they mineral colorant or soil was determined and when possible identified with the help of the particle atlas (*38*). The previously created control mounts were used as comparative materials to the unknown mineral or pigment deposits that were observed in the artifact samples.

Optical Microscopy and Imaging

Lab grade hematite (Fe_2O_3) and copper sulfate (anhydrous and hydrated) were mounted on slides and used as controls to compare to mineral deposits that might have been found adhering to the fibers. Rabbit hair and milkweed that had been colored with an aqueous hematite solution and with an aqueous copper sulfate (blue vitriol) solution were also used for comparison. Fibers removed from each simulated material were mounted in water (Refractive Index (RI) of 1.0), and in Permount (Fisher Scientific) (RI of 1.55). The collected particulate matter and fibers removed from the yarn samples were similarly mounted and examined using optical microscopy.

SEM Imaging and EDS

After having observed and photographed each yarn sample that was placed on carbon tape covered planchette; these were carbon coated. At least one picture at a magnification between 400X and 3000X at an average of 10 keV was taken before the EDS data were collected. The EDS measurements were performed in the differently colored areas as they had been noted during the mounting of the fibers onto the planchettes.

Measurements were obtained from: (1) several fibers of the same sample for average elemental composition, (2) at least two spots of a single fiber to compare elemental composition within one fiber, and (3) material adhering to fiber surfaces to compare the elemental composition from fiber to adhering material. During the course of performing the SEM-EDS, it was noted that some particulate adhering to the fibers tended to build a charge, while the fiber itself was more conductive which is indicative of a chemical difference between the two.

ICP-MS

The SEM-EDS analysis produced a semiquantitative analysis of the elemental composition found in the samples. In spite of the presence of a number of heavy elements, organics were present in significant quantities. Consequently, the inorganic identification route of inductively coupled plasma-mass spectrometry (ICP-MS) was not pursued at this time.

However, when other archaeological samples are analyzed, the digestion process must be modified to facilitate digestion of the elements that are present as determined by EDS. In addition to nitric acid, other acids such HF may have to be incorporated into the process and may need to be incubated under pressure, with ultrasound and at an increased temperature in the water bath environment for total digestion to take place.

Discussion and Implications

The results of this research encompasses two broad areas: (1) the development of a protocol for the examination of archaeological textiles, and (2) the application of the protocol to selected textiles of Seip Mound, Ohio. Detailed descriptions of the data gained from the Seip textiles and its implications will be published elsewhere.

The Protocol and its Implications for the Study of Archaeological Textiles

As the methods of the protocol were applied to the archaeological textiles, the evaluated analytical methods needed to be adjusted to working with a limited amount of material of unknown composition. The protocol includes a sequence of methods designed to minimize invasive sampling of archaeological textiles, and to keep the samples as small as possible. To successfully sample archaeological textiles in a purposeful and selective manner, samples should be taken only from those regions on the textile that indicate differences in chemical signature. Furthermore, the analytical methods must be adjusted to accommodate the small sample sizes and component concentrations.

Visual Examination

A great deal of information can be gained from a visual examination that is done with painstaking attention to detail. The initial visual examinations may have to be done in less than ideal conditions, such as a museum storage facility, but still can give many clues as to what to expect in the research that is to follow.

It was learned that particulate matter in the glass cases holding the textiles included degraded textile fibers mixed with contaminating matter. These fragments were more informative than had been expected; and if found in association with archaeological materials, these should be considered the first bit of information about the textiles' state of degradation. The more particulate matter there is, the more likely the fibers in the textiles themselves will be very fragile, which must be taken into account when choices are made about proceeding in the research.

Furthermore, since this material contains the severely degraded bits and pieces from the textiles, it can also be used (a) to assess the condition of the fibers, (b) to identify the fibers used in the textile as a whole, and (c) to identify the coloration in general. It may even be possible to use this particulate for some of the other analytical methods, besides OM such as infrared (*67*) or Raman spectroscopy, or even ICP-MS.

While collecting and using this particulate matter to study the textile as a whole may be useful, it cannot totally replace direct sampling because particulate cannot be associated directly with particular areas of the textile. It is also possible that the particulate is contaminated with material that is not from the textile. Therefore, when studying coloration, selective sampling of particular colors or areas with different chemical signatures will still need to be done. Thereby, materials of different composition or color are kept separated and each can be analyzed by itself.

Lighting and Sample Selection

Lighting should be controlled when selecting the samples and sub-sampling from these. As shown in the forensic photography, different lighting conditions will reveal or hide details and this can be used to one's advantage. The ideal lighting will allow the researcher to see as much as possible for the initial selection of the suitable textiles from which samples will be taken. Creating the ideal environment to do the initial selection of textiles may minimize the number of textiles selected as representative of an assemblage, and thereby the sampling may be kept at a minimum. After working with several different light sources in this research, it became apparent that the fluorescent white light that was present at OHS when the initial specimen selection took place was not sufficient for comprehensive visual examination.

After pictures were taken in simulated daylight, many of the colored patterns on the textiles and the colors themselves became much more visible than when they were examined in fluorescent white light. As an alternative to the light sources that are typically available in the curation facilities, another light source that will reveal colors more effectively should be brought in. A small lamp creating simulated daylight (6500 K) combined with the photoflood lamp (3200 K) that is needed for the photography would allow for a more precise selection of the initial textile grouping and selection, and thereby decreasing the number of textiles needed for study. In controlling the lighting, more details in coloration are revealed and fewer textiles need to be sampled to capture these details.

Availability of ideal lighting conditions during the actual sample taking is important. Photography may reveal details about colored patterns on the fabrics that may be difficult to recognize and sample if the lighting does not produce similar conditions. This can potentially increase difficulty of sampling, the number of textile specimens, and sub-samples unnecessarily.

Besides controlling lighting conditions, adequate magnification should be available during the work with the textiles. A nominal magnification of 8X was available before and during the actual sampling. Only later in the lab did it become apparent that the magnification should have been greater. The fibers within the yarn structures were very fragmented and mineralized, which was easily recognized under adequate magnification and the proper lighting conditions. This fragility explained the difficulties that occurred during sample taking when the pieces of yarns kept falling apart as soon as they were touched with a pick or tweezers. Besides extra time, patience, a steady hand, and good color vision by the researcher, some flexibility to get usable samples is necessary even if that meant picking up single fibers that had only a second earlier been a yarn, but had fallen apart as they were moved. Hence, extra time must be allocated for the sampling process under magnification to make this task productive.

Time and Flexibility

A certain degree of flexibility must be built into the research design, because the actual conditions of the textiles and the contaminants are not known, and could have been acquired over the course of the textile's lifetime. Time is required to adjust and improvise methods and to overcome difficulties while the researcher must retain a willingness to let the data take the lead. The pursuit of clues that were not anticipated or were initially only thought to be minor might be of the utmost importance; these could lead to greater knowledge about the materials.

For instance, the extensive time that was spent with the optical microscope examining the Seip materials was absolutely essential and could not have been replaced by more reading about the subject, or by any other test method or analysis. The same is true for the time spent with the macroscope.

Sequencing of Methods

Using forensic photography as a precursor to any sample acquisition forms the foundation of the protocol, and allows purposive sampling. EDS should be performed to establish which elements to expect before attempting any quantitative elemental analysis such as ICP-OES/MS. Before working with actual artifacts, a set of replicated materials must be used and a successful trial run using the planned methods of analysis whether ICP-OES/MS, GC-MS or any others, must be achieved, so the methods of preparation can be adjusted properly. To facilitate this, appropriate materials must be replicated, which might mean that plants or minerals must be collected, and dyed or painted comparative standards must be created, so the unknown can be compared to the known. For many of the Old World dye plants these standards already exist. However, for North American dye plants comparative collections are in the early phases and subsequent analysis of colorant constituents have not yet been conducted (*68,69*).

Limitations

The degree of the fiber degradation and mineralization of the Seip materials made them difficult to sample and also limited the degree of visual examination that was ethically justifiable without destruction of the textiles.

After the textiles were turned over, several layers of fabric, fringed fabric, a combination of fabric adorned with leather and two composite pieces were identified. However, these layers could not be separated without jeopardizing the textile. To truly examine these details, the textiles need to be unfolded to describe the structures of all layers underneath or to investigate the method of attaching the fringe or leather decorations.

Due to mineralization, these fibers are brittle and readily fractured. Furthermore, in spite of the high degree of mineralization, the presence of organic compounds was indicated by the results of the EDS. The presence of organic components could not be disregarded, although mineralization does not preclude the existence of remaining organic materials in textiles in addition to copper possibly having formed malachite, a copper carbonate. For that reason, it was decided that the semi-quantitative elemental analysis done with EDS would be sufficient at this time, and the organic path would need to be explored more thoroughly before trace element analysis of the inorganic components. Since comparative materials and the analyses of colorant components need to be done prior to this pursuit, further sampling and analysis of Seip materials must be postponed for now.

Conclusion

In conclusion, the lesson to be learned from this research is an essential one. The material must dictate the sequence of events—what, where, when, and how. The protocol reported herein was developed as a beginning to classify colorants in archaeological textiles found North America. The sequence of methods was designed to gain a maximum of data from the textile while keeping the destruction of the artifact at a minimum.

The protocol was tested on prototypes that simulated expected properties of colored archaeological textiles. When applied to artifacts from Seip Mound, some adaptation was required. It was shown that useful information can be gleaned from particulate shed from these fragile materials as well as from strategically located samples.

Acknowledgments

SEM–EDS were conducted with the assistance of Dr. Sreenivas Bhattiprolu from the Microscopic and Chemical Research Center located at the Department of Geological Sciences at the Ohio State University. The spectrometry was conducted with the assistance of Dr. John Olesik and Anthony Lutton. Furthermore, the authors extend their appreciation to the Ohio Historical Society.

References

1. Morgan, L. H. *The League of the Iroquois*; JG Press, North Brighton, MA, [1851]1954.
2. Densmore, F. *How Indians Use Wild Plants for Food, Medicine and Crafts*; Dover Publication: New York, NY, 1974 [1928].
3. Catlin, G. *Letters and Notes on North American Indians*; JG Press: North Brighton, MA, 1995 [1832], Vol.1+2.
4. Hinkle, K. A. M.A. thesis, Ohio State University, Columbus, OH, 1984.
5. Mills, W.C. *Ohio History* **1909**, *18,* 286.
6. Shetrone, H. C.; Greenman, E.F. *Ohio History* **1931,** *40,* 376.
7. Willoughby, C. C. *Ohio State Archaeol. Quarterly* **1938,** *47,* 451–452.
8. Willoughby, C. C. *Ohio State Archaeol. Quarterly* **1938,** *47,* 273–287.
9. Church, F. *Ohio Archaeologist* **1983**, *33,* 10–15.
10. Church, F. *Midcont. J. Archaeol.* **1984**, *9,* 1–25.
11. Song, C. A. PhD thesis, Ohio State University, Columbus, OH, 1991.
12. Song, C. A.; Jakes, K. A.; Yerkes R. W. *Midcont. J. Archaeol.* **1996,** *21,* 247–266.
13. Kuttruff, J. T. *Am. Antiq.* **1993**, *58,* 125–145.
14. Saltzman, M. et al. *Dyestuffs* **1963**, *44,* 241.
15. Martoglio, P. A.; Bouffard, S. B.; Sommer, A. J.; Katon, J. E.; Jakes, K. A. *Anal. Chem.* **1990**, *62*, A1123–A1128.
16. Martoglio, P. A; Jakes, K. A.; Katon, J. E. In *Proceedings of the 50th Annual Meeting of the Electron Microscopy Society of America*; Bentley, B. W.; Small, J. A., Eds.; Microbeam Analysis Society: San Francisco, CA, 1992; pp 1534–1535.
17. Jakes, K. A.; Katon, J. E.; Martoglio, P. A. In *Archaeometry '90*; Pernicka, E.; Wagner, G. A., Eds.; Birkhauser Verlag: Basel, 1990; pp 305–315.
18. Sibley, L. R.; Jakes, K. A. In *Archaeometry of Pre-Columbian Sites and Artifacts*; Scott, S. A.; Meyers, P., Ed.; The Getty Conservation Institute: Marina del Rey, 1994; pp 396–418.
19. Berns, R. S. In *Billmeyer and Saltzman's Principle of Color Technology*, 3rd ed.; Wiley and Sons Inc.: New York, NY, 2000.
20. Brown, J. A. *The Artifacts*; Spiro Studies, Second Part of the Third Annual Report of Caddoan Archaeology—Spiro Focus Research, University of Oklahoma, 1976; Vol. 4, 464.
21. Boyd, J. *Oklahoma Anthropological Society Newsletter* **1967**, *15,* 4.
22. King, E.; Gardner, J. S. *The Research Potential of Anthropological Museum Collections*; Cantwell, A.; Griffin, J. B.; Rothschild, N. A., Eds.; Annals of the New York Academy of Science, NY, 1981, Vol. 376.
23. Tomlinson, P. *J. Archaeol. Sci.* **1985**, *12,* 269–283.

24. Hall, A. R. *Quaternary Science Reviews* **1996**, *15,* 635–640.
25. Casselman, K. L. *Craft of the Dyer: Color from Plants and Lichen*, 2nd ed.; Dover Publications, Inc.: New York, 1993.
26. Moerman, D.; *Native American Ethnobotany*; Timber Press: Portland, OR, 2000; pp 473–515.
27. Moerman, D. *Native American Ethnobotany: A Database of Foods, Drugs, Dyes and Fibers of Native American Peoples, Derived from Plants*; University in Michigan-Dearborn; URL http://herb.umd.umich.edu, accessed 7–9–2005.
28. Jakes, K. A.; Ericksen, A.G. *Southeastern Archaeol.* **2001**, *20,* 56–66.
29. Klein, C. *The 22nd Edition of the Manual of Mineral Science*: (after James D. Dana) / Cornelis Klein; with continued contribution of Cornelius S. Hurlbut, Jr.; John Wiley: New York, 2002.
30. The Amethyst Gallery, Inc.; URL http://mineral.galleries.com/minerals/property/physical.htm; accessed 5–2005.
31. *The Mineral Database*; URL http://www.mindat.org; accessed 5–2005.
32. Wouters, J. *Studies in Conservation* **1985**, *30,* 119–128.
33. Wouters, J. Rosario-Chirinos, N. *J. Am. Inst. Cons.* **1992**, *31,* 237–255.
34. Schweppe, H. *Handbuch der Naturfarben: Vorkommen, Verwendung, Nachweis*; Nikol Verlagsgesellschaft, Hamburg, Germany, 1993.
35. Duke, J.; *Agricultural Research Services: Dr. Duke's Phytochemistry Database*; USDA - ARS - NGRL, Beltsville Agricultural Research Center, Beltsville, Maryland; URL http://www.ars-grin.gov/cgi-bin/duke/farmacy-scroll3.pl; accessed 7–9–2005.
36. USDA-NRCS Plants Database; United State Department of Agriculture Plant (USDA) and Natural Resources Conservation Service (NRCS) Plants Database; URL http://plants.usda.gov/index.html; accessed 7–9–2005.
37. Andrew, S. R.; Eastop, D. *Conservator* **1994**, *18, 50–56.*
38. Johnston-Feller, R. *Color Science in Examination of Museum Objects: Non-destructive Procedures*; The Getty Conservation Institute: Marina del Rey, 2001; p 207.
39. Hayashi, K. In *International Symposium on the Conservation and Restoration of Cultural Property*; Tokyo National Research Institute on Cultural Properties: Tokyo, 1979; pp 48.
40. McCrone, W. C.; Delly, J. G. *PAE2: The Particle Atlas*; electronic version, 2nd Edition; McCrone Research Institute: Chicago, IL, 1997.
41. McCrone, W. C.; McCrone, L. B.; Delly, J. G. *Polarized Light Microscopy*; McCrone Research Institute: Chicago, IL, 1984.
42. Slayter, E. M.; Slayter, H. S. *Light and Electron Microscopy*; Cambridge University Press: Cambridge, England, 1992.

43. Petraco, N. *Color Atlas and Manual of Microscopy for Criminalists, Chemists and Conservators*; CRC Press: Boca Raton, FL, 2004.
44. Jakes, K. A. In *Beyond Cloth and Cordage: Current Approaches to Archaeological Textile Research in the Americas*; University of Utah Press: Salt Lake City, 2000; pp 31–68.
45. Curl, A. M.; Jakes, K. A. In *Postprints of the Textile Specialty Group*; Am. Inst. Conservation, 2003; pp *53–67.*
46. Gremillion, K. J., Jakes, K. A.; Wimberley, V. *Current Archaeological Research in Kentucky*; Pollack, D.; Gremillion, K. J., Ed.; Kentucky Heritage Council: Frankfort, KY, 2000, Vol. 6.
47. Goldstein, J. I., et al.; *Scanning Electron Microscopy and X-Ray Microanalysis: A Text for Biologists, Material Scientists, and Geologists*, 2nd ed.; Plenum Press: New York, 1992.
48. Chen, H. L.; Jakes, K. A.; Foreman, D. W. *Textile Research J.* **1996**, *66*, 219–224.
49. Chen, R.; Jakes, K. *J. Am. Inst. Cons.* **2001**, *40*, 91–103.
50. Sibley, L. R.; Jakes, K. A. In *Conservation and Characterization of Historical Paper and Textile Materials*; Zeronian, H.; Needles, H., Eds.; Advances in Chemistry Series No. 212; American Chemical Society: Washington, DC, 1986; pp 253–275.
51. Sibley, L. R.; Jakes, K. A.; Swinker, M. *Clothing and Textile Res. J.* **1992**, *10,* 21–28.
52. Rowe, M. *Handbook of Rock Art Research*; Whitley, D. S., Ed.; AltaMira Press: Walnut Creek, CA, 2001.
53. Jakes, K. A.; Mitchell, J. C. *J. Archaeol. Sci.* **1996**, *23,* 149–156.
54. Srinivasan, R.; Jakes, K. A. *J. Archaeol. Sci.* **1997**, *24,* 517–527.
55. Chen, R.; Jakes, K. A. *J. Amer. Inst. Conserv.* **2001**, *40*, 91–103.
56. MURR; *Archaeometry Laboratory at the University of Missouri Research Reactor (MURR)*; URL http://www.missouri.edu/~rjse10/home.htm; accessed 6–2–2005.
57. Goulding, J. In *Forensic Hair Examination of Hair*; Robertson, J., Ed.; Taylor & Francis: London, 1999; pp 175–192.
58. Weinstein-Evron, M.; Ilani, S. *J. Archaeol. Sci.* **1994**, *21*, 461–467.
59. Speakman, R. J.; Neff, H.; Glascock, M. D.; Higgins, B. In *Archaeological Chemistry: Materials, Methods, and Meaning*; Jakes, K.A., Ed.; ACS Symposium Series No. 831; American Chemical Society: Washington, DC, 2002; pp 48–63.
60. Sall, C. A.; Zedeño, M. N.; Speakman, R. J. In *Laser Ablation ICP-MS in Archaeological Research*; Speakman, R. J.; Neff, H., Ed.; University of New Mexico Press: Albuquerque, NM, 2005; pp 155–166.
61. Speakman, R. J. *Laser Ablation ICP-MS in Archaeological Research*; Speakman, R. J.; Neff H., Eds.; University of New Mexico Press: Albuquerque, NM, 2005; pp 167–186.

62. Verhecken, A. In *Dyes in History and Archaeology 20*; Kirby, J., Ed.; Archetype Publications: London, 2005.
63. Wouters, J. In *Dyes in History and Archaeology 20*; Kirby. J., Ed.; Archetype Publications: London, 2005.
64. Baldia, C. M.; Jakes, K. A. *J. Archaeol. Sci.* **2006**, in press.
65. Williams, R. A.; Williams, G. *Medical and Scientific Photography: an Online Resource for Doctors, Scientists and Students*; RMIT University; URL http://msp.rmit.edu.au; 3–22–2005.
66. Thompson, A. J. PhD thesis, Ohio State University, Columbus, OH, 2003.
67. Jakes, K. A.; Baldia, C. M.; Thompson, A. J. In *Archaeological Chemistry: Analytical Techniques and Archaeological Interpretation*; ACS Monograph Series, 2006; this volume.
68. Jakes, K. A. In *Archaeological Chemistry: Organic, Inorganic and Biochemical Analysis*; Orna, M.V., Ed.; ACS Symposium Series No. 625; American Chemical Society: Washington, DC, 1996; pp 202–222.
69. Jakes, K. A.; Sibley, L. R.; Yerkes, R. W. *J. Archaeol. Sci.* **1994**, *21*, 641–650.

Chapter 3

Infrared Examination of Fiber and Particulate Residues from Archaeological Textiles

Kathryn A. Jakes[1], Christel M. Baldia[2], and Amanda J. Thompson[3]

[1]Department of Consumer Sciences, The Ohio State University, Columbus, OH 43210
[2]Department of Human Development and Environmental Studies, Indiana University of Pennsylvania, Indiana, PA 15701
[3]Department of Clothing, Textiles, and Interior Design, University of Alabama, Tuscaloosa, AL 35487

Infrared spectra of particulate material shed from fragile archaeological textiles were compared to the spectra of known materials in order to classify them by composition. Spectra were collected of comparative plant and animal fibers including charred, mineralized, pigmented and dyed examples, and of particulate from fabrics from two archaeological sites in eastern North America. Separation of the comparative materials into groups based on their composition was achieved, as well as some distinctions made in dyed, pigmented, and mineralized samples. Compositional features of the archaeological materials were discerned that can aid in their identification.

Textiles recovered from archaeological sites provide a wealth of information concerning the lifeways of the people of the past who manufactured and used them. Typical studies of these fabrics report fabric structure (e.g., details of twining or weaving), and sometimes also include information on yarn structure (e.g., ply and twist). In fewer cases, fiber content is identified, yet learning information about the type of fiber used and its condition can provide insights into the knowledge held by prehistoric people of plant and animal resources for fibrous products. From fiber analysis, we learn about the skill people had in collecting and processing these materials and producing yarns and fabrics with distinctive properties. The researcher might address questions such as the conditions required for separation of bast fibers from the remainder of the plant stem, or the pliability of yarns required to twine a flexible bag. In addition to craftsmanship and knowledge of materials, we can also make inferences concerning the scheduling of activities for fabric production, as well as culturally determined use of materials. Because there is an ideal period in a plant's growth cycle in which fiber cells are mature, plant collection must be scheduled, while manipulation of the separated fibers to produce yarns or fabrics might be held until a later time. Coloration of fabrics may have cultural implications indicating, for example, clan affiliation or status in the group. Analysis of fibers in textiles provides information on their physical and chemical condition resulting from the burial context or from treatments incurred during the textile's useful lifetime. This condition information is needed for determination of conservation treatment even as it can be used to discern how the material was preserved in the burial. Many questions can be answered from textile and fiber study concerning the technology of manufacturing and the cultural use of the materials.

While it is often stated that few textiles survive from archaeological sites in eastern North America, some examples with preserved features have been uncovered. These materials have survived burial conditions because of the peculiarities of the burial context or due to some feature of their composition that made them less susceptible to microbial degradation. These preserved materials often are very fragile; handling them results in the loss of small particulate material. Rather than discarding this particulate, collection and analysis may provide some clues to the content and condition of the artifact.

Even if the particulate represents a contaminant, it is useful to know the nature of that contaminant. Once the textiles are examined visually and with magnification, examination of the particulate by microscopy should be undertaken. Identifying features of the fiber or of other materials that comprise the particulate may be difficult to see, possibly because they are occluded with soil or were damaged in active use or in long-term degradation. Therefore, additional analyses might be needed.

In this work an infrared spectral database of comparative materials was initiated. Spectra of plant fibers from the Comparative Plant Fiber Collection (CPFC) and of rabbit hair and wool were studied to evaluate whether they could be grouped according to features of their chemical composition. The spectra of the plant and animal fibers charred in the laboratory for different periods of time were examined to determine if any chemistry was retained which could be used in classification. Infrared spectra of laboratory mineralized plant fibers and mineral specimens, sample dyed and pigmented plant and animal fibers were also evaluated for identifying features. Finally, the spectra of some example archaeological materials were examined to explore the usefulness of the database in uncovering identifying features.

Literature Review

Although fewer in number than textiles recovered from the southwestern US, Native American textile materials recovered from eastern North America are no less intriguing. The textiles on which this work is focused were recovered from the Seip Mound Group, a Middle Woodland period site in Ohio, and Etowah Mound, a Mississippian period site in Georgia. Both are ceremonial burial mound sites. Site information and some descriptions of the materials recovered are described elsewhere *(1–9)*. The Middle Woodland materials are twined, and a predominant number are made from bast fiber, although some have rabbit hair included (*6, 7*). Many of the fragments are charred, probably from having been exposed to heat related to the cremations and burning ceremonies conducted at these sites. There is some evidence for preservation by copper mineral impregnation as well. A few display evidence of coloration, a feature that can be anticipated from the comments of early travelers and later ethnographers (*9, 10*).

While some of the Mississippian textiles are of similar structure to the Middle Woodland textiles, others are very complex materials and are lace-like in appearance. Many of the materials from Etowah are preserved by mineralization, and display green-colored deposits on their surfaces. Bast fiber, rabbit hair, and feathers have been identified (*2, 11*). The textiles from these two sites selected for analysis are representative of the complexity of structure and fineness of yarns seen in the materials; they provide evidence of the sophisticated technology of prehistoric people in all phases of fiber collection, processing, yarn spinning, fabric manufacture and, when present, coloration.

Infrared Spectroscopy of Textile Fibers

Infrared spectroscopy is a common analytical method used in industry and research. The reader may refer to basic texts such as (*12–14*) to obtain guidance in the method, instrumentation, sample preparation and interpretation of spectra.

In brief, the absorption of infrared radiation by a sample is recorded. The infrared region of the electromagnetic spectrum extends from 14,000 to 200 cm^{-1} but the mid IR region from 4000–400 cm^{-1} is of particular usefulness in studying organic materials. Since specific functional groups and molecular structures absorb particular wavelengths of infrared radiation, the bands of absorbance can be used to identify the composition of the material under study (*12–14*). Spectra are collected as ratios against the spectrum of the air or purge gas in the chamber. Infrared spectra of textile fibers are used to identify them, or to recognize the presence of finishes and dyes as well as to distinguish features resulting from degradation or defects resulting in manufacturing (e.g., *12, 15*). IR spectra of historic and archaeological materials have been used in the identification of dyes (e.g. *16–18*).

The usefulness of infrared spectra has been recognized by those interested in art and archaeology; the Infrared and Raman User's Group (IRUG) was established "to encourage the exchange of information, develop IR and Raman spectral standards, and distribute comparative spectral data for the study of works of art, architecture and archeological materials" (*19*). The IRUG online database of spectra provides a useful resource for those interested in comparing spectra of a material to those of known materials. An additional resource that includes some IR spectra in the identification of materials is the Conservation and Art Material Encyclopedia Online (CAMEO) (*20*). While neither of these databases contain many fiber spectra at present, additions to the databases increase annually.

The Comparative Plant Fiber Collection

The Comparative Plant Fiber Collection is a comparative collection of plant materials and the bast fibers they yield through different processing techniques *(21, 22)*. The plants collected were typical of those used by prehistoric native Americans of eastern North America in the manufacture of fabrics, particularly those with fine yarns. The CPFC contains representative fibers from 34 genera and species of plants that were collected in Ohio and Georgia in 1991 and 1992, with multiple stems gathered from each plant, and the fiber processed from the plant by four different methods:

- hammering the stems, then hand peeling the fibers
- soaking the stems in water for two weeks to simulate the effects of retting then hand peeling the fibers
- boiling the fibers in demineralized water for six hours, then peeling the fibers
- boiling the stems in demineralized water with potassium carbonate for six hours and peeling

Some of the plants readily release numerous fine, strong, and long fibers that could be spun into fine yarns comparable to those observed in the prehistoric fabrics *(21, 22)*. While coarser splits of wood could have been used in making baskets or cordage, the first focus of the CPFC was the fibers used in textiles that would show advanced fiber processing technology.

Most of the plants collected for the CPFC are dicots, some are woody, and some are herbaceous. The woody specimens are all hardwoods (angiosperms) except for eastern red cedar, which is a softwood (gymnosperm). Herbaceous plants can become "woody" with age, lignifying with time, as occurs in weathered, over-matured plants. Attempts among members of our research group to extract fibers from plant stems that have been allowed to lignify with age have shown that the fibers are not readily released. Plants that originally yielded long strands of fiber became so rigid that it was only possible to extract short pieces of fiber, and these were not suitable for spinning and weaving. In order for fibers extracted from stems to have the flexibility needed for spinning and weaving they need to be extracted from the plant during a certain period in its lifecycle. Therefore this study explored the characteristics of the fibers most suited to spinning and weaving and making fine fabrics.

The CPFC fibers were grouped into classes based on fiber processing characteristics and fiber microscopic characteristics: Group I milkweeds, dogbanes, nettles, mulberry; Group II moosewood; Group III rattlesnake master, cattail, red cedar; Group IV basswood, walnut, willow, paw paw, and slippery elm *(21, 22)*. The research reported herein further investigated the distinctions that can be made between groups based on their infrared spectra. Fibers from the comparative plant fiber collection were employed along with an additional number of fibers to initiate a comparative IR database.

Oxalates

While optical and scanning microscopic studies might show some of the oxalate inclusions that accompany the plant fibers *(23, 24)*, infrared spectroscopy is particularly useful in confirming their presence and composition because the absorbance bands in the regions of 1620 cm^{-1}, 1315 cm^{-1}, and 780 cm^{-1} are clear indicators of the whewellite form of calcium oxalate *(25)*. It might be possible to gain additional confirmation of calcareous composition by x-ray microanalysis *(23)*. While the shapes of the crystal inclusions might be diagnostic, their presence or absence alone also may aid in classification. For example, Scurfield, Michell and Silva *(25)* state that gymnosperms display fewer crystals than angiosperms.

Lignins

Although lignin accelerates photooxidation of plant fibers, the "degrading effects of acids, alkalis and solution of other chemicals is determined by the accessibility of the cellulose in the cell-wall of the fibre....A high lignin content retards microbiological deterioration because penetration of water is hindered." *(26:36)*. Thus, if lignin is one compositional feature that can preserve fibers in the archaeological context, it should be observable in those fibers that have been preserved. In addition, Barker and Owen *(27)*, Owen and Thomas *(28)* and Evans *(29)* reported that the percentage of each of the three major components (cellulose, hemicellulose, and lignin) vary from wood to wood and that hardwoods and softwoods show significant differences in their lignin polymer composition. Softwood lignin (guiacyl lignin) is composed of trans coniferyl alcohol derivatives with a small component based on coumaryl alcohol, while hardwood lignin (syringyl lignin) is composed of both trans coniferyl and trans sinapyl alcohol with small amounts of coumaryl alcohol derivatives (*27*) as shown in Figure 1.

Thus, all softwoods show infrared absorption of aromatic skeletal bands above 1510 cm^{-1}, while those of the hardwoods absorb below 1510 cm^{-1}. The carbonyl absorption in softwood is lower than 1739 cm^{-1}, while that in hardwood is higher than 1738 cm^{-1} *(27)*. Since hardwoods and softwoods are distinguished by the types of lignin they contain, it might be possible to classify the fibers extracted from these plants according to their lignin composition. In support of this concept, Mukherjee *(31)* reports that the IR spectra of jute fiber (an herbaceous dicot angiosperm) shows a peak in the region of 1515–1510 cm^{-1}, attributed to the aromatic ring stretch typical of the guiacyl nucleus.

Charring of Fibers

The recovery of many blackened or charred textiles from Hopewellian sites has been proposed to be related to the cremation rituals of these groups of people (8, 32). Cremation pits and other evidence of cremation ritual are characteristic features in these sites (33–40) and, although, cremation temperatures of a body must be high enough that organics, such as fabric, would be completely combusted, the presence of blackened and charred fabric fragments reflect the possibility that some fabrics were exposed to somewhat lower temperatures. Processes of smothering the flames are described by some archaeologists to explain features of these sites (e.g., 33:133), that would allow material to survive the ritual. It is possible, then, that these charred and blackened fabrics that have been recovered are associated with the cremations. In fact, these materials may

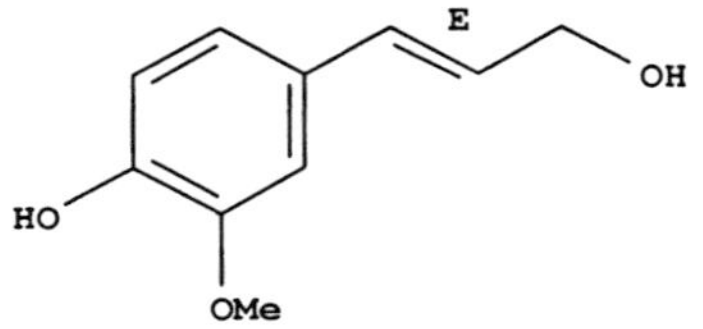

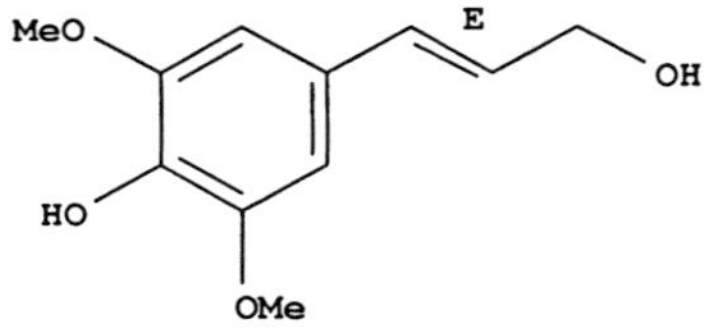

CAS 32811–40–8
trans coniferyl alcohol

CAS 20675–96–1
trans sinapyl alcohol

Figure 1. Structure of trans coniferyl alcohol and trans sinapyl alcohol (reproduced from reference 30. CAS Registry database records duplicated with permission of CAS, a division of the American Chemical Society).

be all the more significant because they were used in these rituals (32). They could contain some of their original composition since it is unlikely that they were exposed to the temperatures that would have completely reduced them to a carbonaceous network of char and yet the charring would make them less susceptible to microbial degradation.

In addition, Thompson *(8)* and Baldia *(9)* have each observed patterns of gradation in shades of black in the patina on some of the charred Hopewellian pieces. These patterns are indicative of the materials having been colored prior to their being charred. Baldia *(9, 41)* conducted forensic photographic evaluation and suggested targeted selective sampling for further dye analysis. The charred materials, then, may hold some essential information aiding in their identification and their composition, including coloration, in the past.

Charring is known to preserve aspects of the physical structure of wood, seeds, and fruit *(42)*. Srinivasan and Jakes *(43)* have shown that in charring some aspects of the physical shape of Indian hemp fiber are retained. In the carbonization of wood, Ercin et al *(44)* report the loss of cellulose, hemicellulose and lignin infrared absorbance bands in the range of 1300–1000 cm^{-1} and the appearance of two new bands at 1250 cm^{-1} attributed to the asymmetric C-O-C and at 1450 cm^{-1} attributed to aliphatic C-H bending.

Mineralization of Fibers

Some archaeological textiles in eastern North America survive due to the process of mineralization that they have undergone (*1, 2, 45–47*). Impregnation by copper salts inhibits microbial degradation and ultimately results in the formation of a pseudomorph of the fibers. Chen, Jakes and Foreman (*45*) report the infrared spectral characteristics of Indian hemp fibers that were artificially mineralized in a laboratory for a 2–year period. The fiber surfaces were covered with brilliant green copper-containing deposits. In comparison to modern bast fibers, the mineralized fibers show sharper, more well-defined individual absorbance bands in the region of 1200–1500 cm^{-1}. Spectral subtraction of the unmineralized fiber from mineralized fiber resulted in a spectrum that was similar to that of malachite, the basic copper carbonate. Infrared spectra of these laboratory-mineralized fibers and samples of mineralized cordage from an archaeological site were similar.

Dyeing of Fibers

Although brilliantly colored fabrics from eastern North American sites are limited in number, a few have been described (*4, 11*) and there is substantial evidence for a dyeing technology to have existed in prehistoric eras. Historic-era travelers and later-day ethnologists commented on the native use of plants as dyes and on the natives' colored garments (*10*). The archaeobotanical record also reflects accumulation of the seeds of sumac and bedstraw, the foci of the work of Jakes and Ericksen (*10*), since these could not be attributed to dietary purposes. While some work has been conducted in the identification of colorants on archaeological materials from eastern North America (*48, 49*) the need for comparative materials was recognized as necessary prior to the sampling of the archaeological materials. Examples of dyed textiles have been prepared, experimentally replicating the dyeing technology thought to have been employed by natives of prehistoric eastern North America (*50–52*). These colored materials provide representations of the color ranges produced by the dye plant and dyeing assistants; they also provide comparative samples on which experimental identification procedures can be attempted. In addition, the blackened charred textiles that display patterns in shades of black (*8, 9*) also are indicative of the application of coloration prior to charring.

Materials and Methods

Fiber Samples Examined

The plant and animal fibers used in this work and the treatments applied to them are summarized in Tables I and II. In addition to fibers obtained from the Comparative Plant Fiber Collection, undyed wool was obtained from Testfabrics (West Pittston, PA), and rabbit hair was obtained from an independent animal breeder. Miscellaneous chemicals that were used were obtained through Fisher Scientific. The infrared spectra of verdigris and malachite that are part of the Infrared and Raman User's Group database *(19)* were used for comparative purposes. Both dyed and mineralized fibers created in other work were used *(45, 50, 51)*. Fibers were charred following a protocol reported previously *(43)*. No correlation between periods of charring (e.g., 10, 20 or 30 minutes) and the period of charring incurred by the archaeological materials is implied. Volume of material and the oxygen accessibility, among many factors, will play a part in the extent of charring. Particulate from textiles from Seip Mound and from Etowah Mound C (EMC) were examined to test application of the infrared information to archaeological materials.

Infrared Spectra

Using a Perkin Elmer Spectrum 2000 Fourier Transform Infrared spectrometer, equipped with a DTGS (deuterated triglycine sulfate) detector, infrared spectra in the region of 4000–400 cm^{-1} were collected from KBr pellets made with finely-chopped and dried fiber. Sixty–four scans with a resolution of 4 cm^{-1} were coadded using Spectrum software. All spectra were baseline corrected, and smoothed with a Savitsky-Golay 9 point smoothing function. Recognizing that fibers from the plant fiber collection displayed variability in color and composition, multiple fibers were selected and chopped in order to represent the range of chemical composition of the bulk of the material in contrast to the particular chemical information that would be learned from an infrared spectrum of a very small location on a single fiber. Subsequently, the spectra of fiber residues and particulates acquired from archaeological textiles were collected and compared to the comparative material spectra. While this work concentrates on the obvious visible distinctions in the spectra, ultimately the spectra will be accumulated in a training set to provide for statistically supported identifications through pattern recognition based on discriminant analysis.

Table I. Comparative Plant and Animal Fibers Examined

Fiber	*Scientific classification*	*Botanical information*
Herbaceous species		
Common milkweed	*Asclepias syriaca* (L.)	Dicot, angiosperm
Indian hemp	*Apocynum cannabinum* (L.)	Dicot, angiosperm
False nettle	*Boehmeria cylindrica* (L.)	Dicot, angiosperm
Wood nettle	*Urtica divaricatum* (L.)	Dicot, angiosperm
Stinging nettle	*Urtica dioica (L.)*	Dicot, angiosperm
Rattlesnake master	*Eryngium yuccifolium* (L.)	Dicot, angiosperm
Grasses		
Cattail	*Typha angustifolia* (L.)	Monocot, angiosperm
Woody species		
Black willow	*Salix nigra* (L.)	Dicot, angiosperm
Basswood	*Tilia americana* (L.)	Dicot, angiosperm
Paw paw	*Asiminia triloba* (L.)	Dicot, angiosperm
Red mulberry	*Morus rubra (L.)*	Dicot, angiosperm
Red cedar	*Juniperus virginiana* (L.)	Conebearing, gymnosperm
Protein fiber		
Wool		
Rabbit hair		

Table II. Treated Comparative Plant and Animal Fibers and Mineral Specimens Examined

Colored fiber replicates and related standards	*Charred fiber replicates*	*Mineralized fiber replicates and related standards*	*Archaeological fibers*
Bedstraw dyed common milkweed	Indian hemp, charred 10, 20 and 30 minutes	Indian hemp fiber mineralized for 6 mos.	Etowah Mound C 840, partially mineralized bast
Iron oxide painted common milkweed, with albumin binder	Common milkweed charred 20 min	Indian hemp fiber mineralized for 24 mos.	Etowah Mound C 842, partially mineralized bast
Iron oxide painted common milkweed with beef tallow binder	Rabbit hair, charred 10, 20, 35 and 45 min.	Cuprite	Seip 5, charred bast, red deposits
Iron oxide painted rabbit hair with albumin binder	Bedstraw dyed common milkweed, charred 20 min	Malachite (spectrum from IRUG)	Seip 36, charred bast, red cast, Fe
Iron oxide painted rabbit hair with beef tallow binder	Iron oxide painted common milkweed, Charred 20 min.	Copper hydroxide	Seip 32, copper "painted", green stains, composite layers
Iron oxide		Verdigris (spectrum from IRUG)	Seip 10, red/brown hair and bast, composite, not charred
Aqueous bedstraw extract			

Results

Infrared Distinctions Between Fibers in the CPFC

All of the fibers in the CPFC display bands typical of cellulose (*53*). These include: a broad band between 3600 and 3200 cm^{-1} attributable to hydrogen bonded OH stretching, a band around 2900 cm^{-1} which is related to CH and CH_2 stretching vibrations, and numerous bands in the region from 1500–1200 cm^{-1} that are related to vibration modes of OH and CH groups. While pure cellulose might not display any bands in the region from 2000–1500 cm^{-1} except one at around 1635 cm^{-1}, the presence of bands around 1700 cm^{-1} is indicative of C=O moieties stemming from cellulose oxidation or esterification. There are many bands in the region of 1300–900 cm^{-1}. This part of the spectra is called the fingerprint region, and small differences in this area can be linked to structural changes in the cellulose. It should be noted also that carbon dioxide in the sample chamber will display a band around 2200 cm^{-1} and adsorbed water may also be seen in the 3600–3200 cm^{-1} region as well. Its presence will add "noise" to the spectra of the material observed, but in some cases adsorbed water may not be readily removed.

The spectra of the CPFC fibers exhibit some distinctions between each other and from that of cellulose alone. While microscopic study might show oxalate inclusions in association with the fibers, the infrared spectroscopy is particularly useful in defining these. The bands in the regions of 1740 cm^{-1}, 1620 cm^{-1}, 1315 cm^{-1}, and 780 cm^{-1} are often very obvious, particularly in the hardwood species (Figure 2, Table III). High amounts of calcium were reported in the study of these same materials *(23)* thus providing corroborating evidence of their composition as calcium oxalate. In contrast to the comment of Barker and Owen *(27)* that crystals occur much less frequently in wood of gymnosperms than angiosperms, the eastern red cedar fibers examined in this work showed extensive crystal formation, observed both microscopically and through infrared spectra.

In examining the lignin regions of the IR spectra (Table III), the fibers from the herbaceous dicot species contain lignins similar to both the softwood and hardwood lignins while the monocot grass cattail contains only the lignin similar to hardwoods. Red cedar shows the absorbances of the softwood lignin, and the hardwoods show those of the hardwood lignins only. Thus, some distinctions between plant fibers based on the absorbance of lignin were achieved.

By considering presence of both the oxalates and the lignins, a further separation of fibers into groups is possible. For example, Figure 3 shows black willow with both lignin and oxalate bands labeled. Figure 4 presents a flow diagram useful for separating the fibers into categories based on the oxalate and type of lignin present.

Table III. Plant Fibers and Relative Quantities of Their Components Observed in the Infrared Spectra

CPFC plant fiber	*Amount and Type of Lignin Present*	*Amount of Oxalate Present*	*Amount of Hemicellulose Present*
Herbaceous species			
Common milkweed	Small* if any, softwood type	Small ***	Small
Indian hemp	Small if any, softwood type	None observed	Small if any
False nettle	Both softwood and hardwood type	Small, not as obvious as in hard woods	Present, sharp, clear peaks
Wood nettle	Both softwood and hardwood type	Small, not as obvious as in hard woods	Present
Stinging nettle	Both softwood and hardwood type	Small, not as obvious as hard woods	Present
Rattlesnake master	Both softwood and hardwood type	None observed	Significant
Grasses			
cattail	Significant **, hardwood type	Small	Significant, very distinct sharp peaks
Woody species			
Black willow	Significant, hardwood type	Significant	Significant
Basswood	Small, hardwood type	Significant, strong peaks	Significant
Paw paw	Small, hardwood type	Significant	Significant
Red mulberry	Small hardwood type	None observed	Significant
Red cedar	Significant, softwood type	Significant, strong peaks	Significant

NOTE: Small* indicates that the quantity of material that is present is relatively small compared to the remainder of the spectrum. Significant **quantity indicates obvious amounts of the component. Where peaks are very outstanding notation is made of distinct sharp peaks. Small quantity of oxalate *** in common milkweed reflects that very little oxalate was observed in IR spectra of the untreated material but charred milkweed displayed oxalate.

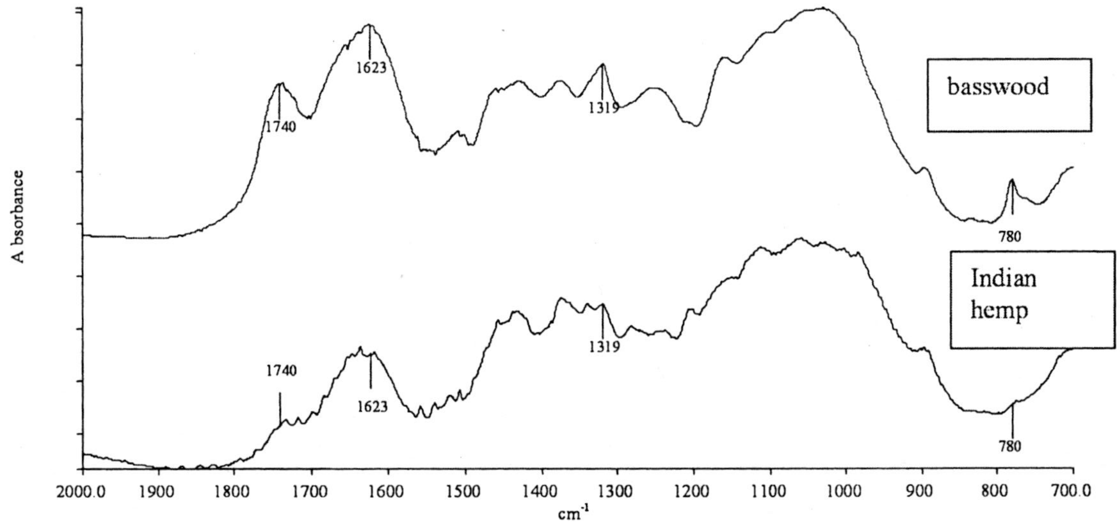

Figure 2. Infrared spectrum of basswood compared to that of Indian hemp. Marked peaks are those typical of calcium oxalate.

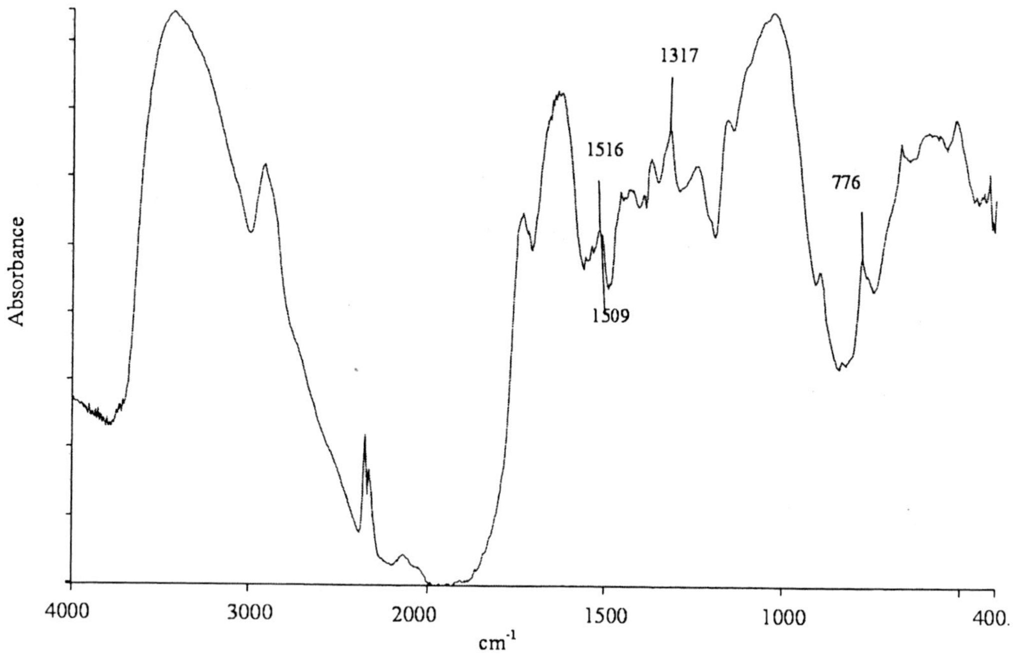

Figure 3. Infrared spectrum of black willow with some bands attributable to lignin and oxalate marked.

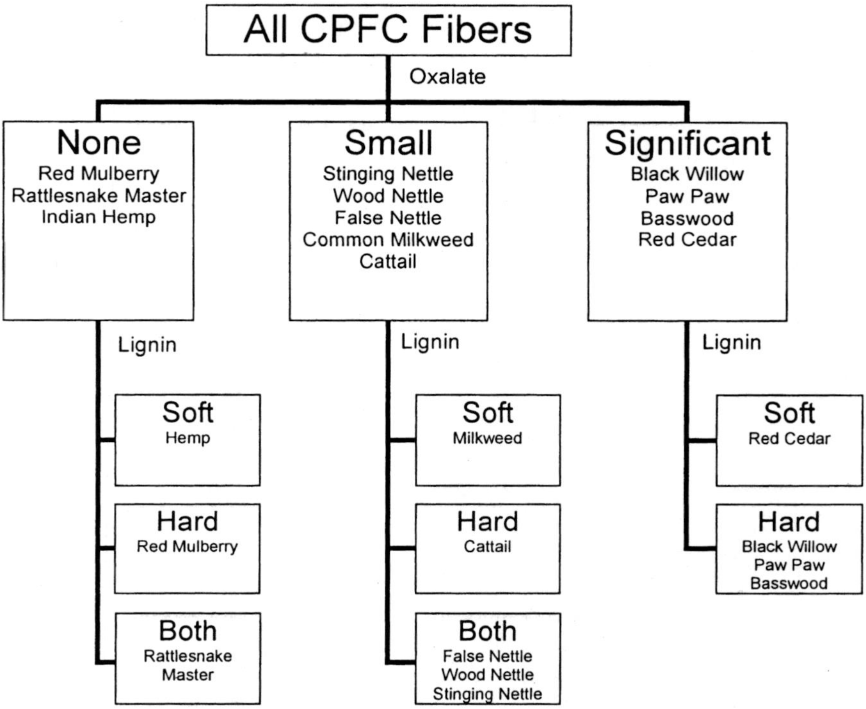

Figure 4. Flow diagram for separation of plant fibers into groups based on infrared spectra.

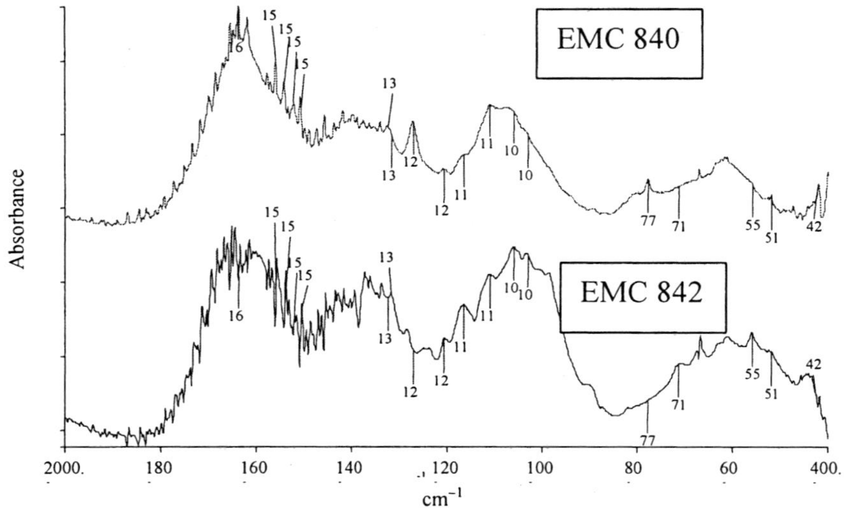

Figure 5. Infrared Spectra of EMC 840 and EMC 842.

Application of CPFC Distinctions to Archaeological Materials

The infrared spectra of two samples from Etowah (EMC 840 and 842) (Figure 5) show lignin absorbances of both the softwood and hardwood types. While no distinct hemicellulose absorbances can be seen, lignin absorbances include both 1505 cm^{-1} and 1519 cm^{-1}, thus indicative of the herbaceous dicots. The oxalate composition is very small and difficult to distinguish. One sample of EMC 840 was identified as a nettle (Urticaceae) by Philip Rury of the Harvard University Herbarium *(2)*, which appears to be partially corroborated by these infrared results.

Distinctions in Infrared Spectra in Protein Fibers

The infrared spectrum of rabbit hair was similar to that of the wool fibers with readily apparent amide bands in the regions of 1650 cm^{-1} and 1530 cm^{-1} (Figure 6).

Application of Protein Infrared Spectra to Archaeological Materials

Although the particulate from Seip 10 was noted to include both bast and hair fibers (9), the infrared spectra of samples studied reflected cellulosic content

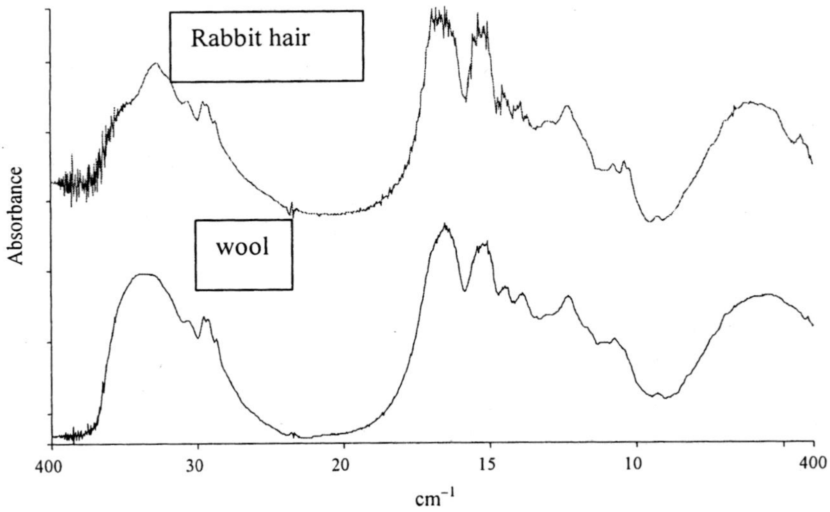

Figure 6. Infrared spectra of rabbit hair and sheep's wool.

only. More samples will need to be examined from a wider range of archaeological textiles.

Charred Fibers

As Indian hemp was charred for increasing periods of time, some loss of chemical structure was incurred (Figure 7). The sample charred for 30 minutes is white and fragile, while the 10 and 20 minute samples are still black and fibrous and appear more like the charred fibers observed in the archaeological textiles. It is obvious that the composition has been altered by exposure to heat. The cellulose is dehydrated with the 2900 cm^{-1} band reduced in each of the levels of charring. The 10 and 20 minute samples are similar to each other, but somewhat different from the uncharred Indian hemp. The 1630 cm^{-1} band in the uncharred fiber is shifted to 1582 cm^{-1} due to dehydration. The new bands in the region of 1450 cm^{-1} and 1200 cm^{-1} in the 30 minute sample are comparable to those noted by Ercin and Yurum *(44)*.

The comparison between charred and uncharred Indian hemp fibers is more readily apparent in Figure 8, which compares only the 10–minute charred sample and uncharred Indian hemp. It can be seen that some features are lost but some sense that the material is cellulosic can still be observed.

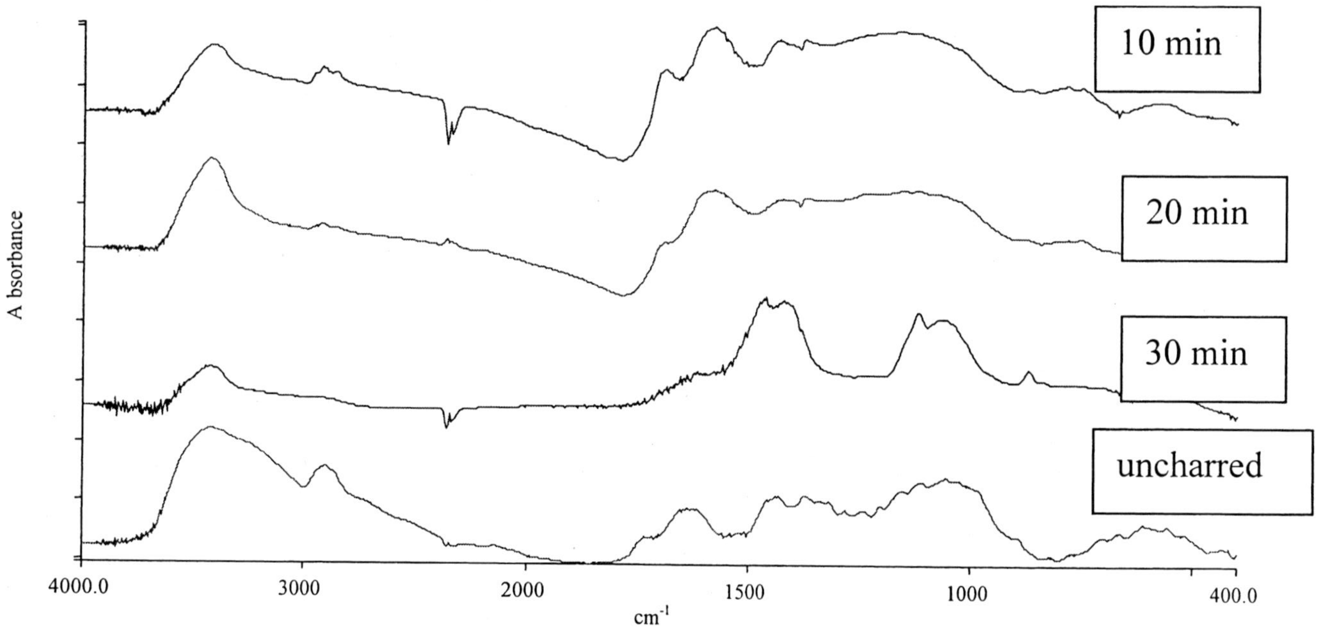

Figure 7. Infrared spectra of Indian hemp charred 10, 20 and 30 minutes compared to uncharred Indian hemp.

Charred back willow displays the distinctive peaks of its oxalate inclusions but the lignins that were observed in the uncharred material are no longer present (Figure 9). Charred red cedar produced similar results with outstanding oxalate bands observed.

In some cases new information may be revealed as a result of charring. The infrared spectrum of charred common milkweed (Figure 10) displays oxalates that are not as apparent in the uncharred sample. In some cases of charred fibers, the spectrum can be said to be cellulosic, but further distinction of the organic structure based on lignins cannot be made.

Charred Protein Fibers

In charring rabbit hair, some compositional features of the infrared spectra were lost (Figure 11). It should be noted that the 35 and 45 minute charred samples were still black and fibrous, yet their IR spectra were considerably different from the materials charred for a lesser period of time. This means that some charred protein fibers might not provide much infrared information. The 1650 cm^{-1} was reduced and shifted a bit to 1664 cm^{-1}, a new C=O band occurs at 1716 cm^{-1}, the N-H amide II band was completely gone after 35 minutes of charring, but was still present in the 10 and 20 minute samples.

Comparison of Charred Fiber Spectra to Archaeological Fibers

Seip 36 is a charred material with a red cast *(9)*; it also has a high iron content. The infrared spectrum of the particulate recovered from Seip 36 (Figure 12) looks like charred cellulose and shows no charred protein. The spectrum is not the same as charred milkweed but it appears more similar to charred Indian hemp (Figure 13), with bands near 1582 cm^{-1} and 1400 cm^{-1}.

Mineralized Fibers

As the Indian hemp fibers were mineralized in the laboratory, not only did they become encrusted with a green-colored deposit, but their interiors were replaced with copper compounds as well. This is supported by the change in the infrared spectra *(45)*. As the fibers became increasingly mineralized, the infrared absorbance peaks became sharper, and the 817 cm^{-1}, 881 cm^{-1} and 1045 cm^{-1} and 1384 cm^{-1} peaks increased in relative size.

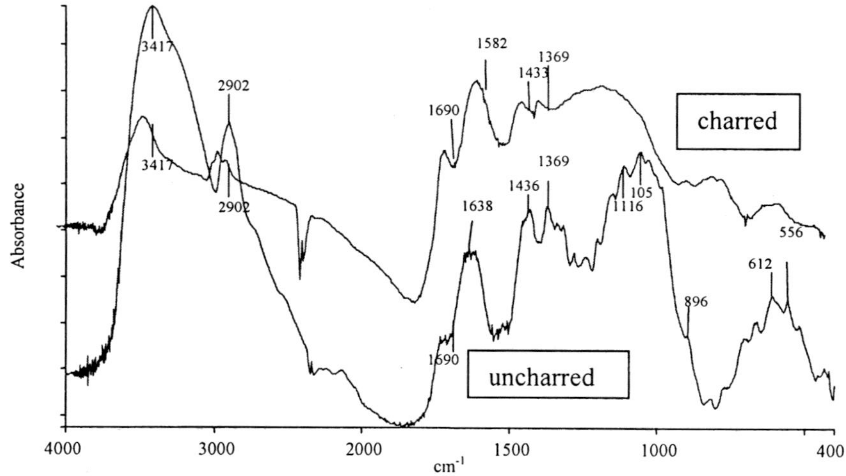

Figure 8. Infrared spectra of Indian hemp charred 10 minutes compared to uncharred Indian hemp.

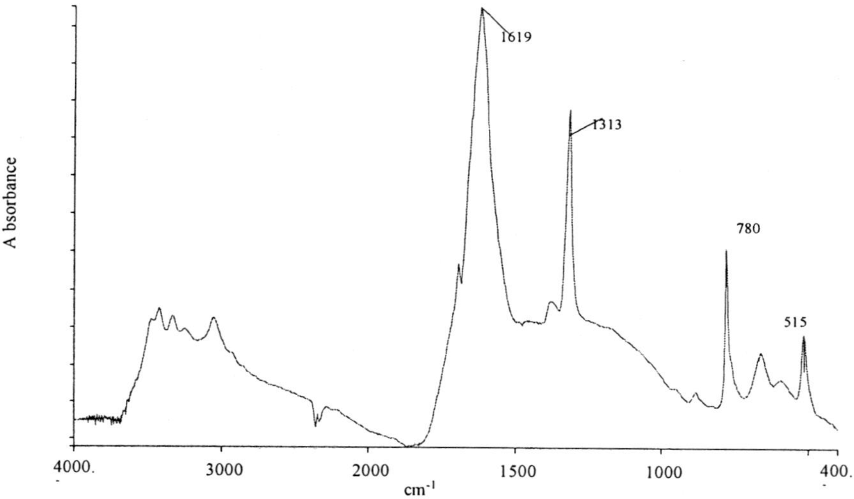

Figure 9. Infrared spectrum of charred black willow, oxalate bands labeled.

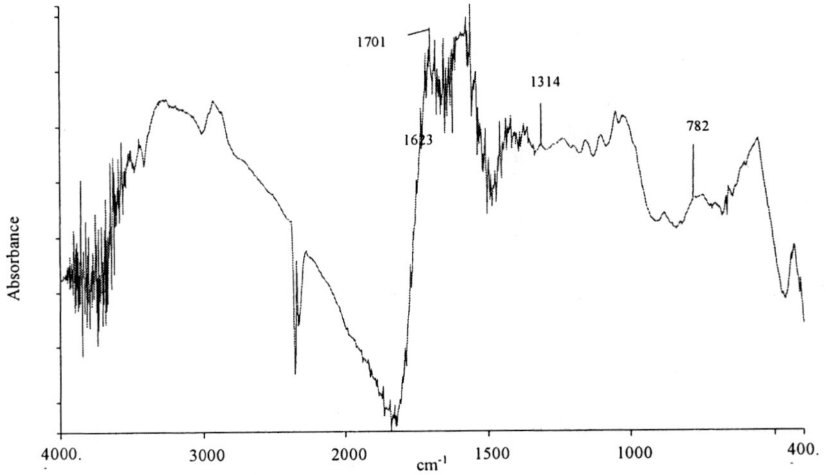

Figure 10. Infrared spectrum of charred milkweed with indicators of oxalates labeled.

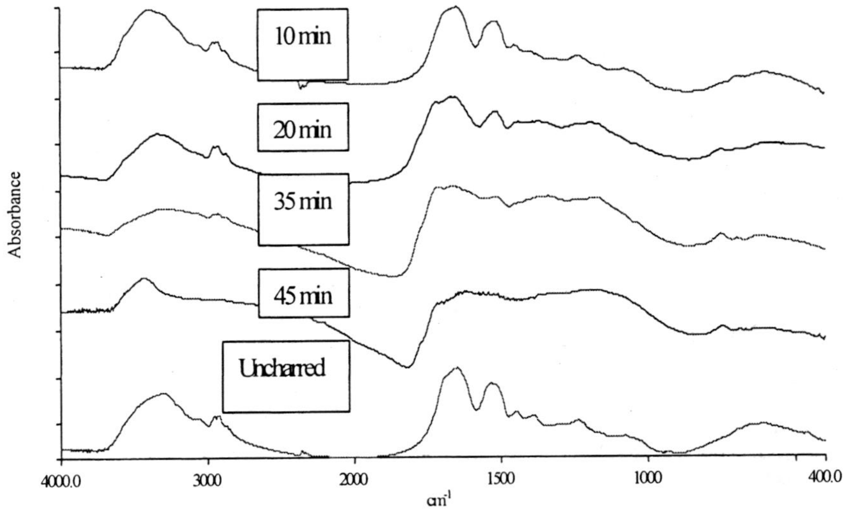

Figure 11. Infrared spectra of rabbit hair charred 10, 20, 35, and 45 minutes compared to uncharred rabbit hair.

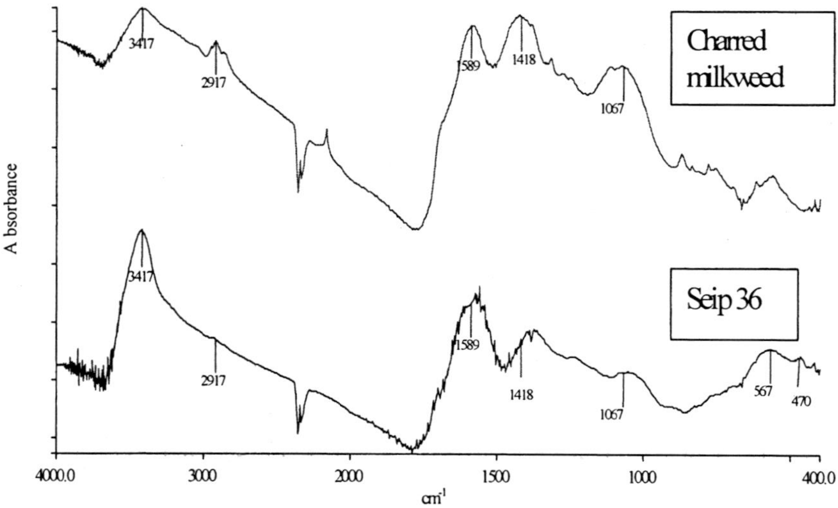

Figure 12. Infrared spectrum of charred milkweed compared to that of Seip 36.

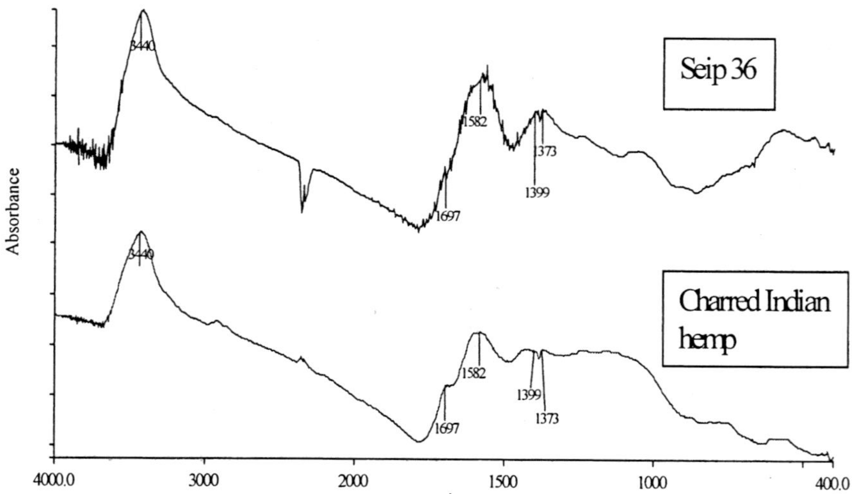

Figure 13. Infrared spectrum of Seip 36 compared to that of Indian hemp charred 10 minutes.

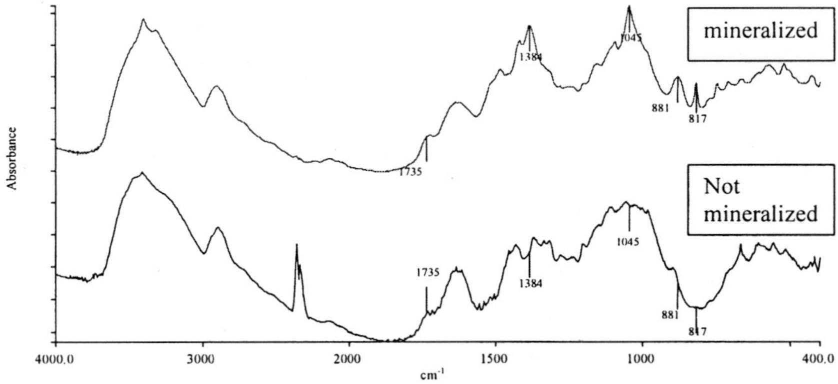

Figure 14. Infrared spectrum of Indian hemp mineralized 2 years compared to Indian hemp, not mineralized

Comparison of Mineralized Fiber Spectra to Archaeological Textiles

Etowah Mound C 842 G is a sample from the green encrustation from a very green mineralized fiber. The infrared spectra of these sample fibers show bands that are similar to the laboratory mineralized Indian hemp fibers, but they are sharper (Figure 15). Similarities in infrared spectra between this sample and malachite (IRUG 0057) *(19)* (Figure 16) indicate that it is primarily malachite.

Seip 32 is a material with deep green colored stains *(9)*. The infrared spectrum of the green colored sample (Figure 17) does not appear similar to malachite (basic copper carbonate) as do the Etowah materials, but it is similar to verdigris (IRUG 00152) *(19)*, which is composed of acetates and diacetates of copper. Thus the mineral replacement of the two fabrics is somewhat different. Whether that difference is due to the environment of mineralization or to the previous chemical composition of the textiles cannot be determined.

Pigments

The infrared spectrum of the iron oxide (Figure 19) used to simulate ocher displays distinctive absorbance bands in the region of 537 cm^{-1} and 465 cm^{-1}. Not only are the bands obvious in fibers pigmented with iron oxide, but the rabbit hair which had been colored with the pigment retained evidence of that pigmentation even when it was charred (Figure 20).

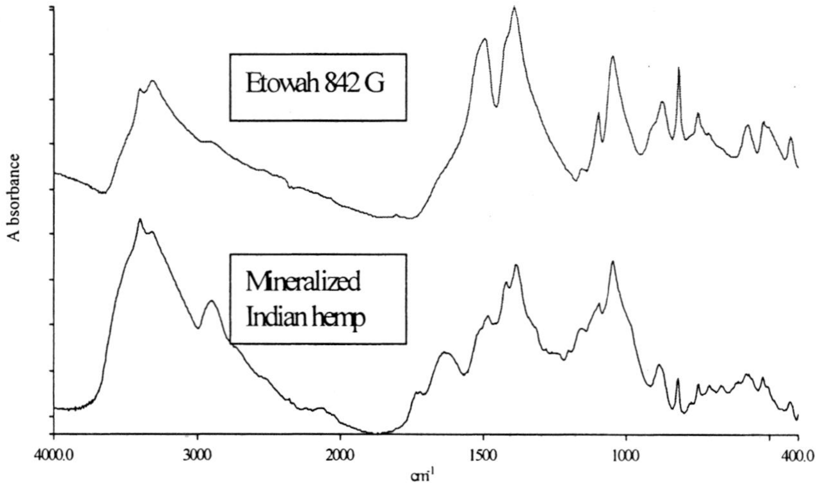

Figure 15. Infrared spectrum of Etowah Mound C 842 G green-colored mineralized fibers compared to that of lab mineralized Indian hemp fibers.

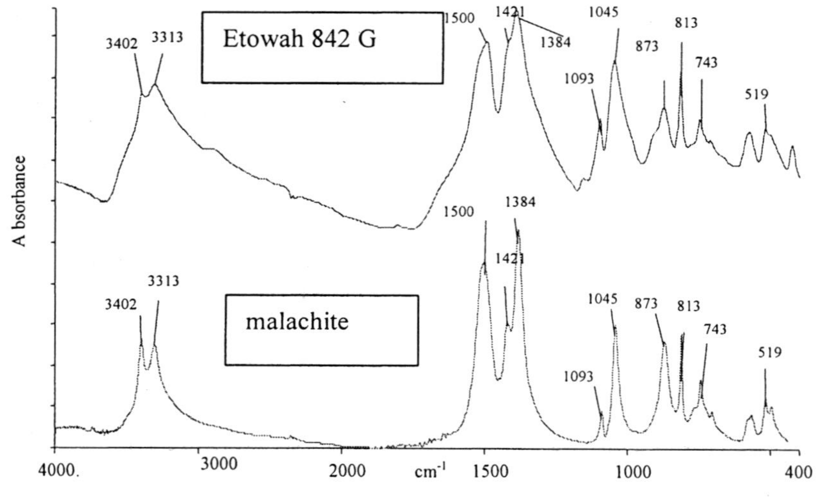

Figure 16. Infrared spectrum of Etowah Mound C 842 G green-colored mineralized fibers compared to that of malachite (IRUG 00057)(19).

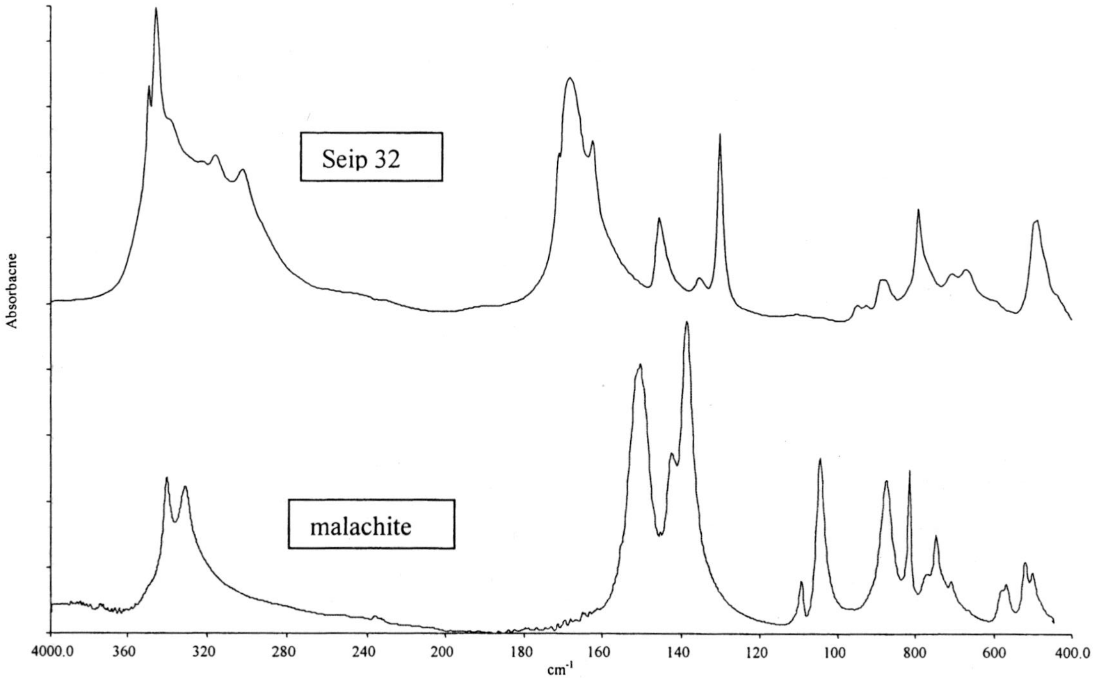

Figure 17. Infrared spectrum of Seip 32 particulate compared to malachite (IRUG 00057)(19).

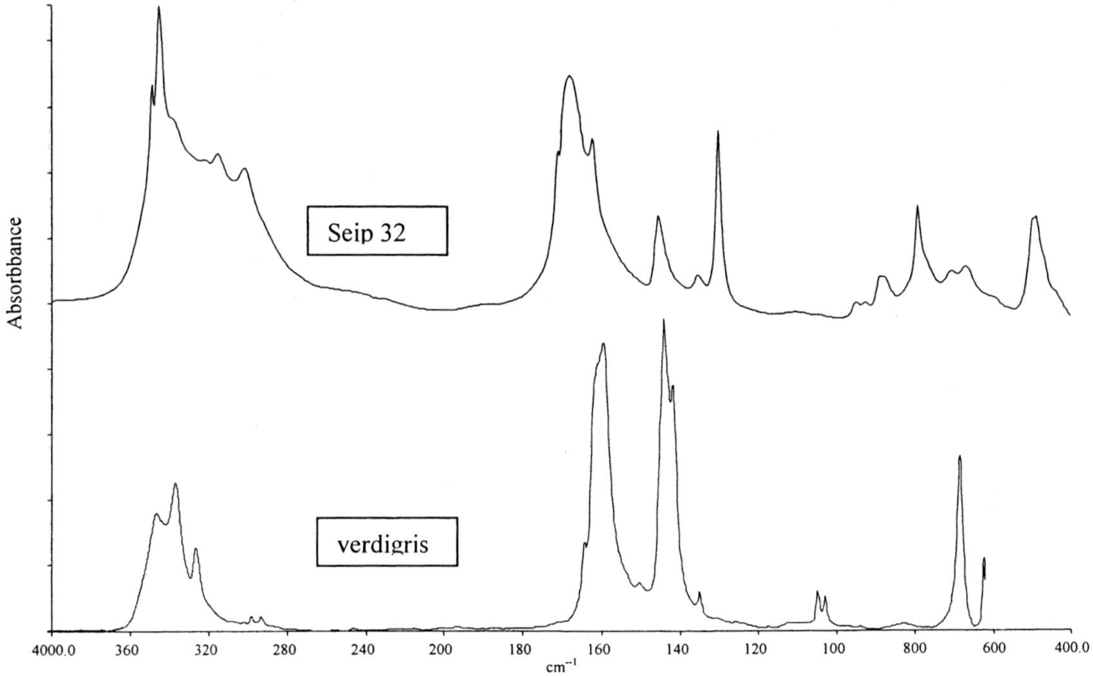

Figure 18. Infrared spectrum of Seip 32 Particulate Compared to verdigris (IRUG 00152) (19).

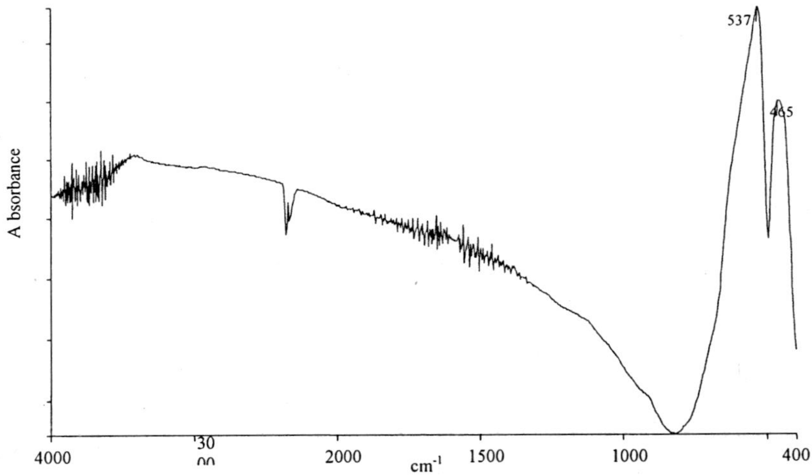

Figure 19. Infrared spectrum of iron oxide. Water and carbon dioxide bands are also visible.

Pigment Possibilities in Archaeological Textiles

Baldia (9) describes the particulate recovered from Seip 36 as bast fibers that are reddish in color and she reports that the fibers contain a significant amount of iron as determined by EDS. Not only does the infrared spectrum of the Seip 36 particulate show cellulose, it also contains absorbances in the region of 567 cm^{-1} and 470 cm^{-1} that reflect its iron oxide content (Figure 12). Seip 5, another material noted by Baldia (9) as charred bast with red deposits displays an infrared spectrum with the bands of cellulose and bands around 590 cm^{-1} and 463 cm^{-1} that can be attributed to iron oxide.

Dyes

While the identification of dye might not be possible by infrared study alone, the alteration of the infrared spectrum might be contributive to such an identification (*16, 17, 18*). The spectrum of bedstraw extract on polyethylene film is presented in Figure 21. In examining the infrared spectra of bedstraw dyed common milkweed (Figure 22), it can be seen that the bands differ as a result of dyeing treatment. While no outstanding extra bands can be seen that can be attributed to the anthraquinones of the dyes, differences in CH absorbance

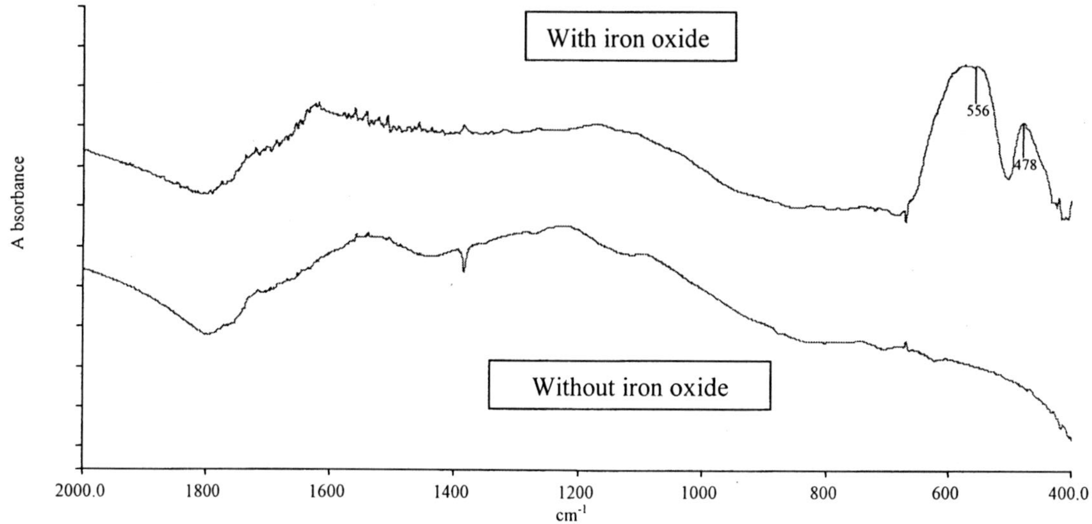

Figure 20. Infrared spectrum of charred rabbit hair with iron oxide compared to charred rabbit hair with no iron oxide.

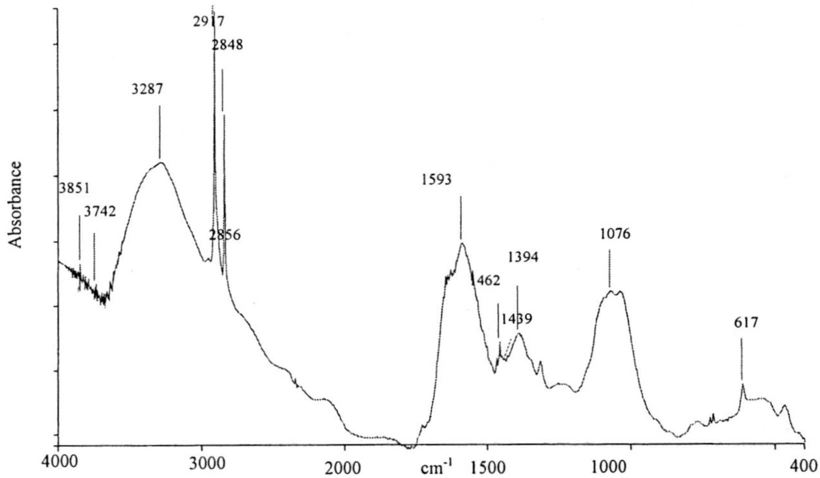

Figure 21. Infrared spectrum of bedstraw extract on polyethylene film

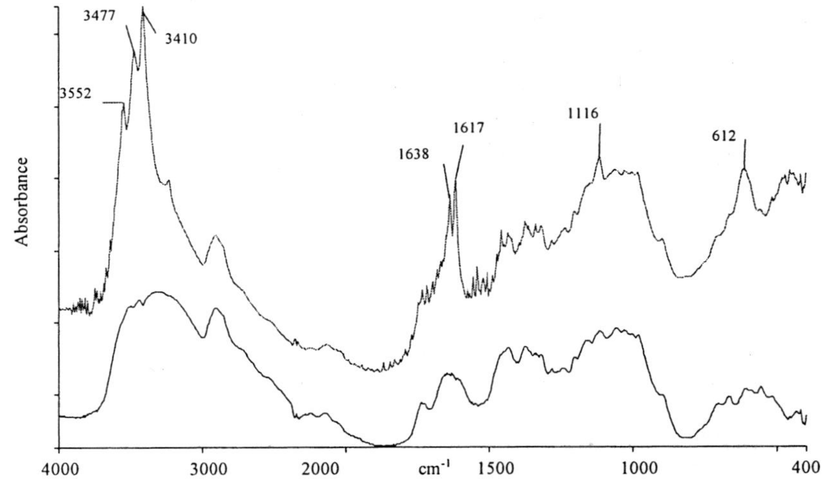

Figure 22. Infrared spectrum of bedstraw dyed milkweed compared to that of undyed milkweed.

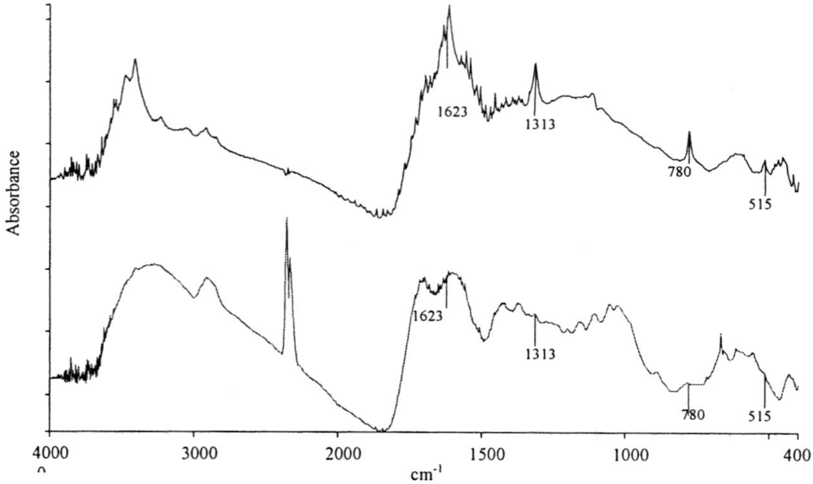

Figure 23. Infrared spectrum of charred dyed milkweed compared to that of charred undyed milkweed

above 3000 cm^{-1}, sharper bands in the 1615 and 1638 cm^{-1} region, and an increased absorbance at 612 cm^{-1} can be noted.

In examining charred dyed milkweed (Figure 23), the oxalates can be seen as before but the features of the dye are not readily apparent. Similarly in examining the infrared spectrum of Seip 36 (Figure 12), no oxalates or lignins can be seen although the overall shape is similar to that of a charred cellulosic. Some iron oxide is noted but no 617 cm^{-1} peak as was noted in fibers dyed with bedstraw. Much more work will need to be done on comparative dye materials, and the effective colorant add-on and the means to identify those colorants.

Conclusions

In examining the few archaeological textiles that survive, often the features of the fiber that might aid identification are difficult to see, whether because they are occluded with soil, or have deteriorated so that the key characteristics are no longer visible. Infrared spectra of the particulate material shed from these fragile materials may provide useful clues to determining their composition. Infrared spectra of comparative plant and animal fibers, including charred, mineralized, dyed, and pigmented examples, yielded multiple results:

- protein can be separated from cellulose, even in partially charred materials
- subgroups of plant fibers can be separated according to lignin and oxalate composition
- while lignins do not persist through charring, oxalates will survive and these can provide an additional means to classify fibers
- spectral characteristics of partially mineralized fibers can be distinguished and can be used to understand differences in their formation, i.e., in the archaeological context
- iron oxide pigment can be identified, even in charred samples, thus indicating that analysis of additional pigments may lead to more information on coloration in charred and uncharred textiles
- characteristics of a particular organic dye may be difficult to discern by infrared analysis alone, although some spectral differences due to the dyeing process may be apparent. In the latter case, the intriguing gradations of shades of black on the charred textiles encourage the attempt to investigate coloration of archaeological textiles further

Continued accumulation of spectra of comparative materials is desirable, with the ultimate goal of gathering enough spectra to create a training set for discriminant analysis and determination of key factors that distinguish archaeological plant or animal protein fibers. Particulate shed from archaeological materials provides a resource that can be examined for information about the textile without incurring damage to the textile.

References

1. Sibley, L. R.; Jakes, K. A. In *Conservation and Characterization of Historical Paper and Textile Materials*; Zeronian, H., Needles, H., Eds.; Advances in Chemistry Series No. 212; American Chemical Society: Washington, DC, 1986; pp 253–257.
2. Jakes, K. A.; Sibley, L. R. In *Archaeometry, Proceedings of the 25th International Symposium*; Maniatis, Y., Ed.; Elsevier: Amsterdam, 1989; pp 237–244.
3. Sibley, L. R.; Jakes, K. A. *Clothing and Text. Res. J.* **1989**, *7*, 37–45.
4. Sibley, L. R.; Jakes, K. A. In *Archaeometry of Pre-Columbian Sites and Artifacts*; D. A. Scott, D. A, Meyers, P. Eds.; The Getty Conservation Institute: Marina Del Rey, 1994; pp 396–418.
5. Sibley, L. R.; Jakes, K. A.; Larson, L. H. In *A Most Indispensable Art: Native Fiber Industries from Eastern North America*; Petersen, J. B., Ed.; University of Tennessee Press: Knoxville, 1996; pp 73–87.

6. Song, C. A. Ph.D. thesis, Ohio State University, Columbus, OH, 1991.
7. Song, C. A.; Jakes, K. A.; Yerkes, R. W. *Midcont. J. Archaeol.* **1996**, *21*, 247–266.
8. Thompson, A. J. Ph.D. thesis, Ohio State University, Columbus, OH, 2003.
9. Baldia, C. Ph.D. thesis, Ohio State University, Columbus, OH, 2005.
10. Jakes, K. A.; Ericksen, A. G. *Southeastern Archaeol.* **2001**, *20,* 56–66.
11. Sibley, L. R.; Jakes. K. A.; Swinker, M. E. *Clothing and Text. Res. J.* **1992**, *10*, 21–28.
12. Berni, R. J.; Morris, N. M. In *Analytical Methods for a Textile Laboratory*; Weaver, J. W., Ed.; American Association for Textile Chemists and Colorists: Research Triangle Park, NC, 1984; pp 265–292.
13. Willard, H. H.; Merritt, L. L.; Dean, J. A.; Settle, F. A. *Instrumental Methods of Analysis;* Wadsworth Publishing: Belmont, CA, 1988.
14. Colthup, N. B.; Daly, L. H.; Wiberley, S. E. *Introduction to Infrared and Raman Spectroscopy*; Academic Press: Boston, 1990.
15. Lang, P. A.; Katon, J. E.; O'Keefe, J. F.; Schiering, D. E. *Microchem. J.* **1986**, *34*, 319–331.
16. Abrahams, D. H.; Edelstein, S. M. *Am. Dyestuff Reporter* **1964**, *53*, 19–25.
17. Gillard, R. D.; Hardman, S. M.; Thomas, R. G.; Watkinson, D. E. *Stud. Cons.* **1994**, *39*, 187–192.
18. Hofenk de Graaff, J. H. *The Colourful Past: Origins, Chemistry and Identification of Natural Dyestuffs*; Archetype Publications: London, 2004.
19. Price, B; Pretzel, B. *Infrared and Raman User's Group, Edition 2000*; Philadelphia, PA. (URL http://irug.org/)
20. *Conservation and Art Material Encyclopedia Online*; Museum of Fine Art: Boston (URL http://www.mfa.org/_cameo/frontend/home.asp)
21. Jakes, K. A.; Sibley, L. R.; Yerkes, R. W. *J. Archaeol. Sci.* **1994**, *21*, 641–650.
22. Jakes, K. A.; Chen, H.-L.; Sibley, L. R. *Ars Textrina.* **1993,** *20*, 157–179.
23. Jakes, K. A. In *Archaeological Chemistry: Organic, Inorganic and Biochemical Analysis*; Orna, M. V., Ed.; ACS Symposium Series No. 625; American Chemical Society: Washington, DC, 1996; pp 202–222.
24. Jakes, K. A.; Mitchell, J. D. *J. Archaeol. Sci.* **1996,** *23,* 149–156.
25. Scurfield, G.; Michell, A. J.; Silva, S. R. *Bot. J. Linn. Soc.* **1973,** *66,* 277–289.
26. Tímár-Balázsy, Á; Eastop, D. *Chemical Principles of Textile Conservation*; Butterworth Heinemann: Oxford, UK, 1998.
27. Barker, B.; Owen, N. L. *J. Chem. Ed.* **1999**, *76*, 1706–1709.
28. Owen, N. L.; Thomas, D. W. *Appl. Spec.* **1989**, *43*, 451–455.
29. Evans, P. A. *Spectrochim. Acta* **1991**, *47A*, 1441–1447.
30. Chemical Abstracts Service Registry database, Scifinder, (March 2006).
31. Mukherjee, A. C.; Mukhopadhyay, U. *Text. Res. J.* **1983** *53,* 473–475.
32. Thompson, A.; Jakes, K. A. *Southeastern Archaeol.*; in press.

33. Mills, W. C. *Ohio Hist.* **1906**, *16*, 113–193.
34. Mills, W. C. *Ohio Hist.* **1909**, *18*, 269–313.
35. Mills, W. C. *Ohio Hist.* **1916**, *25*, 263–398.
36. Shetrone, H. C. *Ohio Hist.* **1926**, *25*, 5–227.
37. Shetrone, H. C.; Greenman, E.F. *Ohio Hist.* **1930**, *40*, 343–509.
38. Willoughby, C. C. *Ohio State Archaeol. and Hist. Quarterly* **1938**, *47*, 273–287.
39. Greber, N. *Midcont. J. Archaeol. Special Paper No. 5*. The Kent State University Press: Kent, OH, 1983.
40. Greber, N.; Ruhl, K. C. *Hopewell Site*; Westview Press: London, 1989.
41. Baldia, C.; Jakes, K. A. *J. Archaeol. Sci.* accepted.
42. Dimbleby, G. W. *Plants and Archaeology*; John Baker Publishers Limited: London, 1967.
43. Srinivasan, R.; Jakes, K. A . *J. Archaeol. Sci.* **1997**, *24*, 517–527.
44. Ercin, D.; Yurum, Y. *J. Anal. Appl. Pyr.* **2003**, *67*, 11–22.
45. Chen, H. L.; Jakes, K. A.; Foreman, D. W. *Textile Res. J.* **1996**, *66*, 219–224.
46. Chen, H. L.; Jakes, K. A.; Foreman, D. W. *J. Archaeol. Sci.* **1998**, *25*, 1015–1021.
47. Jakes, K. A.; Sibley, L. R. In *Archaeological Chemistry III*; J.B. Lambert, Ed.; Advances in Chemistry Series No. 205; American Chemical Society: Washington, DC, 1984; pp 403–424.
48. Martoglio, P. A.; Bouffard, S. B.; Sommer, A. J.; Katon, J. E.; Jakes, K. A. *Anal. Chem.* **1990**, *62*, A1123–A1128.
49. Jakes, K. A.; Katon, J. E.; Martoglio, P. A. In *Archaeometry '90*; Pernicka, E.; Wagner, G. A., Eds.; Birkhauser Verlag: Basel, 1990; pp 305–315.
50. Thompson, A. J. M. S. thesis, Ohio State University, Columbus, OH, 2000.
51. Baldia, C. M. S. thesis, Ohio State University, Columbus, OH, 2003.
52. Thompson, A. J.; Jakes, K. A. *Southeastern Archaeol.* **2002**, *21*, 252–257.
53. Zhbankov, R. G. *Infrared Spectra of Cellulose and Its Derivatives*; Consultants Bureau: New York, 1966.

Chapter 4

Extraction and Analysis of DNA from Archaeological Specimens

Brian M. Kemp[1], Cara Monroe[2], and David Glenn Smith[1]

[1]Department of Anthropology, University of California, Davis, CA 95616
[2]Department of Anthropology, University of California, Santa Barbara, CA 93106

The study of DNA extracted from archaeological specimens is an exciting new avenue of research that can provide unique evidence for addressing archaeological questions. Here we give an overview of notable case studies, some of which were performed by the authors, that used genetic data retrieved from archaeological specimens to make interpretations about the past. Additionally we describe how ancient DNA (aDNA) differs from modern DNA and, thus, why only a few specific genomic markers are usually targeted and why protocols have been developed explicitly for the study of aDNA. We detail an aDNA extraction protocol that we have been developing over the past few years, and a description of genetic screening we routinely perform on DNA extracted from ancient human samples.

Introduction

DNA extracted from archaeological specimens, ancient DNA (aDNA), provides the researcher a window into prehistory, allowing one to say something absolute about the genetic characteristics of someone or something that lived in the past. While modern DNA studies can make predictions about the past based on genetic theory, only aDNA studies can directly test these hypotheses by examining genetic variation at specific temporal and geographic axes. Through the analysis of genetic patterns found in archaeological assemblages, these data can be used, for example, to study human migration, to address the relationships between prehistoric and contemporary populations, to study kinship patterns revealed in burial contexts, to reconstruct prehistoric diets, and to study human behavior. Furthermore, the study of aDNA allows for the observation of molecular evolution over time and, from this, its rate can be estimated. However, these potential benefits must be weighed against the methodological challenges presented by the study of aDNA. The study of aDNA is more difficult than the study of modern DNA primarily because aDNA tends to be degraded, chemically modified, and in low copy number (*1–3*). As such, particular care must be taken in the study of aDNA and a high level of expertise is required to ensure authentic results. Below, we review some studies that have addressed anthropological questions with aDNA data retrieved from a wide variety of archaeological specimens that span both time and geography. We then proceed to outline a protocol that has been developed over the past few years in David Glenn Smith's Molecular Anthropology Laboratory at the University of California, Davis. In particular, this protocol addresses two of the major complications associated with the study of aDNA: modern DNA contamination and polymerase chain reaction (PCR) inhibition.

Examples of Ancient DNA Applications to Topics in Archaeology

The following eight examples are meant to provide a representative survey of the wide range of topics that have been addressed with aDNA evidence. The first three examples are ones that followed the methodology described later in this paper. The additional five examples are ones that we find to be particularly interesting and, in our opinion, were well executed.

Molecular Sex Determination and Human Behavior

Recently, de la Cruz and colleagues (*4*) used aDNA techniques to determine the sex of sacrificial victims that were buried under Temple R, dedicated to the Aztec god of wind and rain, Ehecatl-Quetzalcoatl, at the archaeological site of Tlatelolco. It is believed that these individuals were sacrificed and buried in a single ceremony that signified the founding of the temple. In this case, traditional morphometric techniques were insufficient for sex determination, as most of the individuals were infants, sub-adults, and/or were represented by fragmentary remains. The results of the molecular analyses suggest that most, if not all, of these sacrificial victims were males and provide valuable insight into Aztec ritualistic behaviour.

Coprolites and Population Movement

Kemp and colleagues (*5*) successfully analyzed human mtDNA extracted from eight samples of 700–2,000 year old human coprolites excavated from Fish Slough Cave located in northern Owen's Valley, California. It was discovered that individuals belonging to a rare, derived branch of mitochondrial haplogroup C made five of the fecal samples. A survey of mtDNA variation of Native Americans revealed that the specific haplotype represented by these coprolites is only found today among the Northern Paiute and Washo, who reside in the Great Basin (*6*). This evidence provides a genetic connection between the Northern Paiute and prehistoric Owen's Valley populations and is significant as the Numic sub-branch (including Northern Paiute) of the Uto-Aztecan language family was thought to have spread into the Great Basin (*7, 8*) from Owen's Valley, around the time that the inhabitants of Fish Slough Cave produced the ancient coprolites.

Detection of an Ancient Coastal Migration by Native Americans

Recently, Kemp and colleagues (*9*) analyzed mtDNA from 10,300 year old skeletal remains that were excavated from On Your Knees Cave on Prince of Wales Island, Alaska. The mtDNA of this individual was determined to belong to a newly recognized founding maternal lineage that was carried to the Americas from Asia. It was discovered that a mere 1.4% of Native Americans descend from this founding matriline and that these individuals today live primarily along the western edge of North and South America, being found in, for example the Chumash in Southern California, the Cayapa of Ecuador, and populations of Tierra del Fuego. The evidence is compatible with an ancient, rapid migration of humans along the west coast of the Americas.

Biological Relationships Determined from Burial Context

Keyser-Tracqui and colleagues (*10*) successfully extracted aDNA from 62 individuals from the Eygin Gol Necropolis located in Northern Mongolia, dating to the 3rd century B.C. to the 2nd century A.D. Variation detected in the mitochondrial, autosomal, and Y-chromosomal genomes allowed the researchers to correlate biological relationships of individuals buried in the cemetery to both space and time. Some sections of the cemetery were comprised of related males, possibly indicating an elite patriline that continued over multiple generations. Parent-child relationships were also determined, in some cases indicating that particular portions of the cemetery represent familial groups. Later burials demonstrated a shift in the genetic signature of the population, which is thought to be the evidence of an entrance of Turks in the area.

The Study of Animal Population Demography Over Time

Recently, Shapiro and colleagues (*11*) used Bayesian statistical techniques to model Asian and American bison demographics over a 60,000 year period using aDNA data retrieved from 442 bison remains. Their results support the hypothesis that bison populations began a dramatic decline approximately 37,000 years ago, likely as a result of environment change. Shapiro and colleagues suggest that, if other species were similarly affected, human hunting may not have been the major cause of the extinction of American megafauna.

Dietary Reconstruction

Poinar and colleagues (*12*) successfully extracted DNA from three human coprolites made by Native Americans in Hinds Cave, Texas more than 2,000 years ago. Not only were they able to extract human mtDNA from the coprolites, which were determined to belong to known Native American mitochondrial types, but they also were able to analyze DNA from the plants and animals eaten by these occupants of Hinds Cave. Importantly, they discovered that some of the species of plants and animals identified by molecular techniques were not detected in the same feces by macroscopic techniques (and some macroscopically identified species were not detected by the DNA analysis). The combination of the molecular and macroscopic evidence demonstrated that these three Native Americans had a diverse and well-balanced diet.

Study of Economic Stratification of Prehistoric Populations

Using aDNA techniques, Speller and colleagues (*13*) were able to identify the species of ~1,200 year old salmon remains excavated from various houses and structures at the archaeological site of Keatley Creek, British Columbia. Prior study of the site provided evidence of a socio-economic distinction among the residents of these dwellings based on the distribution of prestige goods and the density and variety of cultural and faunal remains excavated from the features. The researchers sought to test whether this socio-economic distinction resulted in differential access to preferred salmon species. Their results suggest that economic stratification had less effect on the distribution of preferred salmon species within the structures than previously thought. Further, they found no evidence of pink salmon (*Oncorhynchus gorbuscha*) at the site, despite the assumption that this species was a staple in the area.

Study of Disease

Ancient DNA studies have also focused on determining the presence or absence of disease among human, as well as animal populations. Evidence from these studies has contributed significantly to osteological studies of signs of infection with ambiguous etiological origins. The study of ancient disease can also trace the evolutionary trajectory of pathogens and their dispersal through time. For example, Salo et al. (1994) identified tuberculosis (*Mycobacterium tuberculosis*) from a 1,000–year-old Peruvian mummy from the site of Chiribaya Alta, effectively dispelling the notion that tuberculosis was first introduced at the time of European contact. Additional studies of disease include: plagues (*14, 15*), leprosy (*16*), syphlis (*17, 18*), malaria, and gastrointestinal pathogens (*19*).

Problems and Properties of aDNA

Sample Selection

While the first successful study of aDNA was from the hide of an extinct quagga (*20*), the vast majority of aDNA studies since have been performed on bones and teeth, with fewer studies performed on preserved or mummified tissue (for an excellent review see *21*). It is also possible to extract aDNA from coprolites (*5, 22, 23*), quids (chewed wads of plant fiber), and “aprons” (women’s breechcloths) (*24*). Similarly, DNA has also been extracted from ancient pathogens, faunal remains, and botanicals.

The choice of samples will obviously depend on the question that is being posed. For example, plant and animal DNA extracted from human coprolites (e.g., *22*) or DNA extracted from the remains of animals found in house structures (e.g., *13*) would be useful in the study of prehistoric dietary practices. However, sample choice also depends on the availability of material from times and places of interest. In this sense, where human skeletal samples are absent from the archaeological record, are in a poor state of preservation, and/or are simply restricted from study[1], coprolite or quid samples may represent the only material through which ancient population genetics can be studied (*24*). Some aspects of human behavior and prehistory may be inferred from the genetic patterns revealed in domesticates and, therefore, samples of plants and animals may suffice.

Further consideration of sample selection is based on correlations between the type and physical state of the archaeological specimens and the probability of successfully extracting DNA from them. While much of this has been well-covered by review papers (*25–27*), a few points are worth re-emphasizing here, with additional reference. In general, well-preserved remains are more likely to contain well-preserved DNA; sampling material that exhibits signs of degradation due to microbial activity, burning, highly acidic or basic environments, cracked bones or teeth, or teeth with caries should be avoided, if possible. Samples that cannot be radiocarbon dated are also poor candidates for DNA extraction, as DNA degrades in a similar fashion to other biomolecules, and a positive correlation between the preservation of collagen and DNA has been demonstrated in ancient remains (*28*). While, the age of the sample is negatively correlated with DNA preservation, with younger samples having a higher likelihood of containing preserved DNA, environmental conditions probably have a greater effect on DNA preservation than does time alone. That is, older samples from cold, dry climates might contain better preserved DNA than younger samples excavated from hot and/or wet environments. Additionally, DNA has been shown to preserve better in teeth than in bone (*9, 29*), and cortical bone (compact) is preferable over cancellous (spongy) bone for DNA extraction. Lastly, as for yet unknown reasons, DNA has been shown to preserve remarkably well in coprolites (*5, 22, 23*).

Molecular Markers

Due to the high copy number of the mitochondrial genome per cell (>1,000), in comparison to the nuclear genome (two copies or each marker per cell), this system has been the focus of most aDNA studies. In mammals, this genome is passed solely through the female line and, therefore, is only informative about female prehistory. The male analogue of mitochondrial DNA (mtDNA) is the Y-chromosome; markers on this chromosome and others from the nuclear genome are much more difficult to analyze from aDNA extracts due to low copy number. As such, molecular sex identification, addressing paternal relationships, and/or addressing male prehistory is generally not often as feasible, but may be worth considering depending on the archaeological context and the level of DNA preservation in the sample.

Contamination Control

As DNA extracted from ancient remains tends to be in low copy number and is generally degraded into fragments less than 200 base pairs (bp) in length (*1, 2*), aDNA extractions are highly susceptible to contamination originating from modern sources. This is particularly a problem when one attempts the study of ancient DNA from human sources. Modern contaminating DNA can be in higher copy number and more fully intact than the endogenous aDNA and, thus, can better compete than aDNA as a target for amplification during the polymerase chain reaction (PCR) (for those unfamiliar with the process, it is explained very well by *30*). In fact, contamination can completely out-compete aDNA in PCR leading to false positives and aberrant results (*31*). The detection of contamination is dependent on the specificity of the PCR. For example, human studies are particularly prone to contamination because the DNA of any human that has come into contact with the remains can potentially be amplified. This problem is not, however, peculiar to human studies. Bacteria from the soil that are closely related to targeted DNA of pathogen can also contaminate samples studied for the presence or absence of pathogens. Contamination is less of a problem, for instance, if salmon DNA is targeted while the sample is contaminated with human DNA because a properly designed PCR that targets salmon should not be able to amplify human DNA. In spite of this, we cannot stress enough that contamination is a serious concern and all samples, regardless of type, should be treated with the same level of care to minimize contamination.

Ancient DNA extractions can become contaminated via two sources: surface contamination of the archaeological specimens through bare-hands handling of the material or later during DNA extraction and analysis in the DNA laboratory. The former source of contamination can originate at any step of an aDNA study from the time of excavation to the time of DNA extraction. Modern contamination can originate with anyone who has had direct contact with the material, including the archaeologists that excavated the specimens and/or researchers that analyzed (e.g., cataloging, measuring) the specimens. This type of contamination can be minimized by following the advice of Yang and Watt (*32*) for preparing archaeological specimens for DNA analysis. The surfaces of archaeological specimens are especially vulnerable to contamination, and it is crucial that any contamination be removed before DNA extraction begins. Contaminating DNA can be removed from the surfaces of bones and teeth with a treatment of high concentration bleach (NaOCl, sodium hypochlorite) because aDNA is more resistant to the oxidant, than is contaminating DNA (*31, 33, 34*). This is believed to result from the fact that ancient endogenous DNA is protected within crystal aggregates of the bone (*34*), an observation that is consistent with the idea that DNA that is bound to hydroxyapatite aids not only in its long-term preservation, but also protects it from oxidation by highly concentrated bleach

(*31*). A method for the decontamination of non-skeletal remains (e.g. coprolites and quids) has not been developed. While many researchers use ultraviolet (UV) irradiation to "destroy" contamination (see studies reviewed in *31, 33*), no study has demonstrated that this technique is actually effective in removing contamination on archaeological specimens (*33*). One way to avoid potential contamination when extracting DNA from coprolites and quids is to take samples from the interior portions of the specimens.

The latter source of contamination can originate from reagents, labware, PCR carryover, and/or DNA lab personnel. As such, procedures that reduce contamination should be implemented, including using of DNA free lab-ware and reagents, processing ancient materials in a laboratory separated from the one in which modern DNA is examined, and using negative controls in both DNA extraction and amplification to monitor contamination, if present (following the advice of *35*). Contamination can also be removed from PCR cocktails, prior to amplification, with DNAase (*36*).

PCR Inhibitors

Another major problem associated with the extraction of DNA from archaeological specimens is that the procedure often co-extracts impurities that can later complicate, or prevent, the study of the extracted DNA by inhibiting PCR amplification (reviewed by *5*). Commonly encountered inhibitory substances found in aDNA extracted from teeth, bones, mummified tissue, and coprolites include humic acids, fulvic acids, tannins, porphyrin products, phenolic compounds, hematin, and collagen type I (*37–42*). The formation of Maillard products, commonly encountered in coprolite samples, can also prevent PCR amplification by causing DNA to become inaccessibly trapped in these sugar-derived condensation products (*12*). As the negative results in many aDNA studies are attributed to the presence of PCR inhibitors, our extraction method outlined below pays particular attention to the problem and offers a simple test for the presence of PCR inhibitors in DNA extracts.

Ancient DNA Extraction and Analysis

DNA Extraction[2]

The following DNA extraction protocol is one that has been developed and reported in two of our studies (*5, 43*). Remove ≤ 0.5 g from the whole sample.[3]

With the sample held firmly by a vise, portions of bones and teeth can be carefully removed by sawing off portions with one-time use "disposable" hacksaw blades. This is a relatively cost-effective alternative to the use of a Dremmel tool, the blades for which are quite expensive in comparison to hacksaw blades. Moreover, using the hacksaw provides more control over the dispersal of bone and tooth powder during sawing. An alternative method, that works particularly well with rib samples, is to gently snap pieces of bone off the whole, while being held in a closed Ziploc bag. Portions of coprolites and quids can be removed from the whole by using one-time use disposable X-acto blades. Interior portions of coprolites and quids should be sampled, if possible, to avoid sampling any surface material that may have unintentionally become contaminated. Scissors and/or tweezers used to tease apart samples should be cleaned by submerging them in a 50% v/v bleach solution for 5–10 minutes, and then wiping them clean with paper towel. Over time this treatment of the tools can lead to rust, at which time the tools should be replaced.

If the sample is of bone or tooth, submerge it in full strength Clorox bleach (6% NaOCl) for 15 min; otherwise add the sample directly to EDTA, as described below. Rinse the sample well with DNA free water (Gibco) to remove the bleach (1–2 times). In a 15 mL conical tube made of polypropylene (polystyrene will melt if exposed to the mix of phenol:chloroform, below), submerge the sample in molecular grade (DNA free) 0.5 M EDTA, pH 8.0 (Gibco) for > 48 hours. An extraction control to which an equal volume of EDTA is added, but contains no sample, should accompany the extraction and be subjected to all of the following steps. To the sample add 3 mg of Proteinase K (Invitrogen, Fungal Proteinase K) and incubate at 65° C for approximately 4 hours.[4]

Extract DNA from the digested sample using a three-step phenol/chloroform method. First, extract DNA by adding an equal volume of phenol:chloroform:isoamyl alcohol (25:24:1) to the EDTA containing the sample. Thoroughly mix the layers by inverting the tubes vigorously and then centrifuge the tubes at 3100 rpm for 5 min. Remove the aqueous (top) layer, which contains the DNA, and place it in a new tube. Add to this an equal volume of phenol:chloroform:isoamyl alcohol (25:24:1) and re-extract the aqueous phase as just described. Perform a third extraction by adding an equal volume of chloroform:isoamyl alcohol (24:1) to the aqueous phase just removed. Thoroughly mix the layers by inverting the tubes vigorously. Centrifuge the tubes this time at 3100 rpm for 3 min. Remove the aqueous phase and place it into a new tube.

To facilitate removal of co-extracted PCR inhibitors (*5, 44*), precipitate DNA from the solution by adding one half volume of room temperature 5 M ammonium acetate and, to this combined volume, one volume of room temperature absolute isopropanol. Store the solution for a minimum of 7 hours at room temperature (overnight storage at room temperature is acceptable). Centrifuge the tubes for 30 min at 3100 rpm to pellet the DNA. At this point there may or may not be a visible pellet, but this is not an indication or whether the sample contains preserved DNA or not. However, in most cases in which PCR inhibitors have been co-extracted, a pellet is visible as a brown colored smear. Carefully decant the isopropanol from the tube and allow the tubes to air-dry, inverted, for 15 min. Wash the DNA pellet by adding 1 mL of 80% ethanol to the tubes followed by 30 s of vortexing, making sure to dislodge the pellet, if visible, from the side of the tube. Centrifuge the tubes again for 30 min at 3100 rpm. Carefully decant the ethanol[5] and again air-dry the tubes in an inverted state for 15 min.

Suspend the DNA in 300 µL of DNA-free double-distilled water (i.e., ddH_2O) and perform a silica extraction (*45*) using the Wizard PCR Preps DNA Purification System (Promega), following the manufacturer's instructions except that: 1) the "Direct Purification Buffer" need not be added (as it is only used for purifying PCR product from reactions containing mineral oil) and 2) finally elute with 100 µL DNA-free ddH_2O. We recommend the use of the Promega silica extraction kit because they ensure contamination free reagents and it is a relatively inexpensive and quick procedure.[6]

PCR Amplification: Screening SNPs, Molecular Sex Identification, and Sequencing Reactions

Even if contamination is removed from the surface of archaeological specimens (in the case of bones and teeth) or is intentionally avoided by sampling in interior portions of samples (in the case of coprolites and quids), contamination arising during PCR amplification remains a serious problem. We have suggested that using increased amounts of *Taq* polymerase to combat PCR inhibitors (*46–48*) is counterproductive, as it simultaneously increases the sensitivity of PCR reactions to contamination (*5*). In fact, we have countered the hypersensitive nature of PCR (*49*) by reducing the amount of *Taq* polymerase used in our PCR reactions. Additionally, we recommend that small volume (15 µL) PCR reactions be used, when possible, for screening single nucleotide polymorphisms (SNPs) and length polymorphisms, as smaller-sized reactions will ultimately exhaust less DNA extracted from archaeological specimens. We further advocate amplifying small (<200 bp) fragments when working with aDNA, as larger fragments tend to exhibit artificial "mutations" caused by chemical damage (*34*).

Here we describe the PCR conditions that we use routinely to amplify aDNA for screening SNPs and length polymorphisms found in the mitochondrial genome. Set up a 15 µL PCR amplification reaction containing: 0.32 mM dNTPs, 1X PCR Buffer, 1.5 mM $MgCl_2$, 2.4 mM primers, 0.3 U of Platinum *Taq* (Invitrogen), and 1.5 µL of template DNA. At least one negative control, to which no DNA is added, should accompany every set of PCR reactions to monitor the presence of contaminating DNA. A positive modern control, added outside the aDNA PCR set-up laboratory, accompanying the reactions can preclude PCR failure as an explanation for the aDNA samples failing to amplify. Examples of primers that we use for screening the definitive polymorphisms of Native American mtDNA haplogroups A, B, C, D, and X (*50–52*) are provided in Table I. Conduct the PCR as follows: denaturation at 94°C for 3 min, followed by 40 consecutive cycles of 15 second holds at 1) 94°C, 2) the annealing temperature (found in Table I), and 3) 72°C, followed by a final three minute extension period at 72°C. Digest the amplicons with the appropriate restriction enzyme and score sizes as described by Kemp and colleagues (*5*).

In cases when a sample yields particularly well-preserved DNA, molecular sex determination can be performed by screening a portion of the amelogenin gene that exhibits a 6 bp deletion on the X-chromosome relative to its homologue on the Y-chromosome (*53*).[7] The primers for this PCR reaction are found in Table I. Set-up the PCR as just described and run the reaction for 60 cycles. The use of an additional 20 cycles (40 cycles are typically used to amplify mtDNA) helps compensate for the fact that only one pair of sex chromosomes are present in each cell.

As the human mtDNA displacement loop (D-Loop, nucleotide positions 16024–00576) exhibits a great amount of variation, it represents an ideal region to sequence in ancient specimens for comparative purposes with the mtDNA of living and ancient humans. We have designed primer pairs for sequencing the D-Loop in 11 small (<200 bp), overlapping fragments (*9*). These primers are shown in Table II. For sequencing, set up 30 µL PCR amplification reactions containing: 0.32 mM dNTPs, 1X PCR Buffer, 1.5 mM $MgCl_2$, 2.4 mM primers, 0.3 U of Platinum *Taq* (Invitrogen), and 3.0 µL of DNA template. Note that the same amount of *Taq* polymerase is used in these reactions as in the ones described above, notwithstanding that these are twice the volume. Perform touch-down PCR (*54*) using the following cycling conditions: 3 min denaturing at 94°C, followed by 60 cycles of 15 second holds at 1) 94°C, 2) the annealing temperature (found in Table II, decreased 0.1°C after each cycle), and 3) 72°C, followed by a final 3 min extension period at 72°C. A description of how to prepare PCR products for direct sequence can be found in Kemp and colleagues (*5*). Sequence the amplicons in both directions to preclude sequencing errors.

PCR Inhibitors and Repeat Silica Extraction

As mentioned above, impurities co-extracted with DNA from archaeological specimens can make the study of the DNA difficult, if not impossible. Through a

Table I. Primers used to screen the definitive markers of Native American mitochondrial haplogroups A, B, C, D, and X, and those used for molecular sex determination.

Target Region	*Defining Marker**	*Primer*	*Primer Coordinates†/ Sequences (5' to 3')*	*Annealing Temperature*	*Primer Citation*
Mitochondrial Haplogroup A	*Hae*III 663 (+)	611F	00591–00611 ACCTCCTCAAAGCAATACACTG	55°C	(*58*)
		743R	00743–00765 GTGCTTGATGCTTGTTCCTTTTG		
Mitochondrial Haplogroup B	9 base pair deletion	8215F	08195–08215 ACAGTTTCATGCCCATCGTC	55°C	(*59*)
		8297R	08297–08316 ATGCTAAGTTAGCTTTACAG		
Mitochondrial Haplogroup C	*Alu*I 13262 (+)	13256F	13237–13256 ATCGTAGCCTTCTCCACTTC	55°C	(*60*)
		13397R	13397–13419 TCCTCCTATTTTTCGAATATCTT		
Mitochondrial Haplogroup D	*Alu*I 5176 (-)	5120F	05099–05120 CCTAACTACTACCGCATTCCTA	55°C	(*60*)
		5190F	05190–05211 GGGTGGATGGAATTAAGGGTGT		
Mitochondrial Haplogroup X	*Acc*I 14,465, (+)	14440F	14421–14440 CTGACCCCCATGCCTCAGGA	49°C	(*61*)
		14591R	14591–14612 CCATAAATAGGAGAAGGCTTAG		
Amelogenin: Sex Determination	Male=106/112 Female=106/106	Amel-A	CCCTGGGCTCTGTAAAGAATAGTG	55°C	(*53*)
		Amel-B	ATCAGAGCTTAAACTGGGAAGCTG		

* (*50–52*), † Coordinates, numbered according to the Cambridge Reference Sequence (*62, 63*).

Table II. D-Loop sequencing primers and their annealing temperatures.*

Target Region	*Primer*	*Coordinates*[†]/ *Sequence (5' to 3')*	*Annealing Temperature*
D-Loop 1	15986F	15986–16010 GCACCCAAAGCTAAGATTCTAATTT	62°C[#]
	16153R	16132–16153 CAGGTGGTCAAGTATTTATGGT	
D-Loop 2	16106F	16106–16126 GCCAGCCACCATGAATATTGT	62°C[#]
	16251R	16230–16251 GGAGTTGCAGTTGATGTGTGAT	
D-Loop 3	16190F	16190–16209 CCCCATGCTTACAAGCAAGT	58°C[#]
	16355R	16331–16355 GGGATTTGACTGTAATGTGCTATGT	
D-Loop 4	16232F	16232–16249 CACACATCAACTGCAACT	58°C[#]
	16404R	16383–16404 GGTGGTCAAGGGACCCCTATCT	
D-Loop 5	16353F	16353–16372 CCCTTCTCGTCCCCATGGAT	62°C[#]
	16549R	16530–16549 GGGGAACGTGTGGGCTATTT	
D-Loop 6	16470F	16470–16493 GGGGGTAGCTAAAGTGAACTGTAT	62°C[#]
	00106R	00082–00106 CGGCTCCAGCGTCTCGCAATGCTAT	

D-Loop 7	00034F	00034–00058 GGGAGCTCTCCATGCATTTGGTATT	62°C#
	00185R	00160–00185 CCTGTAATATTGAACGTAGGTGCGAT	
D-Loop 8	00112F	00112–00135 CCCTATGTCGCAGTATCTGTCTTT	62°C#
	00275R	00249–00275 CTGTGTGGAAAGTGGCTGTGCAGACAT	
D-Loop 9	00184F	00184–00208 GGCGAACATACTTACTAAAGTGTGT	62°C#
	00356R	00331–00356 GGGGTTTGGCAGAGATGTGTTTAAGT	
D-Loop 10	00255F	00255–00279 GCACAGCCACTTTCCACACAGACAT	62°C#
	00436R	00415–00436 GGGGTGACTGTTAAAAGTGCAT	
D-Loop 11	00369F	00369–00393 CCCTAACACCAGCCTAACCAGATTT	62°C#
	00560R	00538–00560 GGGGTTTGGTTGGTTCGGGGTAT	

*These primers were described by Kemp and colleagues(*9*).

† Coordinates, numbered according to the Cambridge Reference Sequence (*62, 63*).

Touch-down PCR (*54*) used, decreasing the annealing temperature 0.1°C after each cycle.

series of experiments we have developed a technique called "repeat silica extraction" that has demonstrably removed PCR inhibitors from a variety of samples originating from a number of different archaeological contexts (*5*). DNA is introduced into a chaotrophic solution (e.g., 5.4M guanidine thiocynate) that causes it to bind to silica particles, which in the Wizard PCR Preps DNA Porufication System (Promega) are contained within a column. The column is subsequently washed with isopropanol, removing impurities from the sample. Lastly, when water is passed through the column, the cleaned DNA returns to solution. In some cases the silica extraction needs to be repeated multiple times in order to sufficiently remove PCR inhibitors from the DNA extract to permits its amplification (*5*).

If an aDNA extract fails to PCR amplify it should be tested for the presence of PCR inhibitors. This test requires the availability of an authenticated aDNA sample to be used as a "positive" control.[8] Set up side-by-side PCR reactions containing: 1) the template suspected to contain inhibitors, to which is added a volume of the ancient positive control equivalent to that of the template, 2) only the template suspected to contain inhibitors and 3) only the positive ancient control. This side-by-side comparison will allow for the preclusion of PCR failure due to factors other than inhibition (e.g. the stochastic nature of PCR amplification). If the template "spiked" with the positive ancient control (reaction #1) permits its amplification, while the template suspected of containing inhibitors (reaction #2) fails to amplify, the template is likely free of inhibitors and, therefore, does not contain a sufficient amount of DNA for analysis. Alternatively, if the first PCR reaction fails to amplify, whereas the third reaction does amplify, the template is concluded to contain inhibitors. In this case, the silica extraction should be repeated, as described above, and PCR reattempted. Our studies have shown that as may as four repeat silica extractions may be required to sufficiently remove PCR inhibitor from DNA extracts, despite the inherent loss of DNA yield associated with each repetition of the silica extraction (*5*).

Authentication and Assessment of aDNA Results

The strength of aDNA evidence is contingent on the demonstration that positive results are the product of endogenous DNA extracted from the archaeological specimens, and not a product of exogenous contamination. The success of aDNA research is reliant on the collection of flawless and authenticated data. While numerous researchers have provided guidelines for demonstrating the authenticity of aDNA results, we concur with the opinion of Gilbert and colleagues (*55*) that researchers should take a "cognitive and self-critical approach." First, results must be replicable. Minimally, the laboratory that generated the results should demonstrate that at least a portion could be replicated. Depending on the uniqueness of the sample (e.g., a sample of great antiquity) or the data retrieved from it (e.g., a genetic type never before reported), authentication of results may require that separate laboratories generate identical results, for example, from separate skeletal elements of the same individual. The importance of replicating aDNA results is only matched by being able to properly assess the data themselves by asking whether or not the results make sense. Being able to answer this question requires not only experience working with ancient DNA, but also a familiarity with the question being addressed and the subjects being studied [refer to Gilbert and colleagues (*55*) for a more thorough treatment of authentication of ancient DNA results].

Conclusion

As illustrated by the examples above, the analysis of DNA extracted from archaeological specimens can provide unique insights about the past. However, the study of aDNA is methodologically challenging, primarily due to the problematical state of its preservation. As such, protocols developed specifically for the extraction and analysis of ancient DNA have been developed, of which we have highlighted one that we have developed over the past few years. Our protocol offers solutions to two of the most common problems associated with the study of aDNA: the possible presence of contamination on the surfaces of samples and/or the co-extraction of PCR inhibitors. We are optimistic that with attention to proper aDNA protocols, data acquisition, and authentication of results, DNA extracted from archaeological specimens will continue to provide a wealth of information about the past.

Endnotes

[1] Extraction from ancient remains is a completely destructive process and, as such, this process might not be permitted, for example, on extremely rare samples. Likewise, in the United States, Native American human remains that are to be repatriated to their descendants (Native American Graves Protection and Repatriation Act, or NAGPRA) may be unavailable for study.

[2] If available, all reagents should be bought from manufacturers that guarantee them to be contamination free (we have found this to a reliable claim). If not available, for example in the case of 5M Ammonium Acetate used in our protocol, make the reagents with DNA-free water (Gibco) and filter them through 0.2 μm filters (Nalgene, or others).

[3] We do not grind the material into powdered form. Some may prefer to do so as it obviously increases the surface area of the sample. We avoid this procedure as it also increases the opportunity for the introduction of contamination. However, even DNA extracted from powered bone sample appears resilient to bleach and treating powdered bone with bleach also removes some inhibitors (*34*).

[4] The activity of this Proteinase K is optimal at 65°C, but the protein will rapidly denature at higher temperatures. One should take care that the incubation temperature does not exceed 65°C.

[5] If the pelleted DNA is visible at this stage, it will appear gooier than after the isopropanol precipitation and will tend to slide down the side of the tube when decanting the ethanol. Decanting the ethanol very slowly and slightly twisting the tube during the process will cause the pellet to be stuck on the side of the tube. It is a very similar process to pouring a fine wine without causing it to dribble on the tablecloth. If you are primarily a beer drinker, ask your local sommelier for a demonstration of this technique.

[6] At the time of this publication, the Promega silica extraction kit (when bought in bulk) costs about $2 per extraction. Once familiar with the process, each extraction takes about 15 min.

[7]Additional methods for molecular sex determination have been developed (*56, 57*).

[8] Kemp and colleagues (*5*) have suggested that this test is improperly conducted with a modern DNA positive control. For reasons unclear at this time, the effect of PCR inhibition on DNA is influenced by the concentration and/or quality of the DNA template. This aspect of aDNA research has been poorly explored, but a more thorough understanding of this phenomenon will likely improve our ability to study DNA extracted from archaeological specimens.

Acknowledgments

Past members of the David Glenn Smith lab that have contributed to our understanding of how to extract and analyze ancient DNA include: Joseph G. Lorenz, Frederika A. Kaestle, Ripan S. Malhi, Jason A. Eshleman, Beth A. S. Shook, and Deborah A. Bolnick.

References

1. Lindahl, T. *Nature (London)* **1993**, *362*, 709–715.
2. Pääbo, S. In *PCR Protocols: A Guide to Methods and Applications*; Innis, M. A., Ed.; Academic Press: San Diego, 1990; pp 159–166.
3. Gilbert, M. T. P.; Shapiro, B.; Drummond, A. J.; Cooper, A. *J. Archaeol. Sci.* **2005**, *32*, 1053–1060.
4. De la Cruz, I.; González-Oliver, A.; Kemp, B. M.; Román, J. A.; Smith, D. G.; Torre-Blanco, A. *Curr. Anthropol.* Submitted.
5. Kemp, B. M.; Monroe, C.; Smith, D. G. *J. Archaeol. Sci.* In Press.
6. Kaestle, F. A. Ph.D. thesis, University of California, 1998.
7. Bettinger, R. L.; Baumhoff, M. A. *Am. Antiq.* **1982**, *47*, 485–503.
8. Lamb, S. M. *Int. J. Am. Linguist.* **1958**, *24*, 95–100.
9. Kemp, B. M.; Malhi, R. S.; McDonough, J.; Bolnick, D. A.; Eshleman, J. A.; Rickards, O.; Martinez-Labarga, C.; Johnson, J. R.; Lorenz, J. G.; Dixon, E. J.; Fifield, T. E.; Heaton, T. H.; Worl, R.; Smith, D. G. *Am. J. Phys. Anthropol.* In Review.
10. Keyser-Tracqui, C.; Crubezy, E.; Ludes, B. *Am. J. Hum. Genet.* **2003**, *73*, 247–260.
11. Shapiro, B.; Drummond, A. J.; Rambaut, A.; Wilson, M. C.; Matheus, P. E.; Sher, A. V.; Pybus, O. G.; Gilbert, M. T.; Barnes, I.; Binladen, J.; Willerslev, E.; Hansen, A. J.; Baryshnikov, G. F.; Burns, J. A.; Davydov, S.; Driver, J. C.; Froese, D. G.; Harington, C. R.; Keddie, G.; Kosintsev, P.; Kunz, M. L.; Martin, L. D.; Stephenson, R. O.; Storer, J.; Tedford, R.; Zimov, S.; Cooper, A. *Science (Washington, D.C.)* **2004**, *306*, 1561–1565.
12. Poinar, H. N.; Hofreiter, M.; Spaulding, W. G.; Martin, P. S.; Stankiewicz, B. A.; Bland, H.; Evershed, R. P.; Possnert, G.; Paabo, S. *Science (Washington, D.C.)* **1998**, *281*, 402–406.
13. Speller, C. F.; Yang, D. Y.; Hayden, B. *J. Archaeol. Sci.* **2005**, *32*, 1378–1389.
14. Drancourt, M.; Aboudharam, G.; Signoli, M.; Dutour, O.; Raoult, D. *Proc. Natl. Acad. Sci.* **1998**, *95*, 12637–12640.
15. Gilbert, M. T.; Cuccui, J.; White, W.; Lynnerup, N.; Titball, R. W.; Cooper, A.; Prentice, M. B. *Microbiology* **2004**, *150*, 341–354.
16. Donoghue, H. D.; Marcsik, A.; Matheson, C.; Vernon, K.; Nuorala, E.; Molto, J. E.; Greenblatt, C. L.; Spigelman, M. *Proc. R. Soc. B* **2005**, *272*, 389–394.
17. Bouwman, A. S.; Brown, T. A. *J. Archaeol. Sci.* **2005**, *32*, 703–713.
18. Mitchell, P. D. *Am. J. Phys. Anthropol.* **2003**, *121*, 117–124.

19. Zink, A.; Wolf, H.; Nerlich, A. G. *Arch. Pathol. Lab. Med.* **2000**, *124*, 1614–1618.
20. Higuchi, R.; Bowman, B.; Freiberger, M.; Ryder, O.; Wilson, A. C. *Nature* **1984**, *312*, 282–284.
21. Gilbert, M. T. P. Ph.D. thesis, University of Oxford, 2003.
22. Poinar, H. N.; Kuch, M.; Sobolik, K. D.; Barnes, I.; Stankiewicz, A. B.; Kuder, T.; Spaulding, W. G.; Bryant, V. M.; Cooper, A.; Paabo, S. *Proc. Natl. Acad. Sci.* **2001**, *98*, 4317–4322.
23. Sutton, M. Q.; Malik, M.; Ogram, A. *J. Archaeol. Sci.* **1996**, *23*, 263–267.
24. LeBlanc, S. A.; Kreisman, L.; Kemp, B. M.; Carlyle, S. W.; Dhody, A.; Smiley, F.; Benjamin, T. In Preparation.
25. Kaestle, F. A.; Horsburgh, K. A. *Yearbook Phys. Anthropol.* **2002**, *45*, 92–130.
26. Mulligan, C. J. *Am. Antiq.* **2006**, *71*, 365–380.
27. O'Rourke, D. H.; Hayes, M. G.; Carlyle, S. W. *Annu. Rev. Anthropol.* **2000**, *29*, 217–242.
28. Götherström, A.; Collins, M. J.; Angerbjorn, A.; Lidén, K. *Archaeometry* **2002**, *44*, 395–404.
29. Shook, B. A. Ph.D. thesis, University of California, 2005.
30. Palumbi, S. R. In *Molecular Systematics*; Hillis, D. M., Moritz, C., Mable, B. K., Eds.; Sinauer Associates: Sunderland, MA, 1996; pp 205–247.
31. Kemp, B. M.; Smith, D. G. *Foren. Sci. Int.* **2005**, *154*, 53–61.
32. Yang, D. Y.; Watt, K. *J. Archaeol. Sci.* **2005**, *32*, 331–336.
33. Watt, K. E. M.A. thesis, Simon Fraser Univerisity, 2005.
34. Salamon, M.; Tuross, N.; Arensburg, B.; Weiner, S. *Proc. Natl. Acad. Sci.* **2005**, *102*, 13783–13788.
35. Kelman, L. M.; Kelman, Z. *J. Verteb. Paleontol.* **1999**, *19*, 8–20.
36. Eshleman, J.; Smith, D. G. *Electrophoresis* **2001**, *22*, 4316–4319.
37. Tuross, N. *Experientia* **1994**, *50*, 530–535.
38. Kalmár, T.; Bachrati, C. Z.; Marcsik, A.; Raskó, I. *Nuc. Ac. Res.* **2000**, *28*, e67.
39. Hummel, S.; Herrmann, B. In *Ancient DNA: Recovery and Analysis of Genetic Material from Paleontological, Archaeological, Museum, and Forensic Specimens*; Hummel, S., Herrmann, B., Eds.; Springer-Verlag: New York, 1994; pp 59–68.
40. Poinar, G. O.; Poinar. In *Ancient DNA: Recovery and Analysis of Genetic Material from Paleontological, Archaeological, Museum, and Forensic Specimens*; Herrmann, B., Ed.; Springer-Verlag: New York, 1994; pp 59–68.
41. Cooper, A. In *Ancient DNA: Recovery and Analysis of Genetic Material from Paleontological, Archaeological, Museum, and Forensic Specimens*; Hummel, S., Herrmann, B., Eds.; Springer-Verlag: New York, 1994; pp 149–165.

42. Scholz, M.; Giddings, I.; Pusch, C. M. *Anal. Biochem.* **1998**, *259*, 283–286.
43. Kemp, B. M.; Resendez, A.; Román Berrelleza, J. A.; Malhi, R. S.; Smith, D. G. In *Biomolecular Archaeology: Genetic Approaches to the Past*; Reed, D. M., Ed.; Southern Illinois University: Carbondale, IL, 2005; pp 22–46.
44. Hänni, C.; Brousseau, T.; Laudet, V.; Stehelin, D. *Nuc. Ac. Res.* **1995**, *23*, 881–882.
45. Höss, M.; Pääbo, S. *Nuc. Ac. Res.* **1993**, *21*, 3913–3914.
46. Fisher, D. L.; Holland, M. M.; Mitchell, L.; Sledzik, P. S.; Wilcox, A. W.; Wadhams, M.; Weedn, V. W. *J. Forensic Sci.* **1993**, *38*, 60–68.
47. Pääbo, S.; Gifford, J. A.; Wilson, A. C. *Nuc. Ac. Res.* **1988**, *16*, 9775–9788.
48. Edwards, C. J.; MacHugh, D. E.; Dobney, K. M.; Martin, L.; Russell, N.; Horwitz, L. K.; McIntosh, S. K.; MacDonald, K. C.; Helmer, D.; Tresset, A.; Vigne, J.-D.; Bradley, D. G. *J. Archaeol. Sci.* **2004**, *31*, 695–710.
49. Yang, D. Y.; Eng, B.; Saunders, S. R. *Hum. Biol.* **2003**, *75*, 355–364.
50. Schurr, T. G.; Ballinger, S. W.; Gan, Y.-Y.; Hodge, J. A.; Merriwether, D. A.; Lawrence, D. N.; Knowler, W. C.; Weiss, K. M.; Wallace, D. C. *Am. J. Hum. Genet.* **1990**, *46*, 613–623.
51. Forster, P.; Harding, R.; Torroni, A.; Bandelt, H.-J. *Am. J. Hum. Genet.* **1996**, *59*, 935–945.
52. Brown, M. D.; Hosseini, S. H.; Torroni, A.; Bandelt, H.-J.; Allen, J. C.; Schurr, T. G.; Scozzari, R.; Cruciani, F.; Wallace, D. C. *Am. J. Hum. Genet.* **1998**, *63*, 1852–1861.
53. Sullivan, K. M.; Mannucci, A.; Kimpton, C. P.; Gill, P. *Biotechniques* **1993**, *15*, 636–641.
54. Don, R. H.; Cox, P. T.; Wainwright, B. J.; Baker, K.; Mattick, J. S. *Nuc. Ac. Res.* **1991**, *19*, 4008.
55. Gilbert, M. T. P.; Bandelt, H.-J.; Hofreiter, M.; Barnes, I. *Trends Ecol. Evol.* **2005**, *20*, 541–544.
56. Stone, A. C.; Milner, G. R.; Paabo, S.; Stoneking, M. *Am. J. Phys. Anthropol.* **1996**, *99*, 231–238.
57. Schmidt, D.; Hummel, S.; Herrmann, B. *Am. J. Phys. Anthropol.* **2003**, *121*, 337–341.
58. Stone, A. C.; Stoneking, M. *Am. J. Phys. Anthropol.* **1993**, *92*, 463–471.
59. Wrischnik, L. A.; Higuchi, R. G.; Stoneking, M.; Erlich, H. A. *Nuc. Ac. Res.* **1987**, *15*, 529–542.
60. Parr, R. L.; Carlyle, S. W.; O'Rourke, D. H. *Am. J. Phys. Anthropol.* **1996**, *99*, 507–518.
61. Kaestle, F. A. Report on DNA Analysis of the Remains of "Kennewick Man" from Columbia Park, Washington. National Parks Service, website http://www.cr.nps.gov/aad/kennewick/index.htm, 2000

62. Anderson, S.; Bankier, A. T.; Barrel, B. G.; DeBulin, M. H. L.; Coulson, A. R.; Drouin, J.; Eperon, I. C.; Nierlich, D. P.; Roe, B. A.; Sanger, F.; Schreier, P. H.; Smith, A. J. H.; Staden, R.; Young, I. G. *Nature (London)* **1981**, *290*, 457–465.
63. Andrews, R. M.; Kubacka, I.; Chinnery, P. F.; Lightowlers, R. N.; Turnbull, D. M.; Howell, N. *Nat. Genet.* **1999**, *23*, 147.

Chapter 5

Using Archaeological Chemistry to Investigate the Geographic Origins of Trophy Heads in the Central Andes: Strontium Isotope Analysis at the Wari Site of Conchopata

Kelly J. Knudson[1] and Tiffiny A. Tung[2]

[1]Center for Bioarchaeological Research, School of Human Evolution and Social Change, Arizona State University, Tempe, AZ 85281
[2]Department of Anthropology, Vanderbilt University, Nashville, TN 37235

Comparing strontium isotope results from archaeological bone-tooth pairs from individuals buried in mortuary spaces at the Wari site of Conchopata with Conchopata trophy heads shows that the trophy head strontium isotope signatures are more variable. This implies that the individuals transformed into trophy heads likely came from different geologic zones in the Andes, and were more likely to have been victims of raiding or warfare from different parts of the Andes rather than venerated ancestors from Conchopata or the surrounding region. These data also demonstrate the ability of strontium isotope analysis to elucidate individual life histories using archaeological bone-tooth pairs.

Although there is a growing body of bioarchaeological and iconographic evidence that the Wari Empire of central Peru transformed some individuals into trophy heads (*1–3*), there is still much debate about the identity of these individuals. For example, these individuals could have been victims of warfare or raiding, as has been hypothesized for Nasca trophy heads in the Andes (*4–6*). Alternatively, the trophy heads could represent revered ancestors of the local Wari community. For example, in the Andes there is archaeological, ethnohistorical, and ethnographic evidence of the importance of ancestor veneration and the incorporation of skeletal elements into religious rituals (*7–10*). Here, we test the hypothesis that the Wari trophy heads at Conchopata are victims of warfare or raiding taken from outside of the Ayacucho Basin by using strontium isotope analysis of archaeological human tooth enamel and bone from the trophy heads as well as individuals buried in mortuary spaces at Conchopata.

The Conchopata Trophy Heads

During the Andean Middle Horizon (ca. A.D. 500–1000), the Wari Empire exerted substantial influence from its capital at Huari in the Ayacucho Basin. Wari administrative centers like Pikillacta near Cuzco and Wari-style artifacts found throughout the Peruvian Andes testify to Wari political and economic power (*11–17*). Within the Wari heartland, the site of Conchopata is located 12 km south of the site of Huari (*18, 19*). New calibrated radiocarbon dates suggest that the Wari component at Conchopata began around A.D. 600–650 (*20*). In addition to its residential and ritual functions, Conchopata has Middle Horizon burials from at least 242 individuals from numerous tombs and 31 trophy heads from two ritual structures (*1*). The trophy heads were found smashed and burned on the floor of a D-shaped room (EA 72) and a circular room (EA 143) (*1, 21, 22*). Tung has shown that the trophy heads have a standardized preparation with cutmarks on the posterior edge of the mandibular ramus and perforations drilled at bregma (on the superior of the skull), suggesting that the Wari trophy heads were disarticulated and defleshed and then dangled by a cord for display, as shown in Wari iconography (*1, 23*). The combined bioarchaeological and iconographic data suggest that the trophy heads were prepared by highly skilled ritual specialists for their use in the D-shaped and circular ritual structures (*1, 21, 22*).

Strontium Isotope Analyses in Archaeology

In the twenty years since it was first proposed (*24, 25*), strontium isotope analysis of archaeological human remains to examine residential mobility has become increasingly common. Briefly, $^{87}Sr/^{86}Sr$ in a given geologic zone is determined by the age and composition of the bedrock (*26*). The isotopic composition of strontium does not fractionate as strontium moves through the ecosystem, so the strontium isotope ratios in bedrock are then reflected in the soils, plants, and animals of that geologic zone (*27–29*). This isotope signature will also be reflected in the tooth enamel and bone hydroxyapatite from humans who consumed foods grown or raised in a given geologic zone (*30, 31*).

Since tooth enamel does not regenerate after it has formed, strontium isotope ratios in tooth enamel reflect the strontium sources in the diet, and hence the geologic zone or zones in which an individual was living, during enamel formation if local foods provided the strontium in the diet. On the other hand, bone continually regenerates, with turnover rates in cortical bone dependent on the skeletal element analyzed and the age, sex, activity levels and health of the individual (*32–37*).Therefore, the strontium isotope ratios in human bone reflect the strontium source of the last years of life. Comparing the strontium isotope signatures in archaeological human tooth enamel and bone can then reveal changes in place of residence during enamel and bone formation, if local foods were consumed (*24, 25*).

Diagenesis and Strontium Isotope Analysis

When working with trace elemental analyses in archaeological samples, identifying and eliminating diagenetic contamination is clearly necessary. Strontium from the groundwater and burial environment can be incorporated into hydroxyapatite as secondary minerals. For example, calcite ($CaCO_3$) and barite ($BaSO_4$) may be deposited in the pore spaces of the calcium phosphate structure of teeth and bone (*38, 39*). Biogenic apatite could also be altered as trace elements from post-depositional contamination substitute for calcium in calcium phosphate and are converted to hydroxyapatite during recrystallization and crystal growth (*38, 39*).

Because of its mineral composition, the large size of its phosphate crystals and the small amount of pore space, mature tooth enamel is rarely affected by diagenetic contamination (*40–42*). This has been supported by experimental data that show that enamel reliably retains biogenic strontium isotope signatures (*30, 43–48*).

Unfortunately, because of its porosity and large surface area, bone is more susceptible to diagenetic contamination. As in tooth enamel, diagenetic contamination is concentrated on the outer surface of the bone (*49–52*). Therefore, mechanically cleaning the bone samples through abrasion can remove diagenetic contamination on the bone surface (*49, 53*). Although a number of techniques can be used to identify the presence of diagenetic contamination (*54–58*), removing the diagenetic contamination is more difficult. However, washing the bone sample with weak acid removes calcite that accumulated in the bone after burial (*30, 53, 55, 59*). As will be discussed below, the bone samples included in this study were mechanically and chemically cleaned in order to minimize diagenetic contamination.

Field and Laboratory Methodology

Sampling Strategy for Strontium Isotope Analysis

For this project, modern faunal samples were first collected to determine the local strontium isotope signatures in the Ayacucho region. The first author collected six guinea pigs from the Ayacucho market in 2000. Informants stated that the six animals were raised on alfalfa grown outside the city of Ayacucho, so the guinea pig strontium isotope signatures should reflect the strontium isotope signatures in the agricultural fields near Ayacucho, which are the likely food sources for the past inhabitants of Conchopata. Archaeological faunal samples were not used in order to avoid diagenetic contamination.

In addition to the modern faunal samples, the second author collected eleven samples from archaeological human skeletons at Conchopata. Six of these individuals were buried in two tombs at Conchopata. Tooth enamel and bone samples were collected from five adults interred in an undisturbed tomb in Architectural Space 105 (EA 105), and an infant from a looted tomb in Architectural Space 06 (EA 06).

Tooth enamel and bone samples from the five trophy heads came from the circular ritual structure (EA 143). Because the trophy heads and teeth were fragmented, the enamel comes from unspecified first, second or third molars and one right maxillary canine and so covers the period of enamel formation from in utero to 15 years of age (*42*). This sample of five represents approximately 20% of the 24 adult Wari trophy heads that have been discovered in the Andes (*1*). Therefore, this sample size can be used to test the hypothesis that the adult Wari trophy heads found at Conchopata were obtained from individuals taken in warfare or raiding outside of the Ayacucho Basin.

Strontium Isotope Analysis of Archaeological Enamel and Bone

All tooth and bone samples were initially prepared in the Laboratory for Archaeological Chemistry by the first author. Modern faunal samples for strontium isotope analysis were placed in a crucible and ashed at approximately 800°C for 10 hours. The bone samples were then crushed in an agate mortar and pestle. The teeth were removed from modern fauna mandibles after ashing and crushed and stored separately from the bone.

Archaeological teeth samples were mechanically cleaned by abrasion with a Patterson NC-350 dental drill equipped with an inverted-cone carbide burr (White burrs HP-59 type 2 class 2). This removed any adhering organic matter or contaminants as well as the outermost layers of tooth enamel, which are most susceptible to diagenetic contamination (*44, 48*). Approximately 5–10 mg of tooth enamel were then removed with a Patterson NC-350 dental drill equipped with a carbide burr.

Archaeological bone samples were treated for diagenesis before sample analysis. The bone samples were first mechanically cleaned with the Patterson NC-350 dental drill equipped with a carbide burr to remove any organic matter or contaminants. The mechanical cleaning also removed the layers of cortical bone most susceptible to diagenetic contamination, as well as all traces of trabecular bone. The bone samples were then chemically cleaned in an ultrasonic bath. The samples were first sonicated in water for 30 minutes, then rinsed and sonicated in 5% acetic acid for 30 minutes, and finally rinsed and sonicated with 5% acetic acid for 5 minutes (*30, 53, 55, 59*).The bone samples were dried for 1 hour at approximately 80°C. Finally, the bone samples were placed in a crucible and ashed at approximately 800°C for 10 hours.

Strontium isotope ratios were obtained at the Isotope Geochemistry Laboratory in the Department of Geological Sciences at the University of North Carolina at Chapel Hill by the first author under the direction of Paul D. Fullagar. Three to six milligrams of mechanically- and chemically-cleaned bone ash or powdered tooth enamel were dissolved in 15 mL Savillex PFA vials using 500 μL of twice distilled 5N HNO_3 in a class 100 filtered air environmental hood. The samples were then evaporated and redissolved in 250 μL of 5N HNO_3. The strontium was separated from the sample matrix using EiChrom SrSpec resin, a crown-ether Sr-selective resin (50–100 μm diameter) loaded into the tip of a 10 mL BioRad polypropylene column. Total resin volume was approximately 50 μL. The SrSpec resin was pre-soaked and flushed with H_2O to remove Sr present from the resin manufacturing process. The resin was further cleaned in the column with repeated washes of deionized H_2O and conditioned with 5N HNO_3. Resin was used once for sample elution and discarded. The dissolved sample was loaded and washed in 750 μL of 5N HNO_3, and then Sr was eluted with 1 mL of H_2O. Total procedural blanks for Sr are typically 100-to-200 picograms. The sample was then evaporated, dissolved in 2 μL of 0.1 M H_3O_4 and 2 μL of $TaCl_5$ and loaded onto degassed Re filaments. Isotopic ratios were measured by a VG Sector 54 thermal ionization mass spectrometer at the University of North Carolina-Chapel Hill in quintuple-collector dynamic mode, using the internal ratio $^{86}Sr/^{88}Sr$=0.1194 to correct for mass fractionation. Recent $^{87}Sr/^{86}Sr$ analyses of Sr carbonate standard SRM 987 yield a value of 0.710245$\pm$0.000018 (2σ). Long term analyses over approximately 24 months of SRM 987 yielded and average of $^{87}Sr/^{86}Sr$=0.710242. Internal precision for Sr carbonate runs is typically 0.0006 to 0.0009% standard error, based on 100 dynamic cycles of data collection.

Strontium Isotope Results

Expected Strontium Isotope Ratios in the Andes

The Central Andean highlands, where the Wari heartland was located, are predominately composed of late Cenozoic andesites and latites, including the Quaternary Molinoyoc volcanics in the Ayacucho region (*60–62*). Strontium isotope analysis of exposed bedrock in the Ayacucho region has not yet been performed, but geologic data do exist for late Cenozoic andesites in southern Peru. For example, geologic data show that $^{87}Sr/^{86}Sr$=0.7054–0.7067 (n=7) near Moquegua, Peru, and $^{87}Sr/^{86}Sr$=0.7067–0.7079 (n=16) near Arequipa, Peru (*63*). Generally speaking, strontium isotope ratios increase as one travels from north to south in the Cenozoic volcanic zones of the Andes (*60, 63–67*). Therefore, the expected strontium isotope ratios in the Ayacucho Basin andesites should be slightly lower than the observed strontium isotope ratios of southern Peruvian andesites.

In addition to the late Cenozoic andesites and latites in the Ayacucho Basin, there are outcrops of Paleozoic and Mesozoic rocks in the Ayacucho Valley, including the Ticllas red beds as well as other Cenozoic formations such as the Huanta formation, that are composed predominately of silic tuff (*61*). Therefore, the strontium isotope signatures of the late Cenozoic volcanics could be modified as the soils and groundwater in the Ayacucho Basin incorporates strontium from different geologic formations.

However, it also is important to note from where individuals from the Wari heartland would have obtained their food, and specifically their strontium. The agricultural zones that likely provided the food for the inhabitants of Conchopata and other Wari heartland sites are located in the Ayacucho Formation, which predominately consists of Cenozoic dactites and andesites and volcaniclastic lacustrine and fluvial strata (*61*). For example, fields are still in use in the river valleys immediately west and east of Conchopata, and it is likely that Wari agriculture was generally focused on the rivers in the Cachi hydrographic unit, the Huamanga Basin of the Ayacucho Valley (*68–70*).

Strontium Isotope Analysis of Modern Fauna from the Andes

The measured strontium isotope ratios of modern guinea pigs from Ayacucho support the expected strontium isotope ratios based on the geologic literature, as shown in Table I. Small modern fauna can be used as proxies for the biologically available strontium in a given region by providing an average for one small area (*71, 72*). Bone strontium isotope ratios from six modern guinea pigs from Ayacucho are $^{87}Sr/^{86}Sr$=0.707204 (F1229), $^{87}Sr/^{86}Sr$=0.706306 (F1230), $^{87}Sr/^{86}Sr$=0.706555 (F1231), $^{87}Sr/^{86}Sr$=0.711766 (F1232), $^{87}Sr/^{86}Sr$=0.705762 (F1233), and $^{87}Sr/^{86}Sr$=0.705841 (F1234). Five of these samples match the expected strontium isotope signatures of the region around Conchopata, given the geology of the bedrock underneath the agricultural fields. However, one sample ($^{87}Sr/^{86}Sr$=0.711766, F1232) has a much higher strontium isotope signature. It is possible that imported fertilizers were used on the alfalfa fields, or that the informant misreported the source of the guinea pig's diet.

Table I. Strontium Isotope Results from Conchopata

Laboratory Number	*Individual Number*	*Type*	*Sample*	$^{87}Sr/^{86}Sr$
F1218	2095.01	Burial	R fibula	0.706096
F1219	2095.01	Burial	LM2	0.705598
F1220	2095.02	Burial	R fibula	0.705739
F1221	2095.02	Burial	RM2	0.705632
F1222	2095.03	Burial	R fibula	0.705861
F1223	2095.03	Burial	M2	0.705657
F1224	2095.04	Burial	L fibula	0.705663
F1225	2095.04	Burial	LM2	0.705646
F1226	2095.06	Burial	U rib	0.705480
F1227	2095.06	Burial	RM2	0.705739
F1228	2004	Burial	U rib	0.706734
F1784	2907.04	Trophy head	molar enamel	0.708811
F1789	2907.04	Trophy head	cranial fragment	0.707186
F1785	2907.05	Trophy head	molar enamel	0.706270
F1790	2907.05	Trophy head	cranial fragment	0.706483
F1786	2985.10	Trophy head	RC enamel	0.706404
F1787	2985.11	Trophy head	molar enamel	0.710204
F1792	2985.11	Trophy head	cranial fragment	0.709232
F1788	2985.18	Trophy head	molar enamel	0.706259
F1793	2985.18	Trophy head	cranial fragment	0.707289
F1229	A1A	Modern *cuy*	mandible, femur	0.707204
F1230	A2A	Modern *cuy*	mandible, femur	0.706306
F1231	A3A	Modern *cuy*	mandible, femur	0.706555
F1232	A4A	Modern *cuy*	mandible, femur	0.711766
F1233	A5A	Modern *cuy*	mandible, femur	0.705672
F1234	A6A	Modern *cuy*	mandible, femur	0.705841

Determining Local versus Non-Local Strontium Isotope Signatures

Another challenge in strontium isotope analyses in archaeological residential mobility studies is the determination of the local strontium isotope ratio for a particular region. One commonly used method is to define the local strontium isotope signature as the mean of the modern faunal strontium isotope from a given region plus and minus two standard deviations (*73*). Using this definition, the local range for Conchopata is $^{87}Sr/^{86}Sr$=0.7027–0.7118 as defined as the mean of the modern guinea pig samples plus and minus two standard deviations (*71, 72*). Removing the anomalous guinea pig value provides a local range of $^{87}Sr/^{86}Sr$=0.7051–0.7075, which more closely matches the expected values based on Ayacucho geology. In addition, as will be discussed below, the archaeological human bone and enamel values from the burials at Conchopata cluster closely and appear to be local based on their tomb styles, and artifacts (*74, 75*).

Strontium Isotope Results from Human Burials at Conchopata

As shown in Table I, the strontium isotope ratios from the human burials at Conchopata are all very similar ($^{87}Sr/^{86}Sr$=0.7058±0.0003 (n=11, 1σ)). Specifically, enamel samples from five adults buried in the tombs at Conchopata have $^{87}Sr/^{86}Sr$=0.705480–0.705646, while bone samples from the same adults have $^{87}Sr/^{86}Sr$=0.705663–0.706096. The infant bone from EA 06 has $^{87}Sr/^{86}Sr$=0.706734.

The small standard deviation in the enamel and bone strontium isotope signatures from individuals buried in tombs at Conchopata implies that these individuals procured their food from geologic zones with the same strontium isotope signatures during the first and last years of their lives. Since these values correspond closely with the strontium isotope signatures of geological formations near Conchopata and there is scant evidence for large quantities of food being traded into Conchopata, the most parsimonious explanation is that these individuals spent the first and last years of their lives in the Ayacucho region.

Strontium Isotope Results from Conchopata Trophy Heads

While the strontium isotope signatures in archaeological human tooth enamel and bone from tombs at Conchopata are very similar, the five trophy heads have more heterogeneous strontium isotope signatures, as shown in Table I and Figure 1. Three trophy heads exhibit enamel and bone strontium isotope

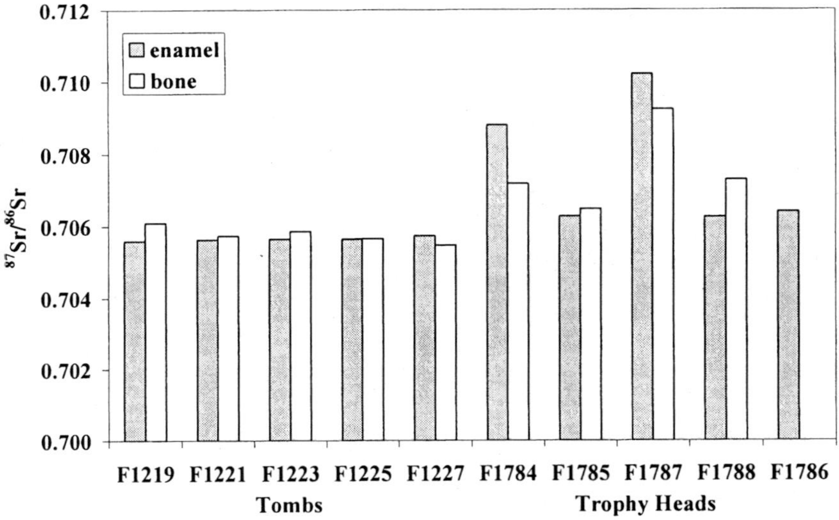

Figure 1. Strontium Isotope Results from Archaeological Human Remains from Conchopata.

signatures that are similar to the strontium isotope ratios of five of the modern guinea pigs and all of the individuals buried in Conchopata tombs.

In contrast, there are two individuals who have enamel strontium isotope ratios that are higher than the others (Individual 2907.4 (Sample F1784, $^{87}Sr/^{86}Sr$=0.708811) and Individual 2985.11 (Sample F1787, $^{87}Sr/^{86}Sr$=0.710204)). The two individuals with high enamel strontium isotope signatures also have high bone strontium isotope signatures (Individual 2907.4 (Sample F1789, $^{87}Sr/^{86}Sr$=0.707186) and Individual 2985.11 (Sample F1792, $^{87}Sr/^{86}Sr$=0.709232)). If these are biogenic strontium isotope signatures, this implies that these two individuals spent the first and last years of their lives consuming foods from different geologic regions than the other individuals analyzed here.

Finally, there is one trophy head that exhibits a low enamel strontium isotope signature (Individual 2985.18 (Sample F1788, $^{87}Sr/^{86}Sr$ =0.706259)) yet a high bone strontium isotope signature (Individual 2985.18 (Sample F1793, $^{87}Sr/^{86}Sr$ =0.707289)). This implies that this individual obtained food from different geologic zones during the first and last years of his life. Interestingly, after being transformed into a trophy head, this individual's head was found at his presumed place of childhood residence, although not his adult residence.

In conclusion, the enamel and bone strontium isotope signatures in the trophy heads exhibit more variability than the enamel and bone strontium isotope ratios in the individuals buried in the Conchopata tombs. A student's t-test demonstrates that the difference in enamel strontium isotope ratios in the two populations is statistically significant ($t = -2.38$, $p < 0.038$). In addition, the only individuals in this study who exhibited a change in strontium isotope signatures, and hence a likely change in residence, between childhood and adulthood are three trophy heads. Since three out of five trophy head samples have high strontium isotope signatures, they likely lived in a region other than near Conchopata for the last years of their lives. This supports the hypothesis that they were victims of warfare or raiding from a different population than the individuals buried in the tombs at Conchopata.

Strontium Isotope Analysis and Diagenesis: A Discussion

Diagenetic contamination of archaeological bone samples is always possible (*76, 77*). Since the trophy heads were fractured and, in some cases poorly preserved, it is possible that the samples were diagenetically contaminated with strontium from the burial environment, even though they had been mechanically and chemically cleaned. However, if the bone samples were contaminated by the burial environment, the diagenetic strontium isotope signature would be close to $^{87}Sr/^{86}Sr=0.706$. Certain trophy head bone samples still show strontium isotope signatures that are above this value. This implies that biogenic strontium isotope signatures were recovered from at least three bone samples at Conchopata. Therefore, it is likely that these individuals did in fact consume food from different geologic regions, and presumably lived in different geologic regions, during bone formation.

In addition to contact with the burial environment, the trophy head samples had been burned, presumably during ritual activities in the structures in which they were found (*1, 21, 22*). Although experimental data has shown that post-mortem burning can change the $\delta^{13}C$ values in bone collagen, this is not the case for strontium isotope ratios (*78*).

Directions for Future Research

The high strontium isotope ratios in the trophy head tooth enamel and bone samples do not match the local strontium isotope signature at Conchopata, and hence are at least partially biogenic strontium isotope signatures. However, independent lines of evidence will also be used in the future to identify contaminated bone samples. More specifically, monitoring the ratio of calcium to phosphorus (Ca/P) and the uranium concentrations in skeletal material can identify samples contaminated with diagenetic strontium (*30, 40, 59, 73*). Diagenetically contaminated skeletal elements have a much higher Ca/P ratio than the biogenic ratio of 2.1:1 if they have incorporated calcium, and likely strontium, from the burial environment. In addition, the uranium concentration in bone may also be used to identify contaminated samples because the mobile ionic form of uranium is incorporated into bone hydroxyapatite instead of remaining in secondary minerals in pore spaces of the calcium phosphate structure (*79*). Therefore, uranium is not removed through weak acid washes, and bone samples with high uranium concentrations may contain contaminated hydroxyapatite while bones with low uranium concentrations are resistant to diagenesis (*40, 73, 80*). These elemental concentration analyses will allow the further identification of diagenetically contaminated samples in this study. In addition to the use of independent lines of evidence to identify diagenetic contamination, the incorporation of more modern and archaeological faunal samples to characterize the strontium isotope signatures of the Ayacucho Basin will provide a more detailed isotopic map of the region.

Conclusion

Strontium isotope signatures in tooth enamel and bone from five adults and one infant buried in two tombs at Conchopata are very similar. This implies that these individuals consumed food from the same geologic zone, or from geologic zones with similar strontium isotope signatures. Since the strontium isotope ratios in these individuals matched the expected Ayacucho Basin strontium isotope signature, it is most likely that these individuals lived in or near Conchopata for the first and last years of life.

On the other hand, the five individuals who were transformed into trophy heads have more variable enamel and bone strontium isotope signatures. Three individuals with variable strontium isotope ratios consumed food from different geologic zones in the first and last years of their lives, and likely lived outside of the Ayacuho Basin. However, even the individuals with high bone strontium isotope signatures may have moved to or been brought to Conchopata shortly before death, since their bone may not have incorporated large amounts of local strontium isotopes. Given the high variability in the trophy head strontium isotope signatures, it is unlikely that these individuals represent venerated Wari ancestors who are local to the Ayacucho Basin. Instead, it is more likely that the Wari trophy heads were victims of warfare or raiding from a variety of different geologic zones.

Acknowledgements

The authors gratefully acknowledge Drs. Anita Cook, William Isbell, and José Ochatoma of the Conchopata Archaeological Project, who provided access to the valuable Conchopata skeletal collection and logistical support in the field.

Drs. James H. Burton, Paul D. Fullagar, and T. Douglas Price generously provided laboratory access and logistical support. Strontium isotope analysis at Conchopata was funded through grants to Knudson (NSF Archaeology Division, BCS-0202329; the Laboratory for Archaeological Chemistry at the University of Wisconsin at Madison) and Tung (NSF Biological Anthropology Division, BCS-0118751; Wenner-Gren Foundation for Anthropological Research, Grant No. 6680; the Fulbright Commission; the Latané-University of North Carolina Research Grant).

References

1. Tung, T. A. Ph.D. thesis, University of North Carolina, Chapel Hill, 2003.
2. Tung, T. A. *Andean Past.* **In press**, *8*.
3. Tung, T. A. In *The Taking and Displaying of Human Trophies by Amerindians*; Chacon, R. J.; Dye, D. H., Eds.; 2006.
4. Silverman, H. *Cahuachi in the Ancient Nasca World*; University of Iowa Press: Iowa City, IA, 1993.
5. Proulx, D. A. In *Ritual Sacrifice in Ancient Peru*; Benson, E. P.; Cook, A. G., Eds.; University of Texas Press: Austin, TX, 2001; pp 119–136.
6. Verano, J. W. In *Tombs for the Living: Andean Mortuary Practices*; Dillehay, T., Eds.; Dumbarton Oaks: Washington, DC, 1995; pp 189–228.
7. Wachtel, N. *EL regreso de los antepasados: los indios urus de Bolivia, del siglo XX al XVI*; El Colegio de México: Ciudad de Mexico, Mexico, 2001.
8. Guaman Poma de Ayala, F. *Nueva crónica y buen gobierno. Edición de John V. Murra, Rolena Adorno y Jorge L. Urioste*; Madrid, Spain, 1613 [1987].
9. Cobo, B. *Inca Religion and Customs*; University of Texas Press: Austin, TX, 1653 [1990].
10. Bourget, S. In *Ritual Sacrifice in Ancient Peru*; Benson, E. P.; Cook, A. G., Eds.; University of Texas Press: Austin, TX, 2001; pp 93–118.
11. Schreiber, K. J. *Wari imperialism in Middle Horizon Peru*; Museum of Anthropology University of Michigan: Ann Arbor, MI, 1992.
12. Tung, T. A.; Owen, B. In *Andean Archaeology III*; Silverman, H.; Isbell, W. H., Eds.; Kluwer Academic/Plenum Publishers: New York, 2005 (In press).
13. *Huari Administrative Structure: Prehistoric Monumental Architecture and State Government*; Isbell, W. H.; McEwan, G. F., Eds.; Dumbarton Oaks: Washington, D.C., 1991.
14. *Pikillacta: The Wari Empire in Cusco*; McEwan, G. F., Ed.; University of Iowa Press: Iowa City, 2005.
15. Schreiber, K. In *Foundations of Power in the Prehispanic Andes*; Vaughn, K. J.; Ogburn, D.; Conlee, C. A., Eds.; Archaeological Papers of the American Anthropological Association: Arlington, VA, 2004; pp 131–150.

16. Isbell, W. H. In *Huari Administrative Structure: Prehistoric Monumental Architecture and State Government*; Isbell, W. H.; McEwan, G. F., Eds.; Dumbarton Oaks: Washington, DC, 1991; pp 293–316.
17. Isbell, W. H.; McEwan, G. F. In *Huari Administrative Structure: Prehistoric Monumental Architecture and State Government*; Isbell, W. H.; McEwan, G. F., Eds.; Dumbarton Oaks: Washington, DC, 1991; pp 1–18.
18. Isbell, W. H.; Cook, A. G. In *Andean Archaeology II: Art, Landscape, and Society*; Silverman, H.; Isbell, W. H., Eds.; Kluwer Academic Press: New York, 2002; pp 249–306.
19. Cook, A. G. *Ñawpa Pacha* **1984–1985**, *22–23*, 49–90.
20. Isbell, W. *Personal Communication, June 2006.* **n.d.**
21. Tung, T. A. Paper presented at the 69th Annual Meeting of the Society for American Archaeology, Montreal, Canada, 2004.
22. Tung, T. A.; Cook, A. G. In *Between King and Commoner: Intermediate Elites in Pre-Colombian States and Empires*; Elson, C., Ed.; University of Arizona Press: Tuscon, AZ, 2005; pp. 68–93.
23. Ochatoma Paravicino, J.; Cabrera Romero, M. In *Andean Archaeology II: Art, Landscape and Society*; Silverman, H.; Isbell, W. H., Eds.; Kluwer Academic/ Plenum Publishers: New York, 2002; pp 225–248.
24. Ericson, J. E. *J. Human Evol.* **1985**, *14*, 503–514.
25. Krueger, H. W. Poster paper presented at Biomineralization Conference, Airlie House, Warrenton, VA, April 14–17, 1985.
26. Faure, G. *Principles of Isotope Geology*; John Wiley: New York, 1986.
27. Faure, G.; Powell, J. L. *Strontium Isotope Geology*; Springer-Verlag: New York, 1972.
28. Blum, J. D.; Taliaferro, E. H.; Weisse, M. T.; Holmes, R. T. *Biogeochem.* **2000**, *49*, 87–101.
29. Aberg, G. *Water Air Soil Pollution* **1995**, *79*, 309–322.
30. Price, T. D.; Johnson, C. M.; Ezzo, J. A.; Ericson, J.; Burton, J. H. *J. Archaeol. Sci..* **1994**, *21*, 315–330.
31. Sealy, J. C. Ph.D. thesis, University of Capetown, Capetown, South Africa, 1989.
32. Carr, T. E. F.; Harrison, G. E.; Loutit, J. F.; Sutton, A. *Brit. Med. J.* **1962**, *2*, 773–775.
33. Parfitt, A. M. In *Bone Histomorphometry: Techniques and Interpretation*; Recker, R. R., Ed.; CRC Press, Inc.: Boca Raton, FL, 1983; pp 143–223.
34. Kulp, J. L.; Schulert, A. R. *Sci.* **1962**, *136*, 619–632.
35. Mulhearn, D. M. *Amer. J. Phys. Anth.* **2000**, *111*, 519–530.
36. Mulhearn, D. M.; Van Gerven, D. P. *Amer. J. Phys. Anth.* **1997**, *104*, 133–146.
37. Wastney, M. E.; Ng, J.; Smith, D.; Martin, B. R.; Peacock, M.; Weaver, C. M. *Amer. J. Physiol.* **1996**, *40*, R208–R216.
38. Sandford, M. K. In *Investigations of Ancient Human Tissue: Chemical*

Analyses in Anthropology; Sandford, M. K., Ed.; Gordon and Breach Science Publishers: New York, 1993; pp 3–57.

39. Radosevich, S. C. In *Investigations of Ancient Human Tissue: Chemical Analyses in Anthropology*; Sandford, M. K., Ed.; Gordon and Breach Science Publishers: New York, 1993; pp 269–332.
40. Kohn, M. J.; Schoeninger, M. J.; Barker, W. W. *Geochim. Cosmochim. Acta.* **1999**, *63*, 2737–2747.
41. Shellis, R. P.; Dibdin, G. H. In *Development, Function and Evolution of Teeth*; Teaford, M. F.; Smith, M. M.; Ferguson, M. W. J., Eds.; Cambridge University Press: Cambridge, UK, 2000; pp 242–251.
42. Hillson, S. *Dental Anthropology*; Cambridge University Press: Cambridge, UK, 1996.
43. Budd, P.; Montgomery, J.; Cox, A.; Krause, P.; Barreiro, B.; Thomas, R. G. *Sci. Total Environ.* **1998**, *220*, 121–136.
44. Budd, P.; Montgomery, J.; Barreiro, B.; Thomas, R. G. *Appl. Geochem.* **2000**, *15*, 687–694.
45. Chiaradia, M.; Gallay, A.; Todt, W. *Appl. Geochem.* **2003**, *18*, 353–370.
46. Grupe, G.; Price, T. D.; Schroter, P.; Sollner, F.; Johnson, C. M.; Beard, B. L. *Appl. Geochem.* **1997**, *12*, 517–525.
47. Lee-Thorp, J.; Sponheimer, M. *J. Anthropol. Archaeol.* **2003**, *22*, 208–216.
48. Montgomery, J.; Budd, P.; Cox, A.; Krause, P.; Thomas, R. G. In *Metals in Antiquity*; Young, S. M. M.; Pollard, A. M.; Budd, P.; Ixer, R. A., Eds.; BAR International Series: Oxford, UK, 1999; pp 290–296.
49. Lambert, J. B.; Weydert, J. M.; Williams, S. R.; Buikstra, J. E. *J. Archaeol. Sci.* **1991**, *18*, 363–383.
50. Waldron, H. A.; Khera, A.; Walker, G.; Wibberly, G.; Green, C. J. S. *J. Archaeol. Sci.* **1979**, *6*, 295–298.
51. Waldron, H. A. *Amer. J. Phys. Anth.* **1981**, *55*, 395–398.
52. Waldron, H. A. *J. Archaeol. Sci.* **1983**, *10*, 35–40.
53. Price, T. D.; Blitz, J.; Burton, J. H.; Ezzo, J. *J. Archaeol. Sci.* **1992**, *19*, 513–529.
54. Edward, J. B.; Benfer, R. A. In *Investigations of Ancient Human Tissue: Chemical Analyses in Anthropology*; Sandford, M. K., Eds.; Gordon and Breach Science Publishers: New York, 1993; pp 183–128.
55. Nielsen-Marsh, C. M.; Hedges, R. E. *J. Archaeol. Sci.* **2000**, *27*, 1151–1159.
56. Nielsen-Marsh, C. M.; Hedges, R. E. *J. Archaeol. Sci.* **2000**, *27*, 1139–1150.
57. Kyle, J. H. *J. Archaeol. Sci.* **1986**, *13*, 403–416.
58. Wright, L. E.; Schwarcz, H. P. *J. Archaeol. Sci.* **1996**, *23*, 933–944.
59. Sillen, A. In *The Chemistry of Prehistoric Human Bone*; Price, T. D., Ed.; Cambridge University Press: Cambridge, UK, 1989; pp 211–228.
60. Klerkx, J.; Deutsch, S.; Pilchler, H.; Zeil, W. *J. Volcan. Geotherm.* **1977**, *2*, 49–71.

61. Mégard, F.; Noble, D. C.; McKee, E. H.; Bellon, H. *Geol. Soc. Amer. Bull.* **1984**, *95*, 1108–1117.
62. Bellido B. E.; Navarez, S.; Simons, F. S., *Mapa geológico del Perú*. La Sociedad Geológica del Perú: Lima, Peru, 1956.
63. James, D. E. *Earth Planet. Sci. Letters.* **1982**, *57*, 47–62.
64. Hawkesworth, C. J.; Hammill, M.; Gledhill, A. R.; van Calsteren, P.; Rogers, G. *Earth Planet. Sci. Letters.* **1982**, *58*, 240–254.
65. Rogers, G.; Hawkesworth, C. J. *Earth Planet. Sci. Letters.* **1989**, *91*, 271–285.
66. McNutt, R. H.; Crockett, J. H.; Clark, A. H.; Caelles, J. C.; Farrar, E.; Haynes, S. J.; Zentilli, M. *Earth Planet. Sci. Letters.* **1975**, *27*, 305–313.
67. Harmon, R. S.; Barreiro, B. A.; Moorbath, S.; Hoefs, J.; Francis, P. W.; Thorpe, R. S.; Déruelle, B.; McHugh, J.; Viglino, J. A. *J. Geol. Soc. London.* **1984**, *141*, 803–822.
68. Schreiber, K. *Wari Imperialism in Middle Horizon Peru*; Museum of Anthropology, University of Michigan: Ann Arbor, MI, 1992.
69. Isbell, W. H.; Brewster-Wray, C.; Spickard, L. E. In *Huari Administrative Structure: Prehistoric Monumental Architecture and State Government*; Isbell, W. H.; McEwan, G. F., Eds.; Dumbarton Oaks: Washington, D.C., 1991; pp 19–54.
70. Pozzi-Escot B., D. In *Huari Administrative Structure: Prehistoric Monumental Architecture and State Government*; Isbell, W. H.; McEwan, G. F., Eds.; Dumbarton Oaks: Washington, DC, 1991; pp 81–92.
71. Price, T. D.; Burton, J., H.; Bentley, R. A. *Archaeometry* **2002**, *44*, 117–135.
72. Bentley, R. A.; Price, T. D.; Stephan, E. *J. Archaeol. Sci.* **2004**, *31*, 365–375.
73. Price, T. D.; Burton, J. H.; Bentley, R. A. *Archaeometry* **2002**, *44*, 117–136.
74. Tung, T. A.; Cook, A. G. In *Intermediate Elite Agency in Precolumbian States and Empires*; Elson, C., Eds.; University of Arizona Press: Tucson, AZ, In press.
75. Isbell, W. H. *Lat. Amer. Antiq.* **2004**, *15*, 3–32.
76. Katzenberg, M. A. In *Biological Anthropology of the Human Skeleton*; Katzenberg, M. A.; Saunders, S. R., Eds.; Wiley-Liss: New York, 2000; 305–328.
77. Sandford, M. K.; Weaver, D. S. In *Biological Anthropology of the Human Skeleton*; Katzenberg, M. A.; Saunders, S. R., Eds.; Wiley-Liss: New York, 2000; pp 329–350.
78. DeNiro, M. J. *Nature (London)* **1985**, *317*, 806–809.
79. Williams, C. T.; Marlowe, C. A. *J. Archaeol. Sci.* **1987**, *14*, 297–309.
80. Bentley, R. A. Ph.D. thesis, University of Wisconsin at Madison, Madison, WI, 2001.

Chapter 6

Interpreting Stable Isotopic Analyses: Case Studies on Sardinian Prehistory

Luca Lai[1], Robert H. Tykot[1], Jessica F. Beckett[2], Rosalba Floris[3], Ornella Fonzo[4], Elena Usai[3], Maria Rosaria Manunza[5], Ethan Goddard[6], and David Hollander[6]

[1]Department of Anthropology, University of South Florida, Tampa, FL 33620
[2]Department of Archaeology, University of Cambridge, Cambridge, United Kingdom
[3]Dipartimento di Biologia Sperimentale, Università di Cagliari, Cittadella di Monserrato, Italy
[4]Laboratorio di Archeologia, Villanovaforru, Italy
[5]Soprintendenza per i Beni Archeologici delle Provincie di Cagliari e Oristano, Cagliari, Italy
[6]Department of Chemical Oceanography, University of South Florida, St. Petersburg, FL 33701

Most archaeological applications of isotopic research aimed at tracing diet in the western Mediterranean focus on human bone collagen and overlook environmental factors, which already have been demonstrated to be relevant. In this biocultural context, marine resources and C_4 crops do not appear to have been as important as the proportion of plant versus animal foods. The corresponding smaller range of recorded isotopic variation makes it necessary to devise specific interpretive tools to distinguish environmental effects from dietary information. In this paper, we explore the option of using bone apatite $\delta^{18}O$.

Stable Isotopes: Principles and Interpretation

In many areas of the world, stable carbon and nitrogen isotopic analyses are only recently becoming part of the standard toolkit of bioarchaeologists interested in investigating diet in prehistoric and historic times. Its use in assessing the proportion of marine food and of C_4 plants is well established, and many successful studies have looked at important nutritional transformations around the world involving radical changes in food procurement. Examples are the transitions from Mesolithic fishing to Neolithic farming in Atlantic Europe *(1)*; from foraging to maize farming in the Americas; and the beginning of millet, sorghum, and rice agriculture in Asia and Africa *(2–4)*.

Since the beginning of dietary research through stable isotopes, the main research focus has changed. In the 1960s and 1970s, following the observation of consistent offsets between radiocarbon dates obtained on C_3 plants versus those obtained on maize, estimating maize contribution to diet was the first important anthropological problem to be addressed *(5–7)*. In the 1980s, a new direction of research was concerned with understanding how environmental effects, primarily humidity and temperature, affected isotopic signatures, and with the implications in assessing diet *(8)*, and the complexity of considering whole ecosystems *(9)*. Studies involving transitions from aquatic to terrestrial diets also became important *(10, 11)*. Between the 1980s and the early 1990s, it seems that understanding the different contributions of macronutrients to the synthesis of different tissues has become a critical issue. The first model (scrambling model) suggested that all macronutrients contributed to the synthesis of collagen, which therefore represented the whole diet, while apatite reflected the energy portion *(12)*. Through experimental efforts it became apparent that in reality collagen tended to preferentially be built from protein intake, whereas apatite included, and therefore reflected, all components *(13, 14)*.

Recently we have witnessed an expansion of the field, both in the number of researchers involved and in the degree of specialization. Investigating both principles and particular archaeological contexts is becoming increasingly difficult due to the complexity of the interactions between dietary signature on one hand, and environmental, physiological, and cultural factors contributing to the existence of diverse combinations at different points in time and space. Therefore, the principles are best investigated through the controlled study of living human and animal populations. On the other hand, the complexity of interpreting isotopic values integrating biologically and culturally specific contexts makes a detailed knowledge of the historically specific conditions a crucial element for a faithful reconstruction. Increased specialization needs to be matched by increased integration of scientific, social and historical perspectives.

Oxygen stable isotopic ratios in animal bone and several materials other than human tissues are used to assess environmental conditions, and are considered a fundamental proxy for paleoclimatic reconstructions. Oxygen isotopes are measured on both phosphate and carbonate *(15)*; for animals that are obligate drinkers, it mainly reflects the isotopic ratios of drinking water, which ultimately derives from rainwater. Although there are several factors contributing to specific $\delta^{18}O$ values, including temperature and the position of air masses, evaporation, and moisture from different sources *(16)*, an overall correlation has been confirmed by studies on the global, continental and local scale *(17, 18)*. Specific archaeological applications using human bone and tooth enamel include the assessment of mobility, residence patterns and the age of weaning. In this paper, we focus on carbon and nitrogen, while oxygen is only used as an interpretive complement.

The principle on which bone chemistry studies are grounded is that, generally speaking, our body composition is made up by the food we consume. The macronutrients we take in are used by the body to build tissues, and in the process they carry along the isotopic signature of their origin. Carbon and nitrogen isotope ratios in tissues (including hair, nails, bone, tooth enamel, flesh and virtually every part of our bodies) change predictably because of differential fractionation. This term refers to the process whereby chemical reactions involving these elements determine the selective uptake of heavier and lighter isotopes in specific ratios. This is due to different atomic mass, which causes the reactions to occur at different rates. Bone tissue is constantly replaced, so that its isotopic composition reflects dietary averages over several years before an individual's death. This means that rather than mirroring the last meals or the last seasons (as do gut contents or isotopes from hair), stable isotopic analysis of bone allows us to assess quantitatively the components of a standard diet for a long period of time.

What is actually measured by mass spectrometry is the ratio between isotopes of the same element ($^{13}C/^{12}C$, $^{15}N/^{14}N$, $^{18}O/^{16}O$), where the lighter isotopes are the most common, and the heavier isotopes occur in minimal proportions. Therefore, isotope measurements are expressed as differences (δ) per mil (‰) relative to standards shared by the scientific community, which are PDB for $\delta^{13}C$, AIR for $\delta^{15}N$, and SMOW for $\delta^{18}O$.

Worldwide, a large isotopic difference in $\delta^{13}C$ that is maintained up the food chain is determined by different plant groups: C_3 and C_4. These groups differ in their photosynthetic pathways, resulting in a very distinct signature, averaging about -26‰ in the former and -12‰ in the latter. From an evolutionary standpoint, C_3 plants are the most common and are dominant in temperate latitudes. C_4 plants are mostly grasses that developed traits adaptive to arid, tropical environments. Utilization by humans caused their spread when they held nutritional significance, as in the case of maize, millet, and sorghum.

It has been shown by controlled-feeding experiments on mammals that the organic and mineral portion of bone (collagen and hydroxyapatite, or apatite) do not reflect the same macronutrients in the same proportion *(13, 19)*. Bone collagen reflects the protein portion of the diet best, because it is mainly synthesized from ingested protein. Bone apatite is a more comprehensive indicator of diet since it is produced from proteins, carbohydrates and lipids. This means that if there is no nutritional imbalance *(20)*, collagen in humans will mostly reflect foods of animal origin, which are much richer in protein, and will reflect plants only if meat, dairy and fish were scarce, so that vegetal proteins were used to synthesize tissues. Apatite will reflect all three macronutrients *(13, 21)*, depending on their relative proportion to each other. However, the details of the mechanisms regulating the isotopic fractionation among diet on one hand, and collagen and apatite on the other, are complex and not fully understood *(22)*. It appears that the lower the amount of protein in the diet, the more carbohydrates and lipids will contribute to collagen composition, so that its isotopic composition *(20, 23)* may in some cases be a combination of both 'scrambling' and selective routing.

Due to physiology, the selective isotopic uptake in consumers varies by tissue type. In bone, the tissue archaeologically most important, $\delta^{13}C$ values shift about +5‰ from the vegetal food source in collagen, and about +12‰ in apatite. Therefore, values for herbivores in C_3 ecosystems should be around -21‰ for collagen and around -14‰ in apatite. Pure C_4-feeders would show $\delta^{13}C$ values around -7‰ in collagen and around 0‰ in apatite. Going up the food chain, $\delta^{13}C$ values show a much smaller difference, not more than 2‰, so that this element is not the best tracer to quantify the importance of nutritional resources from different trophic levels for human diet. Human values for fully terrestrial C_3 ecosystems would typically be around -20‰ (collagen) and -12‰ (apatite), with differences related to the extent of their carnivory. Nitrogen represents the best choice for this purpose, since average differences between consumed and consumer are usually around 3–5‰. Nitrogen is fixed or absorbed by plants, and its values are also passed on up the food chain. Values for plants are therefore around $\delta^{15}N$ 0–4‰, and consequently herbivores' about 4–8‰ and carnivores' about 8–12‰.

Marine ecosystems have much longer food chains, so that the range of variation in $\delta^{15}N$ is much wider. Marine predators show values up to 20‰ in collagen, while fish signatures are generally higher than terrestrial animals *(10, 11)*. Marine $\delta^{13}C$ values also are often enriched, resulting in overlap with the C_4 plants range, so that the best way of setting them apart, and the standard way of presenting the values, is by means of a plot where $\delta^{13}C$ and $\delta^{15}N$ are the x and y coordinates. High $\delta^{13}C$ and low $\delta^{15}N$ indicates C_4 plant-based diets, while high $\delta^{13}C$ and high $\delta^{15}N$ indicates marine protein-based diets.

Lacustrine ecosystems, due to reservoir effects that are particular to individual water bodies, can show remarkable variation, potentially overlapping with both marine and terrestrial values. The reservoir effect applies to some extent also to smaller circumscribed seas such as the Mediterranean.

Stable Isotopes and Diet in Western Mediterranean Prehistory

If we are to correctly interpret the isotopic values in dietary terms in the context of western Mediterranean prehistory, we need to consider the specific resources available and especially those that archaeology has documented as being used as food items. Our focus is on later prehistory.

Food production and consumption between the 6th and the 2nd millennium B.C. is characterized by the spread of the Neolithic suite of domesticated animals and crops, which apparently were adopted in different tempos and combinations in different regional contexts *(24, 25)*. While some areas apparently relied on foraging until relatively late, others seem to have adopted quickly and thoroughly the Neolithic economic patterns. It is believed that in the Mediterranean reliance on fishing did not reach the same importance it did in the Atlantic and Baltic areas *(1)*, although the quality of the evidence is not homogeneous.

Domesticated animals and crops spread from the Near East, mostly through coastal routes along Greece, the Italian Peninsula on both the Tyrrhenian and Adriatic sides, and then down the coast of the Iberian Peninsula and the Corsican-Sardinian complex. Due to environmental factors, on the African side, Neolithicization took the form of a shift to pastoralism rather than agriculture. It seems that at several locations domesticated animals, namely sheep, goat, pig and cattle, were adopted before farming. The importance of agriculture was highly variable in the 6th and 5th millennia B.C., and became more generalized only later, while in some areas pastoralism possibly became more important starting in the 4th millennium. This trend may have been favored by climate change, and/or by the exploitation of secondary animal products such as milk and labor in the form of plow traction *(26, 27)*. The effects of these technological innovations, though, must have varied according to both environmental and cultural contexts.

According to the typical way of reading the values, it seems to be confirmed by isotopic analyses (Figure 1) that in the Late Neolithic through Bronze Age exploitation of seafood was not important. This is not surprising in inland sites in Neolithic Anatolia, such as Çatalhöyük *(28)*, but appears to be the case also in insular settings such as Bronze Age Crete. At Mycenae, only the high status individuals show values that indicate some consumption of marine food, and in Greece as a whole, an explicit comparison of inland and coastal sites spanning the entire Neolithic (late 7th to early 4th millennium B.C.) revealed that the protein component of the diet was relatively homogeneous and based on plants and animals, with no enriched values providing clear evidence for substantial fish intake *(29, 30)*. In the central-western Mediterranean, similar $\delta^{13}C$ and $\delta^{15}N$ values compatible with terrestrial C_3 protein have been documented in coastal environments such as the Maltese archipelago from around 4000 B.C. to the 3rd millennium B.C. *(31)*. This pattern does not change if we look at data from the

Balearic islands between the 4th millennium B.C. to the 1st century A.D. *(32, 33)*, and from several locations on the Italian peninsula dating to Neolithic through Copper Ages *(34)*. While this could be disguised by seafood values specific to the Mediterranean that are much more terrestrial-like *(35–37)*, the general conclusion is that the consumption of marine resources was negligible, and that, given the little range of isotopic values, it is impossible to assess the contribution of animal versus plant foods to the diet analyzing collagen alone.

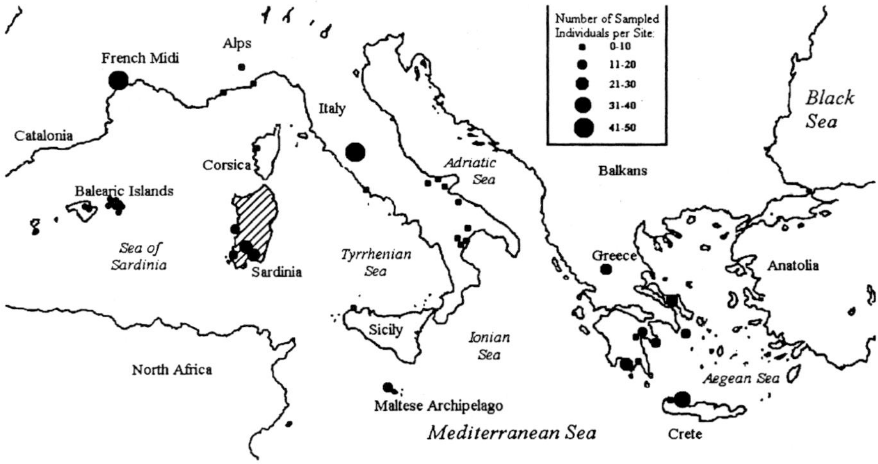

Figure 1. Map of the central-western Mediterranean showing the location of Sardinia and the prehistoric sites that have been analyzed for $\delta^{15}N$ and $\delta^{13}C$ in the whole area

Unlike East Asia, Africa and the Americas, where the presence of C_4 crops makes it easier to detect the beginnings and importance of agriculture, the cereals domesticated in the Near East and spread in the Mediterranean yield $\delta^{13}C$ and $\delta^{15}N$ values similar to wild plants. Since marine resources do not seem to represent a large component of the diet, the majority of western Mediterranean collagen values lie within $\delta^{15}N$ 9.0 ± 1.0‰ and $\delta^{13}C$ 19.3 ± 1.0‰.

Apatite values and the spacing between collagen and apatite are therefore very useful in addressing what becomes the main nutritional question in prehistoric Mediterranean diet and economy: the relative proportion of animal and vegetal foods. This corresponds to the importance of agriculture versus animal husbandry, with all the implications this holds in terms of ethnographically documented correlations involving society and culture. The spacing between collagen and apatite $\delta^{13}C$ can be used, in fact, to 'triangulate' a dietary reconstruction with separate interpretations of collagen and apatite. Larger spacings have been found to correlate with herbivores, smaller spacings with carnivores *(21, 38)*.

Another way of improving the data is by considering environmental factors affecting the isotopic signature. This is particularly useful when a faunal and botanical isotopic baseline is not available. This is the best way of detecting ecosystem-wide shifts, which when comparing human groups can be mistaken for dietary change or variation. It is critical, therefore, to establish whenever possible a baseline that is site-specific, or at least represents the region during a given period. Analyses of faunal remains provide a good estimate both of the animals themselves and the plants they consume. Unfortunately, faunal remains are not always directly associated with human burials, and often there are no faunal remains from sites close enough in space and time, or for various reasons they are not accessible.

Case Studies on Sardinia

Sampling Criteria and Selected Populations

Skeletal collections were selected for sampling to address specific research questions as part of a larger project aimed at understanding variation between coastal and inland environments and change of economic practices over time. In this chapter we use the first results to make some interpretive considerations.

Samples were collected in summer 2003 and fall 2004 from the Ex GIL, a storage facility of the Soprintendenza in Sardara (province Medio Campidano, then Cagliari); in the Old Museum in Cagliari; and at the storage unit of the Dipartimento di Biologia Sperimentale, Sezione Antropologia, in Monserrato (province Cagliari), using a portable drill. Attention was paid to preserving as much as possible the integrity of the remains.

The large majority of skeletal remains from prehistoric Sardinia come from collective burials where skeletal elements were not articulated nor associated in any way deemed significant by the excavators. It was therefore necessary to select one skeletal element in order to insure that the same individuals were not being sampled more than once. Crania were found to be the best choice, since they hold the highest potential to identify the individual's sex, age and pathologies. While we know of different turnover rates in different skeletal elements and within the same element, the specific times have not been documented. Growth speed is believed to affect intra-bone variability mostly in subadults, while by adult age isotopic values should be fairly similar across the skeleton. The possibility of substantial variation has been investigated in humans for $\delta^{15}N$ *(39)*, and no substantial variation has been found, so confirming its value for reconstructing diet change and weaning practices, which is being done routinely by comparing bone and tooth enamel, or through tooth enamel microsampling, less often with bone alone *(40)*. For this study, samples were removed from the cranial vault whenever possible, and in a few cases, from the mandible, maxilla or long bone, to ensure the homogeneity and overall comparability of averages between groups. Authorizations for removal and transport to the United States of America for preparation and analyses were granted according to the regulations, and prepared for analyses by mass spectrometry according to the procedures explained below.

The four sites span roughly 2500 years, between the early 4th and the mid-2nd millennium B.C. Geography and environment show some diversity and similarity (reference for chronology and geographic location are in Figure 2). San Benedetto *(41–43)* is a necropolis of tombs carved into the rock dating to the early part of the Late Neolithic; it is within one day's walk from the coast, but the very steep and rugged terrain in between makes it unlikely that a human group living nearby could have relied on marine food. The group shows very homogeneous morphology, so that the possibility of an endogamic community has been suggested.

Padru Jossu *(44–46)* is a rock-carved tomb, located inland on the outskirts of a wide alluvial plain where a lake was present until the 1800s. Two phases were sampled, dating to the Bell Beaker phase (2500–2200 B.C.) and to the Early Bronze Age (2200–1900 B.C.). Animal remains, interpreted as funerary offerings, were recovered; the relative species frequency in the two represented phases changes from a prevalence of sheep/goat to a higher quantity of cattle and pig specimens, which has been interpreted as an increase in the importance of agriculture as opposed to sheep tending.

Is Calitas *(47–49)* is a pit grave dating to the Early Bronze Age (2200–1900 B.C.) in an inland, hilly area near the coastal plain surrounding the Gulf of Cagliari. It yielded a few articulated skeletons and a large number of scattered bones, with associated grave goods including beads, pottery, stone and metal tools.

Is Aruttas is the only coastal site tested. Skeletal materials were recovered in a cave a few hundred meters from the shore, located in a marshy and sandy area on the central-west coast of Sardinia. The stratigraphic context was disrupted by looters, and pottery sherds found associated were attributed to the final phase of the Late Neolithic. One radiocarbon date, though, shifted the chronology to the Middle Bronze Age (1600–1300 B.C.). From the physical fitness and low caries of the recovered human remains, a well balanced diet has been inferred *(43, 50, 51)*.

Period	Phase	Culture	Date
NEOLITHIC	Middle	Bonu Ighinu	ca. 4700-4000 BC
	Late	Ozieri	ca. 4000-3200 BC
COPPER AGE	Early	Sub-Ozieri, Filigosa, Abealzu	ca. 3200-2700 BC
	Middle	Monte Claro	ca. 2700-2200 BC
	Late	Bell Beaker A	
BRONZE AGE	Early	Bell Beaker B/ Bonnannaro A	ca. 2200-1900 BC
	Middle	Bonnannaro B	ca. 1900-1600 BC
		Nuragic I	ca. 1600-1300 BC

Figure 2. Left: Chronological table of Sardinian prehistory from the Middle Neolithic to the Middle Bronze Age, based on calibrated radiocarbon dates. Right: Map of Sardinia with the location of the four sites where the skeletal remains tested were recovered.

Sample Preparation

Approximately 1 gram of whole bone was selected per each individual, physically cleaned when needed, ultrasonically cleaned and dried. The preparation is a variation of well established procedures *(14, 52)*. After removal of 10 milligrams of bone powder by drilling and milling, collagen was isolated by demineralization with low-concentration HCl in two or more ~24-hour baths, depending on the necessity for more reaction. These baths were preceded and followed by soaking for ~24 hours in NaOH to remove humic acid contaminants. Finally, samples were treated with a 2:1:0.8 mixture of methanol, chloroform, and water to remove lipids, again for ~24 hours. The use of weak HCl (2%) allows the recovery of collagen even from bone that is physically degraded. Sample preservation and reliability can be visually assessed throughout the process, unlike during procedures that involve turning bone into powder *(6, 53)*. Collagen pseudomorphs are the end result; they are oven-dried and weighed to obtain yields. The reliability of collagen samples is commonly assessed using parameters, including the ratio of C:N in the collagen and the quantity of N and C in collagen *(52)*. C:N ratios are not available for this paper so we rely on visual assessment, collagen yields, and consistency of peak area ratios during stable isotopic analysis. Yields under 1% usually indicate poor preservation, and the likelihood that the remaining material may have become fractionated. On the other hand consistency of yields and no relationship between yields and isotopic values suggests that the signal is preserved. Atypical peak area ratios (relative to the majority of other samples in a run) measured during isotopic analysis are also a sign of poor preservation. Measured collagen yields showed more variation between than within sites; there was no substantial stable isotopic difference within each site between lower- and higher-collagen yielding samples, which indicates that diagenesis did not affect isotopic values significantly.

To isolate the apatite, 10 mg of bone powder are treated with a ~72-hour bath in sodium hypochlorite which dissolves the organic portion; non-biogenic carbonate is removed by soaking the sample for ~24 hours in a 1M buffered acetic acid/sodium acetate solution, and attention is paid to be as consistent as possible with sample to solution quantity ratio and soaking times *(54)*. Bone apatite remains less accurate and reliable if compared to collagen and particularly to tooth enamel, because removal of carbonate derived from recrystallization of exogenous carbonates leaked from the soil matrix may not be complete *(55, 56)*. Nevertheless, the procedures used, and the assessment of the integrity of samples through the yields *(57)* measured after each preparation treatment can be used as an indication of reliability. It has been shown that bone mineral has the potential to retain original isotopic signatures, and it has been suggested that in certain conditions recrystallization can even favor this preservation, rather than contaminating the sample *(58)*.

As explained elsewhere *(6, 59)*, one milligram per sample of the end product of collagen preparation is placed in tin capsules and analyzed in continuous flow mode, using a Carlo-Erba 2500 Series II CHN analyzer coupled with a ThermoFinnigan Delta Plus XL stable isotope ratio mass spectrometer. As for the apatite samples, one milligram of purified powder is analyzed on another ThermoFinnigan Delta Plus XL mass spectrometer, in dual-inlet configuration, equipped with a Kiel III individual acid bath carbonate system. Both mass spectrometers are located at the University of South Florida, St. Petersburg campus. Samples of an isotopically known working lab standard (urea for collagen and Carrera marble for apatite samples) are included at the beginning, middle and end of each run of samples. Continuous Flow-IRMS precision (2σ) is typically better than ±0.3‰ for $\delta^{15}N$ and ±0.2‰ for $\delta^{13}C$. Dual-Inlet-IRMS precision (2σ) is typically better than ±0.04‰ for $\delta^{13}C$ and ±0.06‰ for $\delta^{18}O$.

Results

Average collagen values by site range between $\delta^{13}C$‰ = –19.5 at San Benedetto to –18.6 at Is Aruttas, and between $\delta^{15}N$‰ = 9.4 at San Benedetto and 10.8 at Padru Jossu phase A. Apatite $\delta^{13}C$‰ values are between –13.8 at San Benedetto and –10.4 at Is Calitas and Is Aruttas, while the $\delta^{18}O$‰ values are between –4.6 at Is Calitas and –1.2 at Is Aruttas (Figure 3).

Table I. Average Stable Isotopic Values for the Five Analyzed Groups

	Collagen			*Apatite*		
Sites/phases	*n*	$\delta^{13}C$‰	$\delta^{15}N$‰	*n*	$\delta^{13}C$‰	$\delta^{18}O$‰
San Benedetto	16	-19.5±0.2	9.4±0.5	16	-13.8±1.1	-4.3±0.3
Padru Jossu A	9	-18.8±0.2	10.8±0.8	9	-12.0±1.2	-3.2±0.7
Padru Jossu B	13	-18.9±0.4	10.1±0.9	13	-11.3±0.8	-2.6±0.6
Is Calitas	26	-19.2±0.4	10.4±0.9	29	-10.4±0.5	-4.6±0.4
Is Aruttas	11	-18.6±0.3	10.4±0.9	10	-10.4±2.6	-1.2±0.8

By plotting the collagen values with relative standard deviation, we can observe a large overlap. The most distinct group is San Benedetto, which for its more negative values of both $\delta^{13}C$‰ and $\delta^{15}N$‰ should be interpreted as the site with the lowest consumption of animal protein. Is Aruttas shows the most positive $\delta^{13}C$‰ values and fairly high $\delta^{15}N$‰. Since the site is on the coast, we would interpret this difference as due to a limited intake of marine foods, while at Padru Jossu A and Is Calitas this would be more compatible with higher consumption of food of terrestrial animal origin. We must emphasize, in addition to the overlap between the groups, the closeness of their signatures if considered within the larger picture. Figure 4 shows the reference values for the main categories of food for western Europe, and the range covered by Figure 3. Clearly, most protein from all groups came from terrestrial animals, and the seafood contribution may have been limited to none.

Apatite $\delta^{13}C$‰ values, as shown in Figure 5, have more distinct patterns. San Benedetto, the Late Neolithic group, stands out for its more negative values. The chart is again showing a parameter indicating the trophic level of the protein component (collagen $\delta^{15}N$‰), but this time plotted with the value resulting from the whole diet (apatite $\delta^{13}C$‰). The interpretation of apatite values and the resulting $\delta^{13}C$‰ spacing between collagen and apatite is complex and not fully understood *(22)*. One reason for enriched apatite $\delta^{13}C$‰ in ecosystems where C_4 crops are available and important is likely to be their substantial consumption. However, as we have discussed, this seems unlikely in the context of prehistoric Sardinia where all domesticated plants were C_3 until the Iron Age. Large differences cannot be attributed to heavy reliance on marine food, since this would be apparent from collagen as well. Values are indeed consistent with data from several locations in Italy *(34)*. Among possible alternative factors is a substantial difference in lipid content in the diet, and/or a difference in lipid origin, and/or a difference in carbohydrate relative quantities. However, discussing this aspect would require a separate paper and will not be dealt with here. The point we want to highlight is instead the importance of environmental variation in the interpretation of tightly clustered isotopic values, which we discuss below.

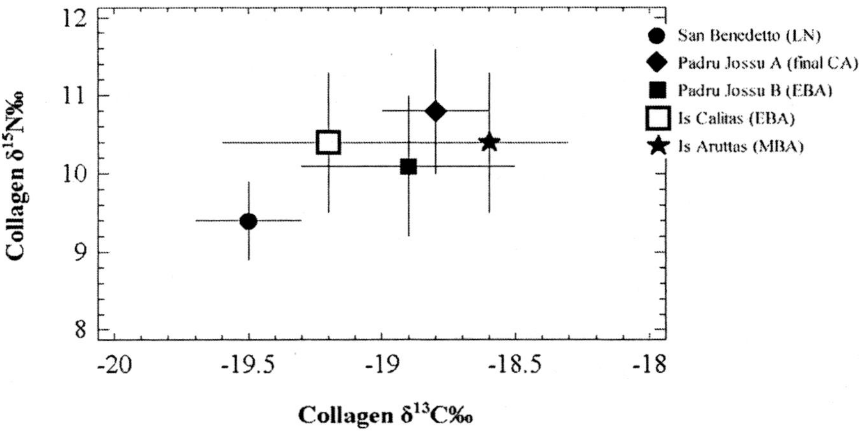

Figure 3.Plot of average collagen $\delta^{15}N$ and $\delta^{13}C$ values for the five human groups analyzed, with standard deviation.

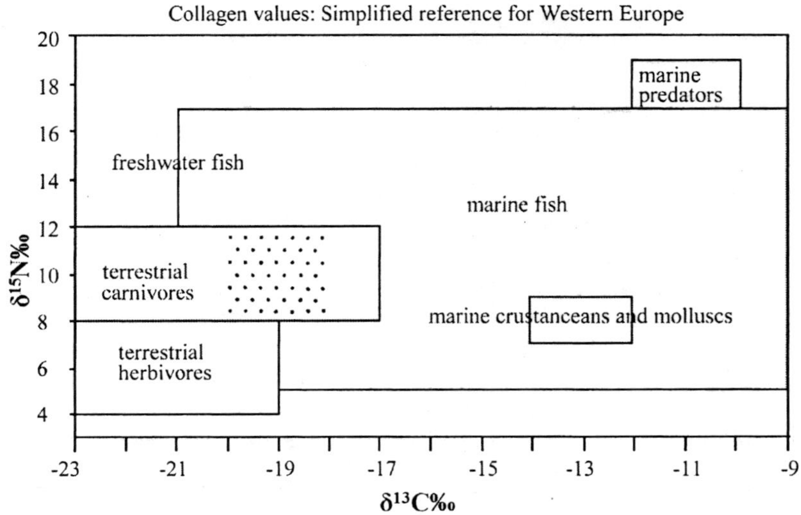

Figure 4 .Plot of reference collagen $\delta^{15}N$ and $\delta^{13}C$ values for western Europe. The range of the plot in Figure 3 is indicated by the dotted texture.

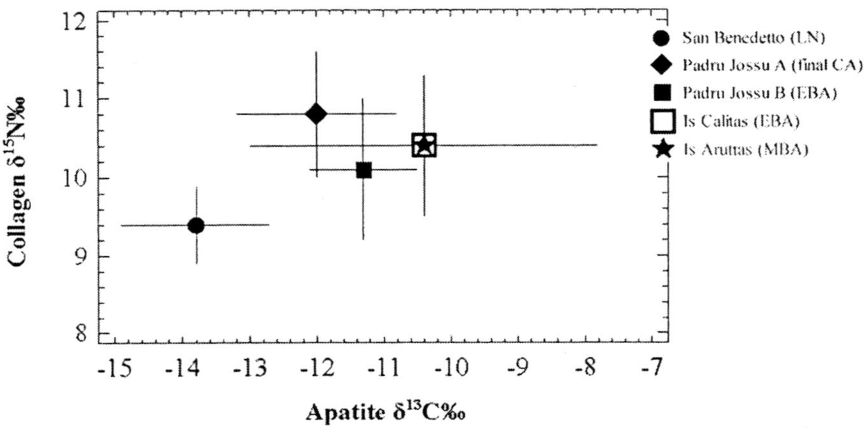

Figure 5. Plot of average collagen $\delta^{15}N$ and apatite $\delta^{13}C$ values for the five human groups analyzed, with standard deviation.

Discussion

It is well established that isotopic values are affected by climate in direct or indirect ways. There are also studies concerning the importance of environmental variation over time and space in interpreting dietary information. Variation in $\delta^{13}C$ in wood charcoal has been found to correlate to elevation within a given region *(14)*. Variation in $\delta^{13}C$ according to climate during the Holocene has been specifically investigated for wide areas from Northwestern Europe to the Near East and North Africa *(60)*. This has demonstrated a clear regional variation, in some cases stronger than temporal variation. When comparing Spain with northern European countries during the entire Holocene, isotopic values were consistently less negative in the former than in the latter, with no overlap for the last 10,000 years. Fluctuations over time corresponded across the studied area, reflecting continent-wide climatic phenomena *(60, 61)*.

Similar variation in $\delta^{13}C$, observed commonly in cave sediments and routinely used along with $\delta^{18}O$ for paleoclimatic reconstruction, has been studied for horse bone collagen and tooth enamel on a larger chronological scale, and found to correlate with climatic trends in the last 40,000 years. It has been argued that this variation may stem from different phenomena: changes in atmospheric CO_2 concentration, and/or secondary effects due to $\delta^{13}C$ depletion in forested environments *(62, 63)*. Regardless of the cause, these isotopic differences would be maintained through the food chain from plants to herbivores (and omnivores), and from herbivores to carnivores (and omnivores).

Studies of isotopic fractionation in several plant species have demonstrated its correlation with water availability *(64–66)*.

Concerning the correlation of $\delta^{15}N$ with climate, there have been observations on specific areas, where an inverse relationship with altitude was documented *(67)*. In the last decade, a substantial body of evidence on $\delta^{15}N$ change over time has being collected, and its relationship with climatic conditions is becoming clear *(62)*. More specifically, the dataset considered by Schwarcz and coworkers is used to explore the correlation with precipitation, which appears to be inverse and statistically strong *(68)*.

What is most important is that all these ecosystem-wide shifts are transferred from the lower trophic levels up to the several groups of consumers. Considering collagen $\delta^{15}N$ and the case of our Sardinian dataset, if we assume an identical diet we should nevertheless expect some degree of variation according to precipitation. This, from a synchronic perspective, involves altitude and geomorphology. Based on present-day rainfall, which is ~750 mm/year at San Benedetto and ~550 mm/year at the remaining sites, if we apply an approximate ratio of ~$\delta^{15}N$ 1‰ per 100 mm/year of rain, dietary interpretation changes substantially. The San Benedetto group, falling above the line describing climatic variation, would show higher consumption of animal protein. On the other hand, $\delta^{15}N$ values for Padru Jossu, Is Calitas and Is Aruttas can be explained by climatic variation alone (Figure 6).

The problem is that we are still dealing with small isotopic differences. Since the groups span over 2,000 years we must also consider climate change. Among the main stable isotopes, $\delta^{18}O$ is the most commonly used for paleoclimatic reconstruction. Its correlation to both temperature and precipitation has been long established. A correlation between biogenic $\delta^{18}O$ in water-dependent mammals and meteoric water $\delta^{18}O$ has been documented *(69, 70)*, so we assume that the isotopic signature should be preserved from the atmosphere to bone carbonate and phosphate, passing through rainwater and drinking water. The relationship itself between $\delta^{18}O$ with $\delta^{13}C$ has been detected in tree leaves *(71, 72)*; in horse teeth through microsampling, which has shown remarkable seasonal covariation *(73)*; and in lacustrine environment on molluscs *(74)*.

Isotopic signatures of groundwater and water reservoirs may alter the $\delta^{18}O$ values of rainwater, and the physiology of different species can be very complex and often not comparable. In addition, their $\delta^{18}O$ signature is derived through different pathways, whereas humans derive the majority of it from drinking water. Despite these specific mechanisms, what is important is the general ecosystem-wide variation. With this premise, and under the condition derived from archaeological evidence that diet at Sardinian prehistoric sites was based on the same resource pool that included C_3 plants and animals but no C_4 plants or marine resources, we can plot collagen $\delta^{13}C$, collagen $\delta^{15}N$ and apatite $\delta^{13}C$ against apatite $\delta^{18}O$ as an environmental proxy, isolating the real dietary signal.

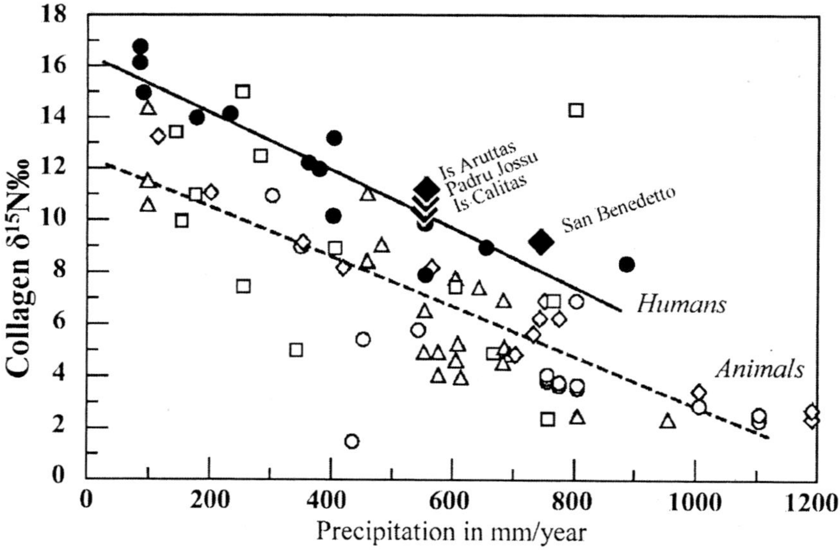

Figure 6. Plot of average human (filled circles) and animal (empty symbols) collagen $\delta^{15}N$ and precipitation values from several sites across the world reprinted after Schwarcz and coworkers (68), Fig. 1, p. 630, with permission from Elsevier. Average values from Sardinia (filled diamonds) have been added to the original plot according to present-day precipitation.

Plotting apatite $\delta^{18}O$ and collagen $\delta^{15}N$ allows us to visually appreciate how much variation is due to climatic factors and by comparison how much is due to diet (Figure 7). The line in the chart is not a regression line, but indicates only the slope of the covariation between $\delta^{15}N$ and $\delta^{18}O$ related to rainfall. Its purpose is to allow a comparison among sites. It is obtained by combining two correlations to precipitation. One is the correlation between collagen $\delta^{15}N$ and precipitation measured by Schwarcz and coworkers *(68)* estimated at ~1‰/100 mm/year. The other one is the correlation between $\delta^{18}O$ and precipitation by Bar-Matthews and coworkers, which is measured on non-human, biogenic carbonates *(75)* estimated by the authors to ~1‰/200 mm/year. This results in a slope of $\delta^{15}N$ ~2‰ = $\delta^{18}O$ ~1‰ which signifies the covariation between $\delta^{15}N$ and $\delta^{18}O$ that is related to climate. The dietary difference between two groups would correspond to the distance from any given slope line. Interpreting the data in this manner, the human group at San Benedetto appears to have roughly as much animal protein in the diet as Padru Jossu A, both having the same distance to the climatic regression line. The values of Is Calitas show enrichment, reflecting a higher animal protein proportion, while those of Padru Jossu B and particularly Is Aruttas are the lowest amount of all groups.

Plotting collagen $\delta^{13}C$ against apatite $\delta^{18}O$ should again set apart the signatures due to dietary protein and climatic variation (Figure 8). The values appear to be linearly correlated and they indicate a small difference in the intake of animal protein. In this case we can read a trophic level effect from $\delta^{13}C$ values by comparing Is Calitas to San Benedetto and Padru Jossu A to Padru Jossu B. The first of each couple shows less animal protein consumption than the second, which is in agreement with the $\delta^{15}N$ interpretation.

Finally, plotting apatite $\delta^{13}C$ against apatite $\delta^{18}O$ (Figure 9) shows a strong linear correlation between the two coordinates, except for the Is Calitas group, which stands out for its enriched apatite $\delta^{13}C$. Since apatite reflects the whole diet, this comparison would lead to the interpretation that this human group is the only one with a substantial dietary distance from the rest. As discussed above, the meaning of apatite values is complex due to its mirroring the signature of all macronutrients, and it can have multiple causes. In accordance with the previous plots, we could argue that a high consumption of animal protein, which is isotopically heavier than plant carbohydrates (~-21‰ vs. ~-25‰), could explain the less negative $\delta^{13}C$. The common explanation that connects smaller $\delta^{13}C$ collagen-apatite spacing to the amount of consumed lipids would in this case contradict the high quantity of animal protein apparent in the previous plots.

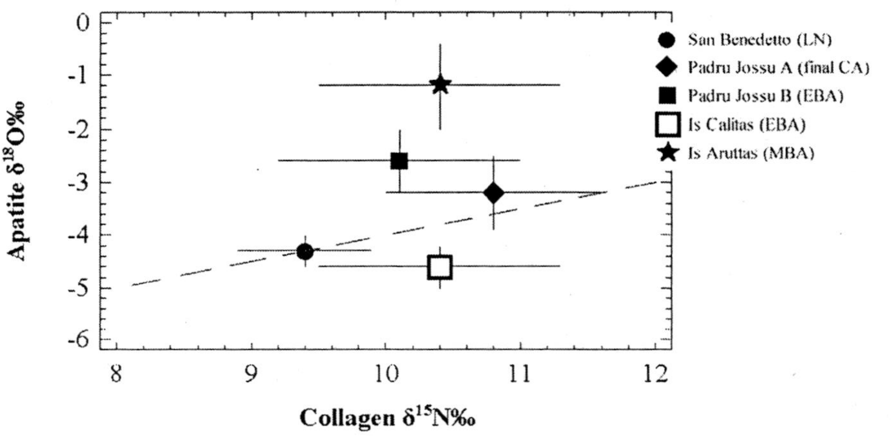

Figure 7. Plot of average collagen $\delta^{15}N$ and apatite $\delta^{18}O$. The regression line is the slope obtained from a rough approximation of the correlation between precipitation and both $\delta^{15}N$ and $\delta^{18}O$. Distance from the line would correspond to dietary variation, with more animal protein to the right and less to the left. As an example, at San Benedetto this quantity would be less than at Is Calitas.

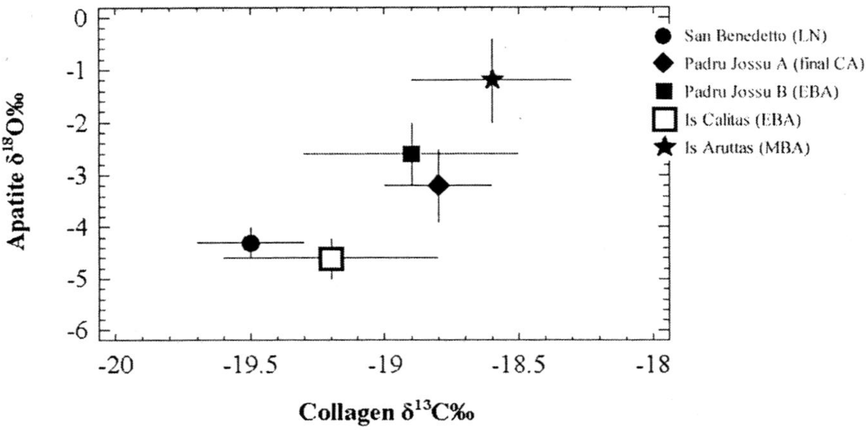

Figure 8. Plot of average collagen $\delta^{13}C$ and apatite $\delta^{18}O$. Assuming the alignment indicates environmentally-related covariation of the two coordinates, $\delta^{13}C$ values that are less negative relative to the alignment would indicate higher consumption of animal protein.

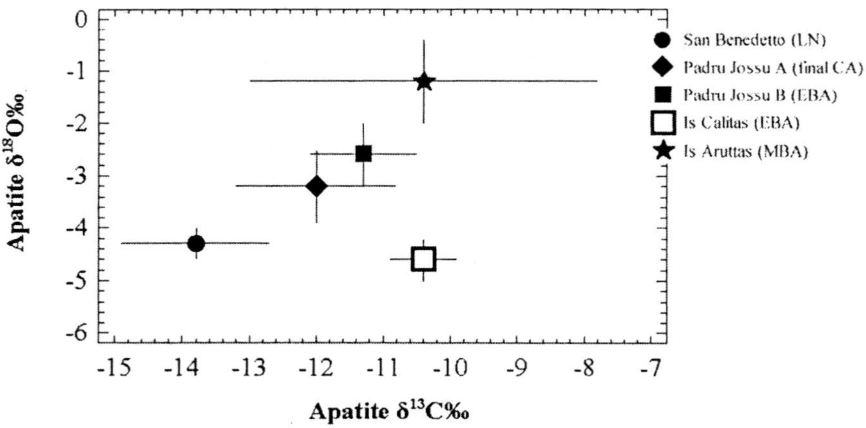

Figure 9. Plot of average apatite $\delta^{13}C$ and $\delta^{18}O$. Assuming that the alignment indicates environmentally-related covariation of the two coordinates as for Figure 8, $\delta^{13}C$ values that are less negative relative to the alignment would indicate dietary difference. In this case, since the plotted dietary indicator is apatite $\delta^{13}C$, the difference involves the whole diet. The group clearly apart would therefore be Is Calitas.

This procedure is to be considered an exploration of alternative directions to detect shifts of climatic origin affecting whole ecosystems and get closer to the actual dietary information, when animal and plant reference values from the same context are not available. We emphasize that the most direct and reliable method remains the isotopic analysis of faunal and botanical remains from the same context, since other factors like the isotopic signature of groundwater and water reservoirs may alter the $\delta^{18}O$ values and therefore the correlation between $\delta^{18}O$, $\delta^{13}C$ and $\delta^{15}N$.

Conclusions

In order to reconstruct human diet from bone tissue, direct isotopic analysis of animal and plant remains from the same archaeological context is the most reliable way to detect isotopic shifts involving the whole ecosystem due to environmental variation. Since this is often impossible for the lack of these control samples, we have explored the use of $\delta^{18}O$ to assess the environmentally induced variation in $\delta^{13}C$ and $\delta^{15}N$ values from collagen and apatite, and assess the dietary information they represent. This can be done assuming a scarce nutritional role of marine resources and the absence of C_4 crops, as seems to be the case in the western Mediterranean and specifically in the Sardinian Neolithic and Bronze Age.

After reconstructing diet by means of plotting collagen $\delta^{13}C$, collagen $\delta^{15}N$ and apatite $\delta^{13}C$ with each other, we plotted each with $\delta^{18}O$. Since all of them partially depend on climatic factors, while $\delta^{18}O$ should reflect climate more faithfully, we would expect a straight line if diet is identical for all groups. Values that are not along this line are interpreted as depending on differences in diet.

From the collagen values we assessed that the protein component of the diet was mainly supplied by terrestrial animal sources among all the populations and periods. Moreover, marine resources were not nutritionally as important, confirming the previous data available for the central and western Mediterranean. Comparing the five groups analyzed so far, San Benedetto appears to reflect the lowest animal protein consumption, Padru Jossu A the highest, and Is Aruttas possible limited consumption of seafood. The apatite $\delta^{13}C$ values showed a more sensible difference between San Benedetto and the remaining sites, to be attributed to more plant carbohydrates and/or lipid consumption.

Interpreting the plots of collagen $\delta^{15}N$, collagen $\delta^{13}C$ and apatite $\delta^{13}C$ with $\delta^{18}O$ leads us to more accurate reconstructions: it seems that the San Benedetto, Padru Jossu B and Is Aruttas groups all had a lower average consumption of animal protein than the Padru Jossu A and Is Calitas groups. Is Aruttas values are not indicative of any animal protein intake (whether terrestrial or marine) higher than the remaining sites, and may instead have been the group with the highest consumption of plant protein. Is Calitas shows the highest consumption of animal products, which affects visibly even the apatite signature when plotted against $\delta^{18}O$.

Drawing conclusions of paleoeconomy requires the integration of several other lines of evidence, which is not in the scope of this paper. Based on the comprehensive interpretation of the isotopic results presented, the Late Neolithic group (San Benedetto), the Early Bronze Age group at Padru Jossu (B), and the Middle Bronze Age group (Is Aruttas) seem to have relied more on farming, while the Late Copper Age group (Padru Jossu A) and the Early Bronze Age group at Is Calitas emphasized reliance on animal husbandry. More specifically, San Benedetto seems to be the most agricultural, and Is Calitas seems to be the most pastoral group. This interpretation is in line with the expected dietary and economic change over time in Sardinian prehistory.

To test these conclusions from both a paleoeconomic and a methodological perspective, analyses of faunal and human samples are critical. As we have highlighted throughout our discussion, documenting isotopic variation over the different trophic levels of whole ecosystems is and remains the best way to reconstruct dietary variation reliably. We suggest the possibility that, in the specific ecological and cultural context of late western Mediterranean prehistory, using $\delta^{18}O$ helps us to disentangle dietary and environmental signature and therefore detect more faithfully nutritional and economic practices.

Acknowledgments

We would like to thank the Soprintendenza ai Beni Archeologici delle Provincie di Cagliari e Oristano, specifically Dott. V. Santoni, Dott.ssa D. Mureddu, and Dott. Carlo Tronchetti for granting the necessary authorizations; and sig.ra Cynthia Ventimiglia, for assistance at the Vecchio Museo during sample collection. We thank sig. Ubaldo Badas, director of the Laboratorio di Archeologia at Villanovaforru, and sig. Giuseppe Garau, president of the Consorzio Gestione Museo di Sardara, for facilitating access to the collections curated at these locations. We express our gratitude to the Sigma Xi Scientific Research Society, the Comune di Soleminis (Sardinia), and the National Science Foundation for each funding a part of the analyses. A personal thank you from the first author to Dr. Tamsin O'Connell and Dr. John Robb for enlightening discussions and precious advice on several issues; to Annette Doying, Jennifer Kelly, Belen Sisay for help in the preparation of samples; to the Farmacia Dott. Giua, Cagliari (Sardinia), for help in first steps towards sample collection; and to Mimmo Casti, Alessandra Lai, Gian Paolo Melis, Marina Melis, and Sharon Watson, for their practical and moral support in many occasions.

References

1. Richards, M. P.; Hedges, R. E. M. *J. Archaeol. Sci.* **1999**, *26*, 717–722.
2. van der Merwe, N. J.; Vogel, J. C. *African Archaeol. Rev.* **1983**, *1*, 33–56
3. Schoeninger, M. J. *Evol. Anthro.* **2005**, *3*, 83–98.
4. Pechenkina, E. A.; Ambrose, S. H.; Xiaolin, M.; Benfer, J. R. A. *J. Archaeol. Sci.* **2005**, *32*, 1176–1189.
5. Ambrose, S. H.; Krigbaum, J. *J. Anthro. Archaeol.* **2003**, *22*, 193–199.
6. Tykot, R. H. In *Physics Methods in Archaeometry. Proceedings of the International School of Physics "Enrico Fermi" Course CLIV*; Martini, M.; Milazzo, M.; Piacentini, M., Eds.; Società Italiana di Fisica: Bologna, Italy, 2004, p 433–444.
7. *Histories of Maize: Multidisciplinary Approaches to the Prehistory, Biogeography, Domestication, and Evolution of Maize*; Staller, J. E.; Tykot, R. H.; Benz, B. F., Eds.; Elsevier, 2006.
8. Sealy, J. C.; van der Merwe, N. J.; Lee-Thorp, J. A.; Lanham, J. L. *Geochim Cosmochim. Acta* **1987**, *51*, 2707–2717.
9. Schwarcz, H. P.; Schoeninger, M. J. *Am. J. Phys. Anthro.* **1991**, *34*, 283–321.
10. Schoeninger, M. J.; DeNiro, M. J. *Geochim. Cosmochim. Acta* **1984**, *48*, 625–639.
11. Schoeninger, M. J.; DeNiro, M. J.; Tauber, H. *Science* **1983**, *220*, 1381–1383.
12. van der Merwe, N. J. *American Scientist* **1982**, *70*, 596-606.
13. Ambrose, S. H.; Norr, L. In *Prehistoric Human Bone: Archaeology at the Molecular Level*; Lambert, J. B.; Grupe, G., Eds.; Springer: Berlin, New York, 1993; pp 1–37.
14. Ambrose, S. H. In *Investigations of Ancient Human Tissue: Chemical Analyses in Anthropology*; Sandford, M. K., Ed.; Gordon and Breach Scientific: Langhorne, 1993; pp 59–130.
15. Iacumin, P.; Bocherens, H.; Mariotti, A.; Longinelli, A. *Earth Planet. Sci. Lett.* **1996**, 1–6.
16. Fricke, H. C.; O'Neil, J. R. *Earth Planet. Sci. Lett.* **1999**, 181–196.
17. Stephan, E. *J. Archaeol. Sci.* **2000**, *27*, 523–535.
18. Longinelli, A.; Selmo, E. *J. Hydrology* **2003**, *270*, 75–88.
19. Tieszen, L. L.; Fagre, T. In *Prehistoric Human Bone-Archaeology at the Molecular Level*; Lambert, J. B., Grupe, G., Eds.; Springer: Berlin, 1993; pp 121–155.
20. Schwarcz, H. P. In *Biogeochemical Approaches to Paleodietary Analysis*; Ambrose, S. H., Katzenberg, M. A., Eds.; Kluwer Academic/Plenum: New York, 2000; pp 189–209.

21. Jim, S.; Ambrose, S. H.; Evershed, R. P. *Geochim. Cosmochim. Acta* **2004**, *1*, 61–72.
22. Hedges, R. E. M. *Int. J. Osteoarchaeol.* **2003**, *13*, 66–79.
23. Hedges, R. E. M.; van Klinken, G. J. In *Biogeochemical Approaches to Paleodietary Analysis*; Ambrose, S. H., Katzenberg, M. A., Eds.; Kluwer Academic/Plenum: New York, 2000; Vol. 5, pp 211–241.
24. Bogucki, P. *American Scientist* **1996**, *84*, 242–253.
25. *Europe's First Farmers*; Price, T. D., Ed.; Cambridge University Press: Cambridge, 2000.
26. Sherratt, A. G. In *Pattern of the Past: Studies in Honour of David Clark*; Hodder, I.; Isaac, G.; Hammond, N., Eds.; Cambridge University Press: Cambridge, 1981, p 261–305.
27. Sherratt, A. G. In *The Oxford Illustrated Prehistory of Europe*; Cunliffe, B., Ed.; Oxford University Press: Oxford, 1994; pp 167–201.
28. Richards, M. P.; Pearson, J. A.; Molleson, T. I.; Russell, N.; Martin, L. *J. Archaeol. Sci.* **2003**, *30*, 67–76.
29. Papathanasiou, A. *Int. J. Osteoarchaeol.* **2003**, *13*, 314–324.
30. Papathanasiou, A. *Am. J. Phys. Anthro.* **2005**, *126*, 377–390.
31. Richards, M. P.; Hedges, R. E. M.; Walton, I.; Stoddart, S.; Malone, C. *Eur. J. Archaeol.* **2001**, *2*, 253–262.
32. Van Strydonck, M.; Boudini, M.; Ervynck, A. In *World Islands in Prehistory. International Insular Investigations, V Deiá Conference of Prehistory*; Waldren, W. H., Ensenyat, J. A., Eds.; British Archaeological Reports: Oxford, 2002; Vol. 1095, pp 189–197.
33. Davis, M. H. L. A. In *World Islands in Prehistory. International Insular Investigations, V Deiá Conference of Prehistory*; Waldren, W. H.; Ensenyat, J. A., Eds.; British Archaeological Reports: Oxford, 2002; Vol. 1095, p 198–216.
34. Tykot, R. H.; Robb, J., unpublished data.
35. Pinnegar, J. K.; Polunin, N. V. C. *Oecologia* **2000**, *122*, 399–409.
36. Pinnegar, J. K.; Polunin, N. V. C.; Badalamenti, F. *Can. J. Fish. Aquatic Sci.* **2003**, *2*, 222–235
37. Polunin, N. V. C.; Morales-Nin, B.; Pawsey, W. E.; Cartes, J. E.; Pinnegar, J. K.; Moranta, J. *Marine Ecology Progress Series* **2001**, *220*, 13–23.
38. Lee-Thorp, J. A.; Sealy, J. C.; van der Merwe, N. J. *J. Archaeol. Sci.* **1989**, *16*, 585–599.
39. Waters, A. L.; Katzenberg, M. A. In *Program of the Seventy-Third Annual Meeting of the American Association of Physical Anthropologists*; Electronic document accessed June 2006 at http://www.physanth.org/annmeet/aapa2004/ajpa2004.pdf , p 204.
40. Balasse, M.; Bocherens, H.; Mariotti, A. *J. Archaeol. Sci.* **1999**, *26*, 593–598.

41. Maxia, C.; Atzeni, E. In *Atti della VIII e IX Riunione Scientifica Istituto Italiano Preistoria e Protostoria, Trieste 19–20 Ottobre 1963, Calabria 6–8 Aprile 1964*, p 123–135.
42. Floris, R. In *La collezione Pistis-Corsi e il patrimonio archeologico del Comune di Iglesias. Mostra archeologica, grafica e fotografica*; Atzeni, E., Alba, L., Canino, G., Eds.; Cooperativa Tipografica Editoriale: Iglesias, 2001; pp 30–31.
43. Germanà, F. *L'uomo in Sardegna dal Paleolitico all'Eta' Nuragica*; Carlo Delfino: Sassari, 1995.
44. Ugas, G. In *Ricerche archeologiche nel territorio di Sanluri (mostra grafica e fotografica, Sanluri 16–26 Giugno 1982)*; Gruppo Archeologico Giovanile Ex Lege 285/77, Comune di Sanluri, Soprintendenza Archeologica per le Provincie di Cagliari e Oristano: 1982; pp 19–26, tables XI-XX.
45. Ugas, G. In *Simbolo ed enigma. Il bicchiere campaniforme e l'Italia nella preistoria europea del III millennio a.C. Catalogo*; Mottes, F. N. E., Ed.; Provincia Autonoma di Trento: 1998; pp 261–332.
46. Germanà, F. *Quaderni della Soprintendenza Archeologica per le Province di Cagliari e Oristano* **1987**, *1*, 49–57.
47. Manunza, M. R. *Quaderni della Soprintendenza Archeologica per le Province di Cagliari e Oristano* **1998**, 59–105.
48. Manunza, M. R., Ed. *Cuccuru Cresia Arta. Indagini archeologiche a Soleminis*; Grafica del Parteolla: Dolianova (Italy), 2005.
49. Buffa, R.; Calo', C. M.; Floris, G.; Marini, E.; Usai, E. *Quaderni della Soprintendenza Archeologica per le Province di Cagliari e Oristano* **2000**, 6–15.
50. Germanà, F. *Archivio per l'antropologia e l'etnologia* **1980**, 343–391.
51. Germanà, F. *Archivio per l'antropologia e l'etnologia* **1982**, 233–280.
52. Ambrose, S. H. *J. Archaeol. Sci.* **1990**, *17*, 431–451.
53. Ambrose, S. H.; Buikstra, J.; Krueger, H. W. *J. Anthro. Archaeol.* **2003**, *22*, 217–226.
54. Koch, P. L.; Tuross, N.; Fogel, M. L. *J. Archaeol. Sci.* **1997**, *24*, 417–429.
55. Lee-Thorp, J. A.; van der Merwe, N. J. *J. Archaeol. Sci.* **1991**, *18*, 343–354.
56. Nielsen-Marsh, C. M.; Hedges, R. E. M. *J. Archaeol. Sci.* **2000**, *27*, 1139–1150.
57. Nielsen-Marsh, C. M.; Hedges, R. E. M. *J. Archaeol. Sci.* **2000**, *27*, 1151–1159.
58. Lee-Thorp, J. A.; Sponheimer, M. *J. Anthro. Archaeol.* **2003**, *22*, 208–216.
59. Tykot, R. H. In *Archaeological Chemistry. Materials, Methods, and Meaning*; Jakes, K., Ed.; ACS Symposium Series No. 831; American Chemical Society: Washington, DC, 2002; pp 214–230.

60. van Klinken, G. J.; van der Plicht, H.; Hedges, R. E. M. *Geophysical Res. Lett.* **1994**, *6*, 445–448.
61. van Klinken, G. J.; Richards, M. P.; Hedges, R. E. M. In *Biogeochemical Approaches to Palaeodietary Analysis*; Ambrose, S. H., Katzenberg, M. A., Eds.; Kluwer Academic/Plenum: New York, 2000; pp 39–63.
62. Stevens, R. E.; Hedges, R. E. M. *Quaternary Science Reviews* **2004**, 23, 977–991.
63. Di Matteo, G.; De Angelis, P.; Scarascia Mugnozza, G. *Forest* **2005**, *4*, 367–377.
64. Araus, J. L.; A. Febrero; M. Catala; M. Molist; Voltas, J.; Romagosa, I. *Global Change Biology* **1999**, *2*, 201–212.
65. Araus, J. L.; Febrero, A.; Buxo, R.; Camalich, M. D.; Martin, D.; Molina, F.; Rodriguez-Ariza, M. O.; Romagosa, I. *Global Change Biology* **1997**, *2*, 107–118.
66. Brugnoli, E.; Scartazza, A.; Lauteri, M.; Monteverdi, M. C.; Máguas, C. In *Stable Isotopes. Integration of Biological, Ecological and Geochemical Processes*; Griffiths, H., Ed.; Bios Scientific: Oxford, 1998; pp 133–146.
67. Ambrose, S. H. *J. Archaeol. Sci.* **1991**, *18*, 293–317.
68. Schwarcz, H. P.; Dupras, T. L.; Fairgrieve, S. I. *J. Archaeol. Sci.* **1999**, *26*, 629–636.
69. Koch, P. L. *Ann. Rev. Earth Planet. Sci.* **1998**, *26*, 573–613.
70. Kohn, M. J.; Cerling, T. E. In *Phosphates. Geochemical, Geobiological, and Materials Importance. Reviews in Mineralogy and Geochemistry*; Kohn, M. J., Rakovan, J., Hughes, J. M., Eds.; Mineralogical Society of America: Washington, DC, 2002; Vol. 48, pp 455–488.
71. Farquhar, G. D.; Barbour, M. M.; Henry, B. K. In *Stable Isotopes. Integration of Biological, Ecological and Geochemical Processes*; Griffiths, H., Ed.; Bios Scientific: Oxford, 1998; pp 27–62.
72. Barbour, M. M.; Farquhar, G. D. *Plant, Cell and Environment* **2000**, *23*, 473–485.
73. Sharp, Z. D.; Cerling, T. E. *Geology* **1998**, *3*, 219–222.
74. Jones, M. D.; Leng, M. J.; Eastwood, W. J.; Keen, D. H.; Turney, C. S. M. *The Holocene* **2002**, *5*, 629–634.
75. Bar-Matthews, M.; Ayalon, A.; Gilmour, M.; Matthews, A.; Hawkesworth, C. J. *Geochim. Cosmochim. Acta* **2003**, *17*, 3181–3199.

Chapter 7

Bitumen in Neolithic Iran: Biomolecular and Isotopic Evidence

Michael W. Gregg[1], Rhea Brettell[2], and Benjamin Stern[2]

[1]**Department of Anthropology, University of Toronto, Toronto, Ontario M5S 3G3, Canada**
[2]**Department of Archaeological Sciences, University of Bradford, Bradford, West Yorkshire BD7 1DP, United Kingdom**

This paper presents the results of the chemical analysis of materials recovered from two of the earliest agricultural villages in southwestern Iran and a late Neolithic pastoral encampment in nearby Khuzistan. Gas chromatography - mass spectrometry (GC-MS) revealed biomarker compounds characteristic of bitumen in residues from ceramic vessels supporting the excavators' contention that the interior surfaces of some vessels were coated with a thin layer of such material and confirmed that 'fragments' collected during excavation were indeed bitumen. Biomolecular and isotopic (δD and $\delta^{13}C$) analysis of the bitumen indicated that the sources utilized lie in the Susa and Deh Luran regions of southwestern Iran.

Bitumen use is well attested in the later Chalcolithic and Bronze Age periods of southern Mesopotamia (*1, 2*), with its earliest reported use in pottery vessels coming from the site of Tell Sabi Abyad in northern Syria, dated to 6100 B.C. (*3*). However, the presence of bitumen has been recorded in both the aceramic and ceramic levels of the early Neolithic sites of Ali Kosh and Chagha Sefid (Figure 1) excavated by Frank Hole, Kent Flannery and James Neely during the early 1960s, the earliest ceramic horizons at both sites dating between 6800 – 7200 B.C. (*4, 5*). Similarly, evidence of 'bituminous earth' was also recovered from the late Neolithic pastoral encampment at Tepe Tula'i (Figure 1) excavated by Frank Hole in 1973, and provisionally dated between 6200 – 5900 B.C. (*6, 7*). Bitumen appears to have been utilized as waterproofing for basketry, reed matting, and dwellings, and as a hafting agent for stone axes and flint tools (*8, 9*). However, Hole et al. also identified residues, presumed to be bitumen, adhering to ceramic fragments. This coating may have functioned as a ceramic sealant and/or repair agent. Alternatively, the primary function of the pottery vessels may have been for the collection and heating of bitumen for use as a waterproofing and adhesive material (*10, 11*).

Materials and Methods

Materials

As part of a wider study into the initial development and use of ceramics in the Middle East, organic residues in pottery from 19 of the earliest villages and pastoral encampments in the Zagros mountains and in the Levant are in the process of being analyzed by gas chromatography - mass spectrometry (GC-MS). Most of these materials were acquired from extant collections with 200 pottery fragments having been examined to date. These included samples from Ali Kosh (Figure 2) and Chagha Sefid, two of the earliest agricultural villages in southwestern Iran. The molecular signatures characteristic of bitumen were identified in 4 sherds from these sites (Table I). Since Hole had reported recovering 'bituminous earth' from Ali Kosh, Chagha Sefid, and Tepe Tula'i (6, *10, 11*), earthen fragments that appeared to be encrusted with bitumen (Figure 2) were subsequently obtained from the Iranian National Museum in Tehran (Table I). One modern bitumen sample for source identification (Table I, CM) was also collected from a seep called Chersh Merghir, (*spring of tar* in Farsi; N 32° 41' 26"; E 47° 19' 56"), where the foothills of the Zagros mountains meet the Mesopotamian plain.

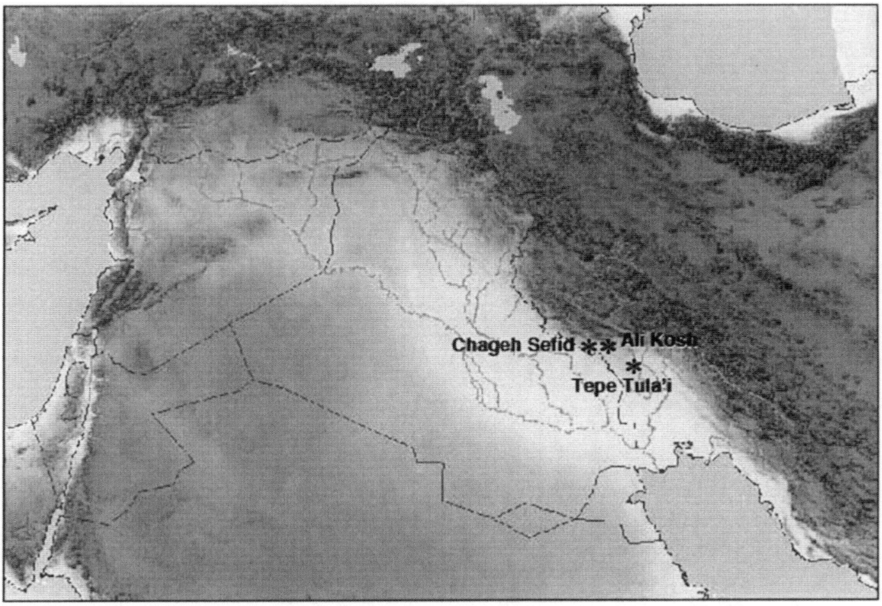

Figure 1. Location of sites in southwestern Iran mentioned in the text; the early Neolithic villages of Ali Kosh and Chagha Sefid and the late Neolithic pastoral encampment of Tepe Tula'i.

Figure 2. left, reed-impressed, bitumen-encrusted, mudbrick fragment (AK) from Ali Kosh; right, pottery fragment (AK1) recovered from the Mohammad Jaffar horizon at Ali Kosh, dated to 7100 B.C.

Table I. Details for the Archaeological and Modern Samples.

	Find	*Description*	*State*	*Color*
Archaeological pottery samples	AK1 INT RES	Pottery fragment with visible residue from Ali Kosh	Sherd	Dark brown
	AK2 INT	Pottery fragment from Ali Kosh	Sherd	Orange brown
	CS1 INT RES	Pottery fragment with visible residue from Chageh Sefid	Sherd	Dark brown
	CS2 INT	Pottery fragment from Chageh Sefid	Sherd	Orange brown
Archaeological bitumen samples	AK	Bitumen recovered from Ali Kosh excavations	Earthy chunks	Brown Cream layers
	TTI	Bitumen recovered from Tepe Tula'i excavations	Earthy chunks	Brown Pink/cream layers
	TTZ	Bitumen recovered from Tepe Tula'i excavations	Earthy chunks	Yellow-brown Cream inclusions
	TTD1.02	Bitumen recovered from Tepe Tula'i excavations	Earthy chunks	Grey/brown
Modern bitumen	CM	Bitumen recovered from modern seep at Chersh Meghir	Viscous liquid	Black

Methods

The pottery sherds from Ali Kosh and Chagha Sefid were subsampled by grinding portions (0.6 g) into a fine powder using a high-speed modeling drill fitted with an abrasive tungsten bit. Subsamples of the archaeological 'bitumen' fragments were homogenized by crushing with a mortar and pestle, and a portion of the tar collected from Chersh Meghir was selected. The lipid fractions were then obtained by repeated (three times) solvent extraction using dichloromethane:methanol (DCM/MeOH, 2:1, v/v) with ultrasonication to aid dissolution. Separation of the solid and soluble fractions was attained by centrifuging at 2000 rpm for five minutes, the solvent soluble fraction being decanted and combined. The solvent was then evaporated on a warm hot-plate under a stream of nitrogen gas to produce the 'total extract'. A subsample of this 'total extract' was then diluted in DCM for analysis by GC-MS. This was conducted using an Hewlett Packard 5890 series II GC, fitted with a 15 m x 0.25 mm id, 0.1 mm film thickness OV1 phase fused silica column (MEGA) connected to a 5972 series mass selective detector. The splitless injector and interface were maintained at 300°C and 340°C respectively. The helium carrier gas was held at a constant inlet pressure of 1 psi. The GC oven was temperature programmed at 50°C for 2 minutes then increased by 10°C per minute to a maximum of 340°C, at which the temperature was held for 10 minutes. The column was directly inserted into the ion source where electron impact (70 eV) spectra were obtained. The resulting chromatograms were examined for the molecular and characteristic fragment ions of various lipid classes including biomarkers of petroleum such as terpanes (*m/z* 191) and steranes (*m/z* 217) (3, *12, 13*).

In order to assist in further characterizing the bitumen samples, bulk isotopic analysis ($\delta^{13}C$ and δD) of the asphaltene fraction was undertaken as this fraction provides the most representative isotopic composition of the sample rather than the 'total extract' as used for GC-MS analysis (*3*). The asphaltene was obtained by repeated (three times) washing of a portion of the 'total extract' in 5 ml of *n*-hexane with ultrasonication to aid dissolution. Separation of the solid asphaltene fraction was assisted by centrifuging, the solvent being decanted. Any remaining solvent was then removed on a warm hot-plate under a stream of nitrogen gas. Bulk carbon isotopic values were obtained using continuous flow isotope ratio mass spectrometry (IR-MS), the samples being flash combusted in a column containing chromium oxide (Cr_2O_3) and silvered cobalt (I) oxide held at a temperature of 1020°C. The resultant gases were then reduced to CO_2 in a column of elemental copper at 680°C and passed through a water trap of magnesium perchlorate before being separated in a GC column for introduction to the MS (Finnigan delta plus XL). The reference CO_2 gas was standardized against the international standard IAEA600 (δ -27.5 ± 0.2) and three methionine standards (δ -26.6) were run as quality control check samples during batch analysis. Hydrogen values were obtained by elemental analyzer-isotope ratio mass spectrometry (EA-IR-MS), the samples being released into a furnace set at 1080°C and thermally decomposed to H_2 and CO over glassy carbon. Any traces of water were then removed by a magnesium perchlorate trap and any CO_2 formed by a Carbosorb trap before the H_2 was resolved by a packed column gas chromatograph held at 35°C. The resultant chromatographic peak was then passed into the ion source of the IR-MS, ionized and accelerated and gas species of different mass separated in a magnetic field. These were simultaneously measured on a Faraday cup universal collector array with masses 2 and 3 being monitored for deuterium.

Results and Discussion

Molecular Characterization

Comparison of the chromatograms obtained from the modern bitumen seep, the archaeological bitumen and the pottery samples demonstrated the essential similarity of the data obtained with the terpane distribution patterns showing parallels in the range and relative abundances of the diagnostic hopanes (examples shown in Figures 3, 4 and Table II). The dominant families present were 17α(H),21β(H)-hopanes, Ts (18α(H)-22,29,30–trisnorneohopane) and Tm (17α(H)-22,29,30–trisnorhopane) and tricyclopolyprenanes, with subordinate molecular classes being the methyl-αβ-hopanes, 17β(H),21α(H)-hopanes and hexahydrobenzohopanes. Thus, the residues in the ceramics and the 'bituminous earth' collected during excavation of these Neolithic sites can clearly be identified as bitumen.

Molecular identification of bitumen sources relies upon assessment of the presence/absence of biomarker hopanes such as oleanane, "a unique genuine chronostratigraphic biomarker" relating oil type to specific geological periods (*13*), and gammacerane, and ratios of their relative abundance (*3, 12*). The archaeological bitumen samples and the modern sample from Chersh Meghir all have minor peaks in the expected location of oleanane; however, this compound was only tentatively identified as the mass spectra do not show the required molecular ion and fragments and there is no clear indication of oleanane in any of the archaeological ceramic samples. These findings are somewhat unexpected as, although oleanane has never been observed in bitumen from Iraq and Syria, it generally occurs in pronounced abundances in sources from Iran, in particular, from Khuzestan and Fars provinces (*13*). This lack of clearly identifiable

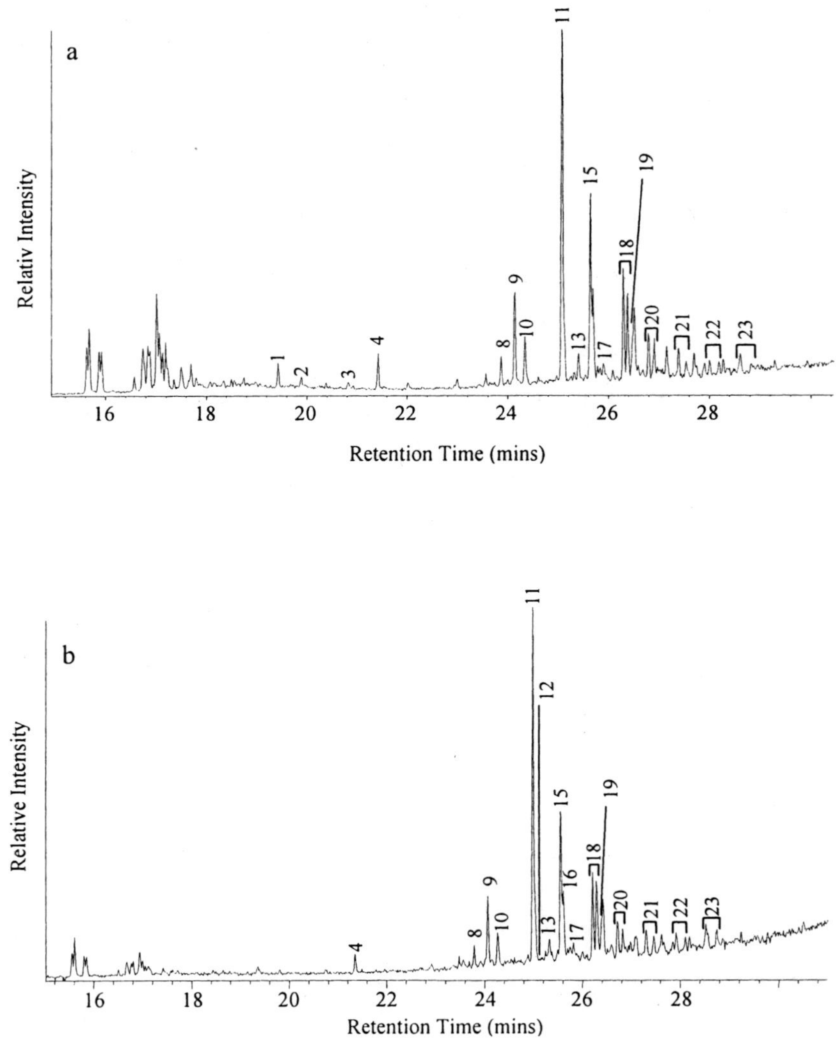

Figure 3. Selected ion chromatogram (m/z 191) for two archaeological ceramic samples: (a) AK2 INTRES and (b) CS1 INTRES. Key to peak identification is shown in Table II.

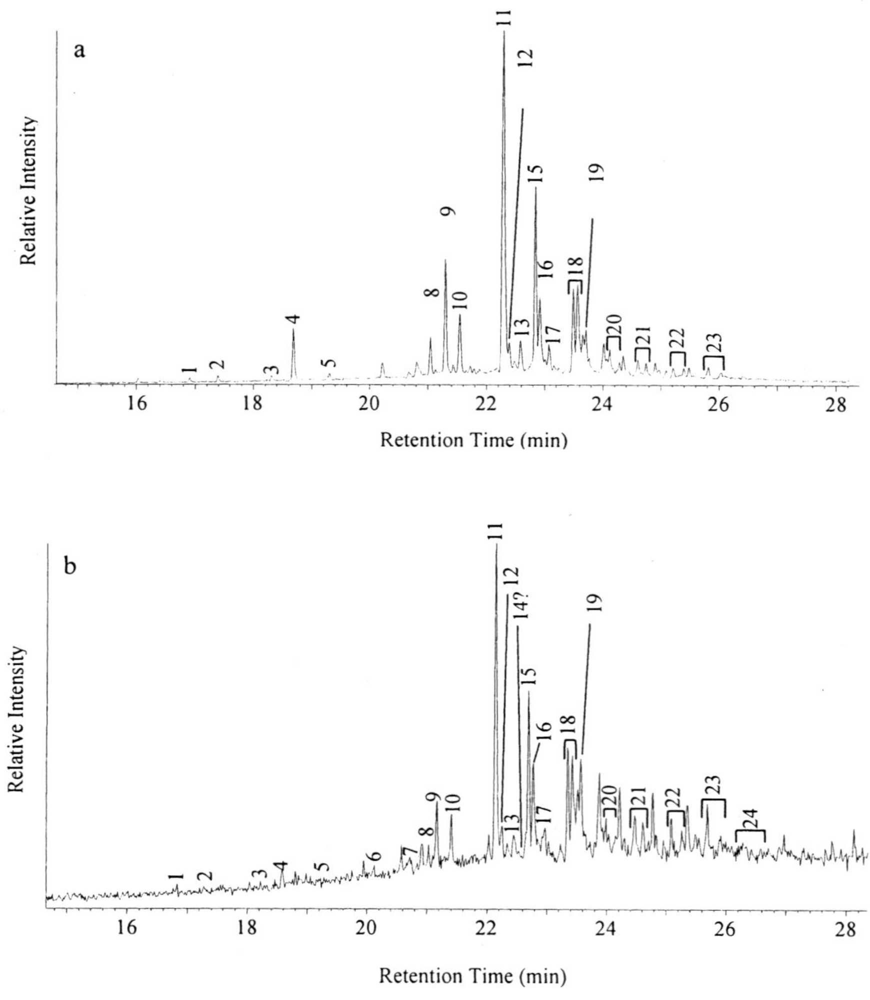

Figure 4. Selected ion chromatograms (m/z 191) of "bituminous earth" samples: (a) AK and (b) TTI. Key to peak identification is shown in Table II.

Table II. Key to the Chromatograms - Terpane Identification.

Peak number	*Brief Identification*	*Name/structure*
1	23/3	C_{23} tricyclopolyprenane
2	24/3	C_{24} tricyclopolyprenane
3	25/3	C_{25} tricyclopolyprenane
4	24/4	C_{24} tetracyclic terpane
5	26/3	C_{26} tricyclopolyprenane
6	28/3	C_{28} tricyclopolyprenane
7	29/3	C_{29} tricyclopolyprenane
8	Ts	18α-22,29,30–trisnorneohopane
9	Tm	17α-22,29,30–trisnorhopane
10	28αβH	17α,21β-29,30–dinorhopane
11	29αβH	17α,21β-30–norhopane
12	30MeαβH	2α-methyl-αβ-norhopane
13	29βαH	17β,21α-30–norhopane
14	?	Possibly oleanane ($C_{30}H_{52}$)
15	30αβH	17α,21β-hopane
16	31MeαβH	2α-methyl-αβ-hopane
17	30βαH	17β,21α-hopane (moretane)
18	31αβHS+R	17α,21β-homohopane (22S) + (22R)
19	Gammacerane	$C_{30}H_{52}$
20	32αβHS+R	17α,21β-homodishopane (22S) + (22R)
21	33αβHS+R	17α,21β-trishomohopane (22S) + (22R)
22	34αβHS+R	17α,21β-tetrakishomohopane (22S) + (22R)
23	35αβHS+R	17α,21β-pentakishomohopane (22S) + (22R)
24	36αβHS+R	17α,21β-sextakishomohopane (22S) + (22R)

oleanane may be due to the degraded nature of the samples or to experimental factors such as peak overlap. More work is required on this matter. However, gammacerane is present in all the residues obtained from the archaeological ceramics as well as the archaeological and modern bitumen samples (Figures 3 and 4). The low abundance of this biomarker in all the samples clearly relates them to patterns observed in Iranian sources and differentiates them from bitumen from any of the Dead Sea sources or from Jebel Bichri in Syria, as does the general pattern of the hopane abundance (*14–16*).

The biomarker fingerprints in all the archaeological samples display significant degradation with the steranes being of particularly low abundance. It was, therefore, not possible to use sterane to terpane ratios for source determination (e.g., 18α-oleanane/17α,21β-hopane vs. Ts/Tm and 27Sdia/29ααR vs. Ts/Tm) (*17*). However, the relationship between the terpanes, also of assistance in distinguishing between sample sources, was ascertained. The ratio of Tm to Ts provides an indicator of the maturity of the sample, the ratio increasing with maturity and demonstrates the relative maturity of the sample source, although it does not provide a secure basis for separation between these samples (Table 3). However, the relationship between gammacerane/17α(H)-21β(H)-hopane and Ts/Tm has been used in this respect and assessment of this parameter (Figure 5) shows that those with gammacerane/17α(H)-21β(H)-hopane ratio less than 0.4 fall within the range of previously analysed materials from Iranian oil seeps, most likely from the Susa region (*3,17*). Other samples (labeled CS2INT, AK1INTRES, TT1; Figure 5) are beyond this range; this may be due to the degraded nature of these samples.

Isotopic Characterization

In order to provide clearer confirmation as to the source of the samples, analysis of $\delta^{13}C$ (‰ vs. PDB) and δD (‰ vs. SMOW) values were undertaken on the asphaltene fraction of each sample (Figure 6). Comparison of results from the ancient pottery fragments and bitumen samples discussed above clearly distinguishes the values obtained from the published isotopic data for bituminous materials from the Dead Sea, Iraq, Lebanon, and Syria and places them within the range of Iranian seeps from the Khuzestan and Luristan (*17*). They also demonstrate a similar range of values to those obtained from bituminous residues identified from the Achaemenid dynasty site at Susa (*16*). The $\delta^{13}C$ value of the modern bitumen sample from Chersh Merghir (labeled CM, Figure 7) is similar to the other samples. However, its δD is considerably lighter. This could be due to the specific degradation conditions of the sample, δD values being heavily influenced by weathering and therefore only partially related to genetic parameters (*12*).

Table III. Numerical Values of the Diagnostic Ratios and Isotopic Data from all Samples.

	Find	GCRN/30 αH	Ts/Tm	$\delta^{13}C$	δD
Archaeological pottery samples	AK1 INT RES	0.57	0.36	-27.2	-70
	AK2 INT	0.29	0.36	-26.9	-70
	CS1 INT RES	0.15	0.27	-25.9	-68
	CS2 INT	0.66	0.32	-26.8	-83
Archaeological bitumen samples	AK	0.22	0.30	-27.2	-67
	TTI	0.50	0.49	-27.2	-65
	TTZ	0.26	0.16	-26.6	-60
	TTD1.02	0.34	0.37	-27.1	-59
Modern bitumen	CM	0.17	0.31	-27.3	-95

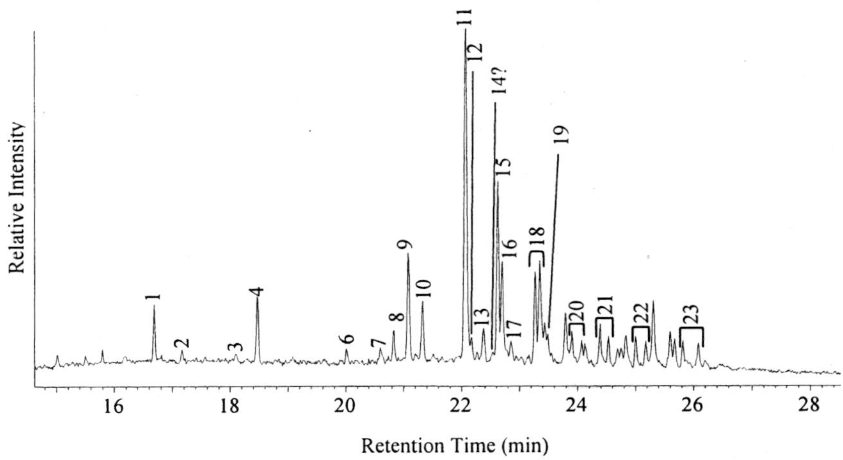

Figure 5: Partial ion chromatogram of the terpanes (m/z 191) from the modern, Chersh Meghir, bitumen seep sample. Key to peak identification is shown in Table II.

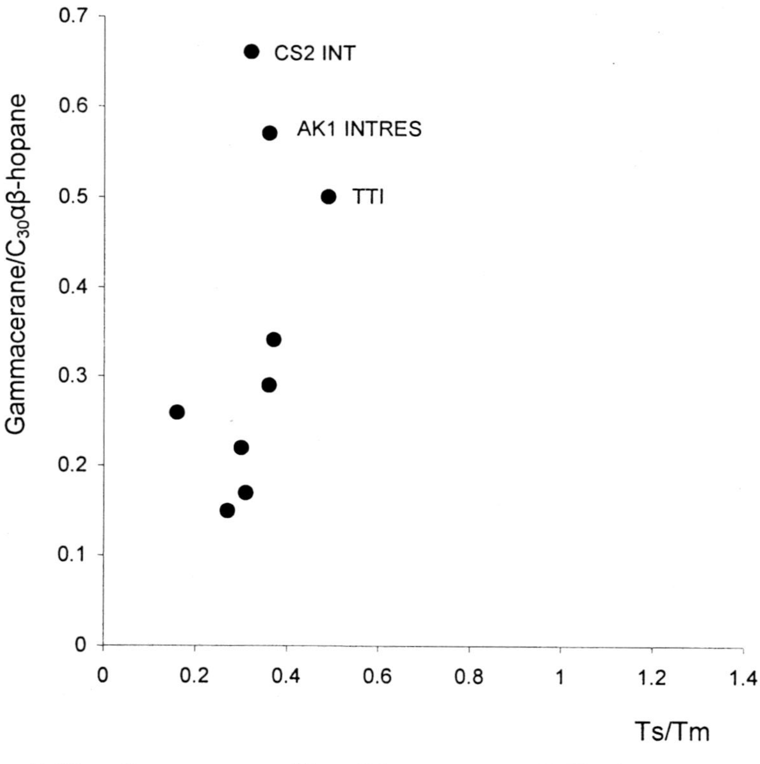

Figure 6. Plot of gammacerane/C_{30} $\alpha\beta$-hopane versus Ts/Tm for comparison of the samples.

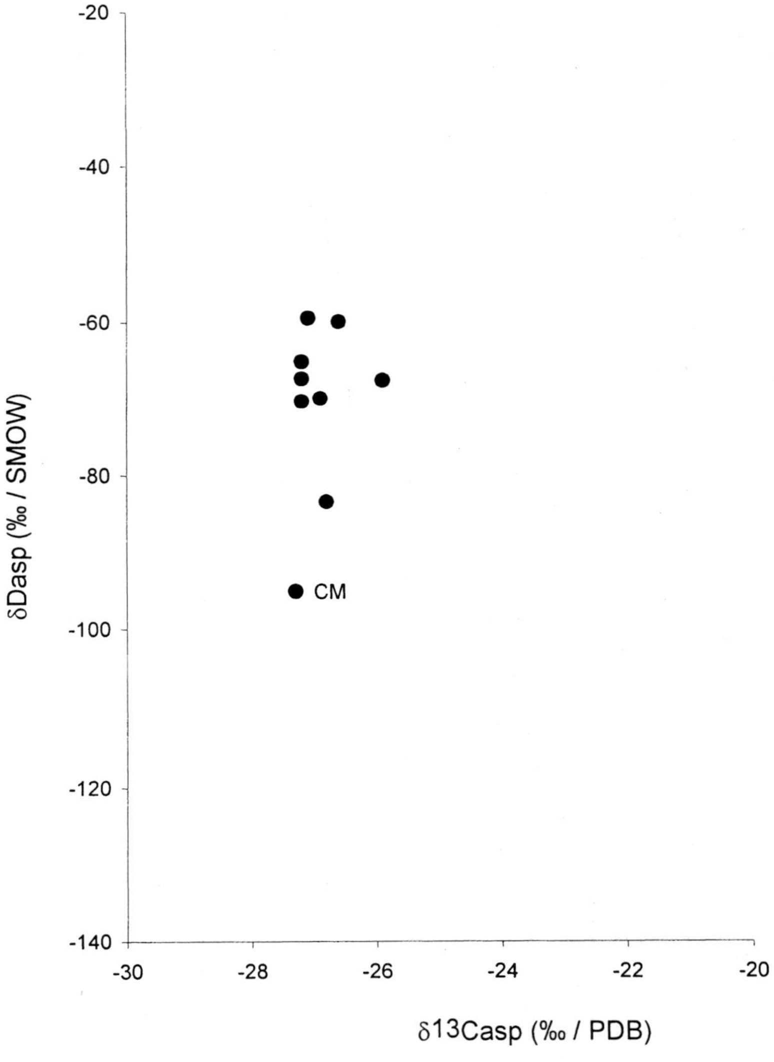

Figure 7: Plot of mean $\delta^{13}C$ (‰ / PDB) vs. δD (‰ / SMOW) for the asphaltene fraction of bitumens from different deposits and the archaeological samples.

Conclusions

The results of this study provide confirmation of the earliest recorded presence of bitumen in pottery vessels (6800 – 7200 B.C.) recovered from primary ceramic horizons in the Middle East. The significance of these residues has yet to be ascertained, with bitumen potentially serving as a ceramic sealant or repair agent or forming the vessel contents (*2*). Evidence from the site excavation records certainly demonstrates the utilization of bitumen as both a waterproofing material and adhesive during this period (*8, 9*). The specific location of the seep/s used as the source of this bitumen has yet to be identified as our results do not adequately match those obtained from the nearest known modern seep at Chersh Merghir, nor unpublished data from three other seeps within the immediate vicinity of the sites (*18*). A comprehesive survey of the region will be required in order to fully address this issue. However, as this paper demonstrates, Neolithic bitumen from these important Iranian sites shares many molecular and isotopic characteristics with 'local' modern sources in Khuzestan and Luristan.

Acknowledgements

We thank the Social Sciences and Humanities Research Council of Canada; the Iranian Centre for Archaeological Research, Tehran; the Iranian National Museum, Tehran; the University of Bradford, United Kingdom; and the University of Toronto, Canada.

References

1. Marschner, R. F.; Wright, H. T. *Papers of the American Chemical Society*, **1977**, *174*, 25–25.
2. Connan, J. *Philosophical Transactions of the Royal Society of London Series B – Biological Sciences* **1999**, *354*, 33–50.
3. Connan, J.; Nieuwenhuyset, O. P.; Van As, A.; Jacobs, L. *Archaeometry* **2004**, *46*, 115–124.
4. Hole, F. *Proceedings of the 1st Annual Symposium of Archaeological Research in Iran*; Ministry of Culture and Arts, Tehran, 1972; p 65.
5. Hole, F.; Flannery, K. V.; Neely, J. A. *Prehistory and Human Ecology of the Deh Luran Plain: An Early Village Sequence from Khuzistan, Iran*; University of Michigan Press, 1969; p 331.
6. Hole, F. *Proceedings of the 1st Annual Symposium of Archaeological Research in Iran*; Ministry of Culture and Arts, Tehran, 1972; p 70.

7. Hole, F. *Paléorient,* **1974**, *2*, 219.
8. Hole, F.; Flannery, K. V.; Neely, J. A. *Prehistory and Human Ecology of the Deh Luran Plain: An early Village Sequence from Khuzistan, Iran*; University of Michigan Press, Ann Arbor, MI, 1969; p 220.
9. Hole, F. *Studies in the Archeological History of the Deh Luran Plain: The Excavation of Chagha Sefid*; University of Michigan, Ann Arbor, MI, 1977; p 223.
10. Hole, F.; Flannery, K. V.; Neely, J. A. *Prehistory and human ecology of the Deh Luran plain; an early village sequence from Khuzistan, Iran*; Museum of Anthropology, University of Michigan Press, Ann Arbor, MI, 1969; p 212.
11. Hole, F. *Studies in the Archeological History of the Deh Luran Plain: The Excavation of Chagha Sefid*; University of Michigan Press, Ann Arbor, MI, 1977; p 43.
12. Killops, S.; Killops, V. *Introduction to Organic Geochemistry* (2nd edn.); Blackwell Publishing: Malden, 2005; pp 196–234.
13. Connan, J. *Philosophical Transactions of the Royal Society of London Series B – Biological Sciences* **1999**, *354*, p 40.
14. Harrell, J. A.; Lewan, M. D., *Archaeometry* **2002**, *44*, 285–293.
15. Haven, H. L.; Rohmer, M.; Rullkötter, J.; Bisseret, P. *Geochimica et Cosmochimica Acta* **1989**, *53*, 3073–3079.
16. Connan, J.; Deschesne, O. *Techne.* **1998**, *7*, 13–16.
17. Stern, B.; Connan, J.; Blakelock, E.; Jackman, R.; Coningham, R. A. E.; Heron, C. *Archaeometry,* submitted.
18. Jacques Connan, personal communication, March 15, 2006.

Chapter 8

Surface Analysis of a Black Deposit from Little Lost River Cave, Idaho

Reshmi Perumplavil and Ruth Ann Armitage

Department of Chemistry, Eastern Michigan University, Ypsilanti, MI 48197

A black coating of unknown origin obscures the pictographs in Little Lost River Cave, Idaho. We have utilized x-ray photoelectron spectroscopy (XPS) to characterize the outermost surface of the black coating, as this material has previously been used to provide a minimum age for the underlying rock paintings. Carbon and oxygen predominated in the XP spectra, whereas nitrogen was detected at varying levels in different samples of the coating. High-resolution carbon 1s XP spectra showed that carbon was present in at least three different forms: hydrocarbon C, carbonyl C, and amide/carbonate C. The N 1s peak was observed at ~399–400 eV binding energy, which is usually attributed to aromatic or amide N. The XPS results are consistent with identification of the coating as a water-deposited layer of humic substances from the overlying soil. As this would be a geologic deposit, the radiocarbon age determined for the coating does not likely relate to the age of the rock paintings.

Little Lost River Cave, in south-central Idaho, is of archaeological interest because of its collection of red and yellow anthropomorphic and zoomorphic rock paintings. The paintings are obscured by a shiny black coating, which was observed in the first investigation at the site in 1954 (*1*), and noted again in excavations at the site in 1990 (*2*). In 2001, this material was sampled in the hope that by determining the radiocarbon age for the coating, a minimum age for the underlying paintings could be determined (*3*). Previous analyses of this material have indicated that it is rich in organic carbon (*4*). Steelman et al. proposed that the coating may be the result of cooking fires within the cave, and obtained for it a radiocarbon age of 2990±50 uncalibrated years B.P. (*5*). Pyrolysis-gas chromatography-mass spectrometry (py-GC-MS) studies (*6*) have shown that the cooking residue hypothesis proposed by Steelman et al. (*4*) is not likely because the material bears a strong resemblance to humic and fulvic acids. Depending on how the coating formed – through deposition of ground water or condensation from smoke and/or cooking – the surface composition of the coating should differ. Thus this work focuses on the formation mechanism and looks at the material from a surface standpoint using X-ray photoelectron spectroscopy (XPS).

Characterization of the surface of the coating is important in validating the radiocarbon date obtained by Steelman et al. (*5*). Plasma-chemical oxidation (PCO) was used in that case to oxidize the organic material at the surface of the coating to carbon dioxide, which was subsequently radiocarbon dated using accelerator mass spectrometry. The PCO sample preparation technique for radiocarbon dating rock paintings was pioneered by Russ et al. (*7*), and has since been proposed as a "nondestructive" technique for dating irreplaceable artifacts (*8, 9*). The low-temperature, low-pressure oxygen plasma discharge selectively oxidizes organic carbon in the presence of inorganic carbon as carbonates and oxalates (*10*). As the interaction with the plasma occurs primarily at the surface of the material, it is important to know that the organic material at that surface stems from an archaeologically relevant event. In this case, that means that the coating surface must not be of a primarily geologic origin if it is to provide a minimum age for the underlying pictographs.

The XPS mechanism, which can be used for quantitative and qualitative chemical analysis of surfaces, is based on the photoelectric effect. A monochromatic soft Mg or Al anode X-ray source is used to irradiate the surface. The absorbed X-rays ionize the core shell, and in response, the atom creates a photoelectron that is transported to the surface and escapes. The ionization potential of a photoelectron that must be overcome to escape into vacuum is the binding energy (BE) plus the work function of the material. The emitted photoelectrons have a remaining kinetic energy (KE), which is measured by using an electron analyzer. Individual elements can be identified on the basis of their BE. The resulting XP spectrum is a characteristic set of peaks for a specific element, with BE as the abscissa and counts per unit time as the ordinate. The BE of an element depends upon its oxidation state and environment. Hence, changes in the chemical state of an element give rise to shifts in peak positions. The quantification of each element depends on peak areas and sensitivity factors that correct for several instrumental parameters.

The surface specificity of XPS is due to the short range of photoelectrons that are excited from the sample; photoelectrons cannot escape without energy loss during transport to the surface. Thus, XPS is sensitive to only the top 3–5 nm of a surface. It can detect all the elements in the periodic table except for H and He, which have very low ionization cross-sections. It can be used in combination with argon ion sputtering to examine the changes in a sample with depth. Ultra high vacuum (UHV) conditions are required to carry out any XPS measurements, and, thus, samples must be vacuum compatible. The surface specificity and ability to distinguish different chemical environments makes XPS a powerful technique in the analysis of thin films and coatings.

XPS has not been widely used in the study of archaeological materials. Ciliberto and Spoto include a chapter on the past uses of XPS in archaeological sciences, describing applications to pottery, metallic objects, paintings, pigments, and the degradation of ancient paper (*11*). Lambert et al. (*12*) provide an extensive review of the application of XPS to archaeological materials. XPS, along with magnetic microanalytical methods, was used to differentiate organic and inorganic materials in black paints on pottery from the American Southwest (*13*). Using XPS and scanning Auger microscopy, Paparazzo (*14*) studied organic carbon-bearing species as the major source of carbon enrichment observed at the surface of a lead pipe (fistula) and a Roman bronze statue. The ability of XPS to provide chemical information from shifts in binding energy and its non-destructive nature makes it a complement to other analytical techniques used in the study of archaeological materials.

XPS will aid in understanding specifically the *surface* of the black deposit covering pictographs in Little Lost River Cave in Idaho. This work will complement other bulk analyses carried out with pyrolysis-GC-MS and thermally assisted hydrolysis /methylation (THM)-GC-MS (*15*). The objectives of this project were to use XPS to qualitatively determine the surface elemental composition of the black residue; semiquantitatively characterize the surface, for comparison with other surface-related materials; and examine the relationship between the chemistry and depth by using Ar^+ sputtering. This, then, will aid in validating the radiocarbon date obtained through plasma-chemical oxidation and accelerator mass spectrometry by Steelman et al. (*5*) .

Methods and Materials

Photoelectron spectra were recorded on a Physical Electronics, Inc. 5100 Series XP spectrometer. Samples were irradiated by use of a Mg Kα monochromatic x-ray source (photon energy = 1253.6 eV) operated at 15 kV and 300 W. The kinetic energies of the photoelectrons were analyzed by a hemispherical electron energy analyzer with a work function of 4.17 eV and constant pass energy. All experiments were performed under ultra high vacuum (UHV) conditions with an overall base pressure of 10^{-9} Torr. A 4–mm diameter analysis area was used. The survey and high-resolution scans were collected using pass energies of 143.05 eV and 44.75 eV respectively. For depth profiling studies, the surface was etched using an Ar^+ gun operated at 20 mA with a chamber pressure of 1×10^{-7} Torr.

The resulting data were then treated by use of standard XPS methods. An asymmetric background correction using the Shirley function (described in detail by Castle and Salvi (*16*)), was used for all spectra, and peaks were fit by using a mixed Gaussian-Lorenzian distribution. Quantitation was performed with the equation $(A_i/S_i) \div \Sigma A_i/S_i$, where A_i is the peak area and S_i is the atomic sensitivity factor for the element i being determined (*17*). Atomic sensitivity factors are empirical constants determined on standards of the elements (*18*). This yields an atom percentage (atom %) for each element at the surface of the material.

Several samples of the coating from Little Lost River Cave were studied using XPS. These materials were collected during three separate trips to the cave: once in 2001, and in 2004 (provided by C. Merrell, Archaeographics), and once in 2005 by one of us (RAA). Because the surface of a sample to be examined using XPS must be flat, only selected samples of the coating were appropriate. Table I describes the samples that were examined. Figure 1 shows the corresponding locations where these samples were collected in the cave.

Standard materials were also investigated for comparison. These included two humic acids, one from Alfa Aesar (CAS#1415–93–6) obtained from an unknown source, and one from the International Humic Substances Society (IHSS) obtained from standard Elliot soil. IHSS fulvic acid from Elliot soil was also used for comparison. Humic acids are ubiquitous geologic contaminants present in archaeological materials that have been exposed to soil or groundwater, and are important comparative materials for surfaces that are to be (or in this case, have been) radiocarbon dated by the PCO-AMS technique.

Results

XPS spectra obtained on the uncoated dolomite sample (Figure 2) showed the presence of carbon, oxygen, and calcium, as would be expected for a calcium carbonate material. Traces of chlorine, probably present as precipitated salts were also observed by electron microprobe analysis in previous studies (*4*). The high-resolution C 1s spectrum, however, showed that carbon was present in at least three different chemical states (Figure 2, inset). The peak occurring at a BE of 284.3 eV is consistent with the presence of adventitious hydrocarbon

Table I. Little Lost River Cave Coating Samples Studied Using XPS

Sample and map location	*Year collected*	*Description*
1	2001	Dolomite fragment with thin layer of shiny black coating
2	2004	Shiny black coating fragment with little dolomite substrate
3	2004	Sooty black coating on dolomite substrate
4	2005	Sooty black coating with thin dolomite substrate
5	2005	Shiny black coating without visible substrate
6	2005	Shiny black coating without visible substrate
7	2005	Shiny black coating on the dolomite substrate
D	2005	Uncoated dolomite from roof spall inside cave

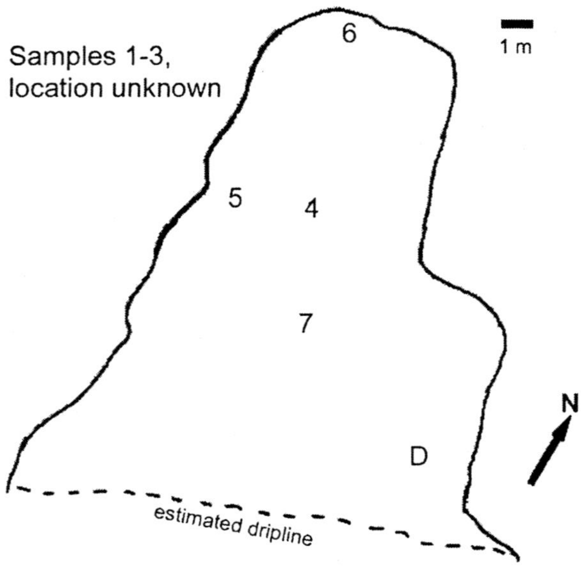

Figure 1. Map of Little Lost River Cave (10BT1), showing sampled locations. Numbers correspond to samples listed in Table I. Map is based on that of Gruhn and Bryan (2).

from the turbomolecular pump, or from hydrocarbon-like material on the surface of the dolomite. The 288.5 eV BE peak corresponds to carbonate carbon; both the carbonate and adventitious carbon peaks were also observed for a clean carbonate standard. These BE values are standard for carbonate and adventitious carbon reported in the literature as well (*17*). However, the third peak at 286.2 eV may correspond to carbonyl moieties at the surface, possibly due to contamination from aeolian dust, which is ubiquitous in the cave. This is consistent with literature values for soils (*19*).

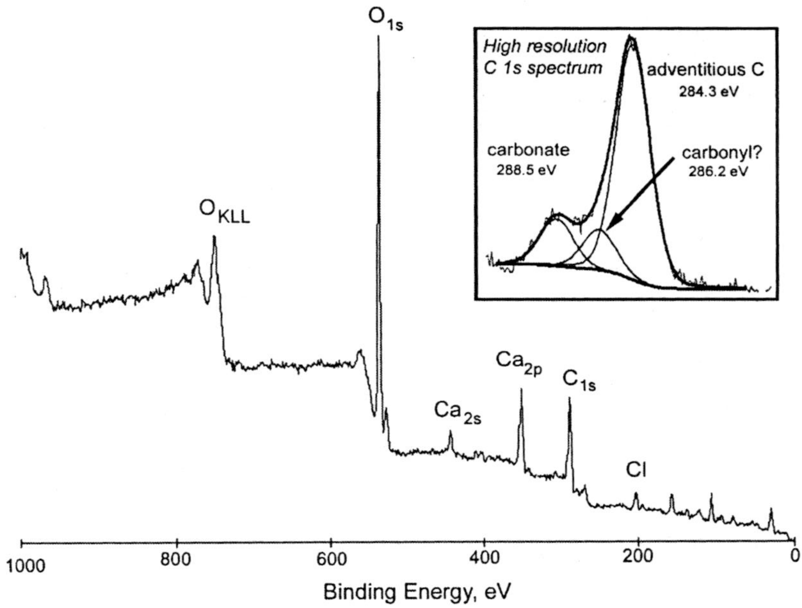

Figure 2. Survey spectrum of uncoated dolomite from 10BT1 (inset: high-resolution C_{1s} spectrum showing presence of at least three chemical states of C).

Survey Spectra

XP survey spectra of the coating samples also showed an elemental composition for the shiny coating which was consistent with that observed by Steelman et al. (*4*) by electron microprobe analysis. The spectra of the shiny coating samples are dominated by carbon and oxygen, with small amounts of nitrogen and calcium and traces of chlorine. Figure 3 shows the survey spectra of the most recently sampled shiny material (Samples 5–7) and a sooty coating

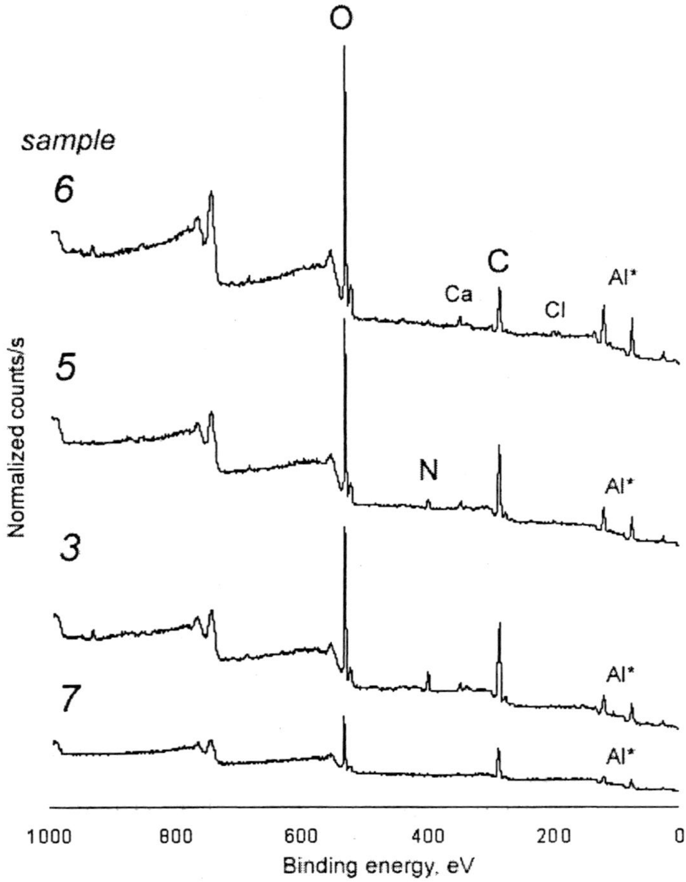

Figure 3. XPS survey spectra for four samples of the coating. All labeled peaks correspond to 1s electrons, except Ca, which is from a 2p electron.

(Sample 3) for comparison. Aluminum peaks were observed from the sample stubs. Nitrogen is found at varying levels in the coating, and does not appear to correspond to the appearance of the material. However, the dolomite survey spectrum also showed significant calcium signals that do not appear in all of the coating samples; when Ca is present, the concentration is less than 1 atom %. This shows that XPS is truly showing the composition of the coating's outermost surface and is not penetrating through to the dolomite substrate. This is significantly different from the earlier microprobe analysis (*4*). Hence the carbon signals in the XP spectra of the samples are specific to the coating, particularly when no Ca signal is observed.

High-Resolution Spectra

XP survey spectra are not particularly useful for identifying a surface-deposited material, as they are purely qualitative. High-resolution spectra are more informative and can be used semiquantitatively. As in the survey spectra, the x-axis corresponds to the binding energy in eV, and the y-axis is counts. Carbon is found in different oxidation states in the coating, as can be seen in Figure 4.

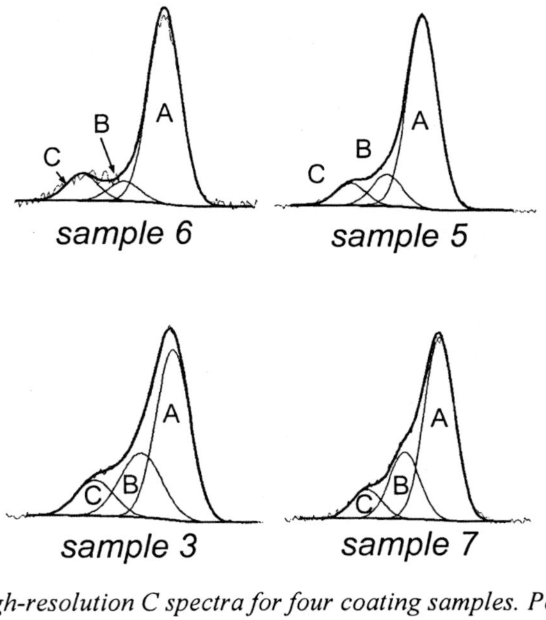

Figure 4. High-resolution C spectra for four coating samples. Peak A (284 eV) corresponds to C-C, C-H (adventitious); Peak B (287 eV) may correspond to C=O; and Peak C (289 eV) may be carbonate or amide C.

For the high-resolution carbon spectra, peak A corresponds to adventitious carbon, typically resulting from the ubiquitous hydrocarbon present in the high-vacuum chamber from pump oil. This peak is often used as a standard by which charging effects can be determined, as it is expected at 284.3 eV (*20*). The coating samples exhibited little charging, exhibiting a shift of only +1–2 eV. For example, peak A was observed at 286 ± 1 eV, compared to the expected 284.3 eV. Peak B, at a charging-corrected value of 287 eV, may correspond to carbonyl or some other functionality wherein C is bonded to O. Peak C, at a corrected BE of 289 eV, is either carbon in an amide functional group, or carbonate.

Identification of Peak C is dependent on the presence of either Ca or N. Sample 3 shows a significant N peak at a BE of ~400 eV. When N is present in such large quantities, this could indicate that the C_{289} signal in Figure 4 corresponds mostly to amide carbon rather than to carbonate (which may also be present, but to a lesser extent). For comparison, the C_{289} signal observed in the dolomite standard (peak C) in Figure 5 must correspond to carbonate because there was no significant nitrogen present on the surface of that sample.

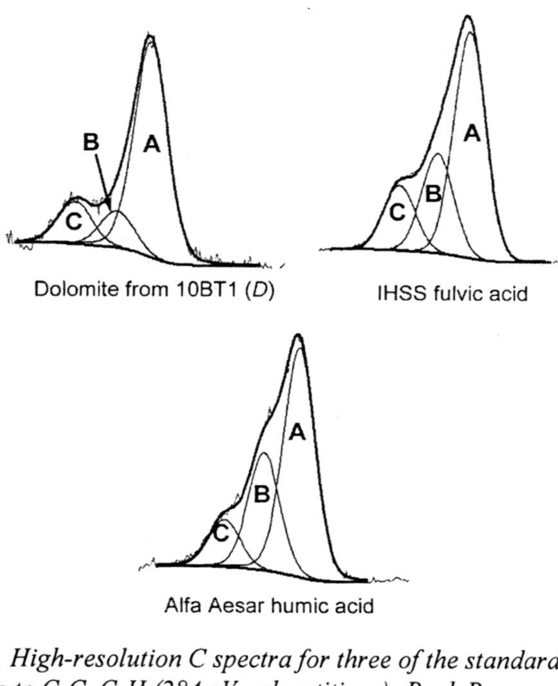

Figure 5. High-resolution C spectra for three of the standards. Peak A corresponds to C-C, C-H (284 eV, adventitious); Peak B may correspond to C=O (288 eV); and Peak C may be carbonate or amide C (289 eV).

Humic and fulvic acids show carbon peaks in Figure 5 at the same three binding energies as were observed in the coating samples. Other authors using XPS have also studied C in humic and fulvic acids from dissolved organic matter, fitting significantly more peaks to the C_{1s} region. Bubert et al. (*21*) fit four peaks with the aid of ^{13}C NMR; as this study used an Al Kα source without an additional monochromator, it is the most comparable to what we present here. A monochromatized x-ray source provides narrower peak shapes in high-resolution XPS. Monteil-Rivera et al. (*19*) were able to fit eight peaks to the C_{1s} region, using a monochromatized source and ^{13}C magic angle spinning solid-state NMR. The high-resolution C spectra presented here were fit with only three peaks; the minimum full width of a high-resolution C peak at half maximum height (FWHM) under the conditions used was approximately 1.7 eV. Thus, the observed range from 284 eV to 289 eV could contain only three peaks at this width.

High-resolution nitrogen 1s signals of the coating samples were observed at 399–400 eV. Nitrogen signals at 399–400 eV are indicative of aromatic or amide nitrogen (*22*), and have been reported in typical soil humic and fulvic acids in studies of soil particles (*19, 22*). The N_{1s} peaks for the IHSS fulvic and humic acid standards were observed at binding energies around 401–402 eV; because +1–2 eV charging was observed for the C_{284} peak, the N_{1s} of these materials also occured at 400 eV, as expected from the literature (*19, 22*).

Nitrogen levels in the samples varied from trace to quantifiable amounts. It was also noticed that nitrogen levels varied irrespective of the physical appearance of the coating: the shiny black samples were not consistently high in N. The presence of nitrogen, too, was consistent with the bulk analysis by py-GC-MS and THM-GC-MS (*6, 15*), which showed significant amide and aromatic nitrogen in the samples (likely derived from the proteinaceous moieties of humic and fulvic acids).

Quantitative Results

Table II shows the quantitative XPS results for all of the samples studied. Not unexpectedly, the dolomite substrate showed the least amount of carbon of all the samples. Two samples, Sample 2 (the shiny coating with little substrate) and Sample 3, had high levels of nitrogen, significantly more than would be expected for a humic or fulvic acid. The atomic compositions of the other coating samples compared well with experimental and theoretical C, N, and O contents for humic acids. The standard humic acid from Alfa Aesar contained a significant amount of K as an impurity. The origin of this particular humic acid is not known, even to Alfa Aesar. Traces of Ca, Cl, Fe, and Si are likely from dissolved salts that are deposited at the cave surface.

Table II. Elemental Composition of Samples and Standards

Material	*C, atom%*	*O, atom%*	*N, atom%*	*Other (atom%)*
Dolomite (D)	36.9	41.8	trace	7.8% Ca, 10.3% Si, 1.2% Cl
Sample 2	59.8	28.1	11.1	trace Ca, Cl
Sample 3	60.1	27.8	12.1	trace Ca
Sample 4	63.3	28.8	trace	trace Ca
Sample 5	54.2	32.8	4.5	trace Ca
Sample 7	67.7	3.1	not detected	trace Ca
Humic acid (Alfa Aesar)	84	10.2	3.6	2.6% K
IHSS humic	64.0	25.3	5.8	
IHSS fulvic	71.3	23.2	4.1	
Calculated humic acid*	64	32	4.4	

*for typical soil humic acid, based on data in Troeh and Thompson (*23*).

The high percentage of carbon in the coating, compared to that of the uncoated (but not "clean") dolomite, indicates the presence of organic carbon, as had first been proposed by Fichter et al. in the elemental analysis done in the 1950s (*1*). However, the origin of this organic carbon is significant. If it was deposited by human activity in the cave (e.g., cooking fires) or a brush fire, and post-dated the red and yellow pictographs, then the coating's age *would* provide a minimum age for the paintings. However, because the coating surface composition seems to be consistent with that of soil humic acids, either from the aeolian dust coating most of the surfaces in the cave, or deposited by penetrating ground water, the age of the coating is more likely not related to that of the paintings.

Depth Profiling

Depth profiling was performed on several of the 10BT1 samples in order to determine how and if the coating changes in composition with depth. Unfortunately, the coating sample surfaces were cracked and uneven, making sputtering of the surface difficult, and the resulting spectra were difficult to interpret. A plot of atom percent vs. sputtering time for the sooty sample

(Sample 3) is shown in Figure 6a. The total C decreases with sputtering, but Ca does not increase. Thus, even sputtering for 170 min cannot remove the sooty coating and reveal the dolomite underneath. Figure 6b shows that the C_{284} signal, typically associated with adventitious carbon, did decrease after 10–20 min of sputtering the surface. Therefore, the aliphatic carbon observed on the surface of the coating samples was truly observed in the coating itself and was not simply an artifact of the high-vacuum system.

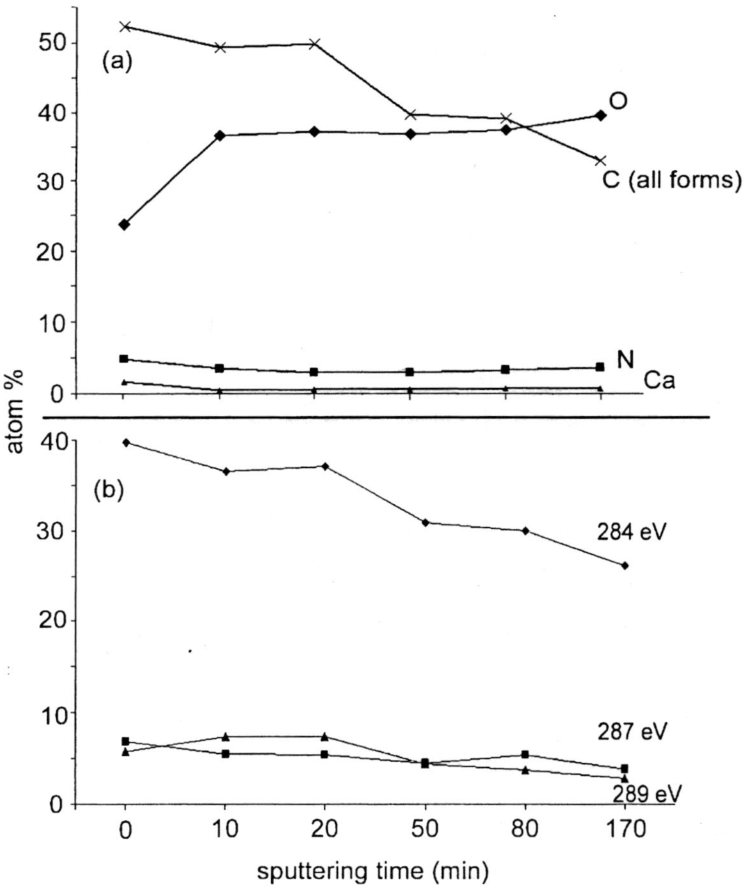

Figure 6. Result of Ar ion sputtering of Sample #1: (a) all elements, (b) carbon only, showing different chemical states present.

Conclusions

The coating samples from different parts of the cave exhibited differences in surface atomic composition and nitrogen levels. High levels of carbon in the coating surface, and varying chemical shifts for the C signals, are indicative of the presence of organic carbon and were also observed during prior analyses of the coating (*4, 5*). Bulk analysis using py-GC-MS and thermally-assisted hydrolysis/methylation-GC-MS also show the coating to be a complex material (*6, 15*). This may indicate that there are several sources of the organic material, some of which (e.g., soot, if it is present) might be useful in dating the underlying paintings, if they could be selectively separated from the coating matrix.

The nitrogen peaks in the high resolution XP spectra at ~399–400 eV BE suggest the presence of aromatic and amide moieties in the coating surface. Hydrocarbon (C-C/C-H) at 284 ± 1 eV BE, carbonyl (C=O) at 287 eV BE and amide/carbonate C at 289 eV BE were observed at the surface of the samples. Nitrogen and carbon results were comparable to soil humic and fulvic acid standards.

XPS is a useful surface technique that complements bulk analyses, and in this case provides an understanding of the organic material that would be oxidized by the plasma-chemical method. Surface and bulk analysis (*6, 15*) suggest that the surface bears a significant resemblance to soil humic acids. Sputtering of the samples did not show significant changes in composition, as might be expected for a coating that formed through smoke and cooking residue depositing on the surface from the air. Instead, it seems to support the idea that the coating formed through the deposition of humic acids as ground water moved through the cave and dried on the wall and ceiling surfaces. The trace elements such as Ca, Si, Cl, and Fe seen in the XP spectra may arise from salts on the surface, which can occur over time through exposure to the elements. The results demonstrate that the radiocarbon date previously obtained from the black coating does not provide a reliable minimum age for the underlying pictographs.

Acknowledgements

Special thanks to Simon Garrett (California State University - Channel Islands) and Todd Bryden (Henkel Technologies, Inc.) for their guidance and instrumentation help with the XPS. Carolynne Merrell, Archaeographics, provided some of the coating samples and additional information on Little Lost River Cave. RAA also thanks Richard D. Hill, Idaho BLM archaeologist, and Daniel Fraser for their aid during the site visit in 2005.

References

1. Fichter, E.; Hopkins, M.; Isotoff, A.; Liljeblad, S.; Lyman, A. R.; Strawn, M.; Taylor, E. A. *Exploratory Excavation in Little Lost River Cave No. 1*; Idaho State College: Moscow, ID; unpublished report, 1954.
2. Gruhn, R.; Bryan, A. *Report on Test Excavation at Little Lost River Cave No. 1 (10BT1) in 1990*; Report on Cultural Resource Permit ID-1–27727; Bureau of Land Management: Idaho Falls, ID, 1990.
3. Merrell, C. L.; Hill, R. D. *Pictograph Documentation from the Little Lost River Cave No.1 (10BT1)*; Report on Agreement No. DAA000103; Bureau of Land Management: Idaho Falls, ID, 2001.
4. Steelman, K. L.; Rowe, M. W.; Guillemette, R. N.; Merrell, C. L.; Hill, R. D. *American Indian Rock Art* **2002**, *28*, 111–120.
5. Steelman, K. L.; Rowe, M. W.; Boutton, T. W.; Southon, J. R.; Merrell, C. L.; Hill, R. D. *J. Archaeol. Sci.* **2002**, *29*, 1189–1198.
6. Fezzey, S. T.; Armitage, R. A. *J. Anal. Appl. Pyrolysis* **2006**, in press.
7. Russ, J.; Hyman, M.; Shafer, H. J.; Rowe, M. W. *Nature* **1990**, *348*, 710–711.
8. Steelman, K. L.; Rowe, M. W. In *Archaeological Chemistry: Materials, Methods, And Meaning*; Jakes, K., Ed.; ACS Symposium Series No. 831; American Chemical Society: Washington, DC, 2002; pp 8–21.
9. Steelman, K. L.; Rowe, M. W.; Turpin, S. A.; Guilderson, T.; Nightengale, L. *Am. Antiq.* **2004**, *69*, 741–750.
10. Rowe, M. W. In *Handbook of Rock Art Research*; Whitley, D. S., Ed.; AltaMira Press: Walnut Creek, CA, 2001; pp 139–166.
11. *Modern Analytical Methods in Art and Archaeology*; Ciliberto, E.; Spoto, G., Eds.; John Wiley & Sons, Inc.: New York, NY, 2000.
12. Lambert, J. B.; McLaughin, D. C.; Shawl, E. C.; Xue, L. *Anal. Chem.* **1999**, *71*, 614A-620A.
13. Stewart, D. J.; Adams, R. K.; Borradaile, J. G.; MacKenzie, J. A. *J. Archaeol. Sci.* **2002**, *29*, 1309–1316.
14. Paparazzo, E. *Archaeometry* **2003**, *45*, 615–624.
15. Brown, J. A. M. S. thesis, Eastern Michigan University, Ypsilanti, MI, in preparation.
16. Castle, J. E.; Salvi, A. M. *J. Vac. Sci. Technol. A.* **2001**, *19*, 1170–1175.
17. Wagner, C. D.; Riggs, W. M.; Davis, L. E.; Moulder, J. F., Eds.; In *Handbook of X-ray Photoelectron Spectroscopy*; Muilenberg, G. E., Ed.; Perkin-Elmer Corporation: Minnesota, 1979; p 189.
18. Wagner, C. D.; Davis, L. E.; Zeller, M. V.; Taylor, J. A.; Raymond, R. H.; Gale, L. H. *Surf. Interface Anal.* **1981**, *3*, 211–225.
19. Monteil-Rivera, F.; Brower, E.B.; Masset, S.; Deslandes, Y.; Dumonceau, J. *Anal. Chim. Acta* **2000,** *424,* 243–255.

20. Johansson G.; Hedman J.; Berndtsson A.; Klasson M.; Nilsson R. *J. Electron Spectrosc. Relat. Phenom.* **1973**, *2*, 295–317.
21. Bubert, H.; Lambert, J.; Burba, P. *Fresenius J. Anal. Chem.* **2000**, 368, 274–280.
22. Abe, T.; Watanabe, A. *Soil Science* **2004**, *169*, 35–43.
23. Troeh, F. R.; Thompson, L. M. *Soils and Soil Fertility*; Oxford University Press: New York, 1993; p 101.

Chapter 9

Shell Bead Sourcing: A Comparison of Two Techniques on *Olivella biplicata* Shells and Beads from Western North America

Jelmer W. Eerkens[1], Jeffrey S. Rosenthal[2], Howard J. Spero[3], Ryoji Shiraki[4], and Gregory S. Herbert[5]

[1]Department of Anthropology, University of California at Davis, One Shields Avenue, Davis, CA 95616
[2]Far Western Anthropological Research Group, 2727 Del Rio Place, Davis, CA 95616
[3]Department of Geology, University of California at Davis, One Shields Avenue, Davis, CA 95616
[4]McClellan Nuclear Radiation Center, University of California at Davis, Davis, CA 95616
[5]Department of Geology, University of South Florida, 4202 East Fowler Avenue, Tampa, FL 33620

We compare two methods for tracking the geographic source of *Olivella biplicata* shell beads along the California and Oregon Pacific coast; bulk element composition by inductively coupled plasma-mass spectrometry (ICP-MS) and stable carbon and oxygen isotopes by isotope ratio-mass spectrometry (IR-MS). Both techniques hold some promise for reconstructing prehistoric trading systems, but neither is perfect. Currently, oxygen isotopes can reliably differentiate recent shell beads (post A.D. 1500) derived from points north versus south of Point Conception. We are working to extend the time depth of this technique. Elemental composition will help to divide the northern isotopic zone into finer geographic sections by providing chemical signatures for certain regions, such as Monterey Bay and extreme Northern California. Additional sampling, particularly on prehistoric specimens that are subject to post-depositional chemical alteration, will be necessary to make this technique reliable for sourcing ancient beads. In concert, the two techniques could provide archaeologists with a dependable means for identifying where the shell used to make a bead was originally collected.

Introduction

Marine shell beads are a regular component of the Holocene archaeological record in many areas of the world. Although they rarely outnumber more common artifacts, such as flaked stone and ceramics, they are sufficiently ubiquitous to have been the subject of considerable archaeological research. Such studies focused on both the symbolic aspects of beads (e.g., wealth display) and the reconstruction of prehistoric trade and/or exchange systems.

Despite the potential to track the spatial movement of marine shell beads, archaeometric attempts at sourcing have been minimal. We are aware of only three other groups that have taken this approach including a neutron activation analysis (NAA) of *Busycon* sp. shells and artifacts in the American Southeast (*1*), the use of strontium isotopes as a measure of geologic age of fossil shells in Southwest Europe (*2*), and the use of oxygen isotope ratios to source *Spondylus* sp. beads in Central Europe (*3*).

These geochemical studies notwithstanding, the most common method to trace the geographic origin of marine-shell artifacts is to identify the shellfish species out of which the beads were fashioned, and then determine the geographic distribution of that species (e.g., *4, 5, 6*). While informative in some cases, there are several potential problems and drawbacks to this approach. First, it is necessary to establish that the modern geographic distribution of a species is consonant with the prehistoric distribution at the time the bead was made. This can be difficult if archaeological and/or palaeontological research along a coastline has been minimal or if variations in Holocene climate have modified regional environmental conditions across millennial timescales. Second, if the shell has been heavily modified through abrasion or other reduction processes it is not always possible to identify the exact species. This can be particularly troublesome when there are several similar species within a genera that have different geographic distributions. Third, if a shellfish species is found across an expansive length of coastline, the geographic accuracy of bead sourcing will be poor. Frequently shells can only be sourced to oceans (e.g., the Atlantic coast), which limits the spatial resolution of data to archaeological questions about the exchange and consumption of marine shell beads.

Sourcing *Olivella* Beads in California and the Great Basin

Building on the geochemical sourcing methods first attempted over 35 years ago, we have been exploring stable isotopic and elemental analytical techniques for sourcing marine shell beads in California. We have focused our efforts on beads manufactured from the shells of the "purple olive snail" (*Olivella biplicata*), by far the most common species exploited for bead-making material in the region. *Olivella biplicata* is endemic to the Pacific coast from Vancouver Island to the northern Baja peninsula. Shells of this species have been turned into beads since the earliest Holocene (9,000–10,000 years before present; ybp) (*7–11*), and are regularly recovered at prehistoric sites across the west, where they are found as far inland as eastern Nevada, Utah and Arizona (*5, 6, 12, 13*). The earliest shell beads from this region are simple spire-ground *Olivella* with little to no additional modification (*7–8, 11*). By the middle Holocene (ca. 5000 ybp), rectangular to oval-shaped beads cut from the body whorl of *Olivella* were traded over a broad region (*13–15*), providing the foundation for a manufacturing industry and exchange network that developed through the late Holocene and culminated in the monetized systems of exchange reported ethnographically (*12, 13, 16–23*).

Centers of prehistoric shell bead production are thought to have existed in the Santa Barbara Channel area of southern California, in central California around Monterey Bay, and at Bodega Bay in northern California (*12*). Despite the ubiquity of *Olivella* shell beads in archaeological sites throughout California and the Great Basin, centers of bead production and exchange outside the Santa Barbara Channel region (e.g., *17, 24, 25*) are poorly documented (*12, 26*). Extensive archaeological research has been conducted along much of the central and northern California coast (e.g., *26–28*), particularly at Monterey Bay (*29–32*) and Bodega Bay (*33*). Yet, *Olivella* bead manufacturing is only occasionally represented by small quantities of manufacturing waste and an occasional bead blank (*12, 26, 31, 34*). This tradition appears to have been small-scale and could not account for the tens of thousands of *Olivella* beads recovered from central and northern California archaeological sites (*12, 14, 35*).

In contrast, sites located on the northern Channel islands produced as much as 150,000 pieces of bead manufacturing refuse and hundreds of production blanks from a single cubic meter of excavation (*18, 20, 36, 37*). As well, a tradition of specialized bead making tools, including micro-blade drills and anvils, is documented on the northern Channel Islands (*16*). Nothing comparable to this level and regularity of production is evident at archaeological sites in central and northern California.

Olivella Sourcing

While it is clear that most *Olivella* beads found in the interior of western North America are ultimately derived from the Pacific Coast, this source zone is over 2000 kilometers long. As a result, a species-based sourcing program is of limited value in the reconstruction of specific exchange networks, and does not allow us to evaluate the indirect evidence suggesting that southern California supplied the vast majority of beads consumed in western North America. Clearly, an alternative method is necessary to test this hypothesis.

To refine potential *Olivella* source zones to more specific sections of the coastline, we have employed two different archaeometric techniques, including determination of elemental composition by inductively coupled plasma-mass spectrometry (ICP-MS) and the use of carbon and oxygen stable isotopes using an isotope ratio-mass spectrometer (IR-MS) (*38*). Each of these techniques is described below, and the final section evaluates and compares their utility in sourcing *Olivella* shell beads.

Compositional Analysis by ICP-MS

Because *Olivella biplicata* snails grow in tidal environments, it was hypothesized that their shell chemistry might be influenced by nearshore seawater chemistry, which would include dissolved minerals from local shoreline deposits. If geology varied enough along the California coast, certain regions might be bathed by seawater with a chemically distinct composition that would be incorporated into the shells.

To test this hypothesis, we collected 40 modern shells from beaches along the California coast and subjected them to ICP-MS analysis. Shells were generally collected dead as they lay on beaches. Specimens were obtained from five main geographic areas, the Channel Islands off the Santa Barbara coast (n = 3), the mainland region from Santa Barbara (n = 3), central California from north of Pt. Conception to Monterey (n = 11), Monterey and Asilomar (n = 12), and regions north of San Francisco (n = 11). We complemented this sample set with 32 shells from coastal archaeological deposits. All were unmodified whole *Olivella* shells thought to come from local seawaters, rather than representing exotic shells brought in through trade. Due to the availability of various prehistoric collections, the sample of archaeological shells could not exactly duplicate the sample of modern shells. Only three of the regions sampled for modern shells were represented by archaeological ones, including the Channel Islands (n = 12), the mainland between Santa Barbara and Kirk Creek (n = 18), and Monterey Bay (n = 2). Prehistoric shells from north of San Francisco were not available to the authors for analysis at the time this study was conducted. Figure 1 shows the location of the sampling areas.

Each shell was washed on its exterior with filtered water, but was not further cleaned. Two samples were extracted from each shell for compositional analysis, one from the shell wall near the mouth aperture and one from the callus. Most prehistoric beads were made from these two parts of the shell or consist of whole shells. Figure 2 shows two typical *Olivella* beads (A and B) as well as an unmodified modern shell (C).

Compositional analyses consisted of 50 milligrams of powdered shell, crushed in an agate mortar and pestle and digested in 70% concentrated nitric acid. These liquid samples were diluted with pure water to 5% nitric acid concentration and introduced to the ICP-MS (an Agilent 7500c). Thirty-three elements were measured for each shell sample, ranging from boron to uranium. Four of these (Ga, Rh, Cd and Ce) were generally below the detection limits of the machine for most samples and were not considered further. The remaining 29 elements were of varying utility in the analysis.

Ca was the only element regularly detected above the ppm range. Some elements, such as Na, Mg, K, and Sr, were frequently above 100 ppm, while others, such as Si, Al, B, were generally above 4 ppm but below 100 ppm. The remaining elements were typically measured in the 10 ppb to 100 ppb range. Some of these elements, such as Sr, Mg, and Ba have been examined for their ability to serve as tracers of various environmental factors, such as salinity and temperature (e.g., *39–44*). These studies have met with varying degrees of success. In some cases temperature and salinity correlate well with certain elements. However, success varies greatly depending on the species used, environment from which those species derive, and geological age of the samples. Unfortunately, no such study has yet been undertaken on *Olivella biplicata* and it is not clear how well such elements track salinity, temperature, or other environmental and/or biological conditions in this species.

Results

Overall, there was only minor geographic patterning within the chemical data. Among modern shells, notable regional differences were observed among some elements. Figure 3 displays ppb readings for two elements, Cr and Sc, relative to %Ca. There is a clear bimodal distribution along the Y-axis for Cr, separating most Monterey samples. Interestingly, there are three non-Monterey samples within this high Cr group (one from Kirk Creek, one from Piedras Blancas, and one from Santa Cruz Island), hinting that the pattern may not be entirely influenced by geography alone. Furthermore, all three of these samples have one anomalously high Cr reading from the wall of the shell paired with a

Figure 1. Map of region, sampling locations, and places mentioned in text.

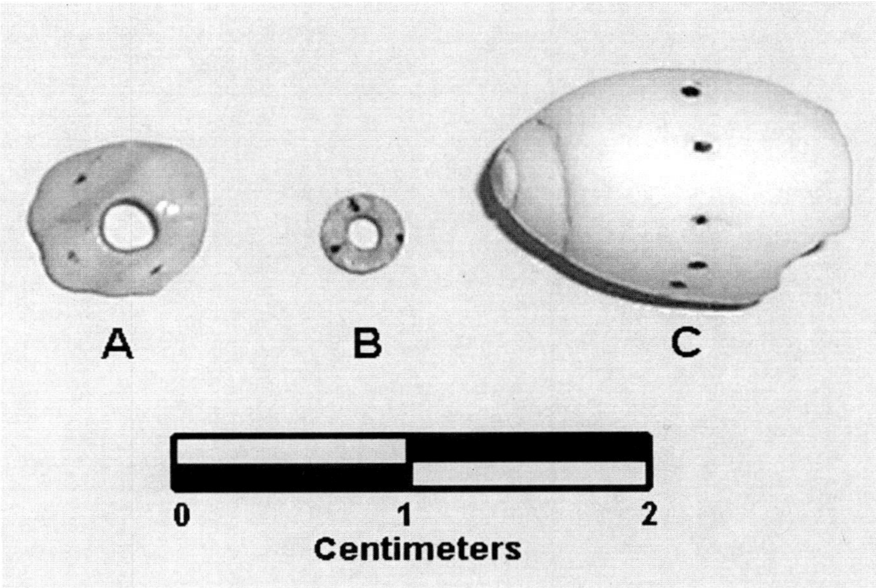

Figure 2. Example of Olivella beads (A and B) and modern shell (C)

normal Cr reading from the callus. This suggests, again, that a non-geographic factor may be involved in determining the bimodal distribution in Cr.

Further evidence that the patterns in Cr are not geographically-influenced is given in Figure 4 which plots all prehistoric samples on the same scales for these two ratios. As seen, Cr values are nearly equal for all four regions represented in the prehistoric sample (note that two prehistoric samples had unusually high ppm values for Cr and are not shown on the graph). Moreover, the Cr values of prehistoric shells all fall within the range of modern Monterey shells. Modern Cr contamination in seawater does not seem to be a likely explanation as most modern samples are *depressed* in Cr concentrations relative to prehistoric ones. While post-depositional alteration to shell chemistry is clear in our samples (see below), such a dilution is also not a satisfactory explanation of the observed patterns in Cr. If Cr was increasing relative to other elements over time as a shell was lying with an anthropogenic midden, as would be suggested by the Channel Island, Santa Barbara, and central California samples, such an increase is not evident in the prehistoric Monterey samples, which remain stable near a ratio between 30 and 50. It is unlikely Cr would systematically increase for a diverse set of samples all over California but not for Monterey samples. Our future research will attempt to pinpoint the explanation for the Cr patterns.

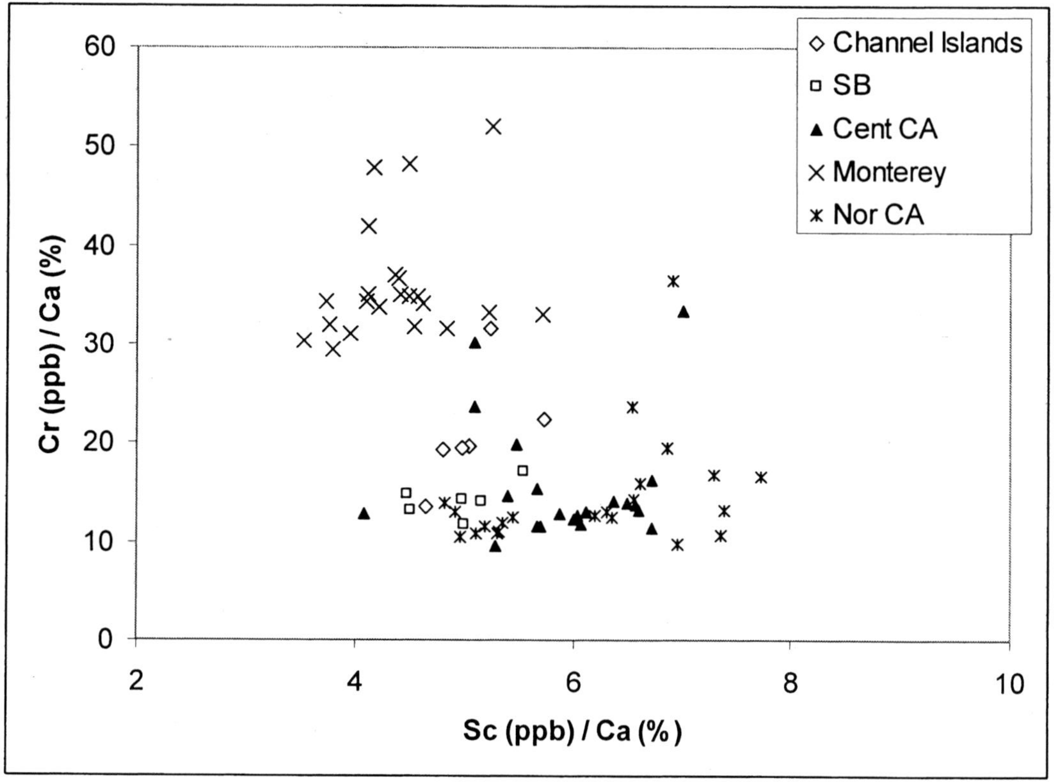

Figure 3. Plot of Cr vs. Sc for modern Olivella shells.

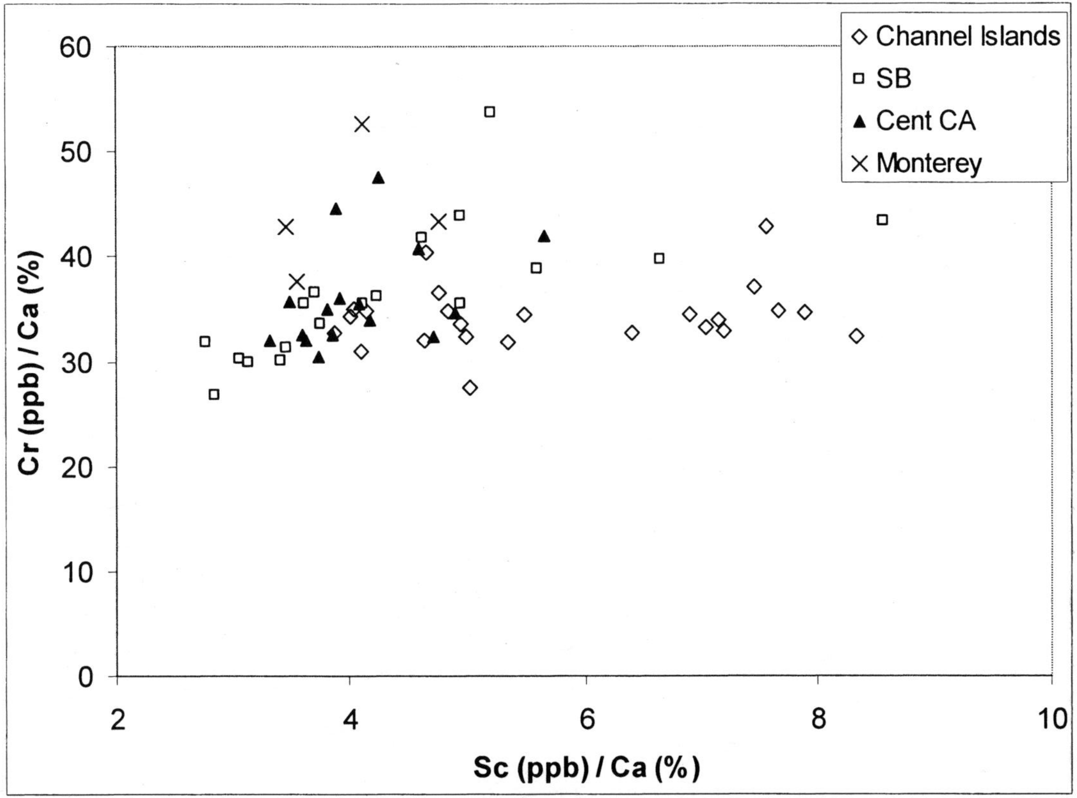

Figure 4. Plot of Cr vs.Sc for prehistoric Olivella shells (scale identical to Figure 2).

Although far less pronounced, patterns in Sc are more promising as geographical signatures. As seen in Figures 3 and 4, Monterey samples, both prehistoric and modern, tend to be slightly depressed in Sc while Santa Barbara and Channel Island samples tend to be elevated. Central California samples are more variable in Sc. More promising in this regards are Al, Sr and Zr concentrations. Shown in Figure 5 are Al (ppm) and Zr (ppb) as ratios of %Ca (note that both are on a logarithmic scale to facilitate display). Monterey samples fall on the upper right hand side of the scatter of points while Channel Islands samples fall on the left and central and northern California samples are intermediate. Prehistoric samples, shown in Figure 6, follow some of these same patterns though variability seems to have increased for some of these regional samples.

Other elements are even less discriminatory geographically, though they occasionally serve to distinguish one particular sampling location within a region where more than one shell was collected. For example, shells from Devereaux Beach in Santa Barbara were frequently higher in heavy metals and rare earth elements than other shells from this region.

The most obvious pattern among all the data concerned the difference between prehistoric and modern samples. A number of elements show either systematic increases or decreases in ppm concentrations between the modern and prehistoric shells. For example, Figure 7 plots Na and Ba as ratios of % Ca. Relative to prehistoric samples, modern ones are systematically enriched in Na and depleted in Ba. Other elements that show such systematic changes include Ti, V, Mn, As, and Cd, where prehistoric samples tend to be enriched in these elements. Enrichment of some elements in prehistoric samples is likely the product of leaching of other elements over time (such as Na), thereby increasing the relative proportion of the former.

Even Ca, the element we used to normalize all our other measures, may be subject to post-depositional alteration. Moreover, such alteration may vary depending on local environmental conditions. Such a factor may account for some of the overlap or blurring we see between regional elemental signatures in our prehistoric samples. Without knowing exactly how much change may have taken place in Ca concentrations in different burial conditions, it is difficult in such sourcing studies to account for this possibility.

In sum, the ICP-MS analyses suggest that *Olivella* shells are subject to fairly significant post-depositional alteration as shells are exposed to midden soils over hundreds to thousands of years. While discouraging, it does not preclude the application of elemental compositional studies for shell-bead sourcing. Some elements, such as Sc, Zr, and Al, among others not discussed in detail here, remain fairly constant when taken as ratios against Ca. It is these elements that should be the subject of provenance analyses and that we will target in future research with *Olivella* shell.

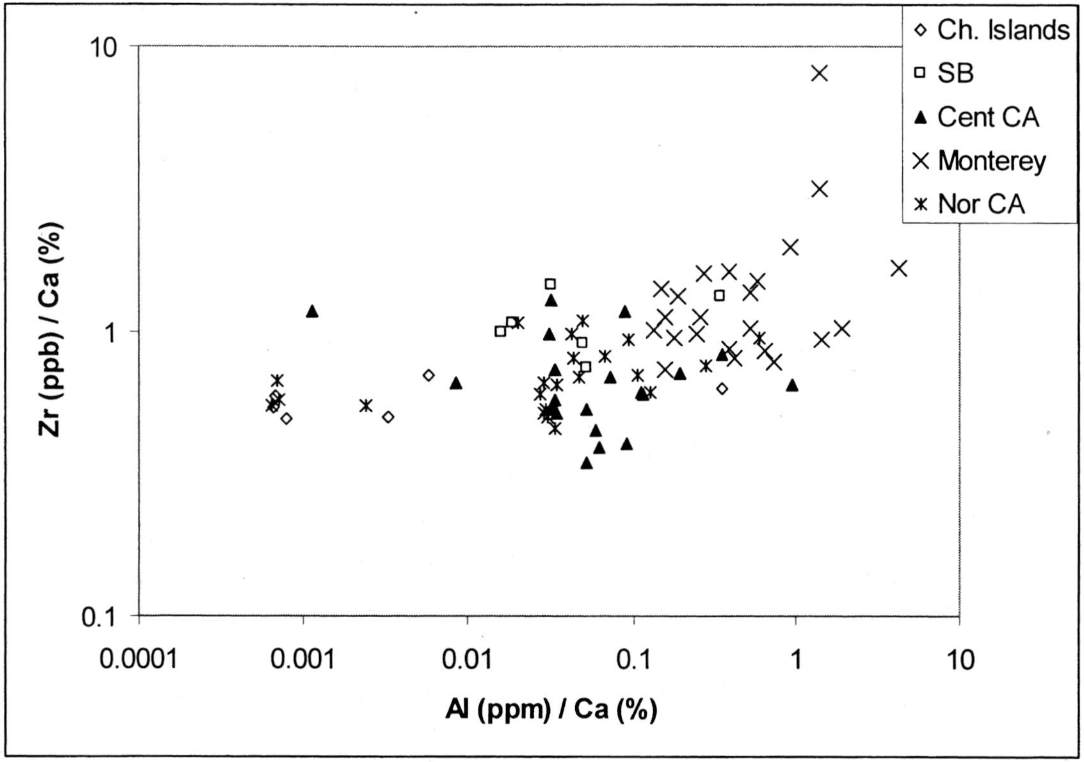

Figure 5. Zr and Al values for modern shell samples

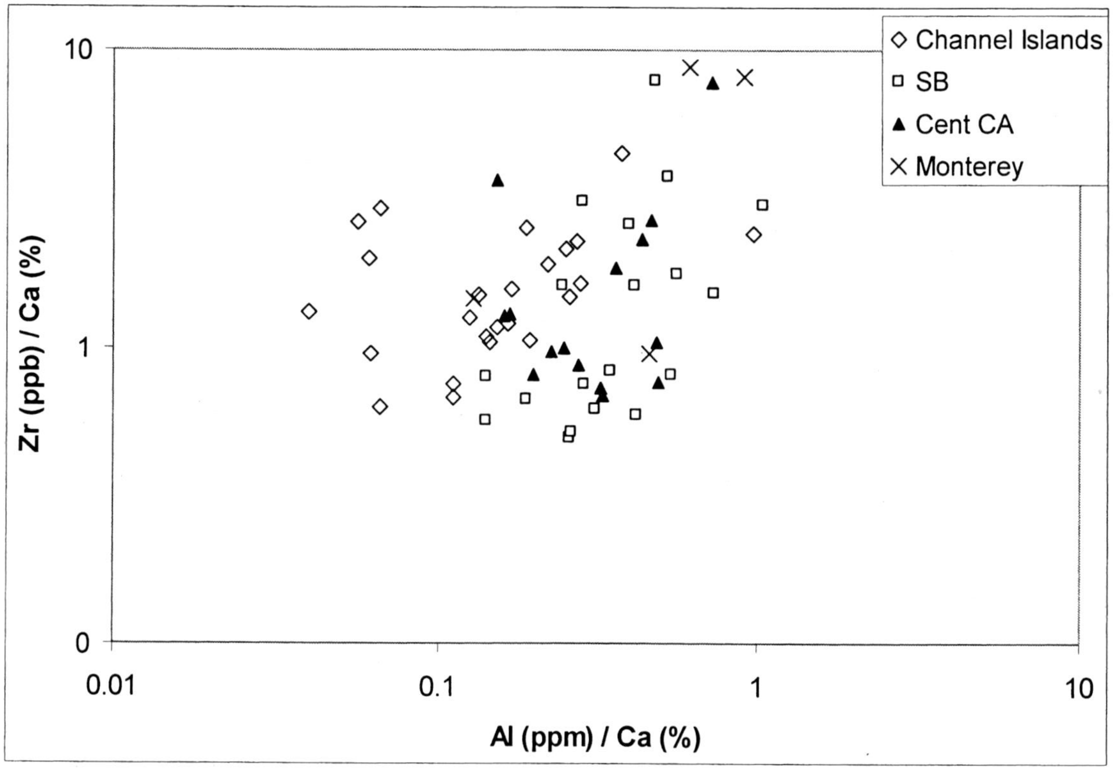

Figure 6. Zr and Al values for prehistoric samples (note that the scale of the X-axis is slightly different than in Figure 4).

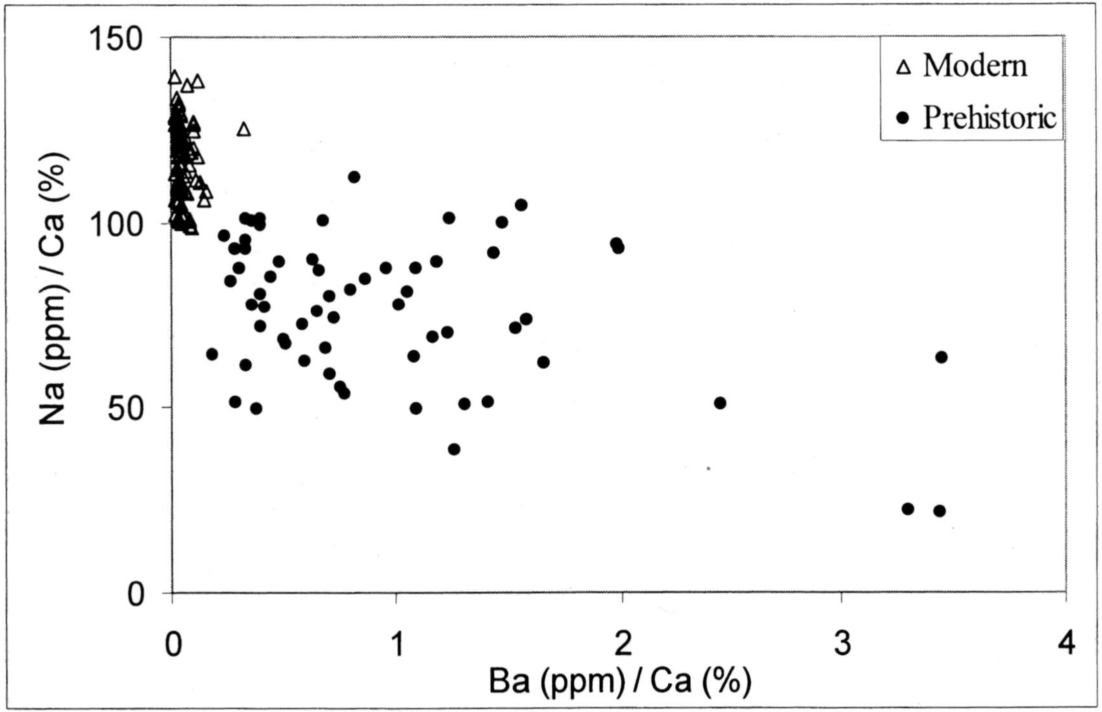

Figure 7. Plot of Na and Ba showing difference between prehistoric and modern shells.

Furthermore, shell compositional studies may be informative for more micro-scale sourcing. While some elements may discriminate large-scale geographic regions (e.g., the California coast south of Point Conception), others could be indicative of micro-habitats (e.g., individual beaches influenced by certain drainages). We did not collect and sample *Olivella* shells at a spatial scale to properly test this possibility, however some data we collected support this notion. For example, in cases where we sampled multiple shells from particular beaches, the chemical composition of these shells were frequently more alike than other shells from the broader geographic region we studied. One potential problem with such an approach to sourcing is that there may not be enough variation between different beaches to individually characterize each. Thus, shells from a beach in northern California may have similar chemical signatures for certain elements to a beach in southern California.

O and C Stable Isotopes

The second approach we have taken to sourcing *Olivella* shells is to exploit differences in oxygen and carbon isotope ratios (*38*). Oxygen isotopes have been particularly useful in this regard due to sea-surface water temperature gradients that exist along the California coast. The oxygen isotope value ($\delta^{18}O$) of biogenic carbonates such as aragonite depend primarily on two factors; the proportion of ^{18}O and ^{16}O in seawater ($\delta^{18}O_{sw}$) which varies as a function of evaporation, precipitation and continental runoff (e.g., salinity), and the calcification temperature in which the organism precipitated its shell (*45–50*). In this study, we focused on the contribution of sea surface temperature, which is a significant factor contributing to oxygen isotopic variability in *Olivella* shells in California.

Along the California coast, two currents influence sea surface temperatures (*51*). The California Current flows southward along the western margin of North America bringing cold, nutrient-rich waters from subpolar latitudes. During spring and early summer, these currents flow along much of the California coast. However, during summer through late fall, the Davidson Counter-Current flows northward from subtropical latitudes, where it transports warmer, nutrient-depleted, waters along the southern California coast. These two currents meet near Point Conception (see Figure 1). As a result, *Olivella* shells growing south of this point record low $\delta^{18}O$ values during most of the year, while those shells north of this region record high $\delta^{18}O$ during the entire year. The only departure from this pattern occurs during El Niño Southern Oscillation (ENSO) events, which bring large amounts of precipitation, especially to Northern California, significantly affecting the salinity of near-shore waters. This influx of freshwater brings Northern California oxygen isotope values in *Olivella* shells within the range of Southern California shells. Fortunately, such events usually last less than two months and are infrequent, on average once every five to seven years. As a result, ENSO events notwithstanding, we can source modern shells with high confidence to either Southern or Northern California (*38*).

A second geochemical tracer is shell $\delta^{13}C$ ($^{13}C/^{12}C$). In mollusks, shell $\delta^{13}C$ records may be influenced by metabolic rate, and reproductive condition, although the processes by which this influence occurs are less well understood (*52–57*). More important is the source of shell $\delta^{13}C$ variation that is related to shifts in the $\delta^{13}C$ of dissolved inorganic carbon (ΣCO_2) in seawater. This environmental parameter is ultimately controlled by the balance between photosynthesis and respiration in the water column. In surface waters, phytoplankton preferentially remove $^{12}CO_2$ during photosynthesis, thereby increasing $\delta^{13}C$ of ΣCO_2. Organic matter sinking below the photic zone is then metabolized by consumer organisms, thereby releasing ^{13}C-depleted CO_2 back into the water column. When these ^{13}C-depleted waters return to the surface during seasonal upwelling events, the shell $\delta^{13}C$ records the ambient decrease in $\delta^{13}C$ of ΣCO_2 (e.g., *49, 58, 59*). In California, upwelling occurs most intensively between April and July (*60, 61*). Thus, shell $\delta^{13}C$ values may be distinctive of certain regions of the California coast that experience more intensive upwelling than others. In combination with $\delta^{18}O$, this information could be used to isotopically fingerprint shells.

These geochemical tracers have been successfully applied to studies of the shells of a variety of marine organisms including bivalve and gastropod mollusks, ostracods, forams, brachiopods and solitary corals (*47, 54, 57, 58*). In the case of mollusk shells, for example, serial microsampling around the spiral whorls from earliest to oldest growth revealed sinusoidal variations in isotope ratios, which result from shell deposition in a seasonal environment (*47, 50, 62–64*). Our previous work showed this to be true of *Olivella* shells as well, where $\delta^{18}O$ levels fluctuate from warm summer temperatures to cold spring and winter temperatures (*38*).

Samples for isotopic analysis were processed at UC Davis on a Micromass Optima isotope ratio mass spectrometer (IRMS). Powdered carbonate samples were drilled from the shell surface in shallow grooves (< 0.3 mm deep) which ran parallel to the growth lines using a 0.5 mm bit attached to a hand-held drill. Powdered carbonate samples ranged from 50 to 80 μg in weight. Sampling began at the *Olivella* shell lip (most recent growth) and continued until the parietal callus was reached (earlier growth), including at least one whole whorl revolution. The analysis of beads required determining the axis of growth and orienting the artifacts in the same way as complete shells. The linear distance between samples ranged from 0.5 mm to 2.5 mm depending on the size of the shell or bead and how intensively it was sampled.

Prior to analysis on the IRMS, powdered aragonite samples were gently heated at 75°C *in vacuo* for 30 minutes to remove adsorbed water and

Table I. Beads Analyzed and Range of Isotopic Values Recorded.

Bead Type	*Site*	*Approx. Date*	*Number Samples Taken*	$\delta^{18}O$ *Range*	$\delta^{13}C$ *Range*
G1	INY-3806	A.D. 850	2	[0.79,0.62]	[1.90,1.74]
G1	INY-3806	A.D. 850	2	[0.83,-0.16]	[2.30,1.91]
G2a	INY-3806	A.D. 850	3	[0.63,0.34]	[2.05, 1.82]
G2b	INY-5207	A.D. 1700	4	[0.40,-0.51]	[2.16,1.36]
K1	INY-5207	A.D. 1700	2	[0.76,0.42]	[2.67,2.33]
A2a	INY-5207	A.D. 1700	8	[0.17,-0.49]	[1.87,1.55]
H1b	YOL-69	A.D. 1810	8	[0.67,-0.27]	[1.74,1.27]
H1b	YOL-69	historic	9	[0.62,0.11]	[0.91,0.20]
H1b	YOL-69	A.D. 1870	3	[-0.05,-0.74]	[2.57,2.45]
H1b	SCL-30	A.D. 1825	3	[0.40,0.16]	[1.53,1.60]
F2a	SCL-732	1518 B.P.	6	[1.63,1.09]	[0.55,0.28]
F2a	CCO-269	1518 B.P.	10	[0.99,0.17]	[1.30,-0.09]
F2a	ALA-413	1512 B.P.	3	[1.32,0.34]	[0.87,0.45]
F2a	ALA-329	1512 B.P.	1	[-0.04,-0.04]	[0.61,0.61]
F2a	SOL-355	1323 B.P.	3	[1.28,0.64]	[1.25,1.11]
F2b	CCO-269	1531 B.P.	9	[1.42,0.96]	[0.87,0.31]
F2b	CCO-269	1399 B.P.	4	[1.15,0.67]	[0.59,0.03]
F2b	ALA-413	1483 B.P.	3	[0.73,0.00]	[0.77,0.67]
F3a	ALA-413	1496 B.P.	5	[1.46,0.85]	[1.36,0.77]
F3a	ALA-329	981 B.P.	5	[1.05,0.75]	[0.94,0.53]
F3a	ALA-46	874 B.P.	4	[1.65,0.43]	[1.18,0.22]
F3a	ALA-46	955 B.P.	10	[0.85,0.18]	[0.66,0.31]
F3a	ALA-329	1075 B.P.	3	[0.65,0.33]	[0.49,0.35]
F3a	ALA-329	1075 B.P.	5	[1.19,0.38]	[0.85,0.18]
F3a	ALA-329	966 B.P.	5	[1.05,0.39]	[0.93,0.36]
F3b	ALA-343	1247 B.P.	2	[1.09,0.55]	[1.10,0.94]
F3b	ALA-343	551 B.P.	5	[0.81,0.38]	[1.21,0.20]
F3b	ALA-343	1180 B.P.	3	[0.46,0.41]	[0.53,0.10]
G2a	ALA-413	1426 B.P.	1	[-3.26,-3.26]	[-7.18,-7.18]
G2a	COL-247	2159 B.P.	2	[0.20,-0.15]	[1.43,0.53]

Table I. *Continued.*

Bead Type	*Site*	*Approx. Date*	*Number Samples Taken*	*$\delta^{18}O$ Range*	*$\delta^{13}C$ Range*
G2b	ALA-413	1531 B.P.	8	[0.67,0.25]	[1.88,1.28]
G2b	ALA-413	1496 B.P.	4	[1.40,1.13]	[1.10,0.92]
G2b	ALA-413	1451 B.P.	4	[1.15,0.42]	[2.13,1.37]
G2b	ALA-413	1518 B.P.	10	[0.76,0.33]	[1.20,0.74]
G2b	ALA-413	1562 B.P.	1	[0.61,0.61]	[0.77,0.77]
G2b	SCl-732	1806 B.P.	6	[-0.83,-3.63]	[-1.94,-4.92]
G2b	SCL-732	1871 B.P.	5	[0.51,0.19]	[0.42,-0.37]
M1a	ALA-42	910 B.P.	3	[0.70,0.39]	[0.73,0.32]
M1a	ALA-42	899 B.P.	4	[0.87,0.47]	[0.81,0.52]
M1a	ALA-42	891 B.P.	3	[0.44,0.14]	[0.94,0.44]
M1a	ALA-42	899 B.P.	5	[1.00,0.70]	[0.77,0.64]
M1a	ALA-329	764 B.P.	6	[1.40,0.94]	[1.85,1.08]
M1a	ALA-329	685 B.P.	8	[1.10,0.52]	[1.36,0.18]
M1a	ALA-42	854 B.P.	2	[1.42,1.03]	[1.60,1.44]
M1a	ALA-42	926 B.P.	6	[1.10,0.67]	[1.52,1.04]
M1a	ALA-329	619 B.P.	2	[-0.90,-1.40]	[-3.22,-3.69]
M2a	YOL-187	548 B.P.	2	[1.09,1.00]	[0.18,0.03]
M2a	ALA-42	493 B.P.	1	[0.40,0.40]	[0.95,0.95]

Note: Type according to Bennyhoff and Hughes (1987).

subsequently reacted in 105% orthophosphoric acid at 90°C using an ISOCARB automated common acid bath system. The resulting CO_2 was purified through a series of cryotraps and introduced into the IRMS through a dual inlet system. Repeated analyses of standards showed that external precision for $\delta^{18}O$ and $\delta^{13}C$ values was ± 0.09 and ± 0.07 respectively (one standard deviation).

Results

We collected 20 modern *Olivella* shell samples from along the California coast to test our hypothesis that $\delta^{18}O$ values vary from southern to northern California as predicted. Each shell was microsampled along growth rings for O and C isotopes, with an average of 15 samples per shell. Figure 8 shows the results of these analyses. As predicted, shells from northern and southern California largely differentiate themselves on oxygen isotopes, shown on the y-axis and expressed as the ratio of ^{18}O to ^{16}O relative to an international standard and multiplied by 1000 ($\delta^{18}O = R_{sample}/R_{standard} - 1] X 1000$) in "per mil" units or parts per thousand. There is some overlap in the central part of the plot, between $\delta^{18}O$ values of 0.6 to 1.7 per mil, correlating to either ENSO events in Northern California, or overlap of summer water temperatures in Northern California and winter temperatures in Southern California. Shells from Santa Cruz Island stray slightly from this pattern, displaying slightly higher $\delta^{18}O$ values than predicted. We also measured $\delta^{13}C$ values to see if this tracer could provide further separation between these regions. Although the Santa Cruz Island samples were also slightly elevated for carbon isotopes, samples were not noticeably different for Northern vs. Southern California. Ellipses in the graph do not represent 95% confidence intervals, but merely subjective "zones" where the majority of samples from particular geographic regions fall.

To date we have analyzed 48 *Olivella* beads from two regions, the San Francisco Bay and Sacramento/San Joaquin Delta area in Northern California (n = 42), hereafter referred to simply as the Delta, and Owens Valley in southeastern California (n = 6). Most of the beads from the Delta region have been independently dated by AMS (*65*). Owens Valley beads comprise three types, saucers (G1/G2), callus cup (K1), and spire-lopped (A2). Delta beads comprise four types following the typology defined by Bennyhoff and Hughes (*12*), including historic needle-drilled (H1b), saucers (G2), saddles (F2/F3), and sequins (M1/M2). We reported on 10 of these beads in an earlier publication, including all six Owens Valley beads and the four historic H1b beads from the Delta (*38*). All ten had oxygen isotopic signatures that were consistent with a southern California source (i.e., at least one isotopic reading lower than 0.6; in fact, most had *maximum* readings equal or less than 0.6).

Prior to the current analyses we had predicted that most of the remaining 38 prehistoric Delta *Olivella* beads would be made from shells procured in northern California. This was based purely on an economic model that minimizes transport distance. In other words, the closest source (by foot) of *Olivella* shell to Owens Valley is southern California, while the closest source of *Olivella* in the Delta region is the coast immediately north and south of San Francisco Bay. Moreover, saddle (F2/F3) and sequin (M1/M2) beads are found exclusively in central and northern California (*12, 13*), again implying a more northern source. On the other hand H1b and G2 beads are found all over California. Results of the carbon and oxygen isotopic analyses are presented in Table II.

Contrary to our predictions, none of the 42 Delta beads display an oxygen isotopic signature consistent with modern shells from northern California (i.e., at least one $\delta^{18}O$ reading greater than 1.7). Instead, the majority of the beads (n = 29; 69%) have oxygen isotopic signatures indicative of a warm water source (i.e., at least one reading less than $\delta^{18}O = 0.6$). All but two of these fall within the range of southern California modern shells. Three shells have $\delta^{18}O$ values

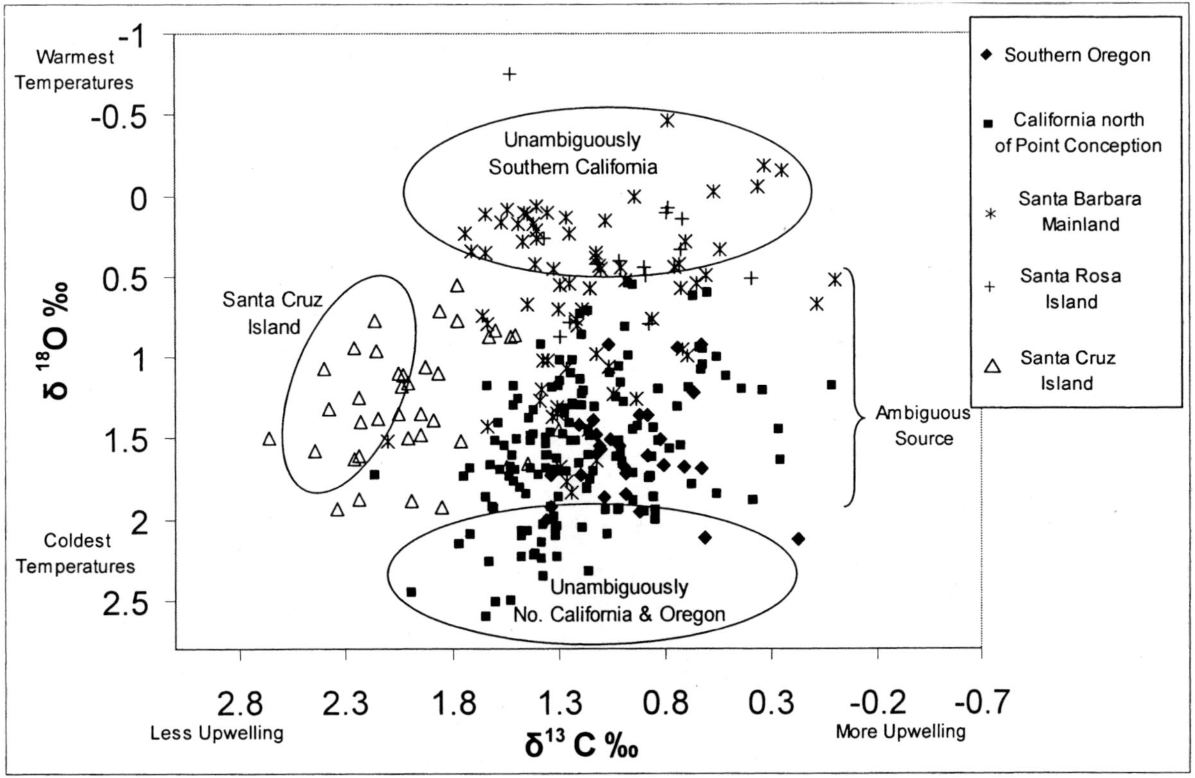

Figure 8. Distribution of Carbon and Oxygen isotope ratios for modern shells.

that are far lower than any values we recorded or predicted for modern shells from southern California (*38*), suggesting a possible source far south of California, perhaps in the Gulf of California, or brackish waters such as in the San Francisco Bay. We are unaware of *Olivella* shells growing in the bay, but our rough calculations suggest that if they did, the lower salinity values in the bay would lead to $\delta^{18}O$ similar to those found on these anomalous beads. However, highly unusual $\delta^{13}C$ values for these shells support the notion they do not come from southern (or northern) California.

The remaining 13 beads (31%) display signatures that are more ambiguous with regard to source. These shells show neither unambiguously northern or southern California isotopic values (i.e., all $\delta^{18}O$ readings are between 0.6 and 1.7). These beads fall in the middle of Figure 8 where the two regional ellipses overlap.

This is a result we found difficult to accept. We would have expected at least *some* of these beads to return oxygen isotopic signatures consistent with the colder waters of northern California. There is some tendency for bead types that we strongly suspect were produced in northern California (F2/F3 and M1/M2 types), to have more ambiguous signatures rather than unambiguously southern California ones. Such a pattern suggests that these beads have slightly lower $\delta^{18}O$ values (indeed, average $\delta^{18}O$ values by type shows this to be true). However, this pattern is far from strong. Thus, while 11 of 29 (38%) F2/F3 and M1/M2 beads have ambiguous signatures, only 2 of 13 (15%) G2 and H1 beads do. Table II shows these results.

Table II. Results of Oxygen Isotope Sourcing for Historic and Prehistoric Delta Beads Based on Modern Shell Data

Type/Source	*F2/F3*	*M1/M2*	*G2*	*H1*	*Total*
SoCal Signature	12	5	5	4	26
Other warm water	0	1	2	0	3
NorCal Signature	0	0	0	0	0
Ambiguous	6	5	2	0	13
Total	18	11	9	4	42

Although our analyses are still in progress, two possible explanations for the isotopic data seem likely at this point. The first is that the majority, if not all, of these beads were in fact produced from shells collected in southern California. If true, this would imply that F2/F3 and M1/M2 beads, which have a geographical distribution limited to central and northern California, were either produced by southern California bead makers exclusively for northern

California consumption, or made by central/northern California bead makers who traded for or directly procured unmodified whole *Olivella* shells from southern California.

Neither of these options is particularly attractive. First, if they were really produced in southern California, we would expect at least some blanks, or broken or lost F2/F3 and M1/M2 beads to show up in archaeological sites. To our knowledge, this is not the case (see also King 1990:32). Second, if they are southern California shells that were reduced into beads in central and northern California, we would expect to see greater evidence for *Olivella* bead manufacturing in archaeological sites in this region. Though some production is evident (e.g., *34*), the scale of such production does not appear to be large enough to account for the tens of thousands of F2/F3 and M1/M2 beads found in archaeological sites.

The second possibility is that prehistoric oxygen isotope signatures for different regions of the coast differ from modern values. Specifically, if water temperatures in northern California had been about 2–3° C warmer than at present when the prehistoric beads were produced (ca. 500–2000 cal B.P.), the northern California ellipse in Figure 8 would be shifted up by approximately 0.4–0.6 ‰ (an increase of ~0.2‰ in $\delta^{18}O$ equates to ~1C reduction in temperature and vice versa). As a result, we would obtain the "ambiguous" isotopic data discussed above even though the beads were produced in northern California.

There are conflicting data on whether northern California sea surface temperatures (SST) were consistently 2–3° C colder from 500–1600 cal B.P. This spans the period when most of our F, M, and G beads were made. Some studies suggest a 1–2 degree SST increase between 700–1300 cal B.P., with a decrease from modern between 500 and 700 cal B.P. (*66*). Such a change could explain a large fraction of the seemingly anomalous data we have collected from our beads. On the other hand, some researchers suggest approximately 1 degree cooler SST, on average, between 700 and 2000 cal B.P., with a slight *increase* between 500 and 700 cal B.P. (*67*). These studies are in slightly different areas, the former based on extreme northern California and near San Francisco Bay, the latter along the Big Sur coast. Information from southern California indicates temperatures there were approximately 1–3 degrees colder between 500 and 1500 cal B.P. (*37*). Yet as discussed earlier, southern California sea surface temperatures are influenced by a different set of oceanic currents than northern California and it is far from clear that the two are necessarily correlated. In any case, there are no clear patterns evident for all of northern California, and indeed, SST may have varied over time and space.

The preceding discussion is an issue we will address in greater detail in the near future. For example, we are in the process of sampling additional prehistoric whole *Olivella* shells from coastal sites in northern California for oxygen and carbon isotopes. At present, we have data on two whole *Olivella* shells collected from archaeological sites near Kirk Creek in northern California. These shells were AMS dated to 1180 and 1755 cal B.P. Both shells display isotopic data wholly consistent with modern samples from that region (e.g., $\delta^{18}O$ readings greater than 1.6), indicating no change in SST. As part of our study, we plan to radiocarbon date the shells to develop a more detailed prehistoric record of sea surface temperature change along the northern California coast over the last 2000 years. Such a program should give us more confidence in sourcing ancient *Olivella* beads, allowing us to extend the time depth over which we can apply this technique.

Discussion

We have applied two distinct techniques, analysis of bulk chemical composition and the ratios of oxygen and carbon stable isotopes, in an effort to determine source provenance of *Olivella* shell beads in Western North America. These techniques produce varying degrees of success.

In our experience, bulk chemical composition holds only minor promise for sourcing specimens where source location is completely unknown. This is due to several complicating factors. Unfortunately, the concentration of most elements in a shell is quite low, limiting the number of variables available for discriminating between possible source zones. For many elements there seems to be high inter-shell variability collected from the same region, suggesting that biological and environmental factors are different for individual snails and have strong influences on shell chemistry. Furthermore, the chemical composition of a shell is quite dynamic following deposition as the aragonite interacts with the soil. Many elements become enriched in prehistoric shells, either due to replacement of existing minerals within the shell, or due to leaching out of other elements (thereby increasing relative concentration of existing ones). Although it may be possible to sample the interior of shells to minimize the effects of such post-depositional alteration and/or only analyze for elements that are more stable within the shell, it will be difficult to control for factors such as depositional environment and the amount of time a shell has been subject to leaching.

Furthermore, even among modern shell specimens there are only a few elements that seem to systematically characterize macro-regions of the California coast. Due to leaching and other issues discussed above, these patterns are less pronounced for prehistoric specimens (though still slightly evident). Thus, our hope that a single or small number of elements would unambiguously allow us to source shells to particular sections of the coast did not prove to be true. Together, these complications make it difficult for us to both define "source zones" based on bulk shell chemistry as well as to assign unknowns to any potential source zones.

Yet we are not without hope for future applications of bulk chemical analyses. More thorough chemical cleaning of the shell to remove inorganic components may reduce some of the variability among shells collected from a region, allowing us to better define potential source zones and assign unknown beads to these source zones.

Moreover, if a general source location is already known or suspected for a shell bead, bulk composition analysis may hold the potential to discriminate between micro-scale environments. For example, shells growing near certain rivers with peculiar geological watersheds within Southern California may have unique chemical composition relative to other rivers within that macro-region. If a bead was then suspected of having been produced in Southern California, bulk chemical composition might rule out or support production near that particular river. Our sampling strategy did not allow us to test this notion, but it is something we hope to investigate in the future.

At present, we believe oxygen and carbon isotopes hold greater promise for sourcing *Olivella* beads. We believe it is now possible to accurately determine the geographic origin of recent (latest prehistoric and historic) beads. We are working to extend this technique to more ancient contexts. Ultimately, such an approach would require knowing the approximate age of a bead before being able to infer geographic source based on the isotopic profile. This is not a problem for *Olivella* beads in California where a detailed chronology has been established (e.g., *12*), but may be more problematical for other bead and pendant types that are poorly dated.

Although the source zones established by stable isotope analysis are quite large, effectively discriminating only between regions north vs. south of Point Conception, such information is still of great anthropological significance in California archaeology. Much has been made archaeologically of the importance of prehistoric bead production, particularly on the Channel Islands off the southern California coast. Bead producers on these islands are thought to have supplied most or all of California with *Olivella* beads. The use of oxygen and carbon isotopes facilitates the testing of this hypothesis. While we are still hesitant to assign the majority of the Delta beads we analyzed to a southern California source, the available evidence clearly does not rule out such a scenario. In future work, we will subject a sample of these beads to ICP-MS analysis in hopes of generating additional data that might inform on geographic provenance.

Conclusions

It is clear that shell-bead sourcing using chemical means is still in its infancy. If shell-bead sourcing is to play a more important role in archaeological studies we will have to better identify criterion that will discriminate among shellfish species growing in different areas. A major hurdle in this respect is the dynamic nature of oceanic environments from which many shellfish species draw their chemical makeup, both temporally and spatially. Oceanic currents distribute the same body of seawater over large areas, exposing shellfish to many of the same conditions. Furthermore, seasonal, inter-annual, and even millennial variations in seawater conditions introduce variability that reduces the geographic accuracy of our source assignments. Moreover, shellfish are dynamic organisms which alter their chemistry depending on local conditions, unlike obsidian and ceramics. These factors still make shell bead sourcing a difficult undertaking.

Combining both bulk elemental composition with C and O stable isotope analysis would provide some degree of cross-checking on shell bead analyses. This is something we will pursue in future research. Indeed, in the future, we expect archaeologists to turn to other techniques to generate complementary data for sourcing shell beads. For example, we believe a useful approach will be the analysis of isotopic ratios of other elements, such as strontium, neodymium, and lead, among others. As our analytical techniques become more accurate and precise, it will be easier to analyze the ratios of rare isotopes that could prove useful in provenance analysis.

Acknowledgements

Funding for this project was provided by the Wenner-Gren Foundation for Anthropological Research (grant #7370) and the National Science Foundation (#0504615). We thank Iain Freckleton, Rowan Gard, Pamela Reynolds, Michelle Gras, and Dave Winter for assistance with sample preparation and mass spectrometry. We further thank Doug Kennett, Michael Glassow, Randy Groza, Linda Hylkema, Mark Hylkema, Randy Milliken, Dustin McKenzie, Allika Ruby, Greg White, and Randy Wiberg for providing us access to shells and beads.

References

1. Classen, C.; Sigmann, S. *Am. Antiq.* **1993**, *58*, 333–347.
2. Vanhaeren, M.; d'Errico F.; Billy, I.; Grousset, F. *J. Archaeol. Sci.* **2004**, *31*, 1481–1488.
3. Shackleton, N.; Renfrew, C. *Nature (London)*, **1970**, *228*, 1062–1065.
4. Kozuch, L. *Am. Antiq.* **2002**, *67*, 697–709.
5. McGuire, R. H. ; Howard, A. V. *Kiva*, **1987**, *52*, 113–146.

6. Nelson, R. S. *Hohokam Marine Shell Exchange and Artifacts*; Arizona State Museum Archaeological Series 179, Arizona State Museum, Tuscon, AZ, 1991.
7. Basgall, M. E.; Hall, M. C. *Archaeology of the Awl Site (CA-SBR-4562), Fort Irwin, San Bernardino County, California*; Far Western Anthropological Research Group: Davis, CA, 1993.
8. Basgall, M. E.; Hall, M. C. *Archaeological Investigations at Goldstone (CA-SBR-2348): A Middle Holocene Occupation Complex in the North Central Mojave Desert*; Far Western Anthropological Research Group: Davis, CA, 1994
9. Erlandson, J. M.; Macko, M. E.; Koerper, H. C.; Southon, J. *J. Archaeol. Sci.* **2005**, *32*, 393–398.
10. Fitzgerald, R. T.; Jones, T. L.; Schroth, A. *J. Archaeol. Sci.* **2005**, *32*, 423–434.
11. Schroth, A. B. Ph.D. thesis, University of California at Riverside, Riverside, CA, 1994.
12. Bennyhoff, J. A.; Hughes, R. E. *Am. Mus. Nat. Hist. Anthro. Papers* **1987**, *64*, 80–175.
13. King, C. D. *Evolution of Chumash Society: A Comparative Study of Artifacts Used for Social System Maintenance in the Santa Barbara Channel Region Before A.D. 1804*; Garland Publishing: New York, NY, 1990.
14. Meyer, J.; Rosenthal, J. S. *An Archaeological Investigation of Artifacts and Human Remains from CA-CCO-637, Los Vaqueros Project Area, Contra Costa County, California*; Anthropological Studies Center, Sonoma State University: Rohnert Park, CA, 1998.
15. Vellanoweth, L. R. *J. Archaeol. Sci.* **2001**, *28*, 941–950.
16. Arnold, J. E. *Craft Specialization in the Prehistoric Channel Islands*; University of California Press: Berkeley, CA, 1987.
17. Arnold, J. E. *Antiquity* **1991**, *65*, 953–962.
18. Arnold, J. E. *Am. Antiq.* **1992**, *57*, 60–84.
19. Arnold, J. E., Colten, R. H., Pletka, S. *Am. Antiq.* **1997**, *62*, 300–318.
20. Arnold, J. E., Munns, A. *J. Field Archaeol.* **1994**, *21*, 473–489.
21. Chagnon, N. A. *University of California Archaeological Survey Annual Report*, **1970**, *12*, 1–25.
22. Hughes, R. E. In *Prehistoric Exchange Systems in North America*; Baugh T. G.; Ericson, J. E., Eds.; Plenum Press: New York, NY, 1994; pp 363–383.
23. Jackson, T. L.; Ericson, In *Prehistoric Exchange Systems in North America*; Baugh T. G.; Ericson, J. E., Eds.; Plenum Press: New York, NY, 1994; pp 385–415.
24. Vayda, A. P. In *Tribal and Peasant Economies*; G. Dalton, Ed; Natural History Press: Garden City, NY, 1967; pp 494–500.
25. Arnold, J. E. *Am. Anthro.* **1995**, *97*, 733–747.

26. Mikkelsen, P.; Hildebrandt, W. R.; Jones, D.; Rosenthal, J. S.; Gibson, R. *Thirty Years After: Reanalysis of Data from the 1974 Excavations at Kirk Creek, CA-MNT-238, on the Big Sur Coast*; Far Western Anthropological Research Group: Davis, CA, 2004.
27. Jones, T. L. *Prehistoric Human Ecology of the Big Sur Coast*; Contributions of the University of California Archaeological Research Facility 61, University of California Press: Berkeley, CA, 2003.
28. White, G. *A Report of Archaeological Investigations at Eleven Native American Coastal Sites, MacKerricher State Park, Mendocino County, California*; Northwest Information Center, Sonoma State University: Rohnert Park, CA, 1989.
29. Cartier, R. *The Saunders Site (CA-Mnt-391): A Littoral Site of the Early Period*; Scotts Valley Historical Society Press: Scotts Valley, CA, 1993.
30. Dietz, S.A. *Archaeological Test Excavations: CA-MNT-101, CA-MNT-298, CA-MNT-929, and El Castillo at the Presidio and City of Monterey, Monterey County, California*; Northwest Information Center: Rohnert Park, CA, 1987.
31. Dietz, S. A.; Hildebrandt, W., Jones, T. *Archaeological Investigations at Elkhorn Slough: CA-MNT-229, a Middle Period Site on the Central California Coast*; Papers in Northern California Anthropology 3, University of California: Berkeley, CA, 1988.
32. Milliken, R. T.; Hildebrandt, W. R.; Mikkelsen, P. *The Moss Landing Hill Site: A Technical Report on Archaeological Studies at CA-MNT-234*; Far Western Anthropological Research Group: Davis, CA, 1999.
33. Kennedy, M. Ph.D. thesis, University of California at Davis, Davis, CA, 2004.
34. Hartzell, L. L. *J. Calif. Great Basin Anthro.* **1991**, *13*, 29–39.
35. Milliken, R. T.; Bennyhoff, J. A. In *There Grows a Green Tree: Papers in Honor of David A. Fredrickson*; White, G.; Mikkelsen, P.; Hildebrandt, W. R.; Basgall, M. E., Eds.; *Center for Archaeological Research at Davis*: Davis, 1993; pp 381–395.
36. Arnold, J. E.; Graesch, A.P. In *The Origins of a Pacific Coast Chiefdom, The Chumash of the Channel*; Arnold, J. E., Ed.; University of Utah Press: Salt Lake City, UT, 2001; pp 71–112.
37. Kennett, D. J. *The Island Chumash: Behavioral Ecology of a Maritime Society*; University of California Press: Berkeley, CA, 2005.
38. Eerkens, J. W.; Herbert, G. S.; Rosenthal, J. S.; Spero, H. J. *J. Archaeol. Sci.* **2005**, *32*, 1501–1514.
39. Bentov, S.; Erez, J. *Geology* **2005**, *33*, 841–844.
40. Lea, D. W.; Pak, D. K.; Spero, H .J. *Science* **2000**, *289*, 1719–1724.
41. Lorrain, A.; Gillikin, D. P.; Paulet, Y.-M.; Chauvaud, L.; Le Mercier, A.; Navez, J.; André, L. *Geology* **2005**, *33*, 965–968.

42. Ramalingam, R.; Saraswati, P. K.; Rogers, K.; Iwao, K. *Marine Micropaleontology* **2005**, *58*, 31–44.
43. Richardson, C. A.; Saurel, C.; Barroso, C. M.; Thain, J. E. *J. Experimental Marine Biol. Ecol.* **2005**, *325*, 55–64.
44. Ries, J. B. *Geochim. Cosmochim. Acta* **2006**, *70*, 891–900.
45. Bemis, B. E.; Spero, H. J.; Bijma, J.; Lea, D. W. *Paleoceanography* **1998**, *13*, 150–160.
46. Böhm, F.; Joachimski, M. M.; Dullo, W. C.; Eisenhauer, A.; Lehnert, H.; Reitner, J.; Wörheide, G. *Geochim. Cosmochim. Acta* **2000**, *64*, 1695–1703.
47. Kirby, M. X.; Spero, H. J.; Soniat, T. M. *Palaios* **1998**, *13*, 560–569.
48. Krantz, D. E.; Williams, F. D.; Jones, D. S. *Palaeogeography, Palaeoclimatology, and Palaeoecology* **1987**, *58*, 249–266
49. Spero, H. J.; Lea, D. W. *Marine Micropaleontology* **1996**, *28*, 231–246.
50. Wefer, G.; Berger, W. H. *Marine Geology* **1991**, *100*, 207–248.
51. Kincaid, E.; Thunell, R. C.; Le, J.; Lange, C. B.; Weinheimer, A. L.; Reid, F. H. H. *Deep-Sea Research II* **2000**, *47*, 1157–1176.
52. DeNiro, M. J. *American Scientist* **1987**, *75*, 182–191.
53. Geary, D. H.; Brieske, T. A.; Bemis, B. E. *Palaios* **1992**, *7*, 77–85.
54. Klein, R. T.; Lohmann, K. C.; Thayer, C. W. *Geochim. Cosmochim. Acta* **1996**, *60*, 4207–4221.
55. McConnaughey, T.; Burdett, J.; Whelan, J. F.; Paull, C. K. *Geochim. Cosmochim. Acta* **1997**, *61*, 611–622.
56. Romanek, C. S.; Grossman, E. L.; Morse, J. W. *Geochim. Cosmochim. Acta* **1992**, *56*, 419–430.
57. Wefer, G.; Killingley, J. S. *Marine Biology* **1980**, *60*, 129–135.
58. Carpenter, S. J.; Lohmann, K. C. *Geochim. Cosmochim. Acta* **1995**, *59*, 3749–3764.
59. Fairbanks, R. G.; Charles, C. D.; Wright, J. D. In *Radiocarbon After Four Decades*; Taylor, R. E., Ed.; Springer Verlag: New York, NY, 1992; pp 473–500.
60. Friederich, G. E.; Walz, P. M.; Burczynski, M. G.; Chavez, F. P. *Progress In Oceanography* **2002**, *54*, 185–203.
61. Korchagin, N. N.; Lozovatsky, I. D. *Okeanologiya* **1998**, *38*, 335–342.
62. Grossman, E. L.; Ku, T. L. *Chem. Geol.* 1986, 59, 59–74.
63. Herbert, G. S. *PaleoBios* **2001**, *21*, 65.
64. Herbert, G. S. *Geological Society of America, Abstracts with Program*, **2003**, *34*, 417.
65. Groza, R. G.; M.A. Thesis, San Francisco State University, San Francisco, CA, 2002.
66. Barron, J. A.; Heusser, L.; Herbert, T.; Lyle, M. *Paleoceanography* **2003**, 1020.
67. Jones, T. L.; Kennett, D. J. *Quat. Res.* **1999**, *51*, 74–82.

Chapter 10

Archaeological Soils and Sediments: Application of Microfocus Synchrotron X-ray Scattering, Diffraction, and Fluorescence Analyses in Thin-Section

W. Paul Adderley[1], Ian A. Simpson[1], Raymond Barrett[2], and Timothy J. Wess[3]

[1]School of Biological and Environmental Sciences, University of Stirling, Stirling FK9 4LA, Scotland, United Kingdom
[2]European Synchrotron Radiation Facility, BP220, 38043 Grenoble, Grenoble Cedex 9, France
[3]Biomaterials Centre, School of Optometry and Vision Sciences, University of Cardiff, King Edward VII Avenue, Cardiff CB10 3NB, Wales, United Kingdom

Archaeological soils and sediments reflect the cultural environment in which they have been formed. Their analysis allows assessment of the nature and intensity of past events. With the results of such analyses playing an increasing role in forming archaeological interpretations, there is a need to verify optical analysis and interpretation of materials and to examine materials that are presently considered amorphous or unknown in conventional optical analyses. This paper discusses the use of microfocus sychrotron X-ray methods and the issues surrounding their application to archaeological soils and sediments.

Soils and sediments are an integral part of archaeological sites. Materials range from soils found in field areas that have been managed in past times to the burial matrix of archaeological structures. Since such archaeological soils and sediments reflect the cultural environment in which they have formed, they can themselves be considered cultural artifacts. Examining these materials can be the key to understanding past agricultural and archaeological site activities thereby allowing interpretation of both the nature and intensity of past events.

The study of soils and sediments has been developed using a variety of methodologies. Most common are procedures derived from geochemical prospecting methods: the elemental characterization of bulk sediment samples (*1, 2*) to geostatistical treatments of the resultant datasets (*3*). These methods, applicable at both site- and regional-scales (*4*), provide data that can be spatially related to large archaeological features. This has led to developments in relating chemical signatures to specific past activities (*5*) in a variety of contexts, particularly the use of elements considered relatively immobile in soils and sediments (*6*). The size of the samples used in these methods relative to the scale of the study can limit extension of these methods to more detailed contexts since, in such instances, analysis of bioarchaeological – sometimes called "ecofacts" - (e.g. bone, wood) and artifactual (e.g. metals, slags, clinker) materials found in these soils and sediments is commonly confounded. At smaller scales, more refined signatures are generally required. There is also an increasing need to understand the nature of the soils and sediment matrix, in terms of spatial organization and of the preservation of the bioarchaeological and artifactual components.

Analyses of microscale features of archaeological soils and sediments through optical thin-section micromorphology can achieve these more refined goals. The scale of analysis and the use of undisturbed samples allows the optical properties of bioarchaeological materials and artifactual materials to be characterized separately at a range of scales from ~5 to 1000 μm. Furthermore, the spatial relationships between these components can be examined, thereby increasing the interpretative value of the methodology. Thin-section micromorphology has become an established method in archaeological studies and has been used in varied on-site contexts including: mixed materials in middens and during site formation (*7-9*); and occupation surfaces (*10, 11*). Past agricultural development surrounding archaeological sites has similarly been examined (*12-15*).

Sample preparation requires undisturbed sampling of the soil or sediment, removal of water in the sample through acetone exchange, resin impregnation with polyester resin, and finally cutting, mounting, and polishing a thin-section on a glass slide. Optical examination is then performed using a polarizing microscope with descriptions and semi-quantitative frequency analyses following standardized international systems (*16, 17*). Such studies sometimes use complementary bulk chemical (*18*) and physical analyses (*8*) to confirm or augment initial optical microscope results. Such validation is essential, especially as the subject area of archaeological soil micromorphology develops, with the consequent expectation of increased interpretative value sought from the methodology.

The logical extension of combining optical observation with chemical and physical measurements is to develop new understandings of the type and nature of the bioarchaeological and artifactual materials in the sediment matrix, extending from detailed chemical characterizations to process-based concepts of material alteration. Knowledge of the use of archaeological materials examined can be greatly enhanced by understanding the processes involved in their formation or degradation. Experimental archaeology methods, where material sources are examined for comparison with archaeological features, can be employed. For example, experimental combustion and examination of the resultant ash and fuel residues has led to major advances in understanding changes in past resource utilization (*7*). This approach requires knowledge of the likely source of the materials studied and of the processes involved in their production. Where this knowledge is absent and materials seen in thin section are of unknown or speculative origin, or, alternatively, where the conditions required to generate the material are not practicable to replicate in a controlled experiment, alternative prospecting methodologies are required.

High-resolution compositional measurements are possible through use of a variety of microanalytical methods. Ideally, these should be non-destructive, can be targeted on small areas of sample, and have low minimum detection limits. Electron-probe X-ray microanalysis (EPXMA) and proton-induced X-ray emission (PIXE) techniques have both been used successfully on archaeological sediment thin sections (*19, 20*). Both techniques yield elemental composition data for a range of elements. EPXMA has the advantage of being non-destructive, whereas PIXE when used on thin-section samples is typically destructive; conversely the detection limit for PIXE is lower than EPXMA.

Several other microanalytical methods in common use potentially have application on soil and sediments section samples. Laser-ablation inductively coupled plasma mass spectrometery (LA-ICP-MS) has been used on soil thin-sections from a controlled field experiment (*21*) but required special resins in the preparation. There is presently (May 2006) no reported use of this method on archaeological soil samples. Likewise, for extremely fine-resolution studies (i.e. <10 μm) with low minimum detection limits and despite difficult calibration, secondary ion microscopy (SIMS) has a potential role in examining archaeological soil thin sections. At even higher lateral resolutions (~100 nm) Auger electron spectroscopy (AES) could also be considered for surface (<5 nm deep) analyses. At present however, the use of these methods in soil systems is limited. SIMS has been focused on biochemical applications (*22*), whereas AES has typically been used to examine metalliferous interfaces (*23*) rather than sediments. LA-ICP-MS, SIMS and AES can all be considered destructive if soil thin sections are being examined.

Each of these methods offers contrasting benefits and weaknesses such as good analytical sensitivity but only over a small number of elements (PIXE), or a wide number of elements but at the expense of large beam sizes preventing accurate targeting (EPXMA). These methods can offer, at most, only limited structural information of the material examined. Since the aim of examining in microscopic detail archaeological soil and sediment thin-sections is to gain an understanding of the nature of both their bioarchaeological and artifactual components, the analytical ideal for examining these would therefore be a method, or combination of methods, offering both structural as well as elemental composition data. X-ray based methods offer this possibility. This chapter outlines three X-ray microanalysis methods (X-ray scattering, diffraction and fluorescence) and discusses issues surrounding their use on archaeological soils and sediments.

Microfocus X-ray Methods of Analysis of Archaeological Thin Sections

X-ray scattering, X-ray diffraction (XRD) and X-ray fluorescence (XRF) techniques are powerful methods of analysis allowing characterization of a large number of material properties at a variety of scales of organization in the material examined. XRD and XRF techniques using conventional X-ray tube sources have become prominent in both academic research and for industrial process control, whereas X-ray scattering is less often utilized. For the three methods to be of use in examining microscale features within thin-sections of archaeological sediment samples there is a need for small areas in the order of a few µm on the sample to be targeted. For examining such small sample targets,the twin requirements are a means of focusing X-rays and sufficient X-ray photons to achieve this usefully. The advent of X-ray focusing methods such as polycapillary optics (*24*) or compound refractive lenses (*25*), has allowed focusing of X-ray beams to micron and sub-micron levels. Such advances, coupled with the use of third-generation synchrotron radiation sources producing high brilliance X-rays, allows the realization of rapid microfocus X-ray methods of analysis.

Microfocus Small-angle X-ray Scattering of Archaeological Sediments

X-ray scattering occurs when an incident X-ray beam interacts at low angles of incidence with the sample target and where there are contrasts in electron density between particles of the material at the nano, i.e. multi-atomic, scale. Given the low angle of the incident X-ray, this method is commonly called small-angle X-ray scattering (SAXS). SAXS (Figure 1a) differs from XRD (Figure 1b) in two ways: in the latter, first the angle of incidence of the X-ray beam relative to emergent X-ray from the sample target is relatively large, and second, the interaction is with the atomic lattice. Since the atomic lattice is typically very regular compared to the contrasts found in electron density at the larger, nano-scale, XRD peaks tend to be sharp and clear, compared to more diffuse or irregular SAXS scatter patterns. These SAXS results, however, can be used to measure the size and characterize the shape and the orientation of materials at nano-scales (*26*). Given that many biological materials show structural hierarchy between scales this method allows biomineralization processes to be studied for a wide range of materials including egg shells (*27*), sea sponges (*28*) and most notably bone (*29*). The typical SAXS apparatus arrangement is shown schematically in Figure 1a. Here a microfocus X-ray beam, through an arrangement of X-ray optics, reaches the sample and is scattered at very small angles; the flat-field energy dispersive detector is positioned at a relatively long distance away from the sample. Depending on the X-ray source and the optics used, the incident wavelength and the sample target spot size can be optimized to allow analyses on a range of materials. By moving the sample relative to the incident beam, sets of spot measurements over meshed areas or along transects on the sample can be examined.

Bone is an obvious bioarchaeological object of study. Key questions surround the nature of the bone material itself regarding its preservation and destruction and the information that it can reveal about the past. Microfocus SAXS studies on thin-sections of bone have shown intra-bone variation in preservation and degradation linked to microbial attack at specific focal points in the bone material (*30, 31*). In archaeological soils and sediments bone is found at various stages of decomposition, its relative preservation depending upon the chemical and microbial nature of the burial environment (Figure 2). In acidic archaeological sediment conditions, calcium-iron-phosphate material infilling the soil matrix is sometimes found (Figure 3). As described in detail elsewhere (*32*), several hypotheses have been made to explain the formation of these materials, but their amorphous nature has limited previous attempts at their analysis. One hypothesis is that they are derived from bone material and, therefore, represent an important bioarchaeological indicator of past site use.

From Vollen in Langenesværet in the Norwegian Vesterålen islands, samples of calcium-iron-phosphate material (Figure 3) were found at depth in an excavated midden at an early fishing site in northern Norway (*20*). Examining these materials in thin-section and reference samples of modern fish-bone materials using SAXS, has allowed the analysis and comparison of their nano-scale properties including size and crystal form (Figure 4). These results show strong similarities between the calcium-iron-phosphate features and modern fish

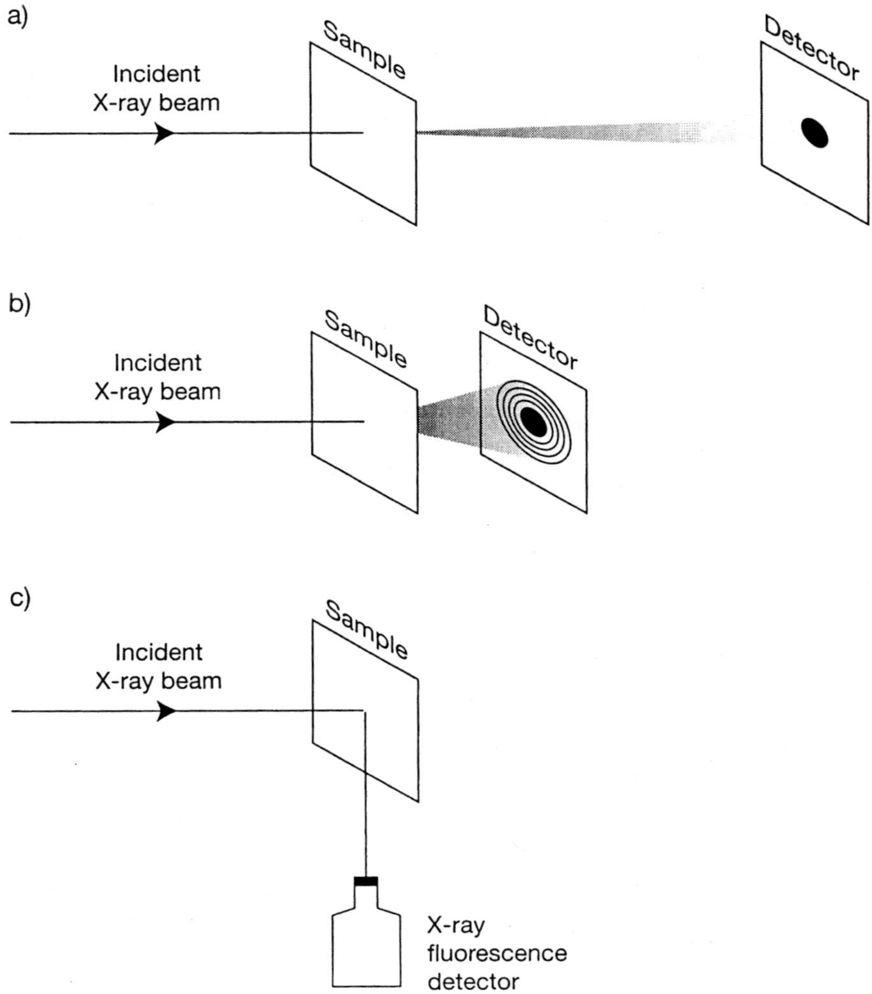

Figure 1. Schematic illustrations of microfocus synchrotron X-ray analysis methods (a) small-angle X-ray scattering (SAXS); (b) microfocus X-ray diffraction (XRD); and (c) microfocus X-ray fluorescence (XRF).

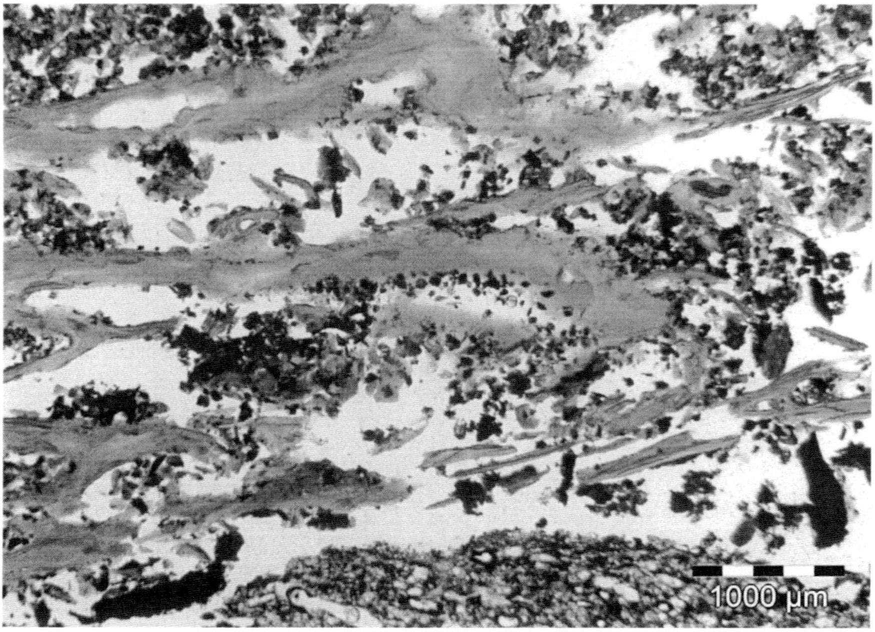

Figure 2. Bone materials in archaeological site contexts. Fractured, iron-stained bone sample from Room D4a, Kongshavn multiroom house complex, Finnmark, northern Norway (transmitted plane polarized illumination).

bone, with their crystallite sizes corresponding (*32*). It is also found that the size and shape of crystallites in the control samples differ with heating pre-treatments, indicating the importance of the collagen matrix in regulating these changes both *in vivo* and *post mortem*. In addition to building knowledge of the processes interacting with buried bioarchaeological materials, this method offers a route towards analyzing optically amorphous materials, and therefore poorly characterized, and hence increasing the interpretative value of these archaeological features.

Microfocus X-ray Fluorescence Analyses and Microfocus X-ray Diffraction of Archaeological Sediments

X-ray fluorescence methods are, in general, ideally suited to prospecting archaeological materials since a wide range of elements at reasonable detection limits can be measured simultaneously and non-destructively. When applied to

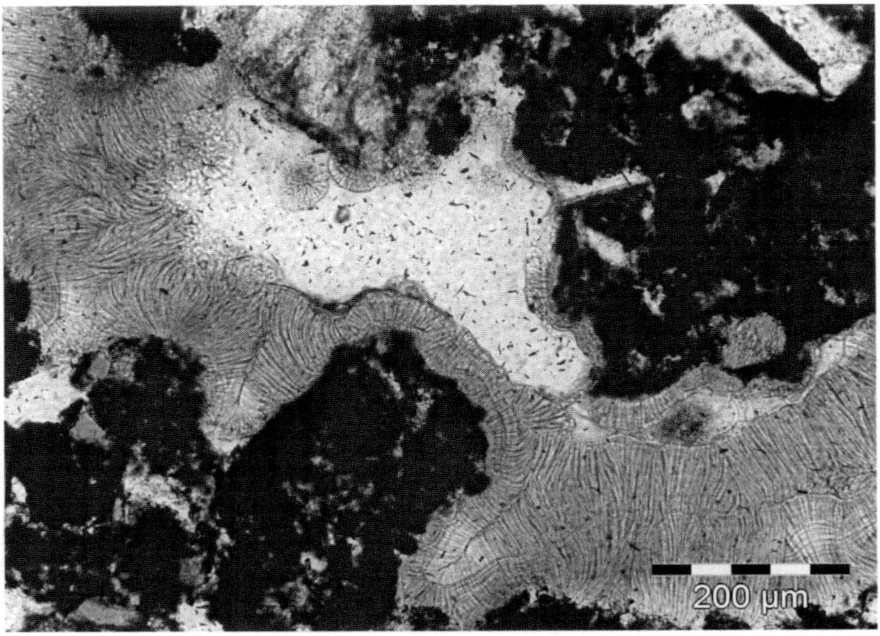

Figure 3. Calcium-iron-phosphate features in thin-section. Ssampled from the acidic portion of a midden profile at Vollen, Langenesværet, Norway (transmitted plane polarized illumination) .

soils, XRF has, typically, been conventionally applied on bulk samples either homogenized in the laboratory or measured *in situ* with portable apparatus. Microfocus synchrotron XRF methods (Figure 1c) have recently become more established in part due to the developments in X-ray optics and synchrotron sources, but also because of the introduction of semiconductor-based detectors; these allow the fluorescent X-ray to be examined at lower intensities thereby usefully lowering detection limits. In the examination of archaeological soil and sediment thin-sections the method therefore offers the possibility of examining not only discrete points of interest on the thin section but also interfaces between artifacts and bioarchaeological materials, and between artifacts or bioarchaeological materials with the surrounding matrix.

In a similar manner to SAXS, microfocus X-ray fluorescence analysis can be targeted at small spot sizes and this can be linked to visual analysis of the archaeological thin-section. Figure 5 shows an example from the Kongshavn archaeological site in Finnmark, Norway (*33*). Here maps of elemental distributions (Figure 5b, 5c) can be linked visually to the assemblage materials

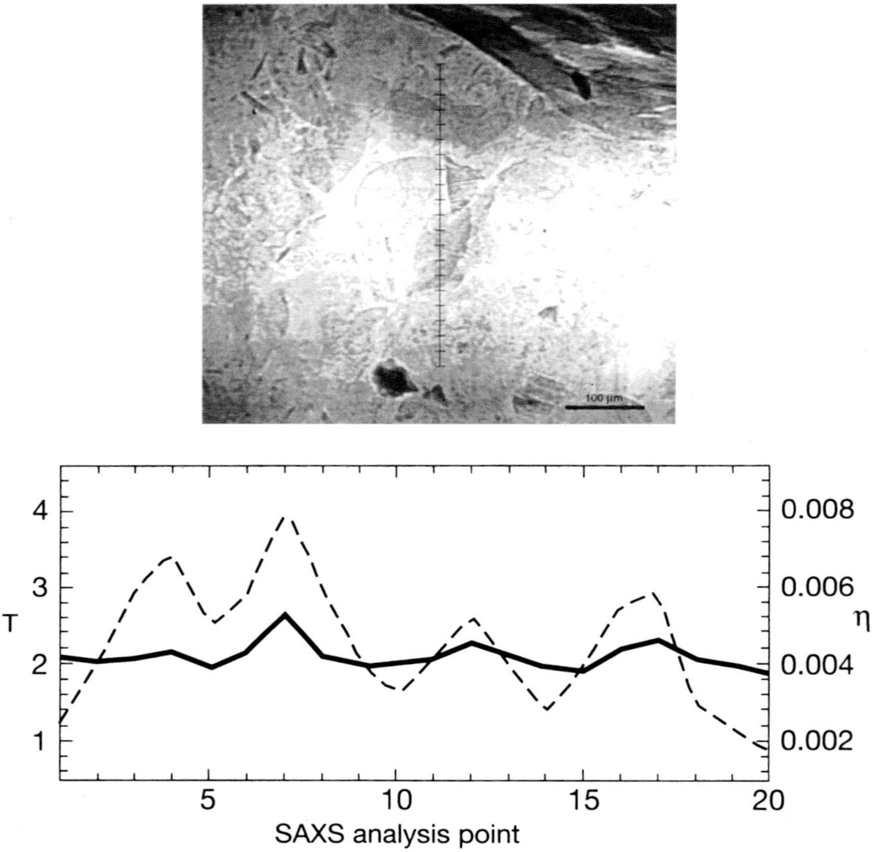

Figure 4. Microfocus synchrotron SAXS measurements: Thickness T (nm) and Shape (η) parameters of calcium-iron-phosphate features from the midden profile at Vollen, Langenesværet, Norway. (Reproduced from Reference 32. Copyright 2004 Elsevier)

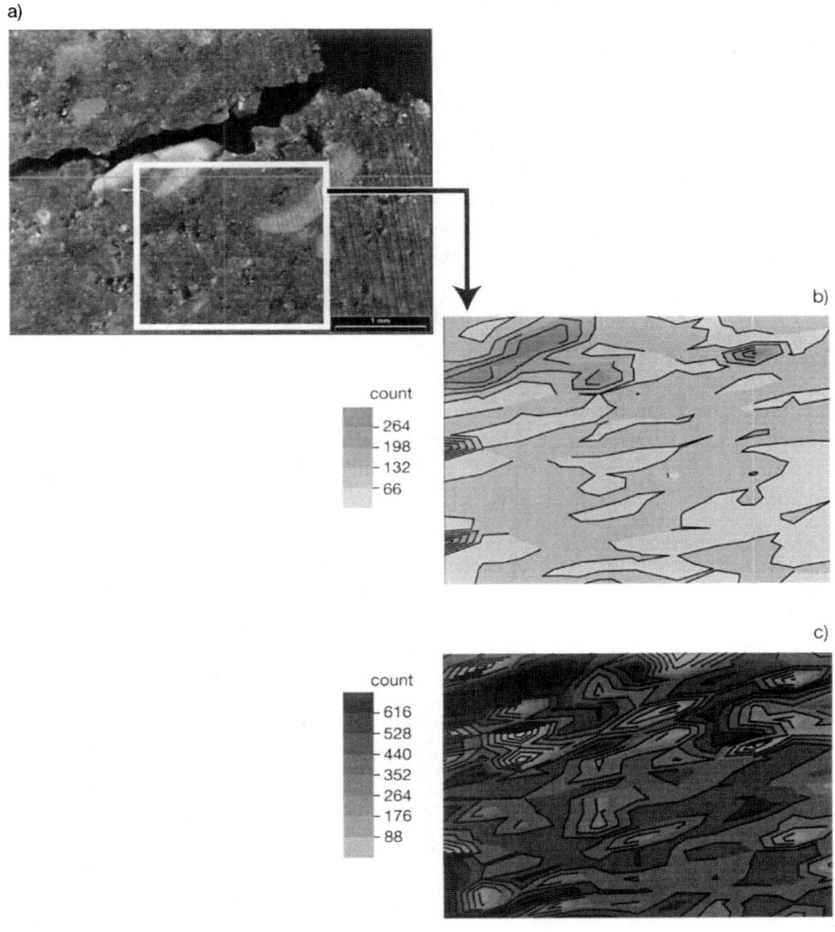

Figure 5. Microfocus XRF analysis of thin-section from Room D4a, Kongshavn multiroom house complex, Finnmark, northern Norway: a) thin-section (200 μm thickness; reflected light illumination); b) calcium (K_α counts x 100); c) iron (K_α counts x 100).

seen in thin-section (Figure 5a), in this example, there is an apparent iron-rich coating on fish bone. Increasing the understanding of the ongoing processes in the burial environment enables interpretation of previously intractable contexts.

Quantitative calibration of microfocus XRF requires both a theoretical and empirical approach to reduce the effects of the measurement environment which include background, sample geometry, and source energy. The approach adopted depends on not only the nature of the sample, but also on the available standards including the range of elements and the support matrix (*34*). When quantitative methods are applied to soil samples, several problems emerge: first, there can be a pronounced inter-sample variation in sample thickness and surface roughness causing geometrical aberrations in measurement, and second, that sample heterogeneity through the vertical thickness of the material may lead to strong self-absorption effects.

Thus, one approach to archaeological soil and sediment analysis is to use the detector counts and compare ratios of what are understood or assumed to be key indicator elements. Another approach is to combine the microfocus XRF analyses with another measurement. This is possible with either typical SAXS or with a microfocus XRD apparatus (Figure 1). Microfocus synchrotron XRD has several common elements with SAXS methodology. Figure 1b shows the schematic experimental arrangement for microfocus synchrotron XRD; whereas the SAXS scatter pattern is examined at very small angles, with XRD the angle is much greater requiring the detector to be positioned close to the sample. Simultaneous microfocus XRD and SAXS measurements are, therefore, not readily realizable. When the method is compared to conventional mineralogical powder diffraction, more commonly used in soils research, there are three key differences. First, the configuration of a microfocus X-ray diffraction with a flat field detector will detect a smaller number of Debye-Scherrer diffraction rings when compared to a conventional moving-goniometer arrangement. Second, when compared to homogenized powders, it is obvious that the soil thin sections will exhibit greater variation. Because of the small spot size examined this effect is extremely dependent on the crystallite sizes of the materials examined. This may be observed in the resultant diffraction spectra as occasional intense peaks (Figure 6) forming a vague ring rather than the complete uniform ring typically found in conventional powder diffraction. To accommodate such heterogeneity, after routine normalization of the spectra to correct for background and geometric effects, the resultant incomplete rings in the diffraction pattern can be integrated.

An ongoing study of sulfur materials taken from an Icelandic Norse-trading site context (*35*) has used this combination of simultaneous co-incident X-ray micro-fluorescence and micro-diffraction analyses. The compositional data from the fluorescence results are used to constrain the multi-phase analysis of the diffraction data. This approach reduces the requirement for accuracy in the

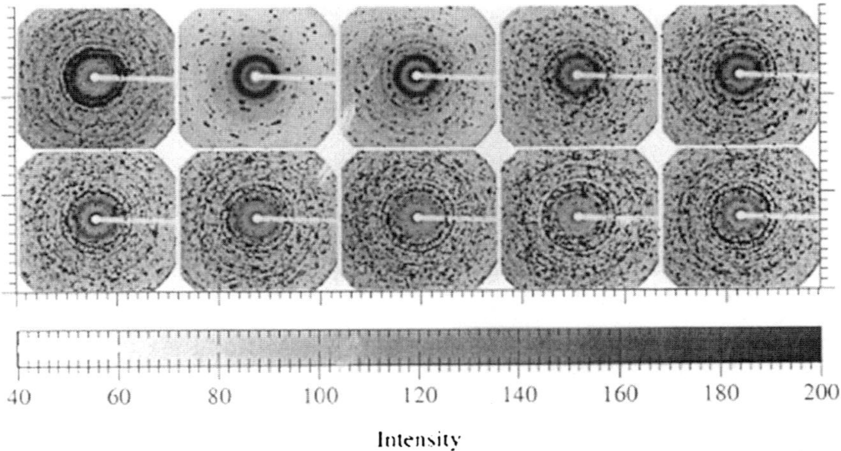

Figure 6. Variations in spectra resultant from large crystallites in heterogeneous soil matrix when examined by microfocus XRD.

quantitative elemental analysis by microfocus XRF. Therefore, this overcomes some inherent issues of microfocus XRF elemental analyses of extremely heterogeneous samples. By producing both structural and elemental compositions, such combination methods can address previously intractable analytical problems in archaeological soil and sediment thin-section analysis.

Common Issues

During the field sampling of archaeological soils and sediments for analysis an obvious key requirement is to gain representative samples of specific contexts. A key factor in examining thin-sections of archaeological soils and sediments by microprobe methods is to again obtain a representative sample. A major advantage of the X-ray methods described is that the sample can be examined by optical methods and target areas sought for X-ray measurements.

A key characteristic of examining materials with X-rays is that the X-rays will penetrate the sample to greater depth compared to electron- (EPXMA) and proton-beams (PIXE). For elemental analyses, this suggests that XRF considers a greater sample depth and, it could be argued, a more representative sample. However, with microfocus XRF, the depth of penetration in the sample (Figure 7) leads to consideration of the issue of self-absorption. This is obviously extremely dependent upon the nature of the sample, particularly the range of

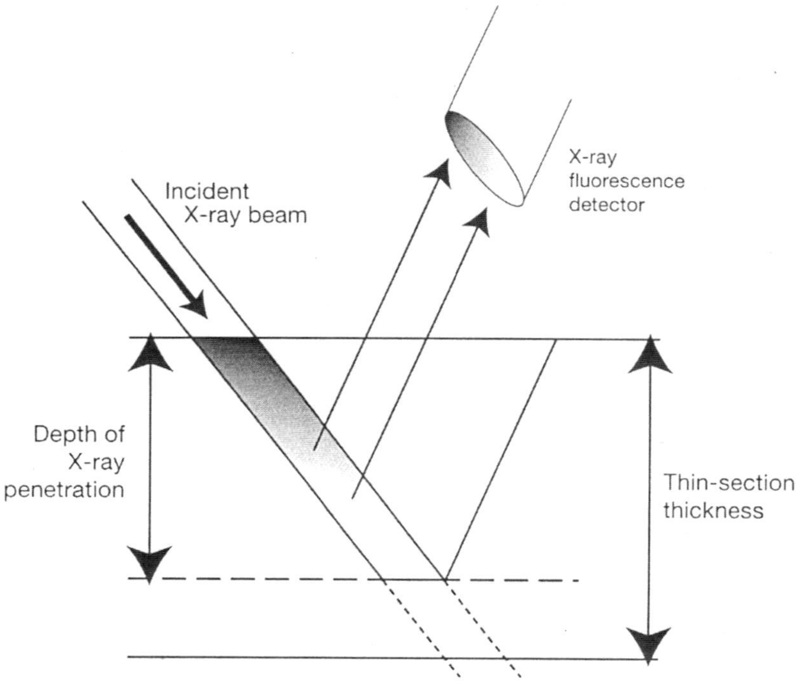

Figure 7. "Information depth" of microfocus XRF analyses.

elements found in the material and the heterogeneity of the material itself. Correcting for self-absorption from a theoretical basis requires *a priori* knowledge of such heterogeneity. Using empirical calibration standards based on a different matrix material (i.e., glass vs. soil) will reduce slightly the accuracy of the resultant data. There is, therefore, a trade-off between the "information depth" of the area sampled, and the accuracy of the results.

For the future, even smaller-scale sample areas can be considered for analysis; as can isotope ratios. It could be envisaged that multiscale prospection analyses could proceed from optical methods to microfocus synchrotron X-ray fluorescence and SAXS and, finally, to destructive isotopic studies with LA-ICP-MS or to finer resolutions with SIMS.

Conclusion

The illustrative studies show the advantages of utilizing combined optical and microfocus synchrotron X-ray based analyses in investigations of archaeological soil and sediment thin-sections. The microfocus synchrotron X-ray small-angle scattering study on calcium-iron-phosphate features has demonstrated that nano-structural features of bone remodeling can be identified at the micron level. The example of the iron-rich coating on the bone in the Finnmark material, illustrates that larger-scale process understandings can be derived for the accumulation of materials in the buried environment. Linked together, the use of microfocus synchrotron X-ray analyses has been demonstrated to further the interpretation of materials found in sites where conditions limit conventional bioarchaeological and artifactual analyses.

Acknowledgements

The authors gratefully acknowledge funding from The British Academy, The Norwegian Science Council and from The Leverhulme Trust for supporting the studies discussed in this paper. An ESRF beamtime award (ME-826) provided synchrotron analyses at ESRF beamline ID18F. From the University of Stirling, School of Biological and Environmental Sciences, Bill Jamieson assisted in production of the figures, and George MacLeod assisted in the thin-section production.

References

1. Jenkins, D. A. *Environ. Geochem. Health* **1989**, *11*, 57–62
2. Hutson, S. R.; Terry, R. E. *J. Archaeol. Sci.* **2006**, *33*, 391–404.
3. Wells, E. C.; Novotny, C.; Hawken, J. R. In *Archaeological Chemistry: Analytical Techniques and Archaeological Interpretation*; ACS Monograph Series, 2006; this volume.
4. Woods, W. I.; Mann, C. C. *Science (Washington, D.C.)* **2000**, *287*, 786–789.
5. Entwistle, J. A.; Dodgshon, R. A.; Abrahams, P. W. *J. Archaeol. Prospect.* **2000**, *7*, 287–303.
6. Cook, D. E.; Kovacevich, B.; Beach, T.; Bishop, R. *J. Archaeol. Sci.* **2006**, *33*, 628–640.
7. Simpson, I. A.; Vésteinsson, O.; Adderley, W. P.; McGovern, T. H. *J. Archaeol. Sci.* **2003**, *30*, 1401–1420.
8. Goldberg, P. *J. Hum. Evol.* **2000**, *38*, 43–90.
9. Simpson, I. A.; Barrett J. H. *J. Archaeol. Sci.* **1996**, *23*, 543–556.
10. Gebhardt, A.; Langhor, R. *Geoarchaeol.* **1999**, *14*, 595–620.
11. Matthews, W.; French, C. A. I.; Lawrence, T.; Cutler, D. F. In *On the Surface Çatalhöyük Excavations 1993–1995*; Hodder, I., Ed.; British

Institute of Archaeology at Ankara and McDonald Institute: Cambridge, 1996; pp 301–342.
12. Adderley, W. P.; Simpson, I. A.; Davidson, D. A. *J. Archaeol. Sci.* **2006**, *33*, 320–334.
13. Simpson, I. A.; Dockrill, S. J.; Bull, I. D.; Evershed, R. P. *J. Archaeol. Sci.* **1998**, *25*, 729–746.
14. Carter, S. P.; Davidson, D. A. *Geoarchaeol.* **1998**, *13*, 535–547.
15. Gebhardt, A. In *Archaeological Sediments and Soils. Analysis, Interpretation and Management*; Barham, A. J.; Macphail, R. I., Eds.; Institute of Archaeology: London, 1995; pp 25–39.
16. Bullock, P.; Fedoroff, N.; Jongerius, A.; Stoops, G.; Tursing, T.; Babel. U. *Handbook for Soil Thin-section Description*; Waine Research Publications: Albrighton, Wolverhampton, 1985.
17. Stoops, G. *Guidelines for Analysis and Description of Soil and Regolith Thin sections*; Soil Science Society of America: Madison, WI, 2003.
18. Simpson, I. A.; Adderley, W. P. In *Halvdanshaugen – Arkeologi, Historie og Naturvitenskap*; Larsen, J. H.; Rolfsen, P., Eds.; University Museum of Cultural Heritage, Oslo, Norway, 2004; pp 329–340.
19. Davidson D. A.; Wilson, C. A.; Meharg, A. A.; Stutter, C.; Edwards, K. J. *Environ. Int.* **2006**, in press.
20. Simpson, I. A.; Perdikaris, S.; Cook, G.; Campbell, J. L.; Teesdale, W. J. *Geoarchaeol.* **2000**, *15*, 743–763.
21. Bruneau, P. M. C.; Ostle, N.; Davidson, D. A.; Grieve, I. C.; Fallick, A. E. *Rapid Commun. Mass Spectrom.* **2002**, *16*, 2190–2194.
22. Cliff, J. B.; Gaspar, D. J.; Bottomley, P. J.; Myrold, D. D. *Appl. Environ. Microbiol.* **2002**, *68*, 4067–4073.
23. Climent-Font, A; Demortier, G.; Palacio, C.; Montero, I.; Ruvalcaba-Sil, J. L.; Diaz, D. *Nucl. Instrum. Methods Phys. Res. B* **1998**, *134*, 229–236.
24. Kumakov, M. A. *Nucl. Instrum. Methods Phys. Res. B* **1990**, *48*, 283–286.
25. Snigerev, A.; Kohn, V.; Snigereva, I.; Lengeler, B. *Nature (London)* **1996**, *384*, 49–51.
26. Fratzl, P.; Schreiber, A.; Boyde, A. *Calcif. Tissue Int.* **1996**, *58*, 341–346.
27. Lammie, D.; Bain, M. M.; Wess, T. J. *J. Synchrotron Radiat.* **2005**, *12*, 721–726.
28. Aizenberg, J.; Weaver, J. C.; Thanawala, M. S.; Sundar, V. C.; Morse, D. E.; Fratzl, P. *Science (Washington D.C.)* **2005**, *309*, 275 – 278.
29. Gupta, H. S.; Wagermaier, W.; Zickler, G. A.; Aroush, D. R.-B.; Funari, S. S.; Roschger, P.; Wagner, H. D.; Fratzl, P. *Nano Lett.* **2005**, *5*, 2108–2111.
30. Wess, T. J.; Drakopoulos, M.; Snigirev, A.; Wouters, J.; Paris, O.; Fratzl, P.; Collins, M.; Hiller, J.; Nielsen, K. *Archaeometry* **2001**, *43*, 117–129.
31. Hiller, J.; Wess, T. J. *J. Archaeol. Sci.* **2006**, *33*, 560–572.
32. Adderley, W. P.; Alberts, I. L.; Simpson, I. A.; Wess, T. J. *J. Archaeol. Sci.* **2004**, *31*, 1215–1224.

33. Simpson, I. A.; Adderley, W. P.; Olsen, B.; Urbanczyk, P. In *Hybrid Spaces? Medieval Finnmark and the Archaeology of Multi-Room Houses*, Olsen, B., Ed.; The Institute for Comparative Research in Human Culture Series B, Novus Press: Oslo, Norway, 2007; (in press).
34. Janssens, K.; Vincze, L.; Vekemans, B. In *Microscopic X-ray Fluorescence Analysis*; Janssens, K. H. A.; Adams, F. C. V.; Rindby, A., Eds.; John Wiley and Sons: Chichester, UK, 2000; pp 155–209.
35. *Excavations at Gásir 2003: An Interim Report (FS238–01075)*. Roberts, H. M., Ed.; Icelandic Institute of Archaeology: Reykjavik, Iceland, 2004; pp 59–60.

Chapter 11

Quantitative Modeling of Soil Chemical Data from Inductively Coupled Plasma–Optical Emission Spectroscopy Reveals Evidence for Cooking and Eating in Ancient Mesoamerican Plazas

E. Christian Wells, Claire Novotny, and James R. Hawken

Department of Anthropology, University of South Florida, Tampa, FL 33620

Recent advances in archaeological soil chemistry provide new ways to examine the use of space where little or no material evidence persists. Drawing on ethnoarchaeological studies, which demonstrate that certain human activities chemically impact soils, this paper compares the results of chemical analyses of soils from two plazas in a prehispanic settlement in Honduras. Chemical compounds were extracted from the surfaces of soil particles using a mild extractant composed of hydrochloric and nitric acids and then characterized using inductively coupled plasma-optical emission spectroscopy. Exploratory data analysis and geostatistics allowed the chemical data to be spatialized across the plaza surfaces. When combined with associated artifact data, the chemical information suggests that the plazas were used, in part, for the preparation and consumption of foods and beverages.

Sixteenth-century Spanish explorers and priests described a wide range of activities—from markets to festivals—that took place in Mesoamerican plazas located at the heart of many Aztec and Maya communities. However, it is often difficult or impossible to evaluate claims about activity patterning in plazas, because these spaces were cleaned of debris and waste after most events. As a result, reconstructing past festive activities often relies exclusively on inferences made from the presence of cooking and serving implements found in the archaeological record. These materials are usually unearthed in secondary deposits, such as middens, which limit our ability to study spatial patterns of culinary practices and the social relations they imply. In this chapter, we synthesize ethnoarchaeological studies from Mesoamerica of how soils are impacted chemically by cooking and eating, with the greater goal of evaluating archaeological inferences of food preparation and consumption. We then present an archaeological example of how these insights might apply to interpreting chemical evidence for plaza activities at Classic period (ca. A.D. 400–1000) Palmarejo in northwest Honduras. We conclude that, by analyzing the chemical residues of food technologies and their byproducts in anthrosols (soils that have been modified by human activity), plaza activities can be studied in primary contexts even where little or no visible material evidence persists.

The Chemical Links Between Food and Soil

Excavation in and around food preparation and serving locales can be extremely informative for determining the kinds of material culture employed in domestic settings. However, activity patterning is more difficult to document in public spaces, such as plazas, because these kinds of areas were often kept clean of refuse for practical purposes, including sanitation. Without *de facto* artifacts or *in situ* assemblages, it is nearly impossible to determine where food-related tasks occurred in relation to other activities.

Recent compositional studies of anthrosols in Mesoamerica suggest that soils from spaces where different kinds of dining activities took place have unique chemical signatures, and that soils contained in refuse dumps can be linked chemically to nearby activity areas (*1–10*). In archaeological studies of soilscapes, the core concept is "soil memory" (*11*), that is, how soils encode the physical, biological, and chemical effects of different human activities. The underlying premise is that certain chemical compounds are associated with particular human activities and their byproducts. For example, the most common compounds examined are phosphates, which indicate the residues of organic substances, including food remains. Since these compounds become fixed to the mineral surfaces of soils and sediments, they remain stable for very long periods in the form of sorbed and complexed ions on clay surfaces, and as insoluble oxides, sulfides, and carbonates, all of which can be studied chemically (*12, 13*). If modern and historical land-use practices, such as farming or livestock grazing, have not destroyed ancient activity areas, variation in the concentrations of chemical elements across a site can be used to infer a range of settlement activities (*14*, *15*).

In Mesoamerica, stuccoed floors, such as those found in plazas and patios, are ideal surfaces from which to extract chemical residues since they tend to be composed largely of lime-based (calcium carbonate) plaster, a calcareous surfacing that has been demonstrated to trap and preserve a variety of chemical compounds over very long periods (*16*, *17*). Inorganic phosphate and potassium compounds, for example, are retained in calcium-rich soils by binding with complex aluminum- and iron-containing minerals, such as apatite clays (*18*). The extremely small size of clay particles (<0.002 mm) correlates with small pore sizes between clay platelets; the high surface tension in these pores results in physical adsorption of phosphates and other chemicals (*19*). Although silt is composed of the chemically inert compound, silicon dioxide, the small size of the particles (0.005–0.002 mm) produces some additional mechanical binding through surface tension.

Compared to the natural background levels of chemical elements in soils unaffected by human occupation, patterns in phosphorus and other elements can assist in determining the locations of activity areas associated with eating and drinking, including processing, consuming, and depositing foods and beverages. It is important to point out that elemental concentrations alone are not reliable indicators of food-related practices or other human activities. Instead, chemical data provide supporting evidence for understanding the organization and spatial distribution of activities. Excavation data are thus essential for inferring activities that generated chemical residues in soils. This kind of "conjunctive research" being carried out over the past few years indicates that, by studying the chemical behavior of anthrosols in addition to excavation data, we can evaluate inferences about the ranges and locations of activities within archaeological contexts where little or no direct material evidence has survived.

Ethnoarchaeological Studies of Cooking and Eating in Mesoamerica

Although elemental phosphorus can often be assigned confidently to activities involving foods and other organic substances, additional major and minor elements—and the activities responsible for their deposition—need to be explored. Since the late 1970s, six important ethnoarchaeological studies have been undertaken in Mesoamerica (Figure 1) with the aim of developing bridging arguments that would allow us to link observed patterning in the chemical data directly with human behavior. These studies are reviewed below.

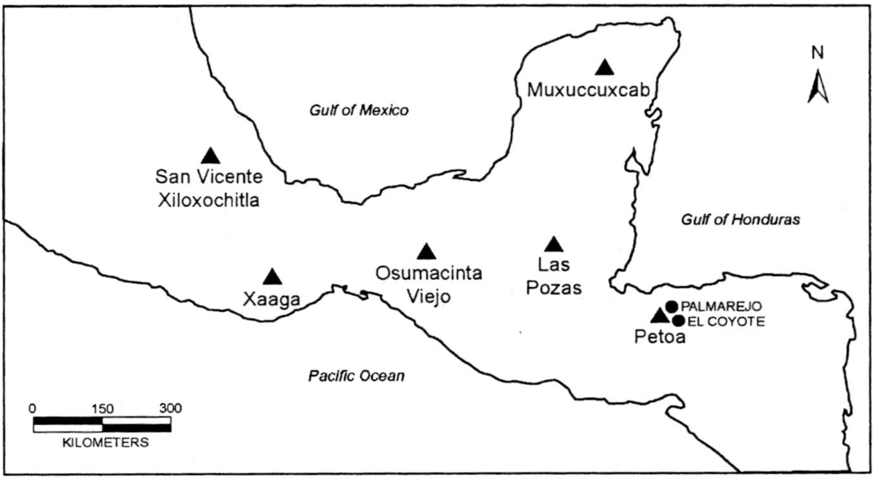

Figure 1. Map of Mesoamerica showing the location of the ethnoarchaeological studies discussed in the text, along with the archaeological sites of Palmarejo and El Coyote, Honduras.

San Vicente Xiloxochitla, Tlaxcala, Mexico

Since as early as the 1970s, Luis Barba and colleagues (*20, 21*) at the Archaeology Prospection Laboratory in Mexico City (part of the Institute of Anthropological Investigations of the National Autonomous University of Mexico) have been developing physical and chemical soil analysis techniques to aid in the study of activity areas at prehispanic archaeological sites in Mexico. In the late 1970s and again in the 1990s, Barba and colleagues studied a modern house in San Vicente Xiloxchitla, in Tlaxcala, Mexico. The house compound had a domicile, a stable for animals, and a separate kitchen building that contained an elevated brick oven. In their study of the soils from the earthen house floor, which derived from underlying volcanic tuff, they found that the area near the oven had high pH values and contained low concentrations of phosphorus (P) along with elevated levels of calcium (Ca) and carbonates, especially in spaces used for processing corn. Food consumption took place in the courtyard in front of the kitchen and deposited very high concentrations of P relative to adjacent areas.

Osumacinta Viejo, Chiapas, Mexico

In the early 1980s, Barba and Denise (*22*) studied volcanic soils from two houses in a recently abandoned village, Osumacinta Viejo, in Chiapas, Mexico. One house was located in the center of town and the other on the outskirts; both were abandoned roughly three years prior to study. Barba and Denise found low P levels around the fireplaces and hearths, but high pH. They also found high concentrations of Ca immediately outside of the houses in spaces used for soaking corn kernels in lime water and iron (Fe) in spaces used for processing agave leaves. These areas contrasted sharply with those where meals were taken, which presented high P concentrations and low Ca and Fe levels.

Muxuccuxcab, Yucatan, Mexico

More recently, in the 1990s, Barba and colleagues (*23*) have been working on soils derived from limestone substrate in a modern house compound in Muxuccuxcab, Yucatan, Mexico located near the prehispanic site of Kabah. Here, the compound was walled in and consisted of one main building and four ancillary structures, one of which served as a kitchen. They found that they could characterize food preparation areas by the presence of protein residues, carbonates, high pH values, low P levels, and burned earth and ash from fires, which produced elevated levels of potassium (K) and magnesium (Mg). Food consumption spaces were distinguished from preparation areas by soils with lower pH values and very high P levels.

Xaaga, Oaxaca, Mexico

In Oaxaca, Mexico, in the early 1990s, Middleton (*24–26*) conducted ethnoarchaeological work at a house compound in Xaaga, a few kilometers from the archaeological ruins of Mitla. The house compound consisted of two living spaces and a cane-walled, thatch-roofed kitchen with compacted earthen floors that had an *horno* (a traditional wood-fire adobe oven) in one corner. The soils derive from fluvial sediments composed of volcanics, conglomerates, and lake deposits, including sandstone and limestone. Middleton found elevated concentrations of Al, Ba, Mn, Na, Sr, Zn, P, Ca, Fe, K, and Mg. Because of the high amounts of wood ash and charcoal, the surface closest to the *horno* contained high levels of K and Mg. Middleton also found high concentrations of Ca and Sr, which he suggested may be associated with the preparation of *nixtamal*—a mixture of dried corn, water, and lime used to prepare *masa*, the dough for making tortillas.

Las Pozas, Peten, Guatemala

Recently in Guatemala, Fabián Fernández and colleagues (*27*) examined limestone-derived soils from two houses in the Q'eqchi' Maya farming village of Las Pozas, located roughly 10 km southeast of the remains of the Classic Maya city of Aguateca. The houses they examined consisted of wooden plank walls, palm-thatched roofs, and earthen floors; one house was abandoned, and the other was still active at the time of study. They mapped the distribution of activities, which included specific-use areas for food preparation and cooking, consisting primarily of grinding, soaking, and boiling corn for making tortillas. Cooking took place on a traditional three-stone hearth on the floor, and dining occurred in the kitchen and near the sleeping and grain storage areas. They found that the kitchen areas contained highly alkaline soils (high pH) with high levels of P, K, Mg, Ca, and Na; heavy metals (Cd, Cu, Hg, Pb, Fe, Ba, Mn, and Zn) were poorly represented.

Petoa, Santa Barbara, Honduras

In Honduras, Wells (*9*) sought to identify some of the behaviors that resulted in chemical signatures of food consumption, especially in the context of feasting and festivals in the rural farming town of Petoa, located about one kilometer south of the prehispanic civic-ceremonial center of El Coyote (*28*). Wells attended a number of plaza events at Petoa—the largest being *feria*, the town's celebration of its patron saint, San Bartolomé—and observed the dining and drinking experiences of the festival's participants. After the party, the municipality subsidized the plaza's cleaning, which was accomplished by some of the town's residents. Wells followed behind and collected soil samples from the plaza's hard-packed earthen surface, where the underlying bedrock is primarily limestone. Foods tended to be consumed in the center of the plaza to the southwest of a church and in front of a makeshift stage on which presentations were made. Foods included tamales and sandwiches as well as a range of sweets, and drinks consisted mostly of carbonated beverages. Overall, consumption areas showed markedly higher concentrations of P, Ca, Fe, and Mn, along with lesser amounts of Al, K, and Mg.

Social Relations from Soil Chemistry

The results of the ethnoarchaeological studies are summarized in Table I. From this, we can observe that soils in food preparation areas involving cooking are often low in phosphates. The alkaline nature of these soils, a result

Table I. Summary of Ethnoarchaeological Soil Studies from Mesoamerica.

Context	P	K	Ca	Mg	Na	Mn	Sr	Ba	Al	Fe	Zn
Food preparation											
San Vicente Xiloxochitla, Tlaxcala, Mexico	low		high								
Osumacinta Viejo, Chiapas, Mexico	low		high							high	
Muxuccuxcab, Yucatan, Mexico	low	high		high							
Xaaga, Oaxaca, Mexico	high	high	high	high	high	high	high	high	high	high	high
Las Pozas, Peten, Guatemala	high	high	high	high	high	low		low		low	low
Food consumption											
San Vicente Xiloxochitla, Tlaxcala, Mexico	high										
Osumacinta Viejo, Chiapas, Mexico	high		low							low	
Muxuccuxcab, Yucatan, Mexico	high										
Petoa, Santa Barbara, Honduras	high	low	high	low		high			low	high	

Note : All concentrations are "high" or "low" compared to adjacent spaces.

of the deposition of wood ash from cooking fires, prevents P from adhering to sediments (*29*). Thus, while P is often deposited in high concentrations in soils near cooking areas, for instance, in the Xaaga and Las Pozas cases discussed previously, ash and charcoal may prevent it (as well as other chemicals) from being retained. While it is generally unconvincing to identify cooking areas by an absence of evidence, wood ash leaves its own chemical signature: high levels of K and Mg. Thus, these studies suggest that, archaeologically, we can identify cooking areas involving the deposition of wood ash by elevated levels of K and Mg and varying levels of P (the variation related to the deposition of ash).

To test this hypothesis, Wells sampled soils from prehispanic hearths (shallow, open pits) and ovens (adobe-lined, enclosed combustion facilities), as well as surrounding floor spaces, in domestic contexts at the archaeological site of El Coyote in northwest Honduras. All samples were processed using a mild-acid extraction procedure (*30*) and analyzed using inductively coupled plasma-optical emission spectroscopy (ICP-OES). We conducted a principal components analysis of the resulting data using a correlation matrix. The results, summarized in Figure 2 as a scatterplot, support the idea that hearths tend to contain high concentrations of K and Mg (and in this case, Al and P as well). However, an unexpected finding is that Na, Ba, and Fe (and to a lesser extent, Sr and Ti) may be useful for distinguishing hearths from ovens.

According to the ethnoarchaeological studies, other kinds of food preparation activities that do not involve wood-burning fires also appear to have chemical fingerprints. For example, Ca and Sr are associated with processing corn, and Fe and P are associated with processing agave. Other food preparation activities, including animal butchering, may be associated with the deposition of Fe, Al, Ti, and K, which represent the chemical makeup of phenocrysts and other inclusions in tools made of obsidian, such as scrapers (*31*).

Finally, food consumption areas tend to be marked by especially high levels of P, along with a variety of other elements, suggesting that we are detecting not only consumption activities, but also evidence for a range of other practices. Here, again, excavation data (even if from nearby midden deposits) are crucial for reconstructing the particular suite of activities associated with the dining experience. Wells (*32*) demonstrates how this line of work can reveal activity patterns related to feasting and craft production in a large ceremonial plaza at El Coyote.

An Archaeological Example from Honduras

We now turn to an archaeological example from Honduras to explore how the ethnoarchaeological studies can inform our understanding of prehispanic cooking and eating practices in plaza spaces. We focus on the site of Palmarejo,

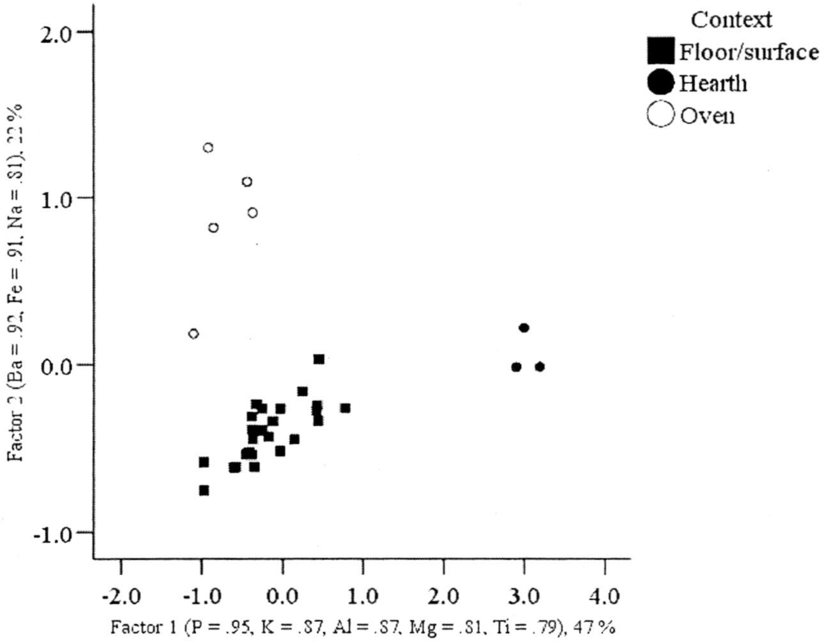

Figure 2. Plot of the first two factors (accounting for 69 percent of the variance) from a principal components analysis of the residential soil sample data (n = 38) from El Coyote using 11 chemical elements (Al, Ba, Ca, Fe, K, Mg, Mn, Na, P, Sr, Ti). Ovens tend to have higher concentrations of Ba, Fe, and Na (which are all highly correlated, r ≥ .7) compared to hearths, while hearths tend to have higher concentrations of P, K, Al, Mg, and Ti (which are all highly correlated, r ≥0 .7) compared to ovens.

a large settlement composed of nearly 100 buildings, representing temples, palaces, and a ball court, which date to the Mesoamerican Classic period, roughly A.D. 400–1000. Some of the tallest buildings surround open plaza spaces at the core of the site. The community appears to have been the capital of a small settlement hierarchy composed of 93 villages and farmsteads (*33*).

The focus of our research has been on investigating the differences between the activities conducted in the main civic-ceremonial plaza and a large elite residential plaza, or "patio" (*34, 35*). The main plaza encompasses roughly 1000 m^2, while the elite patio measures about 2500 m^2. Our excavations in these spaces revealed almost no artifacts. However, excavations on the edges of these locales resulted in a diverse assemblage of materials. The artifacts from the plaza include large grinding stones, presumably used to prepare foods and beverages, large polychrome serving plates, and large effigy censers (for burning incense) in anthropomorphic and zoomorphic forms. In contrast, the materials from areas surrounding the patio include smaller examples of groundstone, small dishes including bowls and cups, and small figurines depicting humans. From this, we can infer that the activities conducted in plaza versus patio spaces used similar kinds of materials but in different ways. For example, each space appears to have been used for cooking and eating as well as religious devotion. However, in the plaza, the size of the artifacts—large cooking implements and large serving wares—suggest that activities involved larger groups of people than in the patio. But how were these activities organized? Where did cooking and eating take place? Were these activities spatially separate from the use of censers?

To answer these questions, we collected a total of 324 samples of soil from the surfaces of the plaza and patio, with the aim of prospecting for activity loci. We sampled these spaces using a lattice grid matrix, collecting samples at regular 2 m intervals in the plaza and 5 m intervals in the patio (Figure 3). All samples were taken at roughly 15 cm below modern ground surface, which in each case represents the approximate level of the ancient plastered surface.

For chemical analysis, we used a mild acid-extraction procedure, based on recent experimental work (*13, 26, 36, 37*). We chose the extraction procedure over total digestion of the sample (*38*) because we are interested specifically in anthropogenic inputs by way of inorganic P and other elements, and not the total compositional chemistry (i.e., anthropogenic and diagenetic inputs). The extractant we used, composed of 20 ml of dilute 0.60 molar hydrochloric acid and 0.16 molar nitric acid, has been experimentally determined to remove soluble and readily labile P and other elements. While the extraction is not always proportional to the total P (or other elements) in soil, for activity area research we are concerned with the spatial patterns of elemental concentrations rather than absolute concentrations, as many variables affect elemental levels in soils (*39, 40*).

For this procedure, we selected a 2.00 g portion from each sample, pulverized it with an agate mortar, mixed it with our extractant in a polyethylene vial, and shook it vigorously for 30 minutes on a platform shaker at 200 rpm. We then filtered the solution using ashless filter paper and decanted it into a clean polyethylene vial. Finally, we diluted the extract with Type II deionized water to bring the concentrations of the elements into the optimal measurement range of the analytical instrument. All samples were analyzed using a Perkin-Elmer 4300DV ICP-OES with Echelle-type grating. For calibration, known solution standards containing the elements of interest in concentrations bracketing the expected concentrations of the sample were run during the analysis. The results were reported in parts per million of the element and subsequently converted to mg element per kg soil using the dry weights of the analyzed portions of the samples. All data were then standardized by conversion to log (base 10) for comparability.

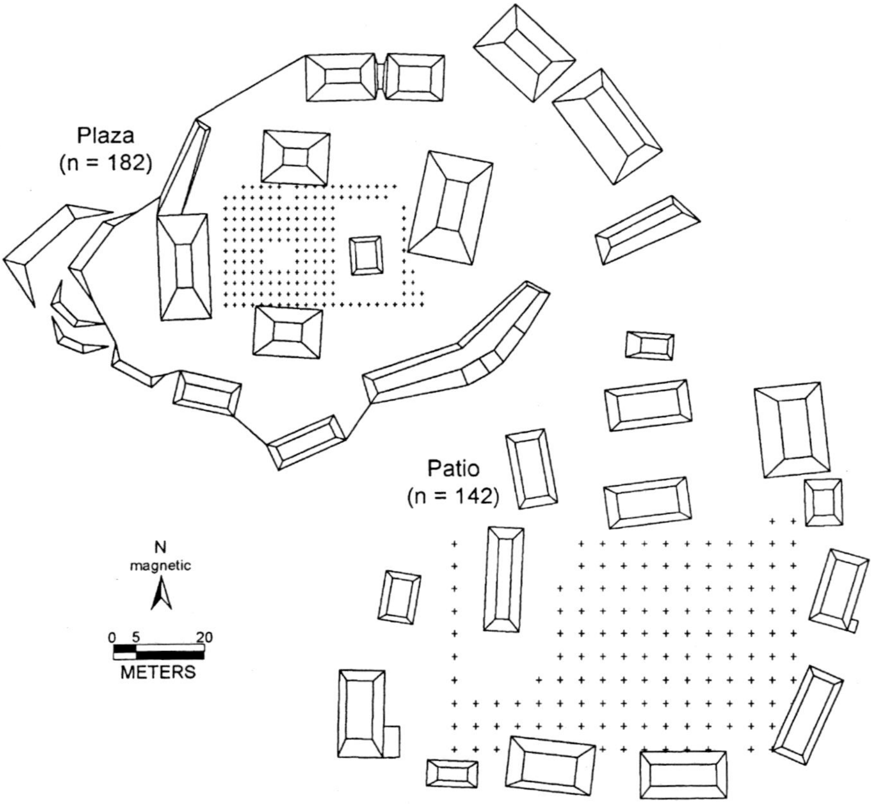

Figure 3. Plan view map of the site core of Palmarejo, showing the location of the sample grids for the plaza and the patio. Modern vegetation prevented sample collection from certain spaces, represented as "empty" areas in the matrixes above.

The data are summarized in two boxplots (Figure 4). What we generally look for are elemental data with large ranges and high standard deviations. This indicates that the deposition of a given element is not homogeneous across the space we are studying. In both the plaza and the patio, Al, K, Mg, and P appear to be good indicators of human activities, because they meet these criteria. Here we are not so much interested in the absolute concentrations as we are in the magnitudes of difference between samples and between elements (as such, we do not discuss the characterization of background samples). For instance, the range of P in the plaza is much larger than that of the patio, indicating that activities involving the deposition of organics in the plaza were more variable than in the patio. From this, we might infer that cooking and eating took place across the entire surface of the patio, but only in certain places in the plaza.

With these patterns in mind, we conducted a principal components analysis of the two datasets using a covariance matrix, since some of the elements have especially high concentrations that could swamp those with lower concentrations in the analysis (Figure 5). Here we can see that, in the plaza, samples from the north half vary by P concentration, while those of the south vary according to levels of Ba and Mg; the reverse is true for western versus eastern samples (not pictured). In the patio, all samples tend to vary along Factor 1, in which Al, Ba, Fe, and Mn account for most of the variance in the data. This suggests that activity loci in the plaza and patio vary by corner or quadrant.

To evaluate this possibility, we conducted a discriminant function analysis of the two datasets (Figure 6). While there is much overlap—which is not surprising as all the soils derive from the same geological substrate—the plot of the plaza data shows that the northwest and northeast corners have different soil chemical signatures compared to the other soils analyzed from this space. This suggests that different activities were carried out in these locales. In contrast, the plot of the patio data suggests that the chemical signatures of soils in the southeast and northeast corners are similar to each other but different from those in the western half of the patio. Thus, in the plaza, activities can be differentiated by north and south, while those in the patio can be differentiated by east and west.

Now that we have a good idea of how soils vary by concentration and combination of chemical elements, it is instructive to examine these differences spatially. The simplest way to do this would be to overlay a graphic depicting the concentration of each sample on maps of the plaza and patio. However, geostatistics offers a more robust set of techniques to examine spatial variation in our data. Here we use Kriging, an empirical model that makes use of spatial dynamics to interpolate unknown values based on known values (*41–43*). The basic idea is that regions close together are more likely to have similar values than regions farther apart.

Kriging is based on the assumption that the variables between two known points follow a stochastic process, which is characterized by a variogram model.

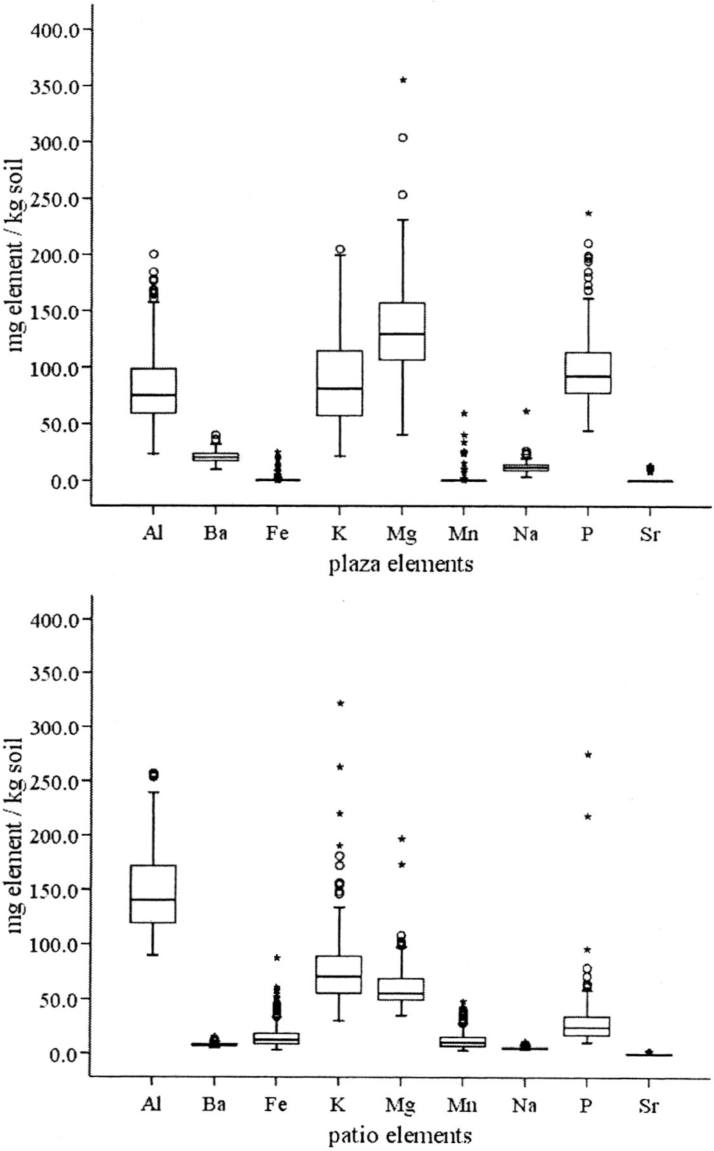

Figure 4. Boxplot summaries of the elemental data. The upper plot shows the plaza data (n = 182) and the lower plot shows the patio data (n = 142).

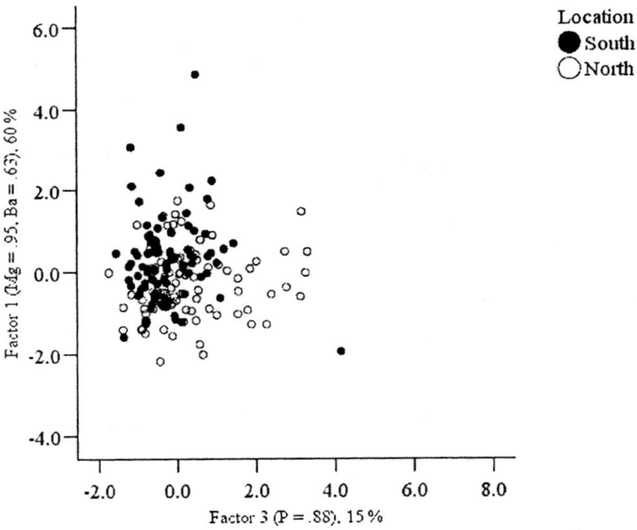

Plaza

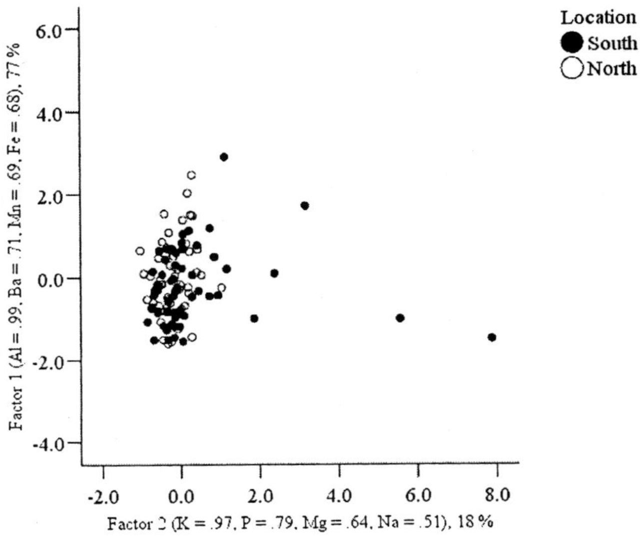

Patio

Figure 5. Scatterplots of the factor scores from a principal components analysis for the plaza (top) and patio (bottom) data. The plots show how the data vary by location (south half versus north half) in the plaza.

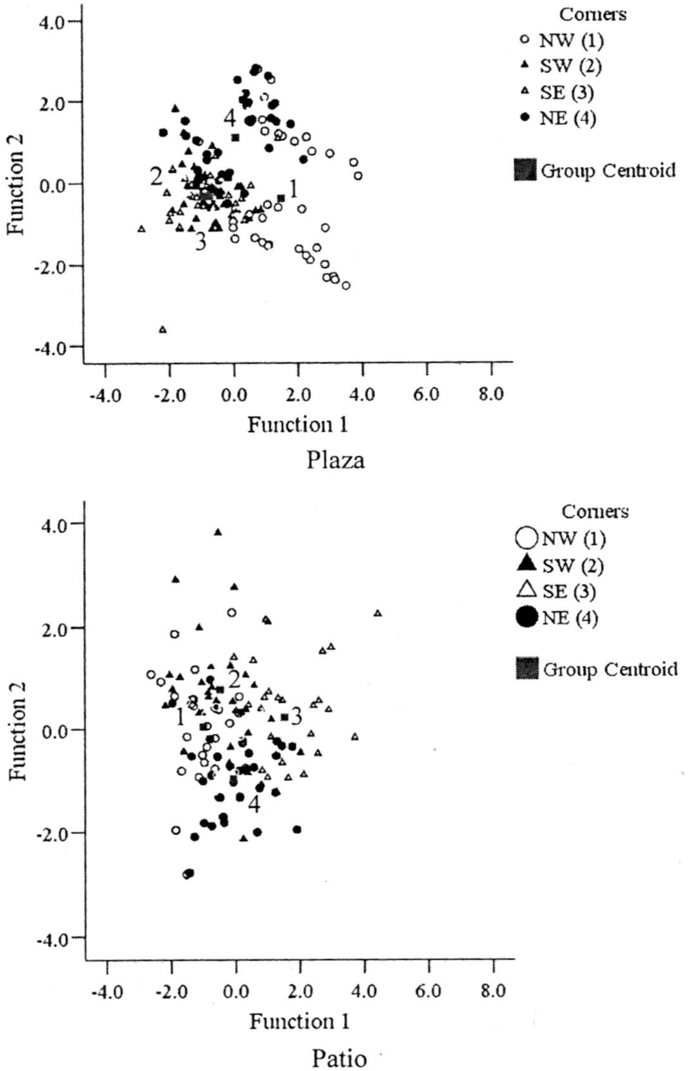

Figure 6. Scatterplots of the first two factor scores from a discriminant function analysis for the plaza (top) and patio (bottom) data. The plots show how the data vary by corner: northwest (NW), southwest (SW), southeast (SE), or northeast (NE).

A variogram model predicts the spatial correlation between two points in such a way that allows the degree of correlation to change according to the spatial arrangement of the samples. Points that are near each other have a certain degree of spatial correlation, but points that are widely separated are statistically independent. Conceptually, variance in Kriging plays the role of a weighting function. In the Kriging method, every known data value and every missing data value have an associated variance. Kriging estimates the missing data values using variance curves as weighting functions when estimating the missing data values. The primary objective in Kriging is to minimize the values of those variances and, by doing so, estimate the "best" data value for each missing data point.

For the current study, we used the computational software program, Surfer, version 8.01 (manufactured by Golden Software, Inc., Golden, Colorado, USA) to construct a linear variogram model (Kriging type = point), where slope = 1 and anisotropy = 1, 0. We used a neighborhood variance of zero and assigned no nugget affect. We chose a linear variance function (*44*) with minimum = 0.00 (at the sample point) and maximum = 1.00 (at 0.25 m from the sample point). Thus, our variogram model can be described as,

$$\gamma\,(h) = C_0 + S \bullet h$$

where the variance [$\gamma\,(h)$] is equal to the sum of the unknown nugget effect (C_0) and the product of the unknown slope (S) and the lag distance (h). The fitted variogram model allowed us to interpolate the data, that is, to predict the chemical concentrations at unsampled locations. We plotted the Kriged data on a regular xyz grid, which we overlayed on a plan view base map of Palmarejo (Figure 7). The result is a visual probability plot of P as it changes over space. The plot shows that the deposition of P, which indicates the use of organic materials, is highly variable in the plaza, while homogeneous over the patio. One possible explanation is that activities in the plaza took place in specific, regularly patterned places, while those in the patio were conducted without fixed locations. Another explanation might be that the plaza and patio were cleaned in different ways, for example, post-activity debris in the plaza could have been picked up, while debris in the patio was swept away—an action that would have resulted in the movement and mixing of soils associated with different activities.

Figure 8 is a Kriged image map that shows how Factor 1 from the principal components analysis varies across space. For the plaza, the variance in Factor 1 is explained mostly by Mg and Ba. For the patio, the variance is explained mostly by Al, Ba, Mn, and Fe. Thus, in each case, particular combinations of elements form chemical signatures of different activity areas. We do not know, as of yet, what these particular activities or sets of overlapping activities were, but thanks to the artifact data, we can surmise that they have something to do with cooking and eating as well as religious devotion. Based on the

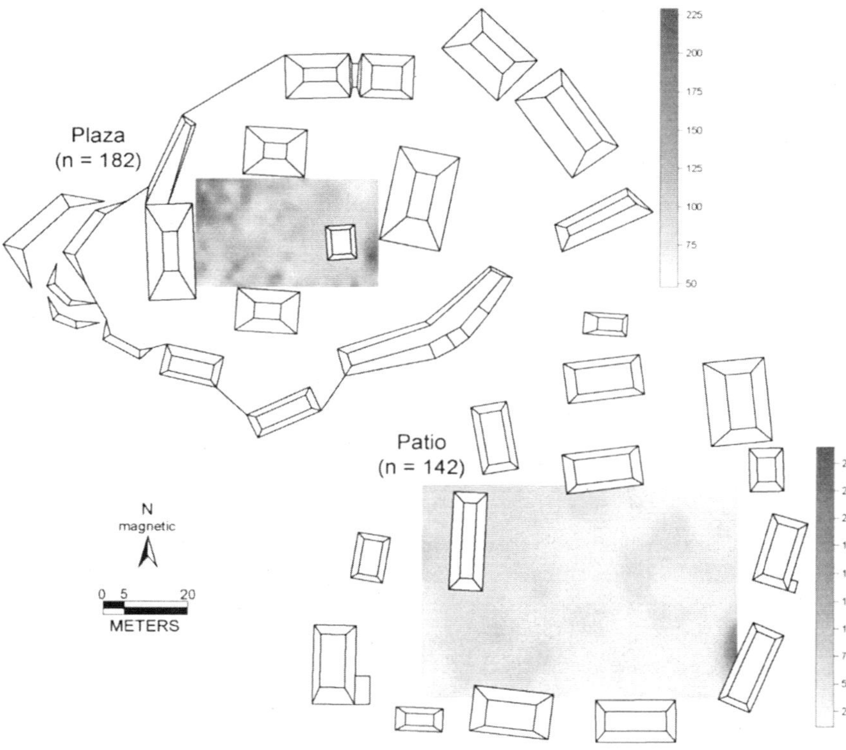

Figure 7. Kriged image map showing the distribution of extractable soil P in mg P/kg soil. Darker hues correspond to higher concentrations of P.

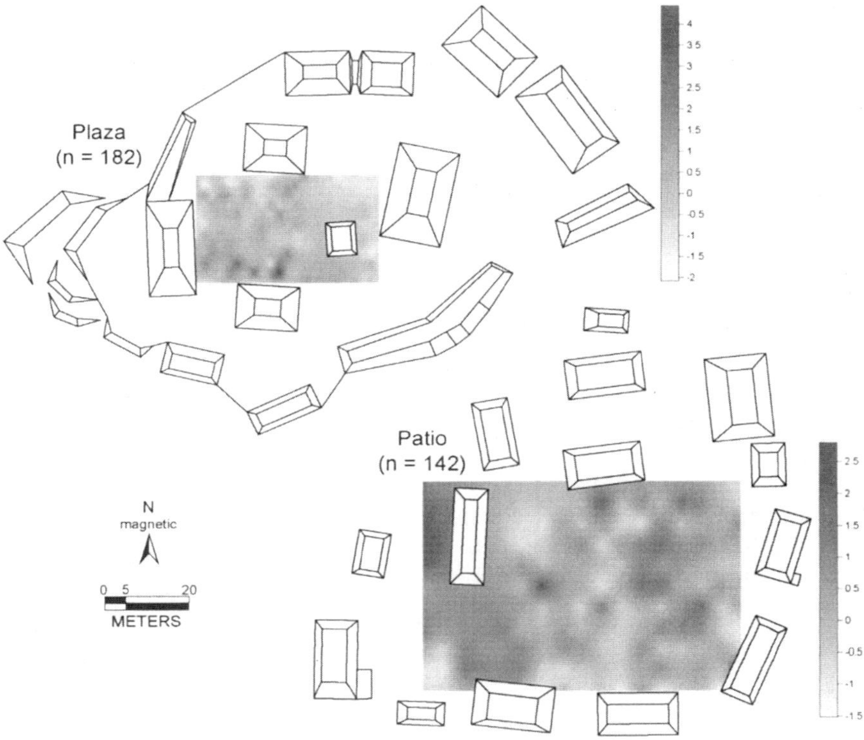

Figure 8. Kriged image map showing the distribution of Factor 1 scores from the principal components analysis (see the caption for Figure 5 for the results of the principal components analysis).

ethnoarchaeological studies, we can speculate that the activities may have resulted in the deposition of corn-based foods or drinks, because the concentrations of P, K, Ca, and Mg are patterned. The presence of groundstone in both the plaza and the patio is consistent with this scenario if the stones were used to grind corn. The next step is to evaluate these quantitative models by correlating artifact type, density, and distribution with the chemical data, as well as to experiment with alternative variogram models.

While the Kriged image plots are useful for thinking about how the chemical patterns might be distributed across space, we would like to point out that here we assume that the chemical concentrations are spatially dependent and that they vary stochastically across space. We make this assumption based on our knowledge of the micro-soilscapes of the plaza and patio, which are highly homogeneous with regard to soil formation processes and substrates. In doing so, however, we assume that our empirical variogram model correctly estimates how chemical concentrations vary across space. This is an important point to keep in mind when interpreting the results, because the sample densities from the plaza and patio are not equal (i.e., individual sample variance may differ between datasets), yet we apply the same variogram model to both datasets (45). In the future, we plan to evaluate our initial observations about the patterns of soil chemistry with different variogram models.

Conclusion

Thanks to ethnoarchaeological work on food residues in anthrosols, we can derive inferences for chemical signatures of certain kinds of food preparation and consumption activities. When this information is combined with data produced from excavation, we can offer compelling arguments for the location and organization of food-related activities in the archaeological record that are supported by different lines of evidence from independent datasets. With these data, we can study the use of space in domestic and other settings in far greater detail than has been the case.

Acknowledgments

We would like to thank Michael D. Glascock for his invitation to prepare this chapter. We greatly appreciate the comments of Karla L. Davis-Salazar, José E. Moreno-Cortés, and three anonymous reviewers, which helped us improve our paper. Work at El Coyote was funded by the National Science Foundation and the Wenner-Gren Foundation for Anthropological Research. The work at Palmarejo was made possible by grants from the National Geographic Society and several funding agencies at the University of South Florida, including the Office of Research and Scholarship, the Office of Undergraduate Research, the Institute for the Study of Latin America and the Caribbean, and the Dr. Kiran C. Patel Center for Global Solutions.

References

1. Barba, L.; Ludlow, B.; Manzanilla, L.; Valadez, R. *Ciencia y Desarrollo* **1987**, *77*, 21–33.
2. Barba, L.; Manzanilla, L. *Antropológicas* **1992**, *1*, 19–46.

3. Barba, L.; Ortiz, A. In *Anatomía de un conjunto residencial Teotihuacano en Oztoyahualco, II: Los estudios específicos*; Manzanilla, L., Ed.; Instituto de Investigaciones Antropológicas, Universidad Nacional Autónoma de México: Mexico City, 1993; pp 595–616.
4. King, S. M. Ph.D. thesis, University of California, Berkeley, CA, 2003.
5. Manzanilla, L.; Barba, L. *Ancient Mesoamerica* **1990**, *1*, 41–49.
6. Parnell, J. J. M.A. thesis, Brigham Young University, Provo, UT, 2001.
7. Parnell, J. J.; Terry, R. E.; Golden, C. Ge*oarchaeol.* **2001**, *16*, 855–873.
8. Parnell, J. J.; Terry, R. E.; Nelson, Z. *J. Archaeol. Sci.* **2002**, *29*, 379–404.
9. Wells, E. C. Ph.D. thesis, Arizona State University, Tempe, AZ, 2003.
10. Wells, E. C. In *XVI Simposio de Investigaciones Arqueológicas en Guatemala, 2002*; Laporte, J. P.; Escobedo, H. L., Eds.; Museo Nacional de Arqueología y Etnología: Guatemala City, 2003.
11. Wells, E. C. In *Function of Soils for Human Societies and the Environment*; Frossard, E., Blum, W. E. H.; Warkentin, B. P., Eds; Geological Society: London, 2006; pp 125–132.
12. Linderholm, J.; Lundberg, E. *J. Archaeol. Sci.* **1994**, *21*, 303–314.
13. Terry, R. E.; Hardin, P. J.; Houston, S. D.; Nelson, S. D.; Jackson, M. W.; Carr, J.; Parnell, J. J. *Geoarchaeol.* **2000**, *15*, 151–166.
14. Entwistle, J. A.; Abrahams, P. W.; Dodgshon, R. A. *J.Archaeol. Sci.* **1998**, *25*, 53–68.
15. Wells, E. C.; Terry, R. E.; Parnell, J. J.; Hardin, P. J.; Jackson, M. W.; Houston, S. D. *J. Archaeol. Sci.* **2000**, *27*, 449–462.
16. Barba, L. In *Etnoarqueología Coloquio Bosch Gimpera*; Sugiura, Y.; Serra Puche, M., Eds.; Instituto de Investigaciones Antropológicas, Universidad Nacional Autónoma de México: Mexico City, 1990; pp 177–200.
17. Ortiz, A.; Barba, L.; López Luján, L.; Link, K. F. In *Materials Issues in Art and Archaeology IV*; Vandiver, P. B.; Druzik, J. R.; Galván, J. L.; Freestone, I. C.; Wheeler, G. S., Eds; Materials Research Society: Pittsburgh, PA, 1994; pp 723–726.
18. Albert, L. E. *Bulletin of the Oklahoma Anthropological Society.* **1986**, *35*, 13–22.
19. Hausenbuiller, R. L. *Soil Science: Principles and Practices*; William C. Brown Company: Dubuque, IA, 1972; pp. 266–270.
20. Barba, L.; Bello, G. *Notas Antropológicas* **1978**, *1*, 188–193.
21. Barba, L.; Ortiz, A. *Lat. Am. Antiq.* **1992**, *3*, 63–82.
22. Barba, L.; Dense, P. *Memorias de la XVII Mesa Redonda de la Sociedad Mexicana de Antropología* **1981**, *2*, 263–277.
23. Barba, L.; Pierrebourg, F. D.; Trejo, C.; Ortiz, A.; Link, K. F. *Revue d'Archéométrie* **1995**, *19*, 79–95.
24. Middleton, W. D. Ph.D. thesis, University of Wisconsin, Madison, WI, 1998.

25. Middleton, W. D. *Archaeometry* **2004**, *46*, 47–65.
26. Middleton, W. D.; Price, T. D. *J. Archaeol. Sci.* **1996**, *23*, 673–687.
27. Fernández, F. G.; Terry, R. E.; Inomata, T.; Eberl, M. *Geoarchaeol.* **2002**, *17*, 487–519.
28. Wells, E. C.; Urban, P. A. In *Materials Issues in Art and Archaeology VI*; Vandiver, P. B.; Goodway, M.; Mass, J. L., Eds.; Materials Research Society: Warrendale, PA, 2002; pp 193–198.
29. Scotter, G. *Forest Chronology* **1963**, *39*, 412–421.
30. Wells, E. C. *Archaeometry* **2004**, *46*, p. 72.
31. Wells, E. C. *Archaeometry* **2004**, *46*, p. 77.
32. Wells, E. C. *Archaeometry* **2004**, *46*, 67–84.
33. Wells, E. C.; Davis-Salazar, K. L.; Moreno-Cortés, J. E. *Proyecto Arqueológico Comunidad Palmarejo: informe preliminar, primera temporada*; Instituto Hondureño de Antropología e Historia: Tegucigalpa, Honduras, 2004.
34. Davis-Salazar, K. L.; Wells, E. C.; Moreno-Cortés, J. E. *Proyecto Arqueológico Comunidad Palmarejo: informe preliminar, segunda temporada*; Instituto Hondureño de Antropología e Historia: Tegucigalpa, Honduras, 2005.
35. Wells, E. C.; Moreno-Cortés, J. E.; Davis-Salazar, K. L. *Proyecto Arqueológico Comunidad Palmarejo: informe preliminar, tercera temporada*; Instituto Hondureño de Antropología e Historia: Tegucigalpa, Honduras, 2006.
36. Burton, J. H.; Simon, A. W. *Am. Antiq.* **1993**, *58*, 45–59.
37. Bethell, P.; Máté, I. In *Scientific Analysis in Archaeology and Its Interpretation*; Henderson, J., Ed.; Institute of Archaeology, University of California: Los Angeles, CA, 1989; pp 1–29.
38. Entwistle, J. A.; Abrahams, P. W.; Dodgshon, R. A. *J. Archaeol. Sci.* **2000**, 27, 287–303.
39. Barber, S. A. *Soil Nutrient Bioavailability: A Mechanistic Approach*; Wiley: New York, 1995.
40. Stevenson, F. J.; Cole, M. A. *Cycle of Soil: Carbon, Nitrogen, Phosphorus, Sulfur, Micronutrients*; Wiley: New York, 1999.
41. Cressie, N. A. C. *Statistics for Spatial Data*; Wiley: New York, 1993.
42. Kitanidis, P. *Introduction to Geostatistics: Applications in Hydrogeology*; Cambridge University Press: New York, 1997.
43. Schabenberger, O.; Pierce, F. J. *Contemporary Statistical Models for the Plant and Soil Sciences*; CRC Press: Boca Raton, FL, 2001.
44. Williams, C. K. I. In *Learning in Graphical Models*; Jordan, M. I., Ed.; MIT Press: Cambridge, MA, 1998; pp 599–612.
45. Haslam, R.; Tibbett, M. *Geoarchaeol.* **2004**, *19*, 731–751.

Chapter 12

Chemical Composition of Song Dynasty, Chinese, Copper-Based Coins via Energy Dispersive X-ray Fluorescence

Jessica Misner[1], Jeffe Boats[2], and Mark A. Benvenuto[1]

Departments of [1]Chemistry and Biochemistry and [2]Mathematics and Computer Science, University of Detroit Mercy, Detroit, MI 48221

Two hundred twenty four Chinese coins (Song Dynasty, ca. 990–1080 A.D.) were analyzed via energy dispersive X-ray fluorescence spectrometry for the following elements: copper (Cu), zinc (Zn), tin (Sn), lead (Pb), iron (Fe), nickel (Ni), manganese (Mn), antimony (Sb), gold (Au), platinum (Pt), palladium (Pd), and silver (Ag). The coins routinely appear to be leaded bronze. However, the amount of lead present in these coins was in many cases significantly higher than expected.

Introduction

The Sung Dynasty (960 – 1280 A.D.) was one link in a chain of imperial dynasties throughout the history of China. At the time of the manufacture of the coins used in this study, established practice dictated that the coins be cast in clay molds, in a traditional manner that was already nearly one thousand years old. Thus, the coins analyzed in this study are all round, with a square hole in the center, have four characters on one side and none on the other, and are said to be of copper, bronze, or brass (*1, 2*). The characters on such coins are read from top to bottom and right to left. The top and bottom characters are the emperor's title, and those on the right and left proclaim the minting authority. A single emperor often had more than one title in the course of his reign. The changes are chronological, and have been summarized in Table I. Schjoth numbers in this table refer to one of the standard reference books for identifying and cataloguing Chinese coins of the various dynasties (*1*). The study numbers that make up the final column of Table I are assigned by the authors as a shorter identifier than imperial titles by which to designate the coins. The coins of three emperors have been examined.

Table I. Imperial Names, Titles, and Identification Numbers

Emperor	*Title*	*Dates (A.D.)*	*Schjoth #(2)*	*Study #*
Chen Tsung	Hsien-p'ing	998–1004	469	SA
	Hsian-fu	1008–1016	473–478	SB, SBB, SC
	T'ien-his	1017–1021	479–482	SD
Jen Tsung	T'ien-sheng	1023–1031	484–486	SE, SF
	Ching-yu	1034–1037	492–494	SG
	Pao-yuan	1038–1039	496–500	SH
Che Tsung	Shao-sheng	1094–1097	582–591	SI, SJ
	Yuan-fu	1098–1100	599–603	SK

Our interest in this series of coins stems from earlier research in which coinage of emperors of the Ch'ing Dynasty (also transliterated "Qing") were examined and found to be impure brass (*3*). We are interested in determining how far back such minting, and the required metallurgical knowledge, went.

Experimental Conditions

A Spectrace QuanX energy dispersive X-ray fluorescence spectrometer was used, with a rhodium target X-ray tube, running on fundamental parameters software, which normalizes the elements examined to 100%, and running on pure element standards. Sample excitation conditions were as follows: 22 kV, 0.10 mA, 100 sec count, Kαβ for Fe, Co, Ni, Cu, Zn, As, Pt, Au, Bi, Pb, followed by: 45 kV, 0.72 mA, 60 sec count, for Pd, Ag, Sn, Sb. Certified brass samples ranging from 36% to 94% copper were run daily prior to each run to ensure instrument accuracy and precision. All samples underwent sonication in hot, soapy water for 15 min, to ensure surfaces were clean and free of dirt or other surface contamination. After sonication, samples were examined visually and via 7–30X microscope to ensure each sample had metal exposed to the X-ray beam, and not surface patination or metal oxides.

Discussion

Modern brass and bronze alloys come in many different compositions, but 2:1 ratios of copper and zinc or tin respectively are used in a number of applications. Ancient and medieval alloys that are called brass and bronze are often mixtures of all three elements, with several other elements present as well. Observing a clean composition of copper and zinc or of copper and tin in any of the samples in this study would be a good indicator of a modern, counterfeit coin. None with such a composition were found.

Previous studies (*3, 4*) and older literature (*5–7*) indicate small amounts of lead are sometimes present and were allowed in Oriental coins, perhaps to lower working temperatures of batches of molten metal. Studies of artifacts of other cultures also indicate that metals such as lead might be present, but such can not be used as a direct indicator for these samples (*8–20*). Beyond this, when initiating this study, we had no specific expectations of the compositions being more than copper, zinc, tin, and lead.

The elemental compositions of the first emperor in the study, Chen Tsung, are represented graphically in Figures 1–3, with each graph representing the coins from one imperial title. A total of 14 coins composed the sample for Figure 1, 36 for Figure 2, and 28 for Figure 3. Figures 4–6 represent the compositions of the coin samples of Emperor Jen Tsung, with 45 coins composing the sample set for Figure 4, 28 for Figure 5, and 13 for Figure 6. The final two, Figures 7 and 8, display the compositions of those coins of Emperor Che Tsung, with 45 coins making up the sample set for Figure 7, and 15 for Figure 8. Each figure thus represents only those coins of a specific imperial title. The multiple Schjoth numbers, listed in Table I, simply delineate the coins of a specific title based on minor stroke differences in the characters on each coin.(*1*).

In each graph and illustrated in Figure 1 is an example, which also serves the following Figures 2 – 8, the diamond icons represent the copper percentage of a specimen, the triangles represent the lead percentages, the circles the tin,

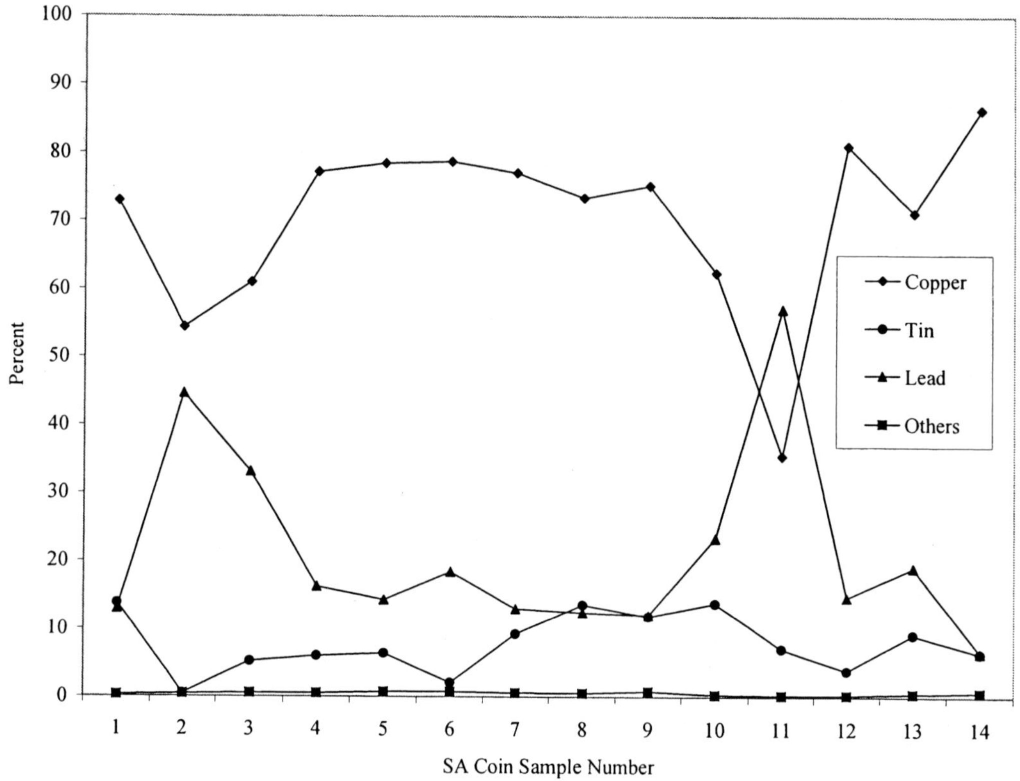

Figure 1. Compositions of Coins of the Emperor Chen Tsung, Imperial Title Hsien-p'ing.

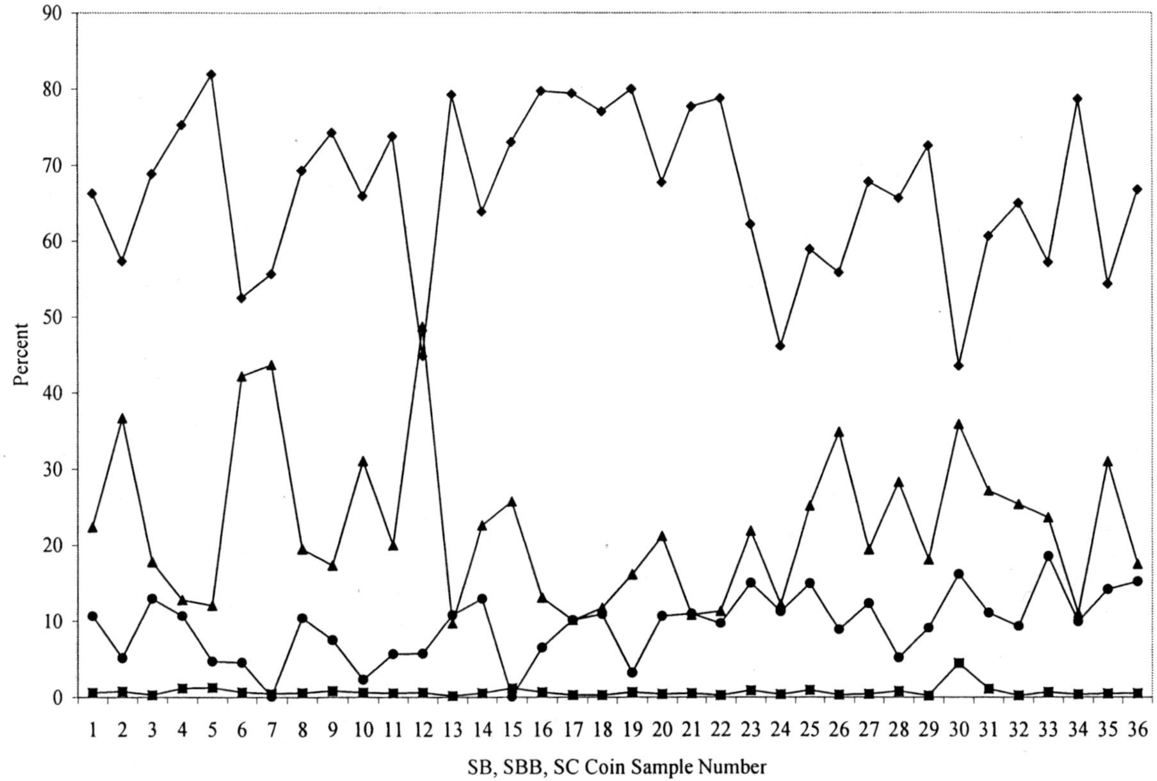

Figure 2. Compositions of Coins of the Emperor Chen Tsung, Imperial Title Hsian fu.

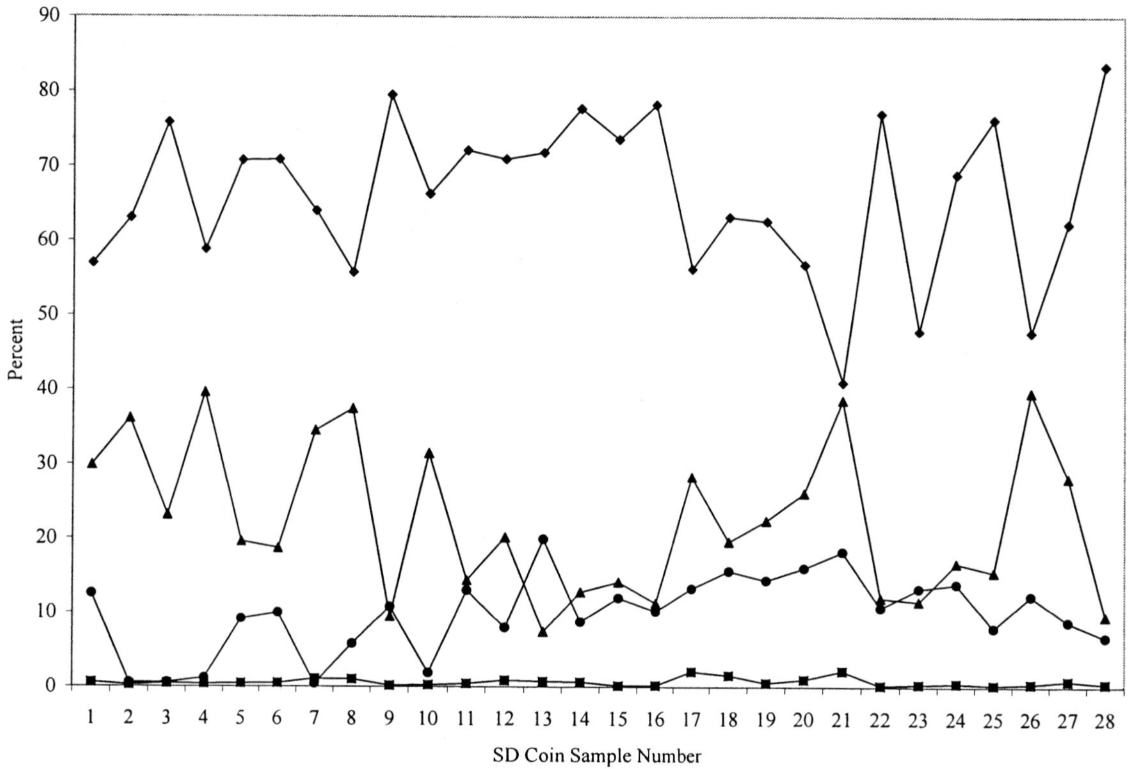

Figure 3. Compositions of Coins of the Emperor Chen Tsung, Imperial Title T'ien-hsi.

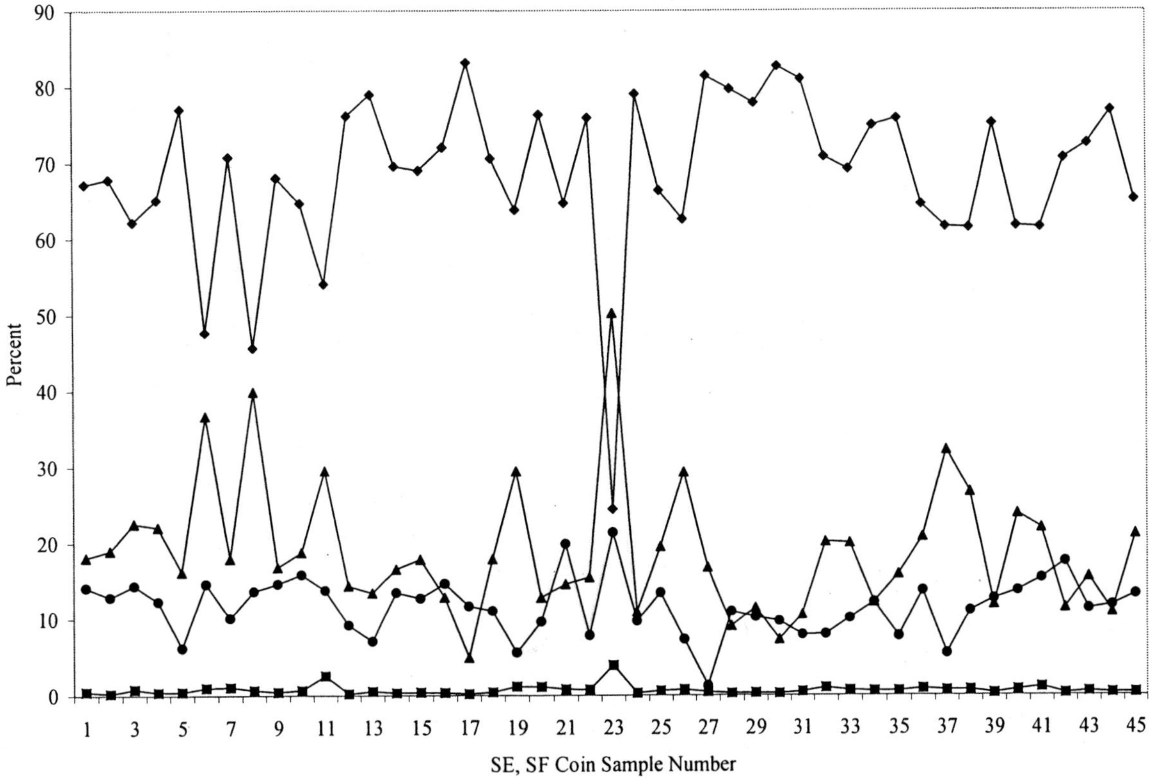

Figure 4: Compositions of Coins of the Emperor Jen Tsung, Imperial Title T'ien-sheng.

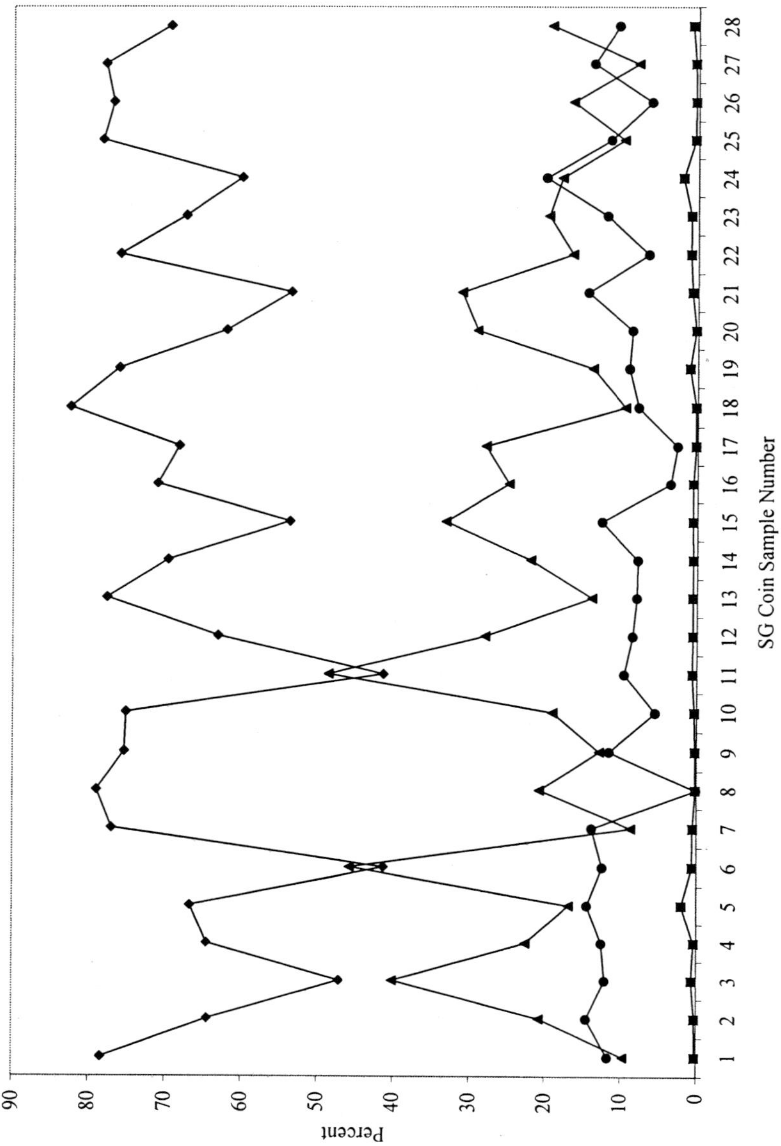

Figure 5. Compositions of Coins of the Emperor Jen Tsung, Imperial Title Ching-yu.

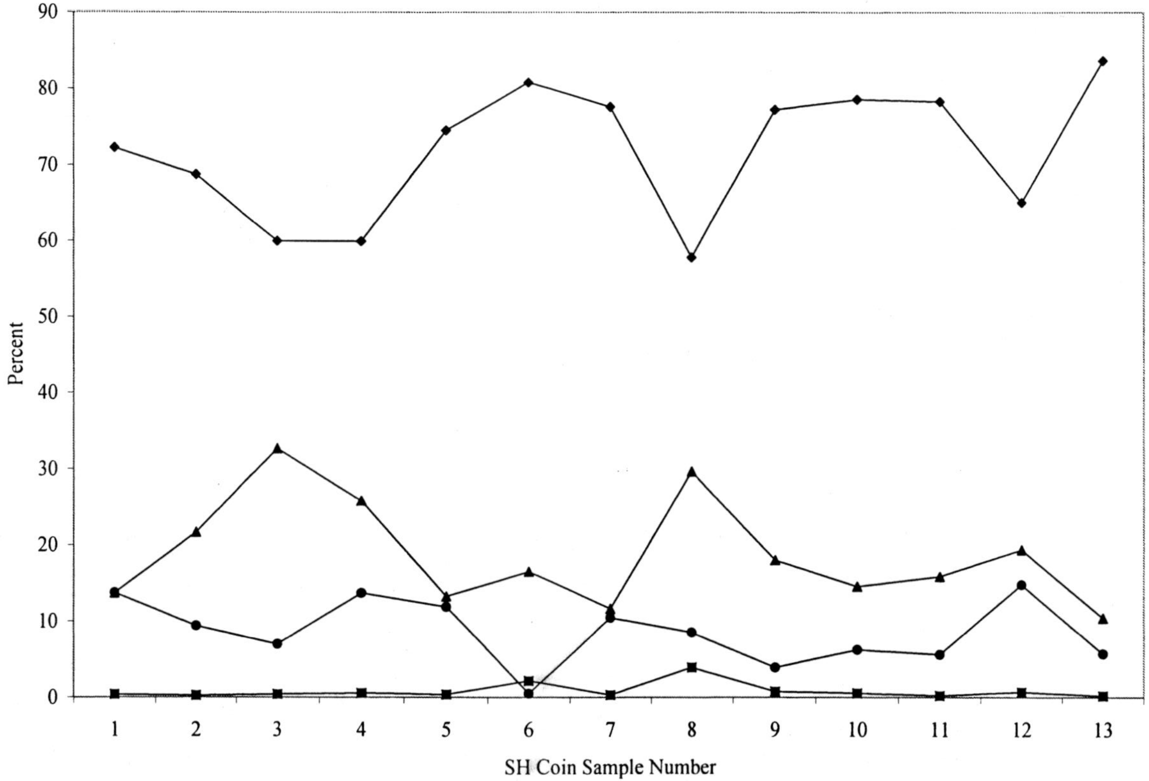

Figure 6. Compositions of Coins of the Emperor Jen Tsung, Imperial Title Pao-yuan.

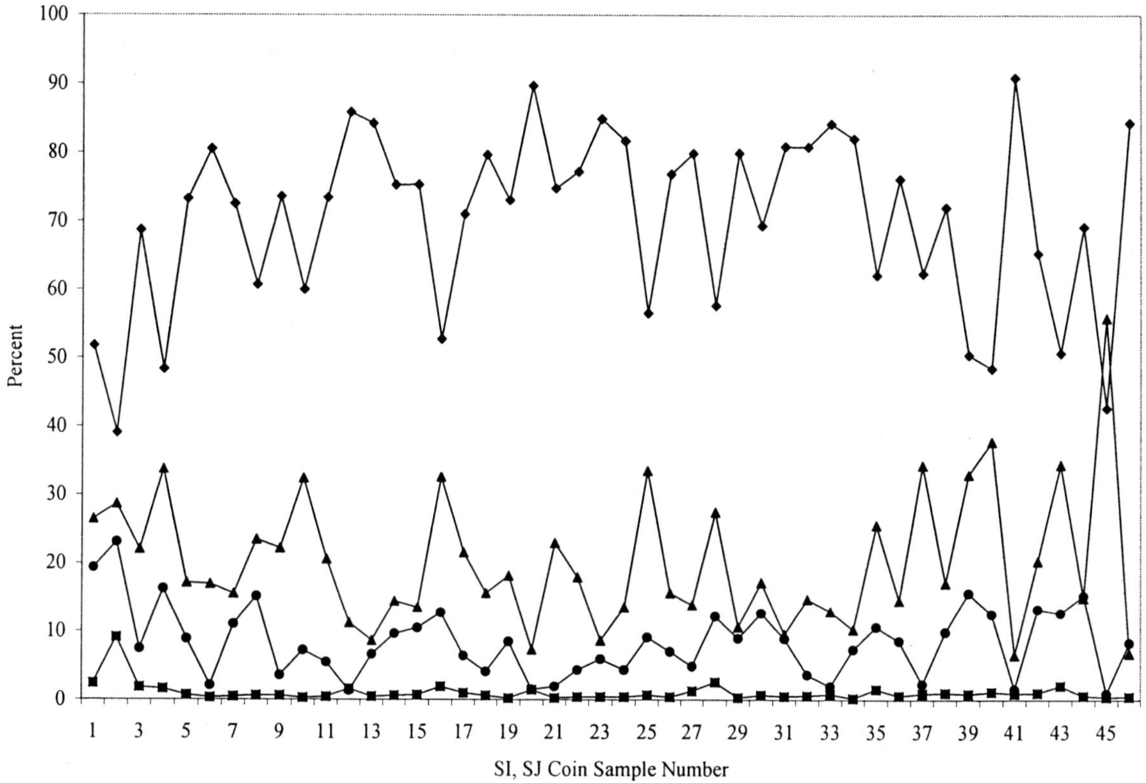

Figure 7. Compositions of Coins of the Emperor Che Tsung, Imperial Title Shao-sheng.

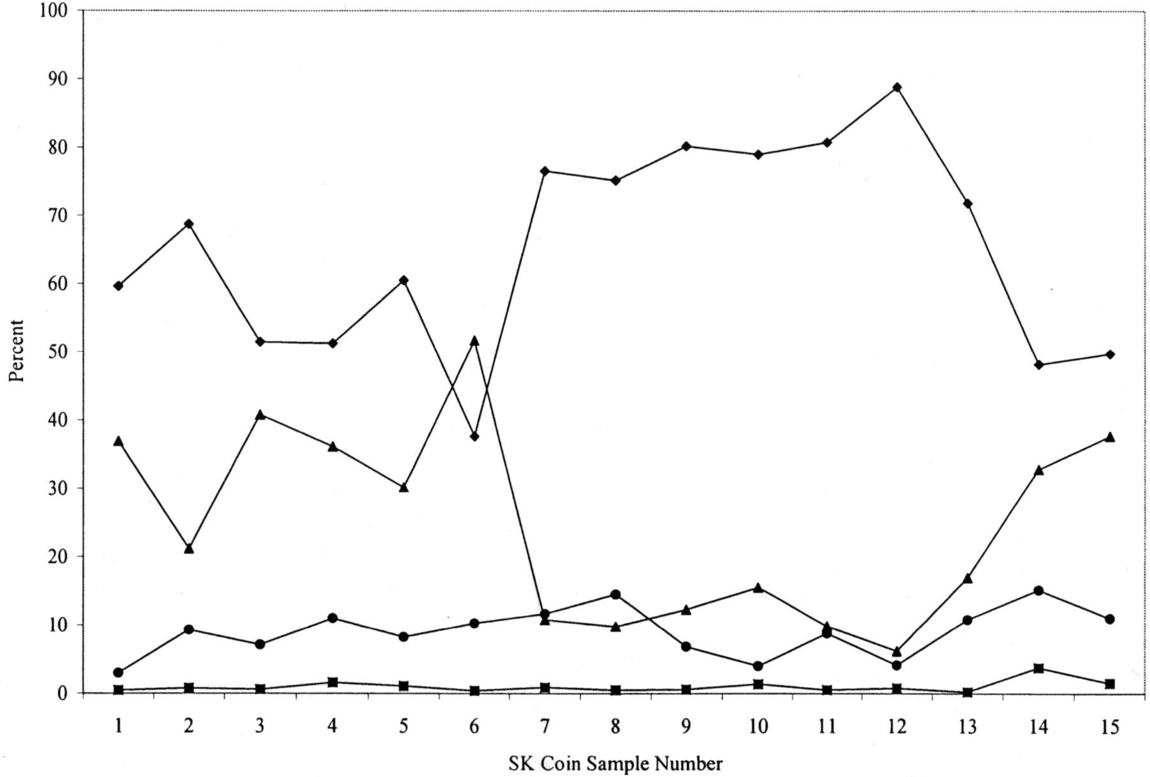

Figure 8. Compositions of Coins of the Emperor Che Tsung, Imperial Title Yuan-fu.

and the square boxes all the remaining elements. Each vertical series of four points thus represents the composition of a single coin.

It is evident that copper appears to be the major component in most of the coins of all emperors and titles. What becomes interesting is that lead is almost always the major component in the percent fraction of elements other than copper, even more so than tin. In Figures 1 – 3, for instance, it appears most prominently that even with the outliers removed, lead is present in greater amounts than tin.

Focusing on the outliers themselves, one realizes that enough lead was occasionally added to the alloy such that the coin could be considered a primarily lead piece. For the most pronounced example, Figure 4, the T'ien-Sheng title of Emperor Jen Tsung, sample number 23 has lead greater than 50%, with copper very close to 25%. This is not a surface enrichment phenomenon, as the density of this sample, and that of other outliers, is significantly greater than the other coins within their respective sets. The most uniform series in terms of standard deviation of lead content percentage is that of Figure 6, although the lot size was small enough – 13 coins – that a larger sample may have revealed similar anomalies.

Zinc is virtually non-existent in the samples, and thus is included with the trace elements, represented as square boxes. In very few instances do we have a trace element combined percentage that rises towards 5% (such as coins SC30, SF23, SH8, SI2, and SK14). Thus, it appears that a change in composition occurs between these coins of the Song Dynasty and later coins of the Qing Dynasty *(3, 5)*. Since certain Qing Dynasty coins have already been found to be leaded brass, it may be worth studying several sets of coins at points between that of the date of manufacture of these coins (ca. 1000 A.D.) and those already examined of the Qing Dynasty time frame (ca. 1750 A.D.) *(3, 5)*, to determine if only one change in composition occurred, if several occurred, or if one occurred slowly over the course of many years.

An interesting trend is seen when the coins are grouped by emperor-title and we consider the total content percentage of additives, that is, the non-copper and non-tin impurities, as shown in Table II. While the mean percentage for additives hovers around 22% in all samples, the standard deviation is seen to decrease approximately linearly with time through the Chen Tsung rule and into the Jen Tsung, before sharply reversing.

The decreasing trend may indicate a steadier, more consistent smelting technique being passed down over time. The reversal could represent some disruption in the normal allocation of mineral resources. It is known, for example, that northern border skirmishes early in Jen Tsung's reign led to mounting military expenditures later on *(21)*. Further study with a larger sample of coins might prove insightful.

Table II. Means and Standard Deviations of Content Percentages

Coin Studies	*Midyear (A.D.)*	*% Copper*		*% Tin*		*% Lead*		*% Additives*	
		Mean	Stdev	Mean	Stdev	Mean	Stdev	Mean	Stdev
SA	1001	70.41	13.19	7.78	4.31	21.29	14.23	21.82	11.70
SB,SBB,SC	1012	66.64	10.72	9.38	4.43	22.44	10.25	23.98	10.78
SD	1019	66.10	10.65	9.84	5.34	22.43	10.34	24.07	9.99
SE,SF	1027	68.64	10.82	11.60	3.80	19.07	8.80	19.76	9.25
SG	1035.5	67.68	11.51	10.04	4.24	21.64	10.74	22.28	10.90
SH	1038.5	71.90	8.71	8.57	4.26	18.71	6.95	19.54	11.03

Conclusions

The coins appear in almost all cases to be leaded bronze. In a number of samples, the outliers contained as much or more lead than copper, and in many instances, the lead content was significantly higher than that of tin.

None of the coins can be considered to be brass. While a few of the many samples did have a small amount of zinc, most had such low concentrations that they were close to the lower detection limit of the instrument.

The bronze compositions are decidedly different from the elemental make up of later series of Qing Dynasty cash from the 18th century, which are brass (*3*). Perhaps obviously, a significant change in coin metallurgy occurred between these two eras. It may be worth investigating when this change occurred, and determining what the reasons were for such.

Finally, paying attention to the outliers of these series, the element lead seems to be added almost at random to all the series of coins, though over time the use of additives seems to have become more consistent, for the most part. There is certainly more lead present than is required to effect lower working temperatures for molten alloys. Other than this variance in lead however, these coins have a notably consistent composition. It appears that the mintmasters and workmen of the foundries had a good understanding and control of copper, lead, and tin in the production of alloys.

References

1. Schjoth, F. *Chinese Currency*; Oslo, 1929; pp 64–66.
2. Lockhart, J. H. S. *The Lockhart Collection of Chinese Copper Coins*; Quarterman Publications, Inc: Lawrence, MA, 1975.
3. Gaines, T.; McGrath, E.; Iduma, V.; Kuzava, R.; Frederick, S.; Benvenuto, M. In *Archaeological Chemistry: Materials, Methods, and Meaning*; Jakes, K. A., Ed.; ACS Symposium Series 831; Washington, DC, 2002; pp.231–244.
4. Gaines, T.; McGrath, E.; Langrill, R.; Benvenuto, M. *Numismatic Chronicle* **2001**, *161*, 308–317.
5. Hartill, D. *Numismatic Chronicle* **1991**, *151*, 67–120.
6. Lutz, J.; Pernicka, E. *Archaeometry* **1996**, *38*, 313–323.
7. Mabuchi, H.; Notsu, K.; Nishimatsu, S.; Fuwa, K.; Iyama, H.; Tominaga, T. *Nippon Kagaku Kaishi* **1979**, *5*, 586–590.
8. Meyers, P.; van Zelst, L.; Sayre, E. V. In *Archaeological Chemistry*; Beck, C.W., Ed.; Advances In Chemistry Series No. 138; American Chemical Society: Washington, DC, 1974; pp 22–33.

9. Gordus, A. A.; Gordus, J. P. In *Archaeological Chemistry*; Beck, C. W., Ed.; Advances in Chemistry Series No. 138; American Chemical Society: Washington, DC, 1974; pp 124–147.
10. Chase, W. T. In *Archaeological Chemistry*; Beck, C. W., Ed.; Advances in Chemistry Series No. 138; American Chemical Society: Washington, DC, 1974; pp 148–185.
11. Chase, W. T.; Ziebold, T. O. In *Archaeological Chemistry II*; Carter, G. F., Ed.; Advances in Chemistry Series No. 171; American Chemical Society: Washington, DC, 1978; pp 293–334.
12. Carter, G. F. In *Archaeological Chemistry III*; Lambert, J. B., Ed.; Advances in Chemistry Series No. 205; American Chemical Society: Washington, DC, 1984; pp. 311–332.
13. Gale, N. H.; Stos-Gale, Z. A. In *Archaeological Chemistry IV*; Allen, R. O., Ed.; Advances in Chemistry Series No. 220; American Chemical Society: Washington, DC, 1989; pp 159–198.
14. Carter, G .F.; Razi, H. In *Archaeological Chemistry IV*; Allen, R. O., Ed.; Advances in Chemistry Series No. 220; American Chemical Society: Washington, DC, 1989; pp 199–212.
15. Moreau, J.-F.; Hancock, R. G. V. In *Archaeological Chemistry: Organic, Inorganic, and Biochemical Analysis*; Orna, M. V., Ed.; ACS Symposium Series No. 625; American Chemical Society: Washington, DC, 1996; pp 64–82.
16. Gordus, A. A.; Henderson, C. E.; Shimada, I. In *Archaeological Chemistry: Organic, Inorganic, and Biochemical Analysis*; Orna, M.V., Ed.; ACS Symposium Series No. 625; American Chemical Society: Washington, DC, 1996; pp 83–93.
17. Carter, G. F. In *Archaeological Chemistry: Organic, Inorganic, and Biochemical Analysis*; Orna, M. V., Ed.; ACS Symposium Series No. 625; American Chemical Society: Washington, DC, 1996; pp 94–106.
18. Carter, G. F. *Am. J. Numismatics* **1997**, *7–8*, 235–250.
19. Kuntz, M.; Ferguson, J.; Iduma, V.; Kuzava, R.; Benvenuto. M. *J. Undergrad. Chem. Res.* **2002**, *1*, 29–37.
20. Ellis, I.; Benvenuto, M. *J. Undergrad. Chem. Res.* **2005**, *4*, 107–112.
21. *Encyclopaedia Britannica*; URL http://www.britannica.com/eb/article-9043514.

Chapter 13

Elemental Compositions of Herodian Prutah, Copper Coins—of the Biblical "Widow's Mites" Series—via Energy Dispersive X-ray Fluorescence

Meghann Mouyianis[1], Jeffe Boats[2], and Mark A. Benvenuto[1]

Departments of [1]Chemistry and Biochemistry and [2]Mathematics and Computer Science, University of Detroit Mercy, Detroit, MI 48221

Thirty six small copper coins issued under the authority of Herod Agrippa I were analyzed using energy dispersive X-ray fluorescence spectrometry for copper, zinc, tin, lead, antimony, iron, gold, silver, and several other elements. This series of coins show significant amounts of lead in the coins, but an otherwise unadulterated bronze composition, with very little in the way of trace elements. The metallurgical make up of the samples and implications of the findings are presented here.

Introduction

The coins of the Herodian Kings, as well as coinage of other subordinate kings and rulers of the Roman Empire, were tolerated in the early imperium, producing an overall coinage system that was Roman throughout the Empire, but that had areas of local production and influence as well. The coins analyzed in this study, prutahs, each display what appears to be three wheat ears on one side, and an umbrella on the other and were minted under the authority of King Herod Agrippa I (ca. 40 A.D.) (*1*). These or their smaller siblings are often referred to as "widow's mites" because of the Biblical reference to them. According to the King James Bible, Mk 12: 41–44:

> "And Jesus sat over against the treasury, and beheld how the people cast money into the treasury; and many that were rich cast in much.
> And there came a certain poor widow, and she threw in two mites, which make a farthing.
> And he called unto him his disciples, and saith unto them, Verily I say unto you, That this poor widow hath cast more in, than all they which have cast into the treasury;
> For all they did cast in of their abundance; but she of her want did cast in all that she had, even all her living."

It is difficult today to determine what value such coins had in ancient times, but it is certainly apparent that these were the small change of their era and locale. Roman coins were available and used throughout the Roman Empire, but these coins were used within the smaller, localized area of influence of the Herodian kings (*2*). Whether Herodian coins were used in other transactions besides those involving the Temple treasury is also difficult to determine exactly. But it seems logical to conclude that there was most likely other local, commercial uses for such coinage as well (*1, 2*).

Older references to these coins simply indicate them to be copper, brass, or bronze (*1, 2*). The experience from our research and that of others indicates that ancient alloys are usually more complex than either a single element, or a simple alloyed mixture (*3–18*).

Experimental Conditions

The sample set consisted of 36 coins, all being of a design representing three wheat ears on one side, and what is generally considered an umbrella on the other. The coins are round, though not perfectly so, with minimum diameters from 14.1 – 18.2 mm (standard deviation 0.890), and maximum diameters of 15.9 – 19.3 mm (standard deviation 0.848), as seen in Table I. Their thicknesses range from 1.6 – 2.5 mm (standard deviation 0.212). While this sample set is not excessively large, it is larger than most previous studies of coins of the Roman Republic or Empire (13, 16), and encompasses a sampling of an overall mintage that is most likely much less than that of the Roman Empire at a comparable time (since the Herodian kings were subordinate to the Empire and had authority over

Table I. Physical Measurements and Chemical Compositions of the Coins

Sample #	*Min D (mm)*	*Max D (mm)*	*Thick (mm)*	*Cu%*	*Pb%*	*Sn%*	*Ag%*	*Fe%*	*Sb%*
WM1	16.1	17.2	1.8	88.3	2.58	8.57	0.06	0.38	0.07
WM2	17.0	18.5	2.0	87.6	8.86	2.72	0.03	0.62	0.10
WM3	16.1	18.8	1.9	81.8	11.01	6.51	0.05	0.43	0.11
WM4	16.4	17.3	2.0	68.7	24.85	5.73	0.02	0.49	0.08
WM5	17.0	17.8	2.5	90.3	4.46	4.06	0.03	0.70	0.09
WM6	15.8	17.9	2.4	90.2	2.80	6.49	0.02	0.30	0.06
WM7	16.6	17.9	1.9	74.2	12.30	12.53	0.06	0.43	0.44
WM8	16.5	17.4	1.9	80.8	7.91	10.48	0.04	0.58	0.11
WM9	14.6	16.3	2.4	91.1	1.30	7.02	0.03	0.41	0.11
WM10	16.0	16.2	2.3	92.7	3.28	3.65	0.01	0.34	0.07
WM11	15.9	17.4	1.8	84.5	10.71	3.94	0.02	0.73	0.10
WM12	15.5	16.2	2.2	80.6	10.31	8.33	0.04	0.61	0.11
WM13	17.0	18.5	1.8	76.5	5.55	16.96	0.05	0.79	0.07
WM14	16.9	17.9	1.8	92.6	1.39	5.64	0.03	0.31	0.05
WM15	16.4	16.8	2.0	77.5	12.20	9.69	0.06	0.33	0.17
WM16	16.3	16.4	2.1	80.1	11.74	7.53	0.01	0.43	0.19
WM17	18.0	18.4	1.9	87.7	2.23	9.70	0.04	0.24	0.09
WM18	17.1	17.2	2.0	68.3	24.03	6.89	0.04	0.66	0.09
WM19	15.1	17.2	2.4	86.2	5.33	7.69	0.04	0.69	0.09
WM20	17.1	17.2	2.0	88.8	8.48	2.09	0.02	0.50	0.04
WM21	15.9	16.3	2.0	93.0	0.61	5.98	0.02	0.29	0.03
WM22	18.2	18.4	2.4	80.2	9.09	9.75	0.03	0.80	0.12
WM23	16.4	17.2	2.2	69.1	19.02	10.72	0.06	0.87	0.15
WM24	16.3	17.3	1.9	82.6	7.40	9.16	0.05	0.58	0.13
WM25	16.8	18.2	2.1	80.5	8.64	10.08	0.04	0.50	0.17
WM26	16.8	18.3	2.0	77.9	14.39	7.10	0.05	0.30	0.19
WM27	14.1	15.9	1.9	71.4	12.87	15.26	0.05	0.23	0.12
WM28	17.0	17.2	2.2	78.6	15.12	5.31	0.04	0.80	0.09
WM29	17.7	18.1	2.4	90.1	7.30	1.87	0.01	0.56	0.13

Table I. *Continued.*

Sample #	*Min D (mm)*	*Max D (mm)*	*Thick (mm)*	*Cu%*	*Pb%*	*Sn%*	*Ag%*	*Fe%*	*Sb%*
WM30	15.8	17.0	2.1	71.8	24.58	2.72	0.03	0.73	0.09
WM31	15.1	16.3	2.2	90.9	5.69	2.79	0.01	0.57	0.06
WM32	17.0	17.1	2.1	74.1	19.61	5.45	0.05	0.66	0.10
WM33	15.6	16.5	2.0	68.1	18.04	13.38	0.04	0.31	0.09
WM34	17.2	19.3	2.1	91.9	3.00	4.57	0.02	0.42	0.08
WM35	17.3	18.1	2.0	91.4	4.97	2.98	0.02	0.53	0.09
WM36	17.1	18.1	1.6	73.3	15.55	10.39	0.05	0.53	0.12
Std Dev	0.89	0.85	0.21	8.1	6.78	3.70	0.02	0.18	0.07

only a small area of the modern day Middle East, whereas the Roman Empire stretched roughly from modern Britain to modern Iraq).

A Spectrace QuanX energy dispersive X-ray fluorescence spectrometer was used, which employed a rhodium target X-ray tube, fundamental parameters software, and pure element standards. Sample excitation conditions were: 30kV, 0.10mA, 100 sec count, Kαβ for Fe, Co, Ni, Cu, Zn, As, Pt, Au, Bi, and Pb, followed by: 50 kV, 0.72 mA, 60 sec count, for Pd, Ag, Sn, and Sb. Certified brass samples were run each day prior to sample runs to ensure instrument accuracy and precision.

Samples underwent sonication in hot, soapy water for 15 minutes, to ensure surfaces were clean and free of any surface contamination. The more elaborate preparation method of Carter (*10, 13*) was not utilized because visual and 7–30X microscopic examination of each coin revealed no remaining encrustation or patination.

The entire set of coins was analyzed twice on each side as a minimum, and the results averaged to ensure reproducibility of the data.

Discussion

Coins and artifacts of the Roman Empire have been studied extensively (*4, 5, 19–39*), but a thorough search of the literature indicates these small, copper coins of the Herodian King Agrippa have been omitted from all previous examinations. The iconography of the coins in this sample set is crude, but all have at least a recognizable wheat ear side or an umbrella side. A worn state or weak design elements are actually not detractors in this case. Rather, one can argue that because of the wear the surfaces being analyzed do not suffer from any elemental enrichment process at a surface that might have resulted from the original method of manufacture.

The coins appear to be essentially a ternary alloy: copper, tin, and lead. Figure 1 displays the copper percentages of the samples in comparison to each other, and Table I shows all copper and other elemental compositions. While the samples with the lowest percentages are near 70%, nearly half of the sample set shows 90% or higher copper. Samples of ancient coinage with this level of purity are rare, but not unheard of (*4*).

Figure 2 illustrates the lead percentage of each coin in the study, again, in comparison to each other. Note that the y-axis peaks at 30%, thus emphasizing the lead differences somewhat in relation to the copper. Coins WM4, WM18, and WM30 are all high in lead, and predictably, are also low in copper. It is noteworthy however, that over half of the samples contain less than 10% lead. This is much lower than in other studies of ancient or medieval coinage (*5, 17, 40, 45*).

Figure 3 shows the percentages of tin in each sample, compared against each other, in similar format to Figures 1 and 2. In this case, the y-axis peaks at 20%, because the samples contain a relatively small amount of tin in relation to copper.

Because it may be difficult to compare the percentages of the three components using only the data as presented in Figures 1–3, Figure 4 shows copper, lead, and tin plotted on the same scale, and Table I reiterates this with the averaged percentage for each coin for each element. While the nuances of the differences in lead and tin compositions may be lessened through the comparison in Figure 4, their relation to copper, the major element, is heightened.

It is immediately evident that there is less tin and more lead on average for the coins – the mean lead percentage is 9.9%, compared with only 7.3% tin. Also noteworthy is the relatively large sample standard deviation of 6.78 in lead percentages, as opposed to 3.70 in tin percentages. The standard deviation can also be compared with the lead mean of 9.9%, indicating a wider variation in lead content. While this large standard deviation does not prove unequivocably that lead was added as an alloying element, it does seem to suggest this to be true. Since these coins were produced for local use, and use of a much more limited extent than the coins of the Roman Empire, it can be deduced that ore sources were most likely local or in the near proximity to the minting site.

The standard deviation of tin content, when compared with the data, suggests that the tin percentages are normally distributed, the expected result for coins made by the same mint or by the same methods. Twenty-four of the thirty-six coins (approximately 67%) had tin percentages within one standard deviation of the mean, and thirty-four of the coins (approximately 95%) fell within two

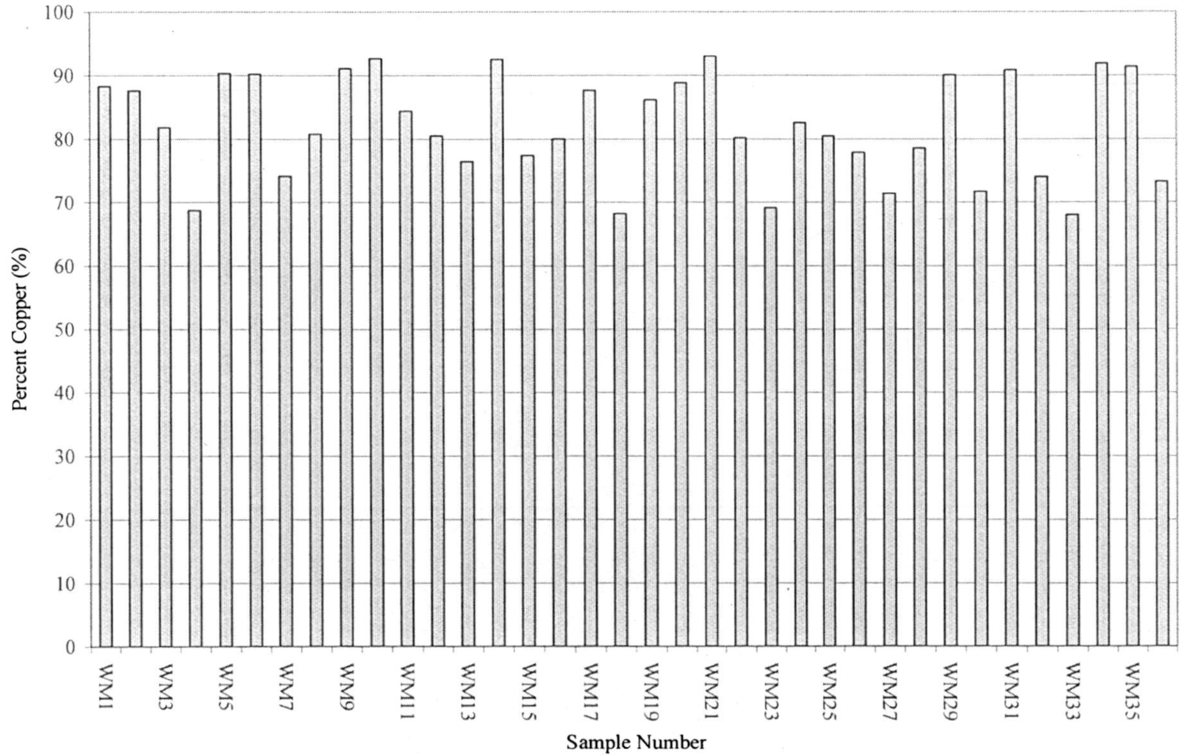

Figure 1. Copper Percentages of Widow's Mites Coins

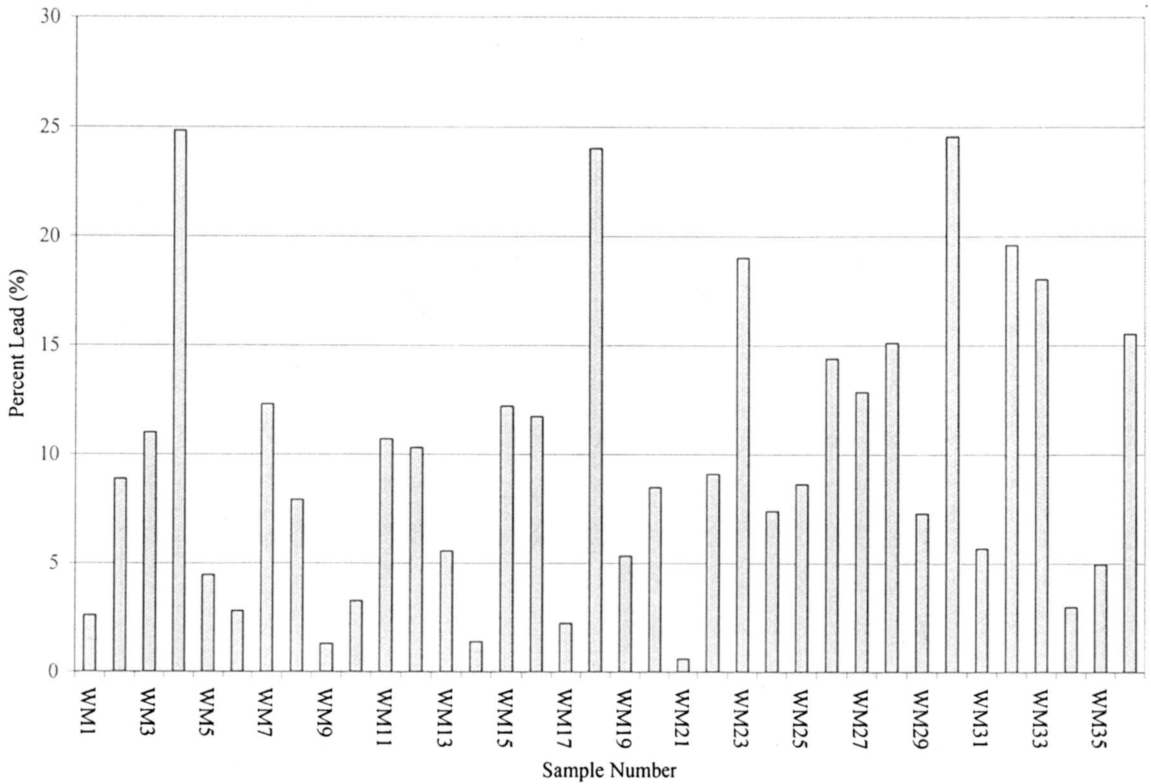

Figure 2. Lead Percentages of Widow's Mites Coins

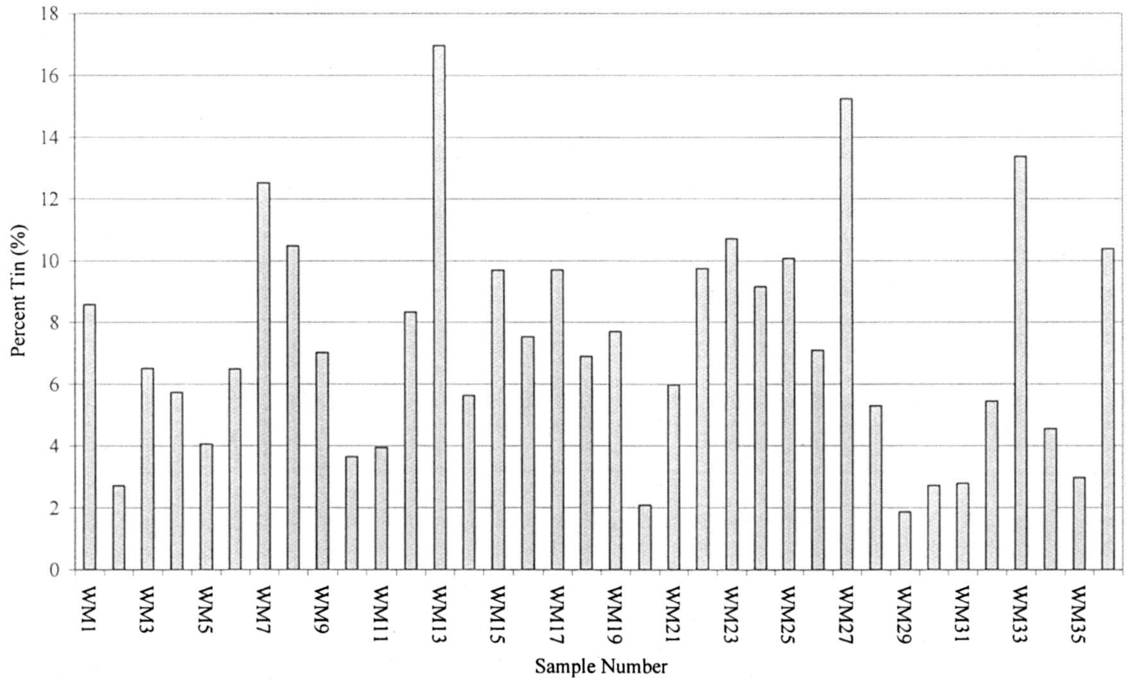

Figure 3. Tin Percentages of Widow's Mites Coins

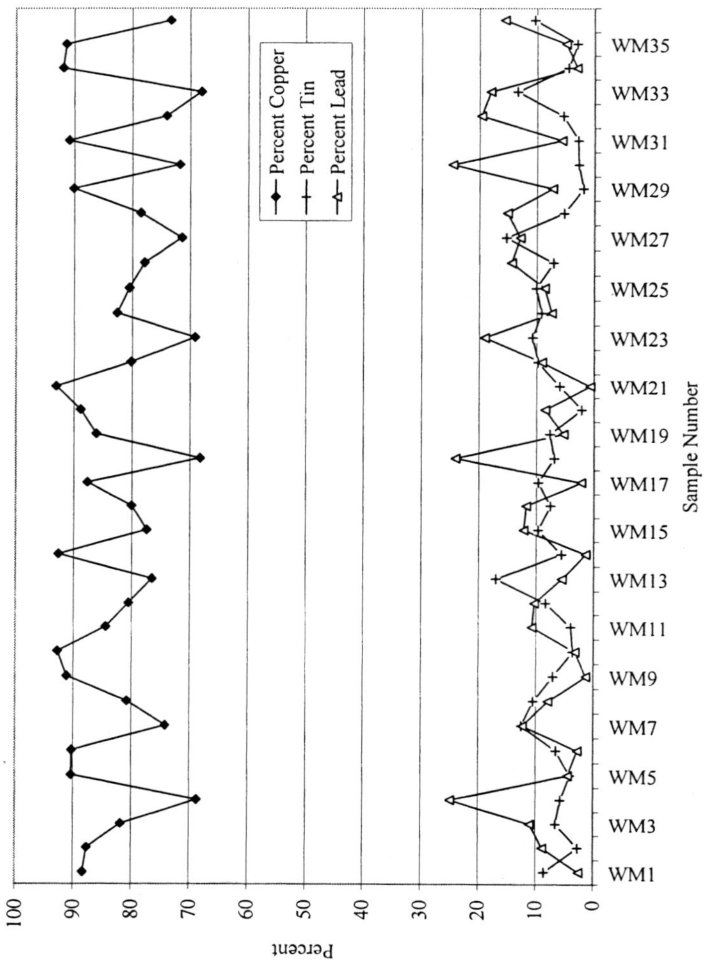

Figure 4. Relative Percentages of Copper, Lead, and Tin in Widow's Mites Coins

standard deviations. The analysis of lead and copper content offer no evidence contrary to this assertion.

Beyond this, iron is present in the range of only 0.2% - 0.9%, which is both quite low and rather uniform. Other studies of coins from different cultures and times show similar, low iron percentages (*4, 5, 17, 40*). In previous studies on Roman coinage of roughly similar dates, Carter found iron concentrations in the range of 0.05 – 0.4% (*10, 13*). There appears to be continued disagreement among scholars as to whether iron in the alloys was present in original ores, is a purposeful addition to the alloy, or is a by-product of the alloy being molten in some iron container (*42*) during the production process. In the case of these 36 coins it remains noteworthy that so little iron was found in any of the samples.

Other elements for which this sample set was analyzed were present in statistically insignificant amounts with the exceptions of silver and antimony. Gold, which is present in numerous copper ores even today (and the recovery of which sometimes pays for the electrical requirements of copper refining), is notably absent from these coins. The small amount of silver present is most likely a natural trace in the copper ores. As just mentioned concerning gold, trace amounts of silver are also recovered from copper ores during refining today, and used to help pay for the copper refining operation. In addition, only trace amounts of zinc appear, approximately at the detection limit of the instrument, and in fewer than one quarter of the coins.

Conclusions

The coins are leaded bronze. The almost complete absence of zinc indicates they are definitely not brass.

The elemental composition shows surprisingly clean make ups overall, considering the technological abilities of the age and culture. This could be an indication that the ancient Judaean foundry workers and minters had access to remarkably pure ore sources, or could be evidence of their ability to refine those less pure ores which were available to them.

The presence of varying amounts of lead in this series of coins appears to be an indicator that the mint workers had some knowledge of leaded alloys and their metallurgy, specifically the knowledge that an alloy was easier to work upon the addition of lead (because the resultant alloy had a lower melting point). Another possibility, that the differing amounts of lead are present in original copper or tin ores in roughly these percentages or ratios simply by chance, seems less likely, because of the variation in lead content from coin to coin.

References

1. Reinach, T. *Jewish Coins*; Argonaut, Inc. Publishers: Chicago, IL, 1966.
2. Sayles, W. G. *Ancient Coin Collecting VI*; Krause Publications: Iola, WI, 1999; pp 111–112.
3. Lutz, J.; Pernicka, E. *Archaeometry* **1996**, *38*, 313–23.
4. Ellis, I.; Benvenuto, M. *J. Undergrad. Chem. Res.* **2005**, *4*, 107–112.
5. Carter, G. F. *Am. J. Numismatics* **1997**, *7–8*, 235–250.
6. Meyers, P., van Zelst, L., Sayre, E. V. In *Archaeological Chemistry*; Beck, C. W., Ed.; Advances in Chemistry Series 138; American Chemical Society: Washington, DC, 1974; pp 22–33.
7. Gordus, A. A., Gordus, J. P. In *Archaeological Chemistry*; Beck, C. W.; Ed.; Advances in Chemistry Series 138; American Chemical Society: Washington, DC, 1974; pp 124–147.
8. Carter, G. F. In *Archaeological Chemistry II*; Carter, G. F. Ed.; Advances in Chemistry Series 171; American Chemical Society: Washington, DC, 1978; pp 347–379.
9. Rapp, G., Jr., Allert, J., Henrickson, E. In *Archaeological Chemistry III*; Lambert, J.B., Ed.; Advances in Chemistry Series 205; American Chemical Society: Washington, DC, 1984; pp 273–310.
10. Carter, G. F. In *Archaeological Chemistry III*; Lambert, J. B., Ed.; Advances in Chemistry Series 205; American Chemical Society: Washington, DC, 1984; pp 311–332.
11. Gale, N. H.; Stos-Gale, Z. A. In *Archaeological Chemistry IV*; Allen, R. O. Ed.; Advances in Chemistry Series 220; American Chemical Society: Washington, DC, 1989; pp 159–198
12. Manea-Krichten, M. C.; Heidebrecht, N.; Miller, G. E. In *Archaeological Chemistry IV*; Allen, R. O. Ed.; Advances in Chemistry Series 220; American Chemical Society: Washington, DC, 1989; pp 199–211.
13. Carter, G. F.; Razi, H. In *Archaeological Chemistry IV*; Allen, R. O. Ed.; Advances in Chemistry Series 220; American Chemical Society: Washington, DC, 1989; pp 213–232.
14. Moreau, J.-F.; Hancock, R. G. V. In *Archaeological Chemistry: Organic, Inorganic, and Biochemical Analysis*; Orna, M.V., Ed.; Advances in Chemistry Series 625; American Chemical Society: Washington, DC, 1996; pp 64–82.
15. Gordus, A. A.; Henderson, C. E.; Shimada, I. In *Archaeological Chemistry: Organic, Inorganic, and Biochemical Analysis*; Orna, M. V., Ed.; Advances in Chemistry Series 625; American Chemical Society: Washington, DC, 1996; pp 83–93.
16. Carter, G.F. In *Archaeological Chemistry: Organic, Inorganic, and Biochemical Analysis*; Orna, M. V., Ed.; Advances in Chemistry Series 625; American Chemical Society: Washington, DC, 1996; pp 94–106.

17. Gaines, T.; McGrath, E.; Iduma, V.; Kuzava, R.; Frederick, S.; Benvenuto, M. In *Archaeological Chemistry: Materials, Methods, and Meaning*; Jakes, K.A., Ed.; Advances in Chemistry Series 831; American Chemical Society: Washington, DC, 2002; pp 231–246.
18. Mabuchi, H.; Notsu, K.; Nishimatsu, S.; Fuwa, K.; Iyama, H.; Tominaga, T. *Nippon Kagaku Kaishi* **1979**, *5*, 586–590.
19. Carter, G. F. *The American Numismatic Society Museum Notes* **1988**, *33*, 91–106.
20. Sheridan, J. A. *The American Numismatic Society Museum Notes* **1988**, *33*, 107–110.
21. Long, J. *The American Numismatic Society Museum Notes* **1988**, *33*, 111–118.
22. Paz Garcia-Bellido, M. *Am. J. Numismatics* **1989**, *1*, 37–50.
23. Metcalf, W. E. *Am. J. Numismatics* **1989**, *1*, 51–70.
24. Kleiner, F. S. *Am. J. Numismatics* **1989**, *1*, 71–78.
25. Hoskins Walbank, M.E. *Am. J. Numismatics* **1989**, *1*, 79–88.
26. Haley, E. *Am. J. Numismatics* **1989**, *1*, 89–116.
27. Newman, R. *Am. J. Numismatics* **1990**, *2*, 37–64.
28. Jones, C. P. *Am. J. Numismatics* **1990**, *2*, 65–76.
29. Damsky, B. L. *Am. J. Numismatics* **1990**, *2*, 77–106.
30. Ermatinger, J. *Am. J. Numismatics* **1990**, *2*, 107–118.
31. Cerutti, S. *Am. J. Numismatics* **1993–94**, *5–6*, 69–88.
32. Kos, P. *Am. J. Numismatics* **1993–1994**, *5–6,* 89–96.
33. McAlee, R. E. *Am. J. Numismatics* **1995–1996**, *7–8*, 113–144.
34. Metcalf, W. E.; Fulco, S. J., W .J. *Am. J. Numismatics* **1995–1996**, *7–8*, 145–54.
35. Butcher, K.; Ponting, M. *Am. J. Numismatics* **1997**, *9*, 17–36.
36. Galst, J. M. *The Am. J. Numismatics* **1998**, *10*, 103–104.
37. Beckmann, M. *Am. J. Numismatics* **2000**, *12*, 119–156.
38. Naiden, F. S. *Am. J. Numismatics* **2003**, *15*, 41–52.
39. Holt, W. C. *Am. J. Numismatics* **2003**, *15*, 61–76.
40. Gaines, T.; McGrath, E.; Langrill, R.; Benvenuto, M. *Numismatic Chronicle* **2001**, *161*, 308–17.
41. Hartill, D. *Numismatic Chronicle* **1991**, *151*, 67–120.
42. Personal communication between M. A. Benvenuto and G. Travis, Numismatic Consultant, NC.

Chapter 14

Chemical Composition of the Isfiya and Qumran Coin Hoards

Michael Notis[1], Aaron Shugar[1,2], Danny Herman[3], and Donald T. Ariel[4]

[1]Archaeometallurgy Laboratory, Department of Materials Science and Engineering, Lehigh University, Bethlehem, PA 18015
[2]Art Conservation Department, Buffalo State College, Buffalo, NY 14222
[3]Unaffiliated
[4]Coin Department, Israel Antiquities Authority, Jerusalem, Israel

The Isfiya and Qumran coin hoards are the largest hoards of Tyrian shekels (tetradrachms) and half-shekels (didrachms) found to date. They represent a very significant source of information concerning the important mint in Tyre and the production technology of silver coinage. However, analysis of coins in large numbers has, until now, been difficult because of accessibility to collections and traditional destructive sampling problems. The ability to perform analysis on large numbers of ancient objects represents a significant change in the way that we are able to view and interpret the meaning and relevance of the scientific results. Two analytical techniques were used to overcome this problem: a new non-destructive analysis method using a handheld X-ray fluorescence (XRF) unit enabled the study of a total of over 700 coins from both hoards (and thus, obtain analytical statistics on an unusually large number of coins); and electron probe micro-analysis (EPMA) was performed on four coins to compare and to confirm the accuracy of the XRF measurements. Based on the results of analysis, it was found that while the coins all show relatively constant high silver levels during the years before and after 44–15 B.C.E., the silver content is erratic and drops significantly for a period of time during these dates. The results are interpreted in the context of social and economic conditions in Tyre and nearby Judea and Syria during that time period. In particular, it appears that interactions between Herod and Augustus may be reflected in the variation of the silver content of the coins.

Introduction

The coin hoards from Isfiya (*1*) and Qumran (*2*) are the two largest hoards of Tyrian shekels (tetradrachms) and half-shekels (didrachms) that have been found to date (*3*). They represent a significant source of information concerning the mint in Tyre and the production technology of silver coinage. However, analysis of the coins in large numbers has before now, been difficult because of accessibility and traditional destructive sampling problems.

As indicated by Meshorer (*4*) the Tyrian shekel, showing the laureate head of Melkart on the obverse side and a standing eagle on the reverse side, was minted continuously for over 190 years starting from 126 B.C.E. By the end of Second Temple period, the Tyrian shekel was the only accepted currency for the yearly census tax, presumably because of its reputation as a coin of high purity silver. [For example, the following texts from the Talmudic era: Mishnah Bekhorot 8:7, also the Babylonian Talmud (BT) Bekhorot 49a-51a, BT Bava Kama 36b, BT Kiddushin 11a, and Tosefta Ketubot 12:6–the very last paragraph in Tosefta Ketubot; in some editions 13.2 or 13.3, not 13.20 as reported in numerous publications]. Furthermore, Meshorer (*4*) and others have noted the presence of a control mark, KP (kappa rho) or KAP on all Tyrian shekels minted after about 17 B.C.E., but the meaning or significance of this control mark has remained in question. It was believed that the present study could shed light on some of these issues.

Experiment

Methodology

Two analytical techniques were used for the analysis of the coins. First and foremost, a non-destructive handheld X-ray fluorescence (XRF) unit was used. Electron probe micro-analysis (EPMA) was also performed on four coins to compare and confirm the accuracy of the XRF measurements.

Source of the Samples

Seven hundred and five coins of the Isfiya hoard and 128 coins of the Qumran hoard, now in the Coin Department of the Israel Antiquities Authority (IAA), were sampled, courtesy of the IAA. This constitutes 16% (705 of approximately 4400 coins) of the original Isfiya hoard (90% of the available coins) and 23% (128 of 561 coins) of the original Qumran hoard (84% of the available coins).

XRF Theory and General Information

XRF analysis involves directing a beam of X-rays at a small area (~ 2 x 5mm) on an artifact or sample and measuring the wavelength and intensity of the secondary X-rays that are 'fluoresced' by the area hit by the primary X-rays. The wavelengths correspond to the elements present, and their intensity is directly related to concentration.

Hand-held portable XRFs, such as the one used for analyzing our coins, have improved considerably over the last few years. The quality of the instruments and their accuracy has increased and has become quite reliable. The development and commercialization of small XRF devices was, until recently, limited by the poor energy resolution of the detectors and by problems associated with transportation of radioisotopic X-ray sources (*5,6*). These shortcomings have now been overcome due to the development of thermo-electrically cooled detectors with improved energy resolution and the production of small dedicated X-ray tubes with good stability (*7*). This offers the ability to analyze elements from Ti(Z=22) to U(Z=92).

The X-rays only penetrate about 20 microns, and as such, this is considered a surface analytical technique. Therefore, it must be assumed that the volume (area × depth) analyzed is representative of the whole object. With coins this is not always the case because the surface is sometimes changed in composition by segregation during solidification or heat treatment, corrosion processes, or by human agency (chemical cleaning). However, Beck *et al.* (*8*) have shown that there is full agreement between surface and bulk methods for Ag content greater than 92% (the solubility limit for Cu in Ag) and although some surface aggregate of dirt was found on some of the coins these few exceptions (see below for more details), were analyzed and the results were recorded. All the coins were cleaned with acetone prior to analysis to remove any other surface contamination.

XRF Instrument and Accuracy of Measurements

The handheld XRF used for this project was an Oxford Instruments X-MET3000–TX. The X-MET3000–TX reads to 0.01 weight percent in bulk analysis; the accuracy and precision are element specific, but in many cases with precision as good as 0.01–0.1% for all elements analyzed. The manufacturer gives the Fe detection limit for Fe in a Cu matrix as roughly 100 ppm = 0.01% (3 sigma, 100s measuring time). The sampling area is an adjustable window set to approximately 3 × 5 mm for analysis. The instrument was run at 40 keV and 6 nA for 70 seconds. The samples were examined for 26 elements. Data collected included the absolute error for each analysis. Elements measured with absolute errors above 50% of the total value were discarded.

Potential Problems with Analysis

The XRF as currently marketed comes with a Ag target which makes analyzing Ag difficult at lower levels; however, Oxford Instruments was able to re-calibrate the instrument to accurately detect Ag above 50 wt%. In addition, the built-in corrections for instrument interference with Fe were removed. This created a slight problem with identifying the presence of Fe in the samples as opposed to interference from the instrument itself or from aggregate soil on the surface of the coins. Coins were analyzed on one side, and if the results showed high levels of Fe, the coin was analyzed on its other side, and both analyses were recorded. The results show that the XRF is susceptible to error from soil concretions on the surface of the coins (see Table I). Along with the issues arising from Fe contamination, the effect of XRF's susceptibility to surface contamination was also seen with the remnants of old glue that was used to hold identifying tags to the coins. Coins that had remnants of glue showed significant increases in their Zn and Fe levels (see Table I). When this occurred, the coin was re-measured on its other side.

Because we were unable to determine if the values for Fe we measured were consistently coming from the built in error in the XRF or from the coins themselves, we were given permission to sample one coin for traditional metallurgical investigation and EPMA analysis. In addition, we analyzed three other coins privately held by one of the authors.

EPMA Analysis

EPMA is a well established technique of analysis for determining chemical composition of metallurgical samples. The accuracy of this technique is very high with reliable results at the 0.01 % level (high PPM range) and is element dependant. Five elements were examined: Ag, Pb, Cu, Fe, and Au. All standards were metallic with an average taken over 6 analyses. The crystal used for Ag, Pb, and Au was PET (Poly-Ethylene Terephthalate) and the crystal used for Fe and Cu was LiF. The EPMA used was a JEOL 733 electron microprobe which is

Table I. Results of surface contamination on XRF analysis (all values are in weight percent). Glue adhering to the surface resulted in higher Zn and Fe levels, while coins that were dirty with aggregates of soil showed significant increases in Fe detection.

	Ti	*Cr*	*Fe*	*Cu*	*Zn*	*Ag*	*Sn*	*Au*	*Pb*	*Bi*	
tyre 8920	0.519	0.354	0.673	1.03	1.32	95.27		0.645	0.188		Glue
tyre 8920b	1.13	0.380	0.358	0.897		95.77	0.651	0.598	0.212		Clean
tyre 9091		0.319	0.695	1.33	1.01	95.57	0.463	0.499	0.114		Glue
tyre 9091b	0.606	0.310	0.433	1.58		95.88	0.570	0.509	0.112		Clean
tyre 9285	0.692	0.237	3.22	1.29		93.99		0.390	0.138	0.041	Dirty
tyre 9285b			0.417	1.11		97.30	0.645	0.322	0.140	0.062	Clean
tyre 9453			11.43	0.712		87.01		0.552	0.298		Dirty
tyre 9453b			0.934	2.40		95.64		0.451	0.570		Clean

Table II. Comparison between XRF and EPMA analysis of Tyre coin 9463 (all values are in weight percent). EPMA analysis of 3 other coins is included for comparison. Results show that the Fe levels detected using the XRF are likely all from detector fluorescence in the instrument. The difference in Pb may well be due to a poorly polished Pb standard in the EPMA.

		Cu	*Fe*	*Pb*	*Au*	*Ag*	*Total*
Tyre 9463	EPMA	2.13	0.014	2.11*	0.555	96.19	101.00
	XRF	2.12	0.669	0.366	0.414	96.43	100.00
N1–19 CE	EPMA	2.18	0.008	0.137	0.649	97.35	100.33
N2–99 BCE	EPMA	3.84	0.008	1.41*	0.417	94.56	100.25
N3–27 BCE	EPMA	2.72	0.019	0.343	0.04	97.52	100.64

optimized for high accuracy in quantitative analysis using wavelength dispersive spectroscopy (WDS). This probe is equipped with four WDS spectrometers. The instrument was run at 20 KeV, 30 nA with 20 second on peak, and 10 second off peak count times. An average of 7 spots was measured on the sample. The Pb standard in the EPMA was not well polished and produced low total counts for the pure metallic Pb. This caused an increase in the Pb levels in the samples and may well explain the wide discrepancy between the XRF and EPMA results for this element in coin Tyre 9463 (see Table II). Three additional coins were measured by EPMA (N1 – N3) but as they were already mounted for electron microscopy analysis, they could not be analyzed by XRF with the same analytical conditions to the other Tyrian shekels; therefore, there is no XRF data on these coins.

EPMA results showed there to be only trace levels of Fe in the coins. Based on these analyses, it was determined that the majority of Fe seen in the XRF analyses was due to instrument interference. It was therefore decided to remove the Fe values from the results and re-normalize the data (see Table II for EPMA vs. XRF results).

Results of XRF Analysis and Discussion

Although all the coins analyzed are presumed to be minted in Tyre, the results of analysis will be given for both hoards separately as well as combined. In addition, the dates covered by each hoard overlap one another with the Qumran hoard dating from 138 B.C.E. to 1 B.C.E. while the Isfiya hoard dates from 34 B.C.E. to 53 C.E. This common mint context, with coins from two different burial locations allowed for an internal check of the analyses by comparing the results from the overlapped periods.

Out of all the coins analyzed–833 total from both the Isfiya and the Qumran hoards–only 782 were able to be accurately dated. For both the Isfiya hoard and the Qumran hoard, of the 26 elements searched for, twelve were found in the coins (Ag, Cu, Au, Pb, Ni, Sn, Ti, Ni, Bi, Pt, Zn and Cr). Results of the analysis show that all the coins are considered to be very pure silver (excluding coin tyre 8823 with 78.69% Ag). A total of 701 of 705 coins in the Isfiya hoard have silver content over 92%, and 704 of the 705 coins have over 90% Ag. Twelve of the 128 coins analyzed in the Qumran hoard have silver values below 92% (88.04–91.75%). The major secondary elements present in the coins are Cu, Pb, and Au. Other trace elements range in value from 0% to 1.452%, but typically are very low with means below 0.45%. The more informative and interesting aspect to consider is the major elements, Ag, Cu, Au, and Pb and their relationship to one another (see Table III for mean values).

When the values of the major elements are plotted against the date of production an interesting pattern emerges. Quite clearly there is a period of time in which there is instability in the level of Ag being used for coin production (see Figure 1). This period corresponds to the dates between 44 B.C.E. and 15 B.C.E.. There is a clear negative correlation between the Ag and Cu indicating that with a decrease of Ag, there is a corresponding increase in Cu; because Cu is the usual element used for debasing Ag, this should not be too surprising. Although there seems to be a general increase in the amount of Au in coins through time, the change in mean values is quite small ranging from between 0.28 and 0.5%. The range of mean values for Pb is relatively small but erratic with the overall value decreasing over time from 0.39 to 0.18.

What may be more interesting is to look at the actual data points for each year rather than the mean. When we plot the data against the date of production, we clearly see a period of instability in silver production with higher ranges of values for each year and a lower overall average silver content. In addition, one can see that there is a prevalence of younger or more recent coins in each hoard as time progresses. This may be related to the number of coins being produced throughout the period, remelting of older coins, or just to the nature and context of the specific hoards. In any case, in both the Isfiya and the Qumran hoards, the instability in silver concentration occurs in the same time period (see Figure 2). The reader should note the one unusually low silver value of 78.7 wt% at 15 B.C.E.

The dates can be arbitrarily divided into three ranges based on the distribution of silver values acquired. We have called these ranges Early, Middle and Late which have the dates 138 – 54 B.C.E., 44 – 15 B.C.E. and 14 B.C.E. – 53 C.E., respectively. While most of the mean values for silver are relatively consistent around 97% (Early – 97.05%, Late – 97.20%), the Middle range of dates has a much lower mean at 94.80% and a range of 20.01 that is

Table III. Mean values of elements (in weight percent) found in the coins from the Isfiya and Qumran hoards.

Location		*Ag*	*Cu*	*Pb*	*Au*	*Sn*	*Zn*	*Ti*	*Bi*
Osfia	Mean	97.10	1.25	0.200	0.495	0.284	0.025	0.467	0.009
	N	705	705	705	705	705	705	705	705
	Std. Deviation	1.18	1.04	0.205	0.170	0.325	0.090	0.376	0.017
	Minimum	78.69	0	0	0	0	0	0	0
	Maximum	99.24	19.43	2.69	1.76	1.05	0.765	1.45	0.172
	Range	20.55	19.43	2.69	1.76	1.05	0.765	1.45	0.172
Qumran	Mean	95.29	3.41	0.325	0.298	0.154	0.012	0.337	0.010
	N	128	128	128	128	128	128	128	128
	Std. Deviation	2.20	2.17	0.152	0.076	0.273	0.024	0.377	0.034
	Minimum	88.41	0.171	0.076	0	0	0	0	0
	Maximum	99.02	10.80	0.794	0.686	0.854	0.173	1.03	0.307
	Range	10.61	10.63	0.718	0.686	0.854	0.173	1.03	0.307
Total	Mean	96.83	1.58	0.220	0.465	0.264	0.023	0.447	0.009
	N	833	833	833	833	833	833	833	833
	Std. Deviation	1.53	1.49	0.203	0.174	0.321	0.084	0.378	0.021
	Minimum	78.69	0	0	0	0	0	0	0
	Maximum	99.24	19.43	2.69	1.76	1.05	0.765	1.45	0.307
	Range	20.55	19.43	2.69	1.76	1.05	0.765	1.45	0.307

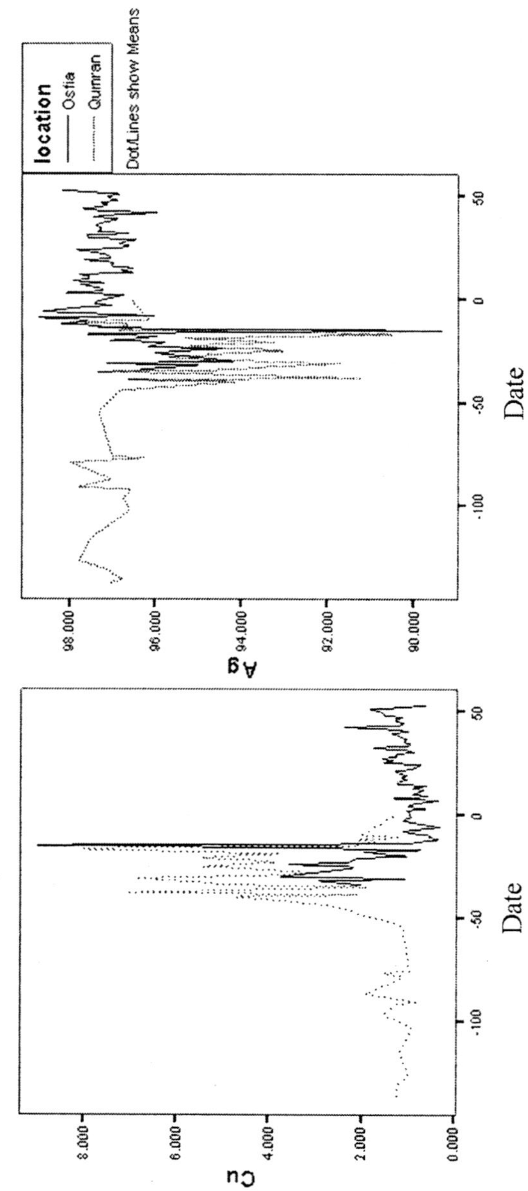
location
Ostia
Qumran
Dot/Lines show Means
Ag
Cu
Date
Date

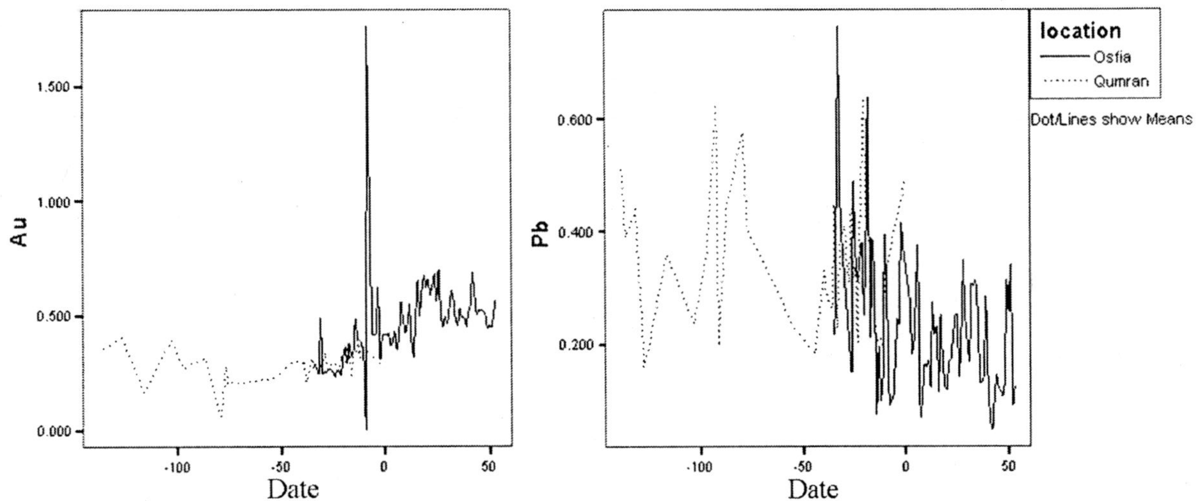

Figure 1. Mean values (weight percent) for Cu, Ag, Au and Pb as plotted against date for the Isfiya and Qumran hoards. There is a high negative correlation between Cu and Ag.

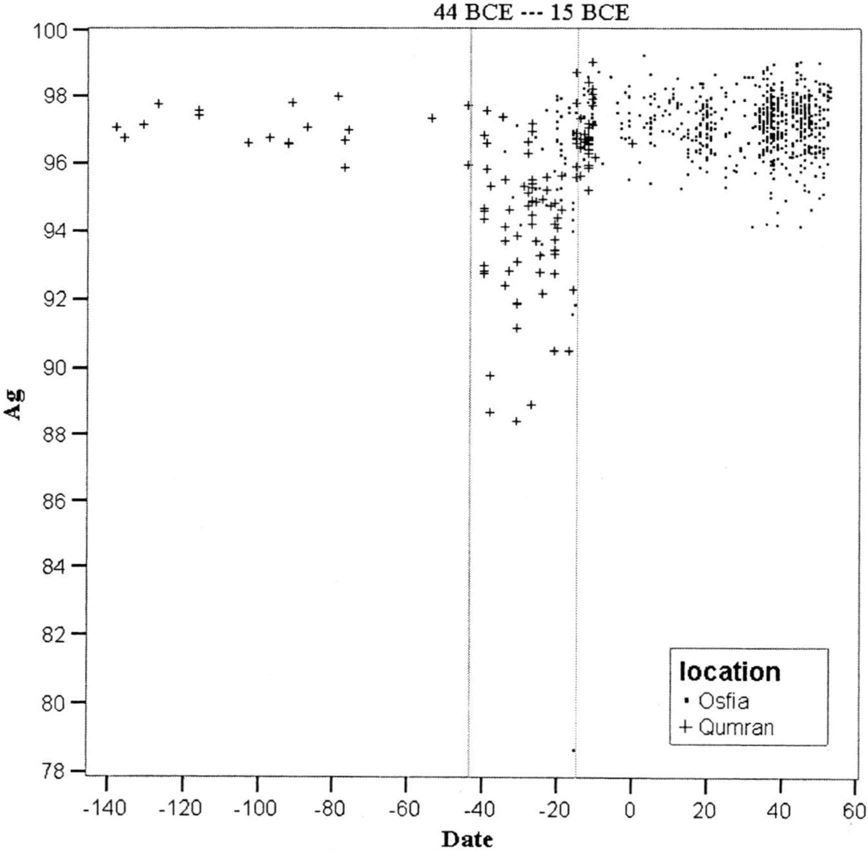

Figure 2. Plot of Ag concentration (weight percent) vs. date of production for each coin. The Early, Middle, and Late arbitrary periods are marked with the grey lines which show the decreased level of silver during the period of instability.Note the one unusually low silver value of 78.7 wt% at 15 B.C.E.

approximately 4 times as high as the Late range, and almost 9.5 times higher that the Early range (see Table IV). Analysis of variance (ANOVA) for these groups supports the clear distinction in composition between these selected dates (see Table V).

The results from the XRF analysis are relatively similar to the published results of Ben-David (*9,10*). Ben-David analyzed 324 Tyrian shekels using Caley's specific gravity method (*11*), but this method is highly inaccurate. Furthermore, it is not clear exactly what coins were measured, as there is no tabulation of the data. Ben-David did not indicate the number of coins, or which specific coins from the hoard that was measured, making a direct comparison impossible. However, the general trend that Ben-David's analysis shows is replicated in our XRF data.

In 1976, Walker (*12*) published a large number of analyses of silver coins using XRF. These analyses have recently received criticism in a series of papers by Butcher and Ponting (*13–18*). They show that the analysis error for severely debased coins of the Roman empire increases with copper content. In our opinion, this is due to a combination of factors including severe segregation during casting (driven by the increasingly higher copper level) and the two-phase nature of the resultant microstructure in the Ag-Cu system when the Ag content is lower than about 92 weight percent. The present XRF study, performed on coins of quite high silver purity, shows excellent agreement with bulk analytical methods and is consistent with the findings of Beck et. al. (*8*). In any case, Walker's results for high silver Tyrian coins (*12*, p.58) are comparable to ours but his values for Ag fall approximately 1–2% lower than our findings. There was also a gap in Walker's data for the years 53–22 B.C.E., almost exactly where we see a drop in silver level.

General Discussion

Tyre was conquered by Rome in 64 B.C., although it remained an autonomous city. However, for the next 30 or so years, it was situated in a region in constant strife.

The first Roman Triumvirate of Pompey, Crassus, and Julius Caesar started in 60 B.C.E., but was the source of almost constant civil war. Julius Caesar was assassinated in 44 B.C.E., and the Second Triumvirate of Antony, Lepidus, and Octavian was formed (43–33 B.C.E.). Continued civil war led to the defeat of the combined forces of Cleopatra and Marc Antony in the naval battle of Actium in 31 B.C.E. Soon after, in 30 B.C.E., Herod met Octavian in Rhodes, and stressed his integrity and loyalty. Octavian promised that Herod would rule "more securely than before" [Flavius Josephus, *Wars of the Jews* (War)1.391],

Table IV. Comparative means (weight percent) for the arbitrarily set Early, Middle and Late ranges of dates. Note that there is a ten year gap between the Early and Middle groups as there were no coins at these dates.

Period		*Ag*	*Cu*	*Au*	*Pb*
Early:	Mean	97.05	1.21	0.278	0.395
138BCE - 54BCE	N	17	17	17	17
	Std. Deviation	0.544	0.282	0.096	0.140
	Minimum	95.85	0.824	0.049	0.159
	Maximum	97.97	1.89	0.406	0.642
	Range	2.12	1.07	0.357	0.483
Middle:	Mean	94.80	3.74	0.295	0.353
44BCE - 15BCE	N	119	119	119	119
	Std. Deviation	2.56	2.57	0.075	0.164
	Minimum	78.69	0.344	0.174	0.047
	Maximum	98.70	19.43	0.686	0.901
	Range	20.01	19.09	0.512	0.854
Late:	Mean	97.20	1.20	0.500	0.184
14BCE - 53CE	N	646	646	646	646
	Std. Deviation	0.824	0.643	0.157	0.173
	Minimum	94.16	0.055	0.000	0.000
	Maximum	99.24	4.16	1.76	1.10
	Range	5.08	4.10	1.76	1.10
Total	Mean	96.83	1.58	0.464	0.214
	N	782	782	782	782
	Std. Deviation	1.52	1.48	0.166	0.183
	Minimum	78.69	0.055	0.000	0.000
	Maximum	99.24	19.43	1.76	1.10
	Range	20.55	19.38	1.76	1.10

Table V. ANOVA with Ag, Cu, Au, and Pb as dependents with the Period (Early/Middle/Late) as the factor

		Sum of Squares	*df*	*Mean Square*	*F*	*Sig.*
Ag	Between Groups	580.197	2	290.099	185.923	0.000
	Within Groups	1,215.485	779	1.560		
	Total	1,795.682	781			
Cu	Between Groups	653.121	2	326.561	242.520	0.000
	Within Groups	1,048.948	779	1.347		
	Total	1,702.069	781			
Au	Between Groups	4.816	2	2.408	112.184	0.000
	Within Groups	16.720	779	0.021		
	Total	21.536	781			
Pb	Between Groups	3.424	2	1.712	58.607	0.000
	Within Groups	22.756	779	0.029		
	Total	26.180	781			

and bestowed additional honors on him [War 1.392–93; Flavius Josephus, *Antiquities of the Jews* (Ant.) 15.195] (*19*, p.171–3).

Octavian was named Augustus in January of 27 B.C.E.. Although the relationship between Augustus and Herod was a delicate balancing act between local internal matters and imperial policy, friendship between Augustus and Herod survived intact for almost 30 years. Herod traveled to Rome numerous times, and Augustus visited Herod in his own territories during Herod's seventeenth year of rule, believed to have been around 20 B.C.E. [Flavius Josephus and Cassius Dio both say that he visited Syria: Ant.15.354, Cassius Dio, *Roman History*, 54] (*19*, p.234).

There were strong economic ties between Herod and Tyre (*20*). Soon after his meeting with Augustus in 30 B.C.E., Herod built temples and marketplaces in Tyre and Berytus [Wars 1:422]. In 20 B.C.E.,Augustus made him procurator of "all Syria" [Wars 1:399]. At about this same time (20–19 B.C.E.) Herod began his most important building project, the complete reconstruction of the Temple in Jerusalem . Although Herod might not have controlled Tyre directly, the New Testament [Acts 12:20] records that Tyre was dependent upon Judea for food.

Harl (*21*) points out that in 30 B.C.E.. Augustus restored the purity of the denarius to its previous high silver level of 97.5–98% fine, and in 15 B.C.E.. established the major Roman mint Lugdunum (Lyon) and secured the silver mines in Spain. Herod's political and personal interactions with Augustus,

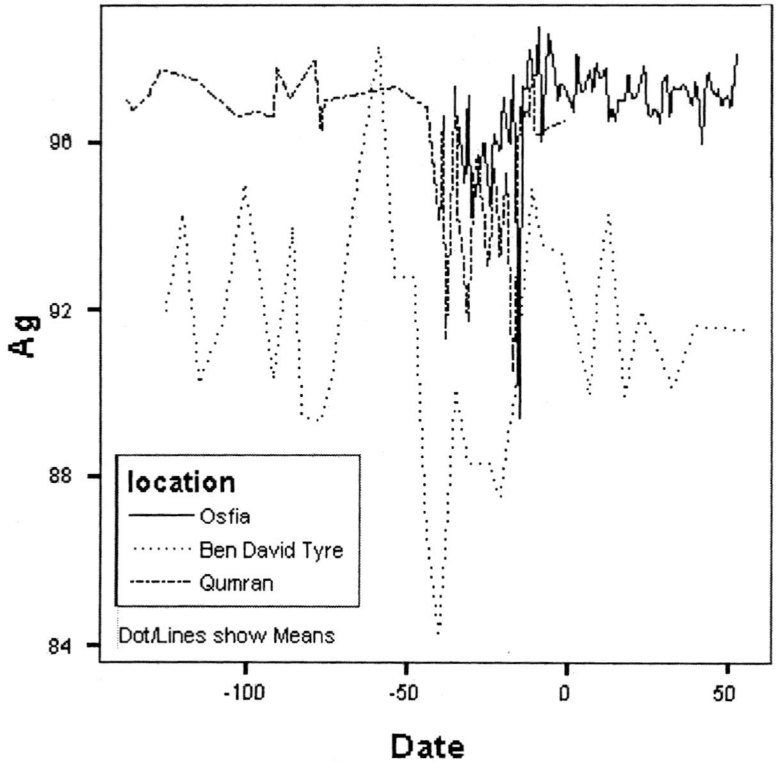

Figure 3. Comparison of the XRF data (weight percent) with the results by Ben David using Caley's specific gravity method.

combined with the need of the Temple in Jerusalem for a stable, quality, silver monetary vehicle would also naturally lead to a demand for and stabilization of the Tyrian shekel.

Based on the results of analysis in this present study, it was found that while the Tyrian coins all show relatively constant high silver levels during the years before and after the period 44-15 B.C.E., during this period the silver content is erratic and drops significantly. The end of this period corresponds closely with the appearance of the KP (or KAP) control mark on Tyrian silver coins, and which lasted until the end of this coinage. The results have been interpreted in the context of social and economic conditions in Tyre and surrounding Judea and Syria at that time period. In particular, it appears that interactions between Herod and Augustus, and between Herod and the Temple in Jerusalem are reflected in the variation of the silver content of the coins.

Conclusions

Based on analysis of over 800 coins, there appears to have been a destabilization of the Tyrian shekel in about 44 B.C.E. and a re-stabilization in about 15 B.C.E. The re-stabilization coincides with the appearance of the control mark KP (or KAP) and which continues throughout the remaining duration of these coins. The destabilization correlates closely with the instability of the Roman Empire prior to the rise of Augustus Caesar, and the re-stabilization correlates to the re-stabilization of Roman mints; and with interactions between Augustus, Herod, and the Temple authorities in Jerusalem.

For coins with high silver content (> 92 weight %Ag), and where surface alteration is minimum, hand portable XRF is a suitable method for analysis, and enables rapid in-situ measurement of large numbers of samples.

Acknowledgement

The authors thank the Rubin-Ladd Foundation for their financial support.

References

1. Kadman, L. In *Congresso Internazionale di Numismatica*; Roma, 1961; Rome, 1965; Vol. II; pp 69–76.
2. Sharabani, M. *Revue Biblique* **1980**, *87*, 274–284.
3. Leonard, Jr., R.D. In *XII.Internationaler Numismatischer Kongress*; Kluge. B.; Weisser, B., Eds.; Berlin, 1997; Vol. I; pp 683–692.

4. Meshorer, Y., In *Studies in Honor of Leo Mildenberg: Numismatics, Art History and Archaeology*; Mildenberg, L.; Houghton. A., Eds.; Wetteren, 1984; pp 171–179.
5. Gigante, G. E.; Cesareo, R. *Radiat. Phys. Chem.* **1998**, *51*, 689–700.
6. Gigante, G. E.; Ridolfi, S.; Visco, G.; Guida, G. In *Archaeometallurgy in Europe*; 2003, Vol. 2; pp 293–302.
7. Thomsen, V.; Schatzlein, D. *Spectroscopy* **2002**, *17*, 14–21.
8. Beck, L.; Bosonnet, S.; Reveillon, S.; Eliot, D.; Pilon, F. *Nucl. Instrum. Methods Phys. Res.* **2004**, *B226*, 153–162.
9. Ben-David, A. *Jerusalem und Tyros: Ein Beitrag zur palaestinensischen Munz und Wirtschaftsgeschichte*; Tübingen, 1969; p 13.
10. Ben-David, A In *Proceedings of the 5th World Congress of Jewish Studies*; Peli, P.; Shinan, A., Eds.; Vol. 4, Jerusalem: World Union of Jewish Studies, 1973; pp 253–267.
11. Caley, E. R. *The Composition of Parthian Coins*; American Numismatics Society, 1955.
12. Walker, D. R. *The Metrology of the Roman Coinage-Part I. From Augustus to Domitian*; BAR Supplementary Series 5; Oxford, 1976.
13. Butcher, K.; Ponting, M. *Oxford J. Archaeol.* **1995**, *14*, 63–78.
14. Butcher, K.; Ponting, M. In *Internationales Kolloquium zur kaiserzeitlichen Münzprägung Kleinasiens., Nomismata: Historische-numismatische Forschungen 1*; J. Nollé; Overbeck, B.; Weiss, P., Eds.; Milan, 1997; pp 167–171.
15. Butcher, K.; Ponting, M.; Chandler, G. *Am. J. Numismatics* **1997**, *9*, 17–36.
16. Butcher, K.; Ponting, M. In *Metallurgy in Numismatics – Volume 4*; Oddy, A.; Cowell, M., Eds.; London: Royal Numismatic Society Special Publication No. 30, 1998; pp 308–334.
17. Butcher, K.; Ponting, M. *Oxford J. Archaeol.* **2005**, *24*, 163–197.
18. Butcher, K.; Ponting, M. *Schweizerische Numismatische Rundschau* **2005**, *84*, 93–124.
19. Richardson, P. *Herod-King of the Jews and Friend of the Romans*; Columbia, SC, 1996; pp 171–173 and 234.
20. Gabba, E. In *Greece and Rome in Eretz Israel. Collected Essays*; Kasher, A.; Rappaport, U.; Fuks, G., Eds.; Jerusalem, 1990; pp 160–168.
21. Harl, K. W. *Coinage in the Roman Economy, 300 B.C. to A.D. 700*; Baltimore, 1990; pp 75–80.

Chapter 15

Selected Applications of Laser Ablation Inductively Coupled Plasma–Mass Spectrometry to Archaeological Research

Robert J. Speakman[1], Michael D. Glascock[1], Robert H. Tykot[2], Christophe Descantes[3], Jennifer J. Thatcher[4], Craig E. Skinner[4], and Kyra M. Lienhop[1]

[1]Research Reactor Center, University of Missouri, Columbia, MO 65211
[2]Department of Anthropology, University of South Florida, Tampa, FL 33620
[3]Directorate of Public Works, United States Army, Garrison, HI 96857
[4]Northwest Obsidian Research Laboratory, Corvallis, OR 97330

Use of inductively coupled plasma-mass spectrometry (ICP-MS) coupled to a laser-ablation sample introduction system (LA-ICP-MS) as a minimally destructive method for chemical characterization of archaeological materials has gained favor during the past few years. Although still a relatively new analytical technique in archaeology, LA-ICP-MS has been demonstrated to be a productive avenue of research for chemical characterization of obsidian, chert, pottery, painted and glazed surfaces, and human bone and teeth. Archaeological applications of LA-ICP-MS and comparisons with other analytical methods are described.

In recent years, laser ablation (LA) systems coupled to state-of-the-art inductively coupled plasma-mass spectrometers (ICP-MS) have gained increased popularity in archaeological science for chemical analyses of a variety of inorganic and organic matrices. In archaeology, LA-ICP-MS has facilitated research concerning provenance, trade, and technology through the analysis of metals, rocks, ceramics, pigments, and other archaeological materials (*1–10*). In addition, analyses of human teeth and bone by this technique have been used to make inferences regarding nativity (*11, 12*) and diet (*13*). LA-ICP-MS also has been used in attempts to identify chemical signatures in archaeological wood samples that might be useful for dating prehistoric volcanic eruptions (*14, 15*).

As an ICP-MS sample introduction technique, laser ablation provides a viable alternative for ICP-MS characterization studies that traditionally have required digestion of solid samples using a combination of heat and/or strong acids—a time consuming and unpleasant task. Laser ablation was first applied to ICP in the late 1970s (*16*), but it was not until the mid-1980s that a laser ablation system was coupled to an ICP mass spectrometer (*17*). The coupling of laser-ablation with ICP-MS has resulted in the development of extremely sensitive microprobes capable of determining most elements of the periodic table. LA-ICP-MS offers several advantages over other analytical methods, including low detection limits, rapid analytical time, low cost per sample, high sample throughput, and minimal damage to the sample. The range of materials that can be characterized by LA-ICP-MS (rocks, ceramics, glasses, pigments, fauna and other organics) and types of analyses (bulk, surface, and microprobe) are unsurpassed by most other analytical techniques. The fact that in situ analyses can be conducted by LA-ICP-MS suggests less chance of contamination resulting from sample preparation in that the sample remains intact within its original matrix until the analysis. Although potential problems exist with data calibration, spectral interferences, and fractionation, these problems can be ameliorated such that any negative impacts to the analysis are minimized. LA-ICP-MS has tremendous potential for providing chemical characterizations of archaeological materials, permitting questions regarding prehistoric production, trade, interaction, and manufacturing technology to be addressed. The examples presented below illustrate a few of the potential applications of LA-ICP-MS to archaeological characterization studies.

Analytical Methods

Data for all case studies presented below were generated using a VG Axiom (high-resolution, double-focussing, single-collector) ICP-MS coupled to a Merchantek Nd:YAG 213–nm wavelength laser ablation unit. The laser can be targeted on spots as small as 5 µm in diameter. The small spot size and the high sensitivity of magnetic-sector ICP-MS to a wide range of major, minor, and trace elements make LA-ICP-MS a very powerful microprobe. Moreover, laser ablation is virtually non-destructive to most samples considering that the ablated areas are often indistinguishable with the naked eye. Unlike instrumental neutron activation analysis (INAA), X-ray fluorescence (XRF), or ICP-MS of solutions which produces a bulk elemental characterization of the entire matrix, LA-ICP-MS provides a point specific characterization of the ablated area of the sample. Relatively homogeneous samples, such as obsidian and to a certain extent cherts, paints, and glazes, are ideally suited for LA-ICP-MS given that spatial variation is minimal in these materials. ICP-MS can generate compositional data for 50–60 elements, whereas, other techniques typically generate compositional data for about 30 (or less) different elements. Some elements such as lead and phosphorus which cannot be measured by INAA but can be measured by LA-ICP-MS may prove important for separating materials into different compositional groups. For many elements LA-ICP-MS has lower detection limits than other instrumental techniques (e.g., Sr, Sb, Ba, and Zr).

In LA-ICP-MS, the sample is placed inside a sample holder or laser cell where ablation takes place. Ablation areas vary in size depending on the sample matrix, but the analyzed area is usually smaller than 1000 x 1000 μm and less than 30 μm deep. During analysis, the laser beam ablates and vaporizes the area of interest on the sample. The ablated material is transported from the laser cell using a 0.9–1.5 l/min flow of argon and/or an argon/helium/nitrogen-mixed carrier gas through Tygon tubing and introduced into the ICP-MS torch, where argon gas plasma capable of sustaining electron temperatures between 8000 and 10,000 K is used to ionize the injected sample. The resulting ions pass though a two-stage interface (sample and skimmer cones) designed to enable the transition of the ions from atmospheric pressure to the vacuum chamber of the ICP-MS system. Once inside the mass spectrometer (in this case a high-resolution, double-focussing, magnetic sector ICP-MS), the ions are accelerated by high voltage and pass through a series of ion optics, an electrostatic analyzer (ESA), and finally the magnet. By varying the strength of the magnet, the ions are separated according to mass/charge ratio and passed through a slit into the detector, which records only a small atomic mass range at a given time. By varying the magnet and flight-tube settings, the entire mass range can be scanned within a relatively short time.

Although laser-ablation sample preparation and analysis are conducted with relative ease, quantification of data can prove challenging. With liquid samples, the amount of material introduced into the ICP-MS remains relatively constant, and instrument drift is usually corrected through the use of internal standards. However, in LA-ICP-MS, conditions such as the texture of the sample, ablation time, the location of the sample within the laser cell, surface topography, laser energy, and other factors significantly affect the amount of material that is introduced to the ICP torch and thus the intensity of the signal monitored for the various atomic masses of interest. As a result, researchers have grappled with normalization methods that permit accurate quantification of LA-ICP-MS data (*6, 7, 10, 18–22*). In the examples below, we present both qualitative (ratios) and quantitative approaches to data interpretation.

Determining Obsidian Provenance

Obsidian is an ideal archaeological material for examining resource procurement patterns and exchange networks because the artifacts can usually be linked to sources with a high degree of reliability. Obsidian has several advantages over a majority of other archaeological materials, especially ceramics, which are found in even greater abundance. First, obsidian sources are restricted to areas where volcanic activity occurred or to locations where secondary deposits were created by other geologic processes. Second, obsidian sources are more often than not chemically homogeneous, and at the same time the individual sources have chemical characteristics that make them different from one another. Measurements of trace element abundances have demonstrated that individual sources can be differentiated from one another, although the characteristic elements are likely to differ for each suite of sources involved in the comparison (*23*). With sufficient field and laboratory work, the spatial extent of a particular geochemical type of obsidian can be established such that a "source" can be defined. Finally, obsidian artifacts are nearly indestructible in most archaeological contexts. Only by the extremely slow process of hydration which attacks the surfaces of artifacts does the artifact gradually get smaller, but the bulk composition remains unchanged. The latter process takes many tens of thousands of years to totally destroy an artifact. Thus, it is possible to compare the compositional fingerprints of artifacts to those of sources and successfully determine the correct source for each artifact with nearly 100% confidence.

In most obsidian provenance studies, the ability to employ compositional differences to discriminate between sources depends, to a certain extent, on the number of elements measured. Because instrumental neutron activation analysis (INAA) is capable of measuring 25–30 elements in obsidian with excellent precision, numerous combinations of trace and major elements are available for comparing differences between sources. The main requirements for success are that all sources have been located and analyzed, and that the internal variation measured within the sources be smaller than the compositional differences measured between the sources (*23*).

Sourcing Obsidian Artifacts in the Western Mediterranean

Until recently, there was no systematic survey, documentation, and chemical and physical analyses of western Mediterranean obsidian sources. Recently, Tykot completed an extensive survey and documentation of western Mediterranean obsidian sources on the islands of Sardinia, Palmarola, Lipari, and Pantelleria (*24–27*) for a more detailed discussion. Samples from these sources were analyzed at MURR by INAA and/or XRF and LA-ICP-MS. As expected, INAA (and XRF and LA-ICP-MS) of geologic samples from these sources demonstrated that obsidian from each island had a unique chemical signature(s). In the case of Sardinia, six compositional groups were identified. Because of the analytical cost and semi-destructive nature of INAA, artifacts were analyzed by LA-ICP-MS rather than INAA. XRF would have provided a viable analytical alternative, but many of the artifacts were smaller than the minimum size required for this analysis on a standard laboratory-based stationary XRF instrument.

Given our extensive analyses of the geologic source samples, the range of chemical variation both within and between western Mediterranean obsidian sources was known in advance. Consequently, it was not necessary to analyze the artifacts for a full suite of elements. Instead a few elements were identified that best separate the various island sources and sub-sources, i.e., iron, cesium, samarium, and barium, and few other elements (Figure 1). The laser was set to ablate along a line, approximately 600 μm in length, over a flat area on the sample. The laser was operated at 80% power using a 100-μm diameter beam operating at 20 Hz. The laser was set to scan across the raster area at a speed of 30 μm/s. Ten measurements were made for each of the isotopes measured. Ratios of the blank-subtracted isotopic-abundance-corrected counts were used to discriminate between the compositional groups for the sources. It is important to note that because this is essentially a "standardless" measurement, that samples of obsidian from the known sources must be analyzed on a daily basis. This is because the operating parameters of the LA-ICP-MS system can change on a daily basis, thus affecting the instrument mass bias. Therefore, data generated for one day's experiment cannot always be readily compared to data generated on a subsequent day. Fortunately, the changes in mass bias do not affect the accuracy of source identification given that samples of known provenance are analyzed with each batch of artifacts.

Although we attributed artifacts to each of the four major western Mediterranean sources, our focus here is on artifacts attributed to Monti Arci, Sardinia. Figure 2 illustrates the use of elemental ratios to discriminate the major island obsidian groups by INAA. By projecting these data as logged ratios of samarium/barium on the X-axis and iron/cesium on the Y-axis, we have maximized the differences between the various Sardinian subgroups in a manner

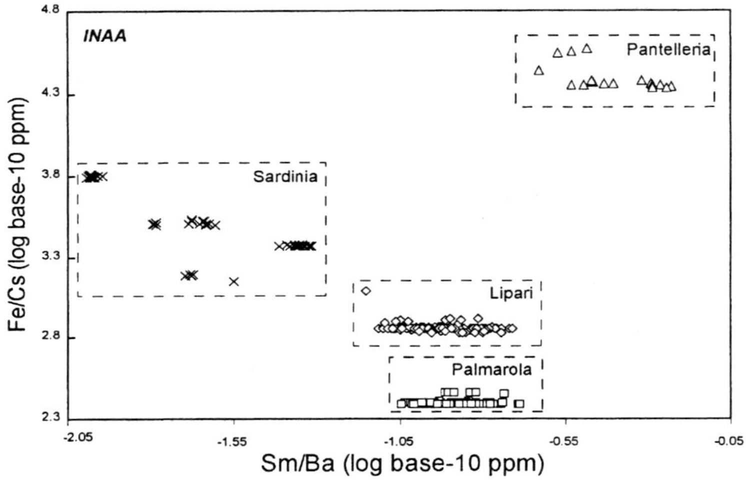

Figure 1. Comparison of INAA elemental ratios for the four major western Mediterranean Island obsidian sources. Only geologic source samples are plotted.

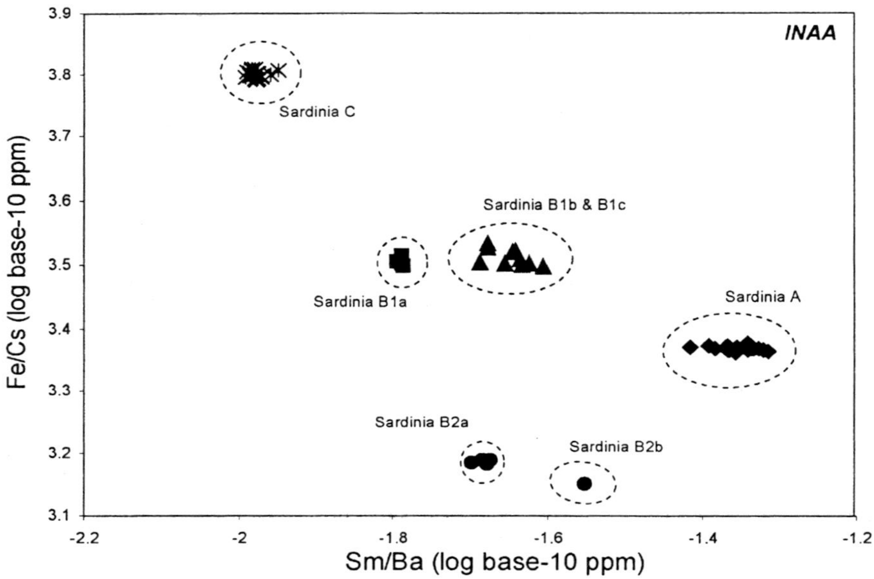

Figure 2. Comparison of INAA elemental ratios for the six Monti Arci (Sardinia) obsidian subgroups. Only geologic source samples are plotted.

that facilitates comparisons with LA-ICP-MS data. As demonstrated in Figure 3, the configuration of the LA-ICP-MS is virtually identical to that observed with the INAA data presented in the previous figure. The analysis of a few geologic source samples permits us to confidently attribute the artifacts to specific sub-source deposits on Monti Arci. As an added advantage, it is possible to analyze between 60 and 100 artifacts per day using this approach.

Discriminating Obsidian Sources When Compositional Variation is Minimal: Sycan Marsh and Silver Lake, Oregon

The Silver Lake/Sycan Marsh obsidian source domes, located in the Fort Rock region of south-central Oregon, were intensively utilized throughout the prehistoric period. Artifact obsidian from this source is found throughout central Oregon, southwest Oregon, northeast California, and southwest Washington, and ranges in age from the Clovis era to the early historic period. Although the two domes that make up the Silver Lake/Sycan Marsh source are usually considered as a single geochemical group, but XRF and LA-ICP-MS studies suggest that their trace element content can be used to distinguish the northern from the southern dome.

The two source domes, located 16 km apart, are separated by a significant physical and ethnographic divide. Obsidian from the northern dome (Silver Lake) is found within the closed Fort Rock Basin that lies at extreme northwestern edge of the Great Basin. Glass from the southern dome (Sycan Marsh) occurs in secondary deposits in the upper reaches of the Klamath Lake Basin. Obsidian from the southern dome would have been available for direct procurement by the Klamath-Modoc groups who occupied the eastern margin of the Klamath Basin. The Northern Paiute groups inhabiting the Fort Rock Basin would have had direct access to geologic material originating from the northern dome. Obsidian from these two culture areas would have been available for use or trade to very different geographic and ethnographic areas. The ability to distinguish artifacts from a specific dome allows examination of the prehistoric use and distribution of the glass with considerably greater archaeological resolution than when considered as a single combined geochemical source.

XRF analyses of Silver Lake and Sycan Marsh obsidian source samples suggested the possibility that the two sources could be differentiated based on small differences in strontium concentrations. However, when the standard error for strontium was taken into account, both groups overlapped at one standard deviation. Because of higher instrumental detection limits for strontium, NAA could not discriminate between the two sources. LA-ICP-MS analyses were conducted to determine if the sensitivity and precision of this analytical technique was sufficient to confirm the existence of the two compositional groups. For this experiment, the laser was set to ablate along a line, approximately 600 μm in length, over a flat area on the sample. The laser was operated at 80% power using a 100 μm diameter beam operating at 20 Hz. The laser was set to scan across the line at a speed of 30 μm/s. Ten measurements were made for each of the five isotopes measured (^{138}Ba, ^{85}Rb, ^{30}Si, ^{88}Sr, ^{66}Zn). Samples were analyzed in a random order to ensure that instrumental drift would not bias the results of the analysis. A ratio of blank subtracted counts for each isotope to silicon provided a means for examining the differences between the two groups.

Data generated by XRF and LA-ICP-MS are presented in Figure 4. Although the X-axis and Y-axis scales differ for the LA-ICP-MS data (ratios) and the XRF data (ppm), it is clear that the separation suggested by XRF is in fact real when the LA-ICP-MS data are taken into account. Future research will include the analysis of artifacts, thereby enabling archaeologists to examine prehistoric human resource procurement patterns and the spatial distribution of these obsidians.

Bulk Analysis of Ceramic Pastes

For more than three decades INAA has been the primary analytical technique for bulk chemical characterization of archaeological ceramics. Provenance studies of ceramic materials permit archaeologists to examine raw material selection and pottery distribution across wide geographic areas. The rapid proliferation of ICP-MS during the last decade has resulted in compositional studies of ceramics being conducted at numerous institutions throughout the world, rather than a few key facilities. We welcome this shift, but maintain that INAA is still the best analytical method available for bulk characterization of prehistoric ceramic pastes. Nonetheless, we recognize that ICP-MS of solutions, and in some cases LA-ICP-MS of solids, can be used to generate data that are comparable to data generated by INAA.

The increased number of LA-ICP-MS applications to studies of archaeological materials has raised the question *can in situ bulk analysis of ceramics be conducted by laser ablation*? The answer in most cases is no. It is not possible to generate bulk compositional data given that a laser-ablation system is a microprobe that permits specific areas of a sample to be targeted, ablated, and introduced to the ICP-MS. Because ceramic pastes are heterogeneous, it is difficult to sample an area with the laser that is representative of the entire ceramic matrix (in most pottery). Additionally, and perhaps more importantly, it has yet to be demonstrated that LA-ICP-MS has the long-term (months to years) replicability necessary to generate large

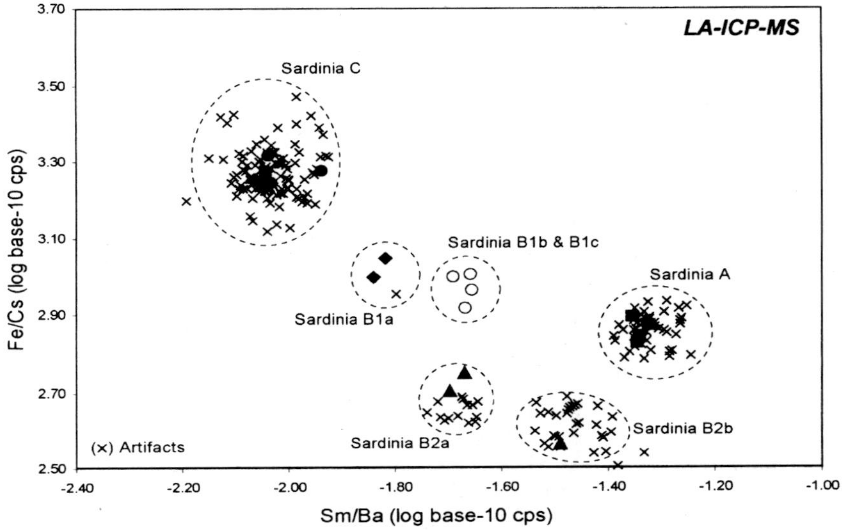

Figure 3. Comparison of LA-ICP-MS elemental ratios for the six Monti Arci (Sardinia) obsidian subgroups. Artifacts are represented by an X. All other symbols represent known source samples that were included in the analyses to verify the accuracy of the assignment of artifacts.

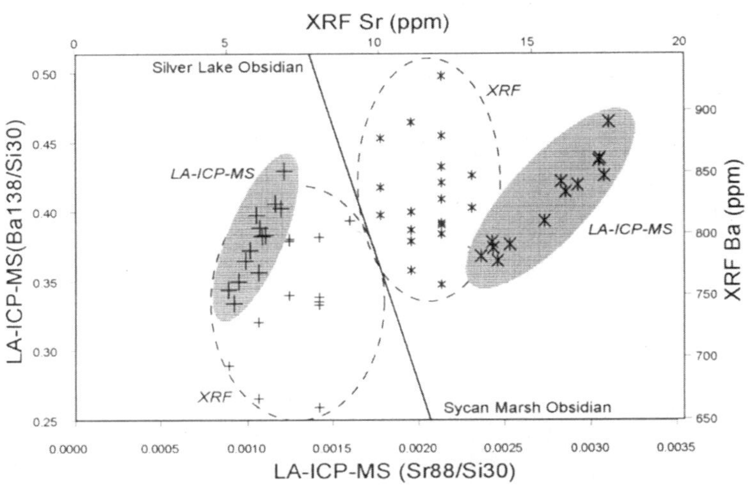

Figure 4. Comparison of LA-ICP-MS and XRF data for Sycan Marsh and Silver Lake obsidian source samples.

compositional databases. Finally, questions remain concerning inter-laboratory comparability of data.

Despite the potential problems with using LA-ICP-MS to generate bulk compositional data for ceramics, there are exceptions where LA-ICP-MS can be effectively employed. The main requirements for ceramic studies of this nature are that the provenance postulate must be fulfilled (*28*)—that differences between groups must be greater than the differences within groups.

As a case study we analyzed the pastes of Dragon Jars housed in the Guthe or Philippine Expedition Collection at the University of Michigan Museum of Anthropology (UMMA). Dragon Jars are large brown- and/or green-glazed stoneware storage jars that were widely produced and distributed throughout China and mainland southeast Asia. Many were decorated, using a variety of techniques, with representations of dragons, botanical elements, lions, or demons. Chemical studies of these types of vessels have been minimal (*29–31*), yet data generated by INAA (Figure 5) suggests that chemical differences in pastes are such that alternative analytical techniques such as XRF or LA-ICP-MS might be used to identify different compositional groups.

A subset of samples previously analyzed by INAA were reanalyzed by LA-ICP-MS. A line approximately 600 μm long was placed over a flat area on the sample. The laser was operated at 80% power using a 100 μm diameter beam operating at 10 Hz. The laser was set to scan across the raster area at a speed of 30 μm/s. Each sample was pre-ablated prior to data acquisition to remove possible surface contamination. Data were calibrated using an approach suggested by Gratuze (*6, 7*). As expected, similar results were obtained by the LA-ICP-MS analyses (Figure 6). We point out, however, that the chemical differences observed for this particular study are not typical of most INAA studies of pottery. Hence, the probabilities of obtaining similar results by LA-ICP-MS were greatly magnified.

Although INAA and LA-ICP-MS Groups 2 and 3 are easily differentiable from one another, examination of the data in bivariate and multivariate space suggest that they are chemically closer to one another than to the other groups. The chemical similarity of these two groups may indicate derivation from separate production centers within the same region, a topic we return to below.

Analysis of Glazes and Paints

One of the potentially more productive avenues of research using LA-ICP-MS involves the characterization of paints and glazes used in the decoration of pottery. Just as bulk analysis of clays by INAA, XRF, ICP-MS, and other analytical methods have demonstrated to be a productive avenue of research for making interpretations regarding past cultural systems, chemical characterization

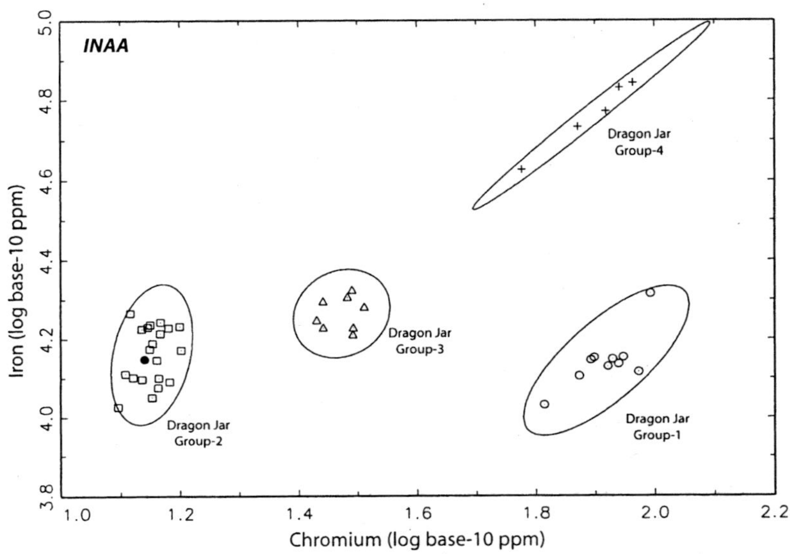

Figure 5. Bivariate plot of chromium and iron base-10 logged concentrations for Dragon Jar paste samples analyzed by INAA. Ellipses represent 90% confidence levels for group membership.

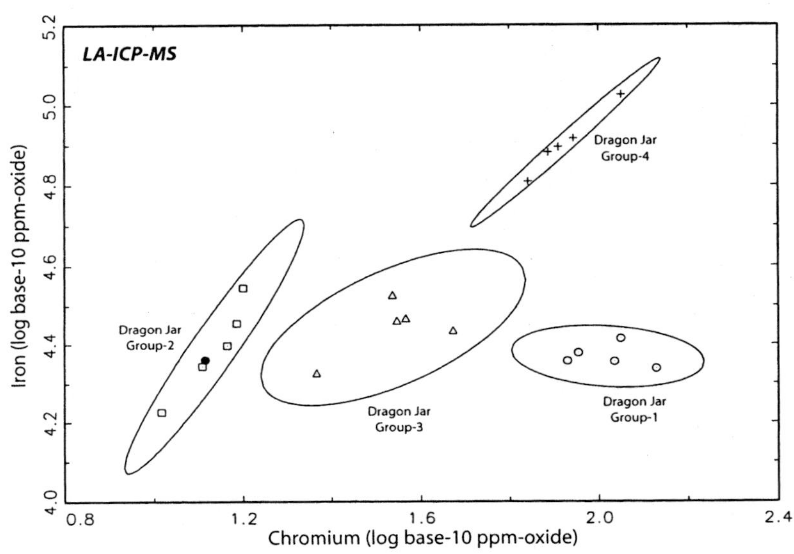

Figure 6. Bivariate plot of chromium and iron base-10 logged concentrations for Dragon Jar paste samples analyzed by LA-ICP-MS. Ellipses represent 90% confidence levels for group membership.

of paints and glazes may prove just as important. Historically, compositional analyses of paints and glazes have been restricted due to limitations imposed by the available analytical techniques. Wet chemistry, physical tests, and other instrumental techniques have been used to characterize paints and glazes with varying degrees of success. However, with INAA, for example, the painted or glazed surface must be separated from the ceramic matrix for analysis. Additionally, analysis of paint by INAA requires a sample size that would result in the destruction of large quantities of paint or glaze. With instrumentation that uses X-rays, it is difficult ensure the underlying clay matrix is not contributing to the analysis. Using LA-ICP-MS, painted and glazed surfaces can be analyzed without having to remove the painted or glazed decoration from the ceramic sherd. By adjusting laser settings the underlying clay matrix can be avoided thus permitting the painted and glazed surfaces to be characterized. Surface contamination can also be removed by using the laser to ablate the decorated surface prior to data acquisition. Additionally, minimal sample preparation time, high sample throughput, and high instrument sensitivity and accuracy make LA-ICP-MS ideal for characterizing this sample matrix.

Characterization of Dragon Jar Glazes

Glazes are an intricate part of the Dragon Jar production process and can contribute independent information on the manufacturing technology of Dragon Jars. In order to investigate the relationships between bulk-paste composition and decoration, (e.g., the glazes), LA-ICP-MS was used to analyze a subset of the 50 samples analyzed by INAA. The number of samples analyzed by LA-ICP-MS is smaller than the number of samples analyzed by INAA given that several of the Dragon Jar specimens (n=13) in the MURR archival collection did not have intact glazes.

As expected, compositional analysis of the Dragon Jar glazes by LA-ICP-MS produced results similar to the INAA study of the pastes (see above) and the study of decorative attributes (see below). Unlike the INAA characterization of the pastes, three groups were identified rather than four. Glaze groups 1 (n=6) and 4 (n=5) correspond directly to INAA paste groups 1 and 4; Glaze Group 2/3 (n=20) is comprised of specimens assigned to the INAA paste groups 2 and 3. The compositional patterning in the glazes is illustrated by a bivariate plot of the cobalt and barium oxides (Figure 7). Seven specimens are unassigned. The pastes from two of the unassigned glazes are assigned to INAA Group 1, but the glaze composition differs from other samples in glaze Group 1. Group 1 Dragon Jar glazes are enriched in Cr, Cs, Li, and Sb; Group 2/3 jar glazes are enriched in Ba, Eu, La, Mg, Sm, Sr, and Th; Group 4 Dragon Jar glazes are enriched in Ag,

Co, Cd, and Pb. The composition of Groups 2 and 3 pastes were found to resemble one another more than either of the other groups. The fact that that glazes found on Group 2 and Group 3 pastes are chemically indistinguishable supports the INAA of the pastes which suggested that Paste Group 2 and 3 may derive from separate production centers within the same region.

Characterization of Mesa Verde Painted Pottery

In a recent project, the black paints on 253 pottery sherds from the Mesa Verde Region of the American Southwest were analyzed to determine the effectiveness of LA-ICP-MS for discriminating between the two major ceramic types, Mancos and Mesa Verde Black-on-white (*32*). Classification of these ceramics is based primarily on their paint composition. Mancos Black-on-white is typically classified as having a mineral-based paint derived from an iron-manganese ore. The other ceramic type, Mesa Verde Black-on-white is usually described as having an organic-based paint. Temporally, Mancos Black-on-white precedes Mesa Verde Black-on-white, thus these two ceramic types are used as temporal markers and serve to test a variety of hypotheses surrounding technology, production, distribution, and social organization. Accurate classification of these pottery types provides a foundation for testing these hypotheses.

Prior to data acquisition, samples were pre-ablated using the laser to remove possible surface contamination. Power settings for the laser were adjusted to prevent the laser from burning through the paint during analysis, ensuring that the material introduced to the ICP-MS was actually pigment and not the underlying clay matrix. Ablation parameters included a 200 μm diameter beam operating at 20 Hz and a laser scan rate of 30 μm/s. Data were calibrated using an approach suggested by Gratuze (*6, 7*). The basic structure of the dataset suggests the existence of at least four different compositional groups, or paint recipes—a mineral group, an organic group, a mixed mineral and organic paint group, and an organic paint group characterized by higher concentrations of lead (Figure 8). Paints have different textures, colors, and appearances, and it is difficult if not impossible to accurately categorize all paints using visual criteria, especially when there are multiple chemical groups possible. Fortunately, LA-ICP-MS removes the subjective component from pigment classification. Given that archaeologists use classification of paint types to explore temporal trends, accurate determination of pigment types is important because errors in classification can result in inaccurate archaeological interpretations. In contrast, chemical characterization of pigments will not only permit accurate classification but may also lead to the discovery of previously unknown chemical variation.

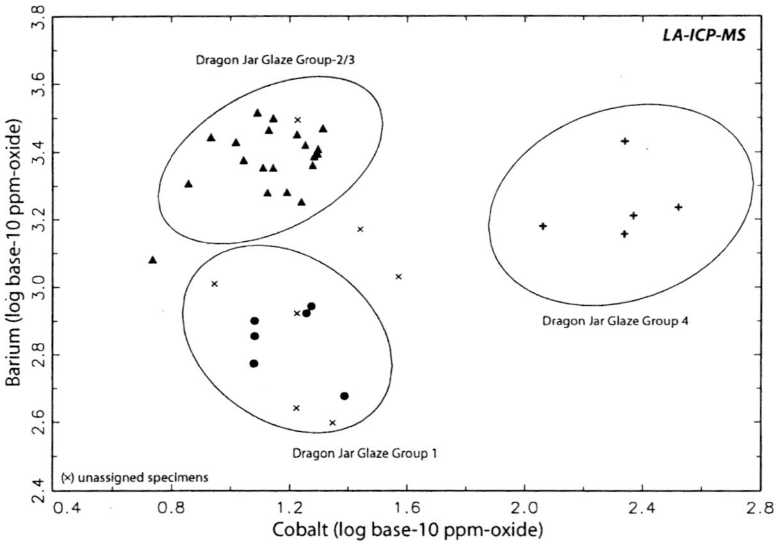

Figure 7. Bivariate plot of cobalt and barium base-10 logged concentrations for Dragon Jar glaze samples analyzed by LA-ICP-MS. Ellipses represent 90% confidence levels for group membership.

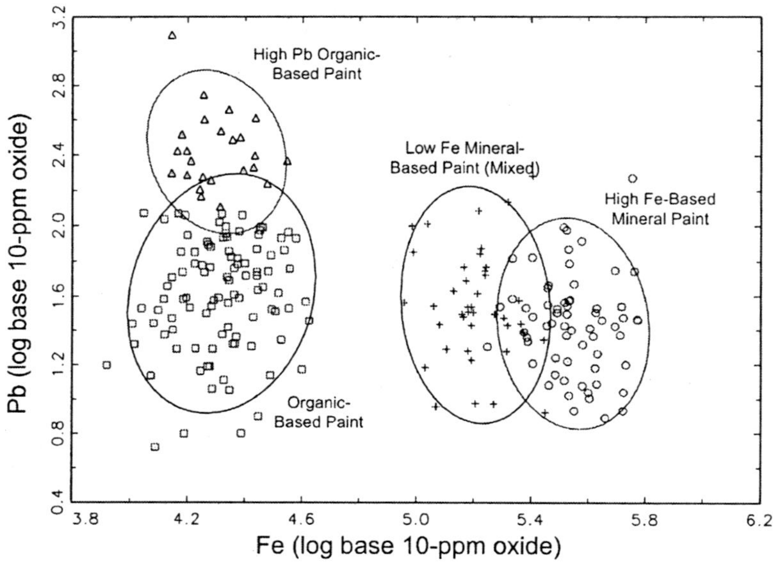

Figure 8. Bivariate plot of iron and lead log base-10 ppm oxide concentrations showing subgroup variation within the mineral and organic-paint groups. Ellipses represent 90% confidence interval for group membership. Unassigned specimens are not shown.

Elemental Contour Maps

One advantage to LA-ICP-MS is that elemental maps of a sherd's surface (and other materials) can be produced by generating data at different areas on a sample and importing these data into a mapping program such as Surfer (Golden Software). Figures 9–11 illustrate differences in paint concentrations on a mineral-painted and two organic-painted sherds. Data for iron, copper, and manganese were generated on an arbitrary grid at 1 mm intervals. The data generated for these experiments were imported into Surfer and manipulated to produce a chemically derived contour map for the surface of each sherd. The highest counts for each element were set to correspond to the color black, the lowest counts to the color white. Darker areas in the figures therefore correspond to higher elemental concentrations. Figure 9 shows the elemental surface map for sample WJJ122, a Mancos B/w sherd. On the left of this figure is a digital photograph of the area analyzed (the black area is paint, the white and gray areas are slip) followed by maps of iron, copper, and manganese. In this case the paint on sample WJJ122 is derived from an iron-based mineral. As a result, the iron and manganese maps show close resemblance to the digital photograph, whereas the copper is somewhat randomly distributed across the surface of the sherd. In contrast, sample WJJ080, an organic-painted sherd classified as Mesa Verde B/w, shows that copper correlates with the painted areas of the sherd whereas the iron and manganese concentrations show little relation to the painted areas (Figure 10). Given that iron and manganese are not expected to occur in significant quantities in organic paint, little agreement between the painted areas and the elemental contour maps is expected. The agreement between copper and the painted area suggests that copper, as well as the other metals (e.g., Zn, Ag, and Pb) found to be elevated in the organic paint, may contribute to the black color in the painted areas as previously discussed. In contrast, sample DMG013, also an organic-painted sherd classified as Mesa Verde B/w, produces a completely different distribution of elements (Figure 11). The distribution of copper in DMG013 does not correlate with the painted area as observed in the copper elemental contour map for WJJ080 (Figure 10). Likewise, the distribution of iron in DMG013 does not correlate with the painted area observed in the iron elemental contour map for WJJ122 (Figure 9). However, there is a suggestion that manganese is slightly enriched in this sample, as the manganese elemental contour map correlates somewhat with the painted area of the sherd.

The elemental contour maps serve to demonstrate that measurable differences between the painted and the slipped areas exist on pottery from the Mesa Verde region. Enrichment of certain metals (Mn, Cu, Ag, Pb) in the organic-painted sherds clearly does not result from diagenesis (post depositional alteration) but instead reflects differences in paint that are related to one or more

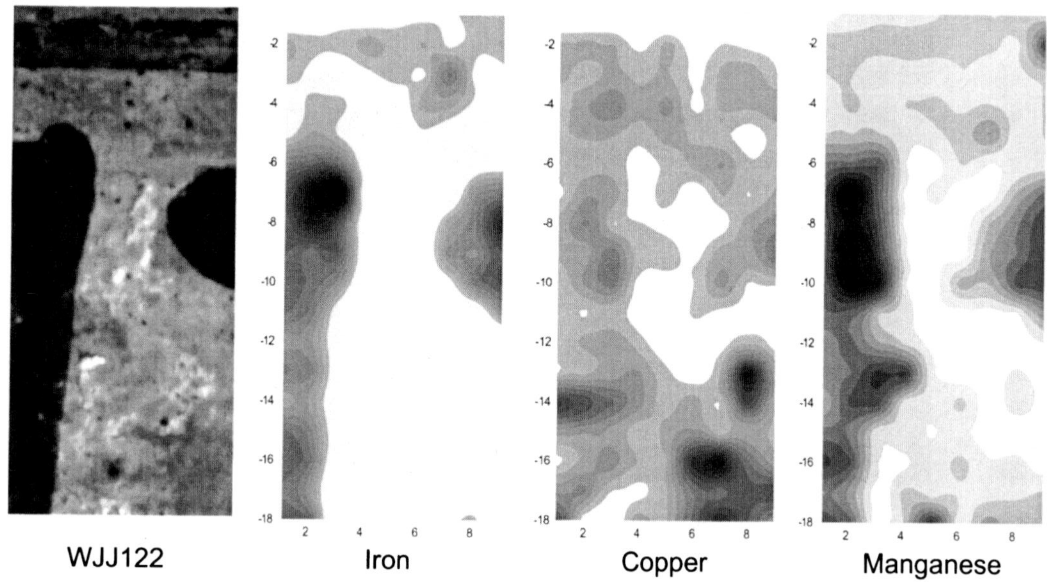

Figure 9. Left: digital photograph of a selected area of sample WJJ122, followed by elemental contour maps of iron, copper, and manganese concentrations on the surface of the sherd. Darker areas on the contour maps indicate higher elemental concentrations for that element. The analyzed area is 10 x 18 mm. Sampling was conducted at 1 mm intervals.

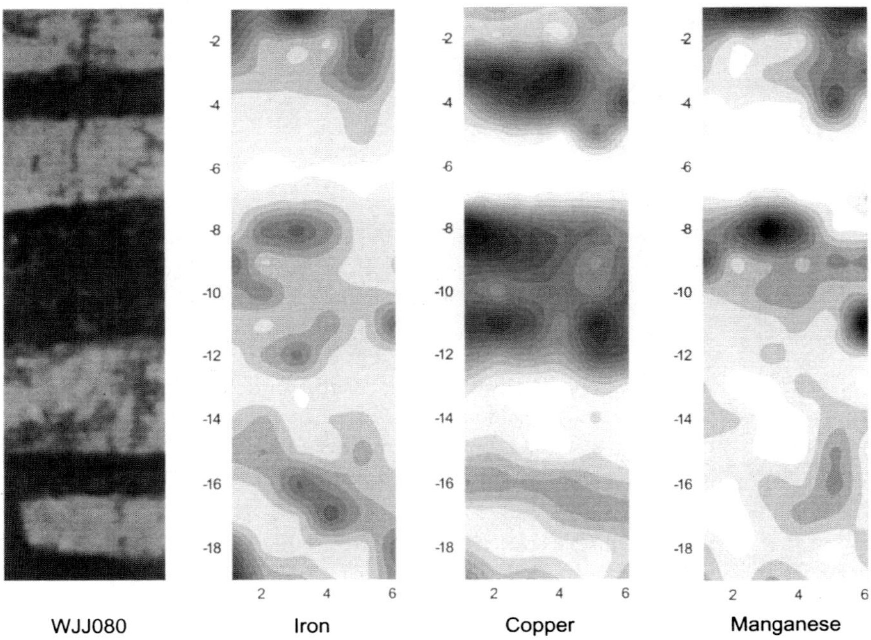

Figure 10. Left: digital photograph of a selected area of sample WJJ080, followed by elemental contour maps of iron, copper, and manganese concentrations on the surface of the sherd. Darker areas on the contour maps indicate higher elemental concentrations for that element. The analyzed area is 6 x 18 mm. Sampling was conducted at 1 mm intervals.

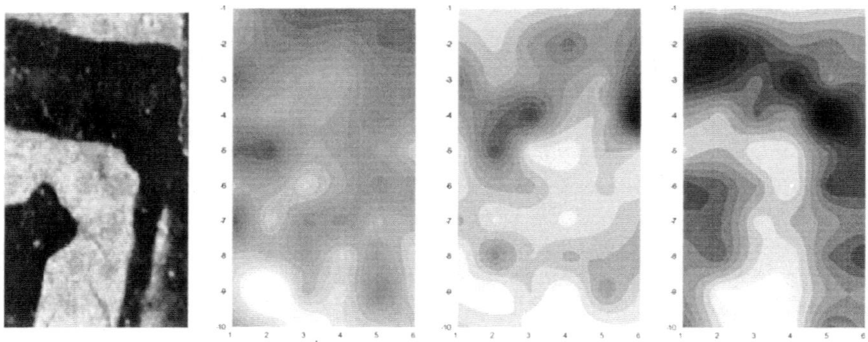

DMG013 Iron Copper Manganese

Figure 11. Left: digital photograph of a selected area of sample DMG013, followed by elemental contour maps of iron, copper, and manganese concentrations on the surface the sherd. Darker areas on the contour maps indicate higher elemental concentrations for that element. The analyzed area is 5 x 9 mm. Sampling was conducted at 1 mm intervals

of the following: the plant species used to manufacture the pigments, the soil substrate where the plant originated, or the water used to make these plants. If diagenesis were the cause for the enrichment, then the copper contour map for sample WJJ080 and the manganese contour map for sample DMG013 would not correlate as closely to the painted areas of the sherd. Finally, the differences between the two organic-painted sherds further support the argument made above that differences exist in the organic paint.

Human Bone and Teeth

Trace elements in human teeth and bone can be used to reconstruct dietary patterns in prehistoric populations. Several methods have been used to generate chemical data for prehistoric human bone. Among these methods INAA of solid bone and ICP-MS and ICP-ES of solutions have been used most often (*33–35*). With INAA, portions of bone or teeth are cleaned, sealed in vials, and irradiated to provide data for 8–10 elements. Samples analyzed by ICP-MS are digested in acid prior to analysis. In both cases, sample preparation is cumbersome.

An inherent problem with chemical characterization of bones is that diagenesis may confound the results of the analysis. One possible way to avoid diagenesis and contamination is through the use of a microprobe to sample specific areas of the bone. LA-ICP-MS may prove to be ideally suited for this application given that point-specific analyses can be conducted in which very small areas are targeted, thereby avoiding mineral inclusions and other potentially contaminated areas. Furthermore, by pre-ablating the target, the laser effectively cleans the sample area immediately before data collection begins, reducing the chances of measuring contamination.

In an attempt to evaluate the applicability of LA-ICP-MS to the study of human bone, we analyzed teeth and bone obtained from seventy-six individuals recovered during excavation of the Paloma archaeological site in central Peru. Paloma was occupied from approximately 5850–3750 B.C. and is located along the coastal plain of central Peru (*36, 37*). The only moisture this area receives is in the form of dense fog which is deposited on the landscape between June and December. Consequently, inhabitants of the site would have had limited access to terrestrial food resources and would have greater dependence on marine resources from the nearby ocean for the majority of their nutritional requirements.

Barium is an alkaline-earth metal incorporated into bone through the intestinal tract. In terrestrial environments, barium and strontium are approximately equal in abundance. In marine environments, barium forms an insoluble precipitate as a result of the high sulfate content in salt water. Formation of this compound effectively removes barium from seawater. As a result, Ba/Sr ratios reflect diet and are an indicator of trophic position. In human populations, individuals with diets high in marine-based food resources typically have low Ba/Sr ratios in their bone and teeth. Populations who consume large amounts of terrestrial-based food resources tend to have higher Ba/Sr ratios.

Research by Burton and Price (*38*) demonstrated that Ba/Sr ratios generated by ICP emission spectroscopy (ICP-ES) can be used to infer the diet (marine verses terrestrial) of prehistoric populations. In this experiment we duplicate results obtained by Burton and Price for the Paloma samples. Our results show that the Ba/Sr ratios obtained by LA-ICP-MS are comparable in precision and accuracy to ICP-ES data (Figure 12). Although it is not unexpected that a coastal population would rely heavily upon marine resources, there are applications where this type of research would have value. What we have done here is demonstrate the efficacy of LA-ICP-MS to this line of research by demonstrating that it is possible to generate results similar to those obtained by other analytical techniques.

Final Remarks and Further Developments

In this paper we have presented several applications of LA-ICP-MS to archaeological materials. Although our discussion has centered upon characterization of obsidian, ceramics, pigments, and bone, LA-ICP-MS has

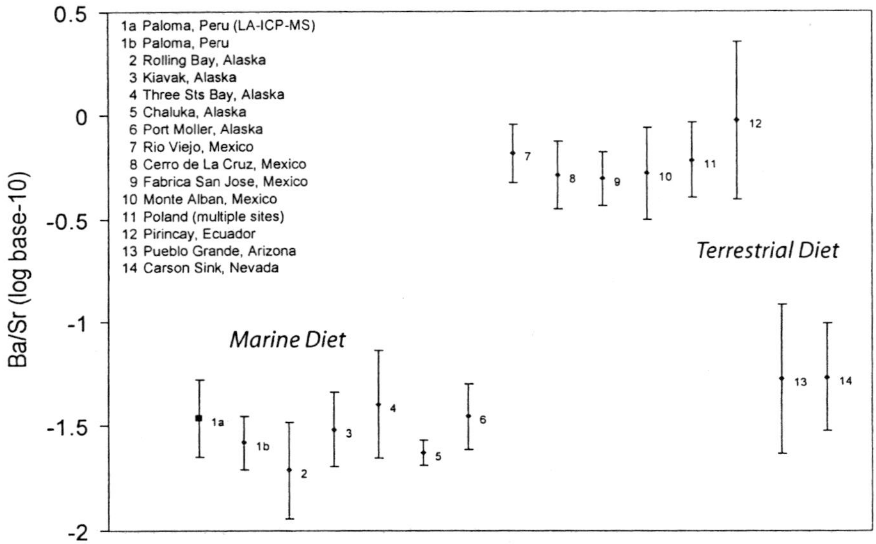

Figure 12. Comparison of barium-strontium element ratios generated by LA-ICP-MS (1a) and ICP-ES (1b) for a sample of individuals from the Paloma, Peru archaeological site, and with other ICP-ES data (2–14). All ICP-ES data were generated by Burton and Price (38).

numerous applications that are not discussed herein. As instrumentation, software, and matrix-matched standards continue to be developed, it is not unreasonable to expect that LA-ICP-MS will evolve into one of the primary analytical techniques employed by researchers for archaeological characterization studies.

Acknowledgments

The research reported here was funded in part by National Science Foundation grants for equipment (grant No. 9977237) and laboratory support (grant No. 0102325). We thank the following individuals and organizations without which this project would not have been possible. Dragon Jar samples were provided by Carla Sinopoli. Robert Benfer provided access to the skeletal material from Paloma. Donna Glowacki, Jim Judge, and Crow Canyon Archaeological Center provided access to the Mesa Verde and Mancos pottery. Hector Neff provided invaluable comments on various aspects of the projects discussed above.

References

1. Devos, W.; Moor, C.; Lienemann, P. *J. Anal. At. Spectrom.* **1999**, *14*, 621–626.
2. Devos, W.; Senn-Luder, M.; Moor, C.; Salter, C. *Fresenius J. Anal. Chem.* **2000**, *366*, 873–880.
3. Kennett, D. J.; Neff, H.; Glascock, M. D.; Mason, A. Z. *The SAA Archaeological Record* **2001**, *1*, 22–36.
4. Neff, H. *J. Archaeol. Sci.* **2003**, *30*, 21–35.
5. Pollard, A. M.; Heron, C. *Archaeological Chemistry*; Royal Society of Chemistry: Cambridge, 1996.
6. Gratuze, B. *J. Archaeol. Sci.* **1999**, *26*, 869–882.
7. Gratuze, B.; Blet-Lemarquand, M.; Barrandon, J. N. *J. Radioanal. Nucl. Chem.* **2001**, *247*, 645–656.
8. James, W. D.; Dahlin, E. S. C., D.L. *J. Radioanal. Nucl. Chem.* **2005**, *263*, 697–702.
9. Mallory-Greenough, L. M.; Greenough, J. D.; Dobosi, G.; Owen, J. V. *Archaeometry* **1999**, *41*, 227–238.
10. Speakman, R. J.; Neff, H.; Glascock, M. D.; Higgins, B. J. In *Archaeological Chemistry: Materials, Methods, and Meaning*; Jakes, K. A., Ed.; ACS Symposium Series No. 831; American Chemical Society: Washington, DC, 2002; pp 48–63.
11. Cucina, A.; Neff, H.; Blos, V. T. In *Laser Ablation ICP-MS in Archaeological Research*; Speakman, R. J.; Neff, H., Eds.; University of New Mexico Press: Albuquerque, NM, 2005; pp 187–197.
12. Dolphin, A. E.; Goodman, A. H.; Amarasiriwardena, D. *Am. J. Phys. Anthropol.* **2005**, *128*, 878–888.
13. Song, R.-J. Ph.D. thesis, University of Massachusetts, Amherst, MA, 2004.
14. Pearson, C.; Manning, S. W.; Coleman, M.; Jarvis, K. *J. Archaeol. Sci.* **2005**, *32*, 1265.
15. Sheppard, P. R.; Ort, M.; Speakman, R. J.; Anderson, K.; Elson, M. In *Geological Society of America Abstracts with Programs*; Salt Lake City, 2005; Vol. 37, p 111.
16. Ambercrombie, F. N.; Silvester, M. D.; Murray, A. D.; Barringer, A. R. In *Applications of Inductively Coupled Plasmas Emission Spectroscopy*, Barnes, R. M., Ed.; Franklin Institute Press: Philadelphia, 1978; pp 121–145.
17. Gray, A. L. *Analyst* **1985**, *110*, 551–556.
18. Neff, H.; Cogswell, J. W.; Ross, L. M. J. In *Patterns and Process: A Festschrift in Honor of Dr. Edward V. Sayre*; van Zelst, L., Ed.; Smithsonian Center for Materials Research and Education: Washington, DC, 2003; pp 201–226.

19. Pearce, N. J. G.; Westgate, J. A.; Perkins, W. T.; Eastwood, W. J.; Shane, P. A. R. *Global Planetary Change* **1999**, *21*, 151–171.
20. Pereira, C. E. d. B.; Miekeley, N.; Poupeau, G.; Kuchler, I. L. *Spectrochim. Acta Part B* **2001**, *56*, 1927–1940.
21. Speakman, R. J.; Neff, H. *Am. Antiq.* **2002**, *67*, 137–144.
22. Hill, D. V.; Speakman, R. J.; Glascock, M. D. *Archaeometry* **2004**, *46*, 585–606.
23. Glascock, M. D.; Braswell, G. E.; Cobean, R. H. In *Archaeological Obsidian Studies: Method and Theory*; Shackley, M. S., Ed.; Plenum Press: New York and London, 1998; pp 15–65.
24. Tykot, R. H. *J. Mediterranean Archaeology* **1996**, *9*, 39–82.
25. Tykot, R. H. *J. Archaeol. Sci.* **1997**, *24*, 467–479.
26. Tykot, R. H.; Ammerman, A. J. *Antiquity* **1997**, *71*, 1000–1006.
27. Tykot, R. H.. *Acc. Chem. Res.* **2002**, *35*, 618–627.
28. Weigand, P. C.; Harbottle, G.; Sayre, E. V. In *Exchange Systems in Prehistory*; Earle, T. K.; Ericson, J. E., Eds.; Academic Press: New York, 1977; pp 15–32.
29. Descantes, C.; Neff, H.; Glascock, M. D. In *Geochemical Evidence for Long-Distance Exchange*; Glascock, M. D., Ed.; Bergin and Garvey: Westport, CT-London, 2002; pp 229–256.
30. Sinopoli, C. M.; Dueppen, S.; Brubaker, R.; Descantes, C.; Glascock, M. D.; Griffin, W.; Neff, H.; Shoocongdej, R.; Speakman, R. J. *Asian Perspect.* **2006**, *45*, in press.
31. Grave, P.; Lisle, L.; Maccheroni, M. *J. Archaeol. Sci.* **2005**, *32*, 885–896.
32. Speakman, R. J. In *Laser Ablation ICP-MS in Archaeological Research*, Speakman, R. J.; Neff, H., Eds.; University of New Mexico Press: Albuquerque, NM, 2005; pp 167–186.
33. Edward, J.; Fossey, J. M.; Yaffe, L. *J. Field Archaeol.* **1984**, *11*, 37–46.
34. Edward, J. B. Ph.D. thesis, University of Missouri, Columbia, MO, 1987.
35. Burton, J. H.; Price, T. D.; Middleton, W. D. *J. Archaeol. Sci.* **1999**, *26*, 609.
36. Farnum, J. F. M.A. thesis, University of Missouri, Columbia, MO, 1996.
37. Benfer, R. A. *Lat. Am. Antiq.* **1990**, *1*, 284–318.
38. Burton, J. H.; Price, T. D. *J. Archaeol. Sci.* **1990**, *17*, 547–557.

Chapter 16

Evaluating the Precision Requirements for Isotope Ratio Determination of Archaeological Materials Using Laser Ablation–Time-of-Flight–Inductively Coupled Plasma–Mass Spectrometry

Increasing Ratio Precision

John V. Dudgeon[1,4], Hector Neff[2,4], Andrew "Flynn" Saint[3], and William Balsanek[3]

[1]Department of Anthropology, University of Hawaii, Honolulu, HI 96822
[2]Department of Anthropology, California State University, Long Beach, CA 90840
[3]GBC Scientific Equipment, Dandenong, Victoria, Australia
[4]Institute for Integrated Research in Materials, Environments, and Society, Long Beach, CA 90840

High precision isotopic determination of archaeological materials using laser ablation ICP-MS remains an elusive goal. Practical, nearly non-destructive, rapid and cost effective, isotope ratio measures using LA-ICP-MS suffer from transient effects, laser-induced fractionation, and generally lower precision than other methods of sample introduction. However, in contrast to high precision measurement requirements in biological systems, lower levels of analytical precision may still meet threshold levels for isotopic discrimination in archaeological materials, depending on the specificity of the research questions. We here report on our efforts to reduce the spread of replicate isotopic measurements of archaeological materials using pre and post-laser sample treatment, and evaluate the degree to which lower levels of precision affect group attribution in specific cases.

Some recent attention on the desirability of using laser ablation sample introduction for the fingerprinting and compositional analysis of archaeological materials has touted the rapid, low cost and minimally destructive result possible with laser ablation techniques (*1*). It is unfortunate that the strides made in multi-element compositional group formation have not permeated into the realm of isotope ratio analyses on some of the same material types. Isotope ratio analysis of biologically available elements like strontium have been shown to be sensitive life history indicators of residential mobility, and also for distinguishing migrants within populations. This is determined by measuring the variation in strontium isotopes in bones and teeth, which are a consequence of soil and groundwater isotope differences in natal regions or areas of long-term residence (*2, 3*). Alternatively, lead isotope analysis is useful to the archaeologist because lead is a widely used mineral in additive technologies such as prehistoric glasses and frits as well as ceramic glazes, and it is used as an alloying agent in ancient metallurgy (*4, 5*).

Of primary interest in this study is the effort to increase the precision of isotopic measurement using inductively coupled plasma time of flight mass spectrometry (ICP-TOF-MS) and to determine if variation in absolute accuracy, relative to well-characterized NIST standard reference materials, is a function of fractionation in the laser ablation chamber, sample transport, or merely a mass bias artifact of instrumental operating conditions and other adjustable parameters in the time of flight mass spectrometer. Several authors have argued the merits of improving isotopic ratio techniques in ICP-MS (*6–8*), but most have come at the problem from disciplines where ultra-precise ratio measures are not only desirable, but critical to understanding biological uptake and incorporation of radionuclides, or isotopic fractionation of exotic particles in analytical chemistry. Taken as a whole, this literature can be divided into two main camps. The first is concerned largely with sample introduction mechanisms (*7, 9, 10*), and the concomitant effort to understand the effect of sample introduction relative to precision and possible fractionation of the ion stream, whether liquid or gas sample carrier is used. The second camp has focused its efforts on describing the instrumental parameters and physics of ion counting in different mass analyzers (*8, 11*), in an effort to estimate precision limits and to explain the deviation from expected isotopic ratios that affect instrumental accuracy. Both of these literatures are technical and perhaps esoteric outside of the realm of analytical chemistry, but they provide us some expectations for experimental procedure that we can test for our own real world samples and projects.

Isotope Ratios for Archaeology

In the remainder of this chapter, we describe some of our lab's efforts to increase the precision and accuracy for replicate measurements of the isotopes of two elements of particular interest to archaeologists: lead (Pb) and strontium

(Sr). Over the past 30 years, lead and strontium isotope ratios have been measured with thermal ionization mass spectrometry (TIMS). Elemental salts are deposited on a filament heated to produce ionized particles, which are then sent into a mass spectrometer where they are detected by multiple Faraday cups arrayed such that ions of several masses are collected simultaneously. TIMS is capable of high precision isotope discrimination, but the instruments tend to be large and expensive, and extensive sample preparation is required prior to sample introduction. Newer ICP-MS-based technologies like multi-collector ICP-MS (especially laser ablation) circumvent some of the sample preparation issues while exploiting the precision of simultaneous mass discrimination, but they are still limited by the number and configuration of ion collectors.

Archaeologically, isotope ratio measures provide a means to identify the location of origin of artifacts or osteological remains. However, ambiguity of provenance based on isotopic signatures may be partially resolved by incorporating trace element signatures that improve the likelihood of source identification. One example currently under study is the comparative analysis of isotope ratios and potential discriminator elements, such as the simultaneous collection of strontium or lead isotope ratio data and differential concentration of rare earth elements within large samples of archaeological skeletal remains (*12*).

In contrast to multi-collector TIMS or ICP-MS based instruments, the GBC Optimass orthogonal time-of-flight ICP-MS installed in our lab utilizes a single discrete dynode ion detector (ETP Pty. Ltd., Ermington, Australia). Instead of measuring the characteristic mass/charge of an ion as a consequence of its deflection along a magnetic flight tube, time of flight mass spectrometry measures the characteristic flight time of an ion as it traverses a parabolic flight path with an effective 1.4 meter flight distance. Traversing this distance allows temporal separation of ions by mass/charge (*13*) and renders essentially synchronous signal counting over the entire mass spectrum. Synchronous data acquisition has the potential to vastly improve transient signal discrimination. Transient signals are produced by sampling spatially or temporally heterogeneous materials, and generally are the rule rather than the exception in laser ablation sample introduction. While not central in the present context, the ability to discern heterogeneity within the sample matrix or from surface contamination in real time allows correction and resampling, and offers insights to the problem of 'noisy data'. More importantly in the present context, the ability to sample the ion stream very rapidly can be exploited to achieve improvements in isotope ratio analyses.

Analytical Parameters

The purposes of this paper are to report on our efforts to evaluate the real world precision of laser ablation time of flight ICP-MS (LA-TOF-ICP-MS) replicate measurements on NIST standards and archaeological materials, and to explore the sources and magnitudes of mass bias and/or mass fractionation in our instrumental setup. Some authors (*7, 14*) have speculated on the causes of laser induced fractionation in laser ablation sample introduction systems. While not fully understood, laser-induced fractionation seems to be a complex interaction at the ablation site of laser energy differentially heating and volatilizing sample particles, resulting in a heterogeneous distribution of elements and isotopes across ablated particle size.

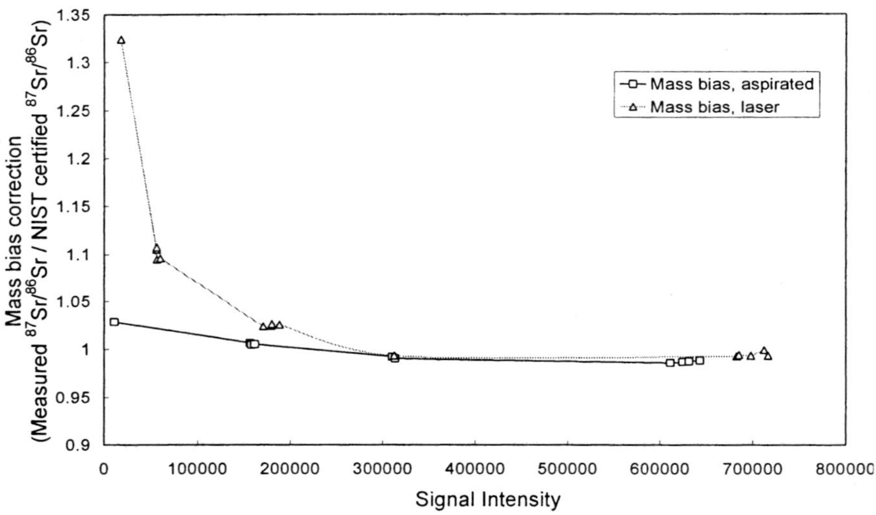

Figure 1. Comparison of mass bias by sample introduction type.

To test for the possibility of laser-induced fractionation, we compared NIST standards in the TOF with both laser and nebulizer aspirated solutions (Figure 1). Our results indicate that while a small amount of bias correction difference exists between laser and aspirated sample introduction at lower signal intensities, no significant difference in isotopic ratios between ablated and aspirated samples exists when signal intensities are maintained at 300,000 cps and higher.

To mitigate laser induced fractionation at different laser energies, we operated the laser at 100% of maximum output energy on all samples (Table I), and modulated the argon and aerosolized sample entering the plasma torch using a simple tabletop rotometer. Controlling the gas flow with the rotometer allows us to standardize the measured signal intensity across samples of widely variable concentrations without altering plasma or detector settings, which might introduce conditions generating differential mass bias.

Table I. Instrument Parameters for this Experiment

Laser Ablation Device	New Wave UP213 Nd:YAG
Wavelength	213 nm
Energy	0.6 mJ
Spot size	100 μm
Fluence	7.35 J cm^{-2}
Rep rate	20 Hz
Depth/pass	5 μm
Scan speed	50 μm/s
ICP-MS	GBC Optimass 8000 TOF ICP-MS
Detector	discrete dynode electron multiplier
RF power	700 W
Plasma Ar gas flow rate	10.0 l min^{-1}
Auxiliary Ar gas flow rate	0.5 l min^{-1}
Sample Ar gas flow rate	1.2 l min^{-1}
Data acquisition mode	time resolved, 1 second/acquisition

Mass bias—the condition in which ions of different masses are sampled by the detector at frequencies varying from their expected proportions—is thought to be a result of space-charge effects in the high pressure region immediately downstream of the plasma torch (*15*).

In the TOF-ICP-MS, space-charge produces varying effects on different isotopic masses, as ions with higher atomic masses maintain higher inertia (*16*). This results in an increased transmission efficiency of heavier ions and an increase in scattering of the lighter ions from the axial path as they encounter the high pressure region between the plasma flame and the ion optics. This effect should be more noticeable for lighter ion isotope species, because of the relatively smaller ratio of mass to mass difference between adjacent isotopes. Our experiments have demonstrated that, irrespective of the magnitude of mass bias when counting isotopes, by standardizing the time-resolved signal intensity within the linear portion of the mass bias count rate curve, mass bias corrections have small variations with respect to rather larger signal differences.

On the other hand, at high signal intensities approaching 50 million analog counts per second, saturation of the detector results in suppression of actual counts for the most abundant isotope, producing a spurious mass bias correction. While the 256 ion counting channels of the detector software are theoretically capable of measuring 64 million counts per second without ion stacking (*13*), our empirical result shows that significantly lower count rates can result in ratio suppression on certain isotopes. Consequently, our strategy was to generate count rates that inhabited the flattest portion of the mass bias correction curve, without detector saturation. Problematically, isotope ratios that were far from percentage equilibrium (e.g., $^{204}Pb/^{206}Pb$) may exhibit both problems: too low count rate for ^{204}Pb to generate the flattest linear mass bias correction, and too high for ^{206}Pb, causing partial detector saturation. A similar problem was found for $^{84}Sr/^{88}Sr$ and $^{86}Sr/^{88}Sr$, which precludes these invariant ratios from being used as an internally-standardized mass bias correction for the measured $^{87}Sr/^{86}Sr$ (*17*).

Experimental Procedure

Pb Isotopes

We began our experiment by taking fused glass beads impregnated with NIST Common Lead Isotopic Standard SRM 981 and making successive passes over the bead surface with the laser, continuously ablating over a pre-ablated raster pattern. To control the ion count rate, we utilized the tabletop rotometer to adjust the argon sample gas flow up or down to normalize the count rate at the detector to around 5–10 million counts per second. At this count rate, we achieved the greatest linear stability of the isotope ratio $^{207}Pb/^{206}Pb$. Using this data as our isotopic ratio mass bias correction, we analyzed several archaeological lead glasses and glazes previously analyzed by Robert Brill (Corning Museum of Glass) using TIMS (Figure 2).

Our results indicate that at standard count rates across lead glasses with a range of lead oxide weight percents, we are able to closely approximate the TIMS-reported $^{207}Pb/^{206}Pb$ and $^{208}Pb/^{206}Pb$ ratios, after correcting for mass bias using NIST SRM 981 standard isotopic lead. As anticipated, the mass bias correction for the $^{208}Pb/^{206}Pb$ ratio was of greater magnitude than the $^{207}Pb/^{206}Pb$, an indication that space-charge effects were of greater magnitude for isotopes two mass units apart than for adjacent masses.

Sr Isotopes

A similar set of experiments was performed to test our ability to precisely characterize strontium isotopes. For the archaeological chemistry of biological remains, strontium isotopes are useful for testing dietary hypotheses and for examining differences in place of birth and place of residence. To test the precision of strontium isotope ratio measurement using TOF-ICP-MS, we made glass fusion beads using NIST Strontium Carbonate Standard SRM 987 and

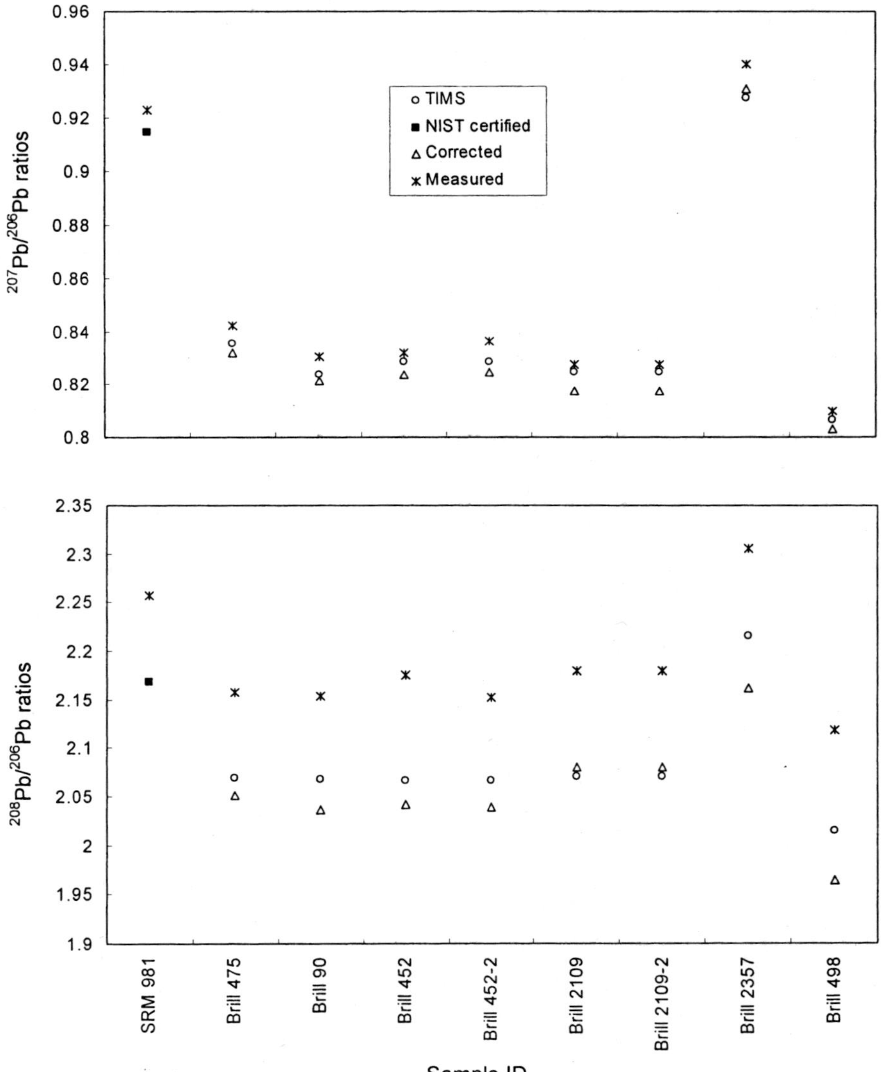

Figure 2. Comparison of TIMS and time of flight ICP-MS Pb isotope ratios from Brill glasses.

examined the effect of varying concentrations of strontium on our precision. Unlike the measurement of lead isotope ratios in pottery glazes or metal artifacts, strontium concentration occurs at lower levels in biological tissues, usually from about 10 to 1000 parts per million (*19*). While this concentration is well within the analytical range of high precision aspirated liquid ICP-MS, it is generally too low to achieve high precision isotope ratio measures using laser ablation, due to lower ion counting rates.

One technique to remedy this problem involves taking several hundred independent, one second 'snapshot' measures of strontium isotope ratios and averaging the individual ratio measures to build higher statistical precision. The method has two important advantages: first, it makes use of the TOF-ICP-MS ability to simultaneously measure multiple masses and generate isotope ratios 'on the fly'; and second, plots of the one second instantaneous isotope ratio deviations from expected ratios can be visually checked for laser ablation artifacts or sample contamination during data acquisition. It is expected that this method will help identify the source of 'noisy' data that sometimes affect isotope ratio determinations using TIMS or other averaged sample introduction methods.

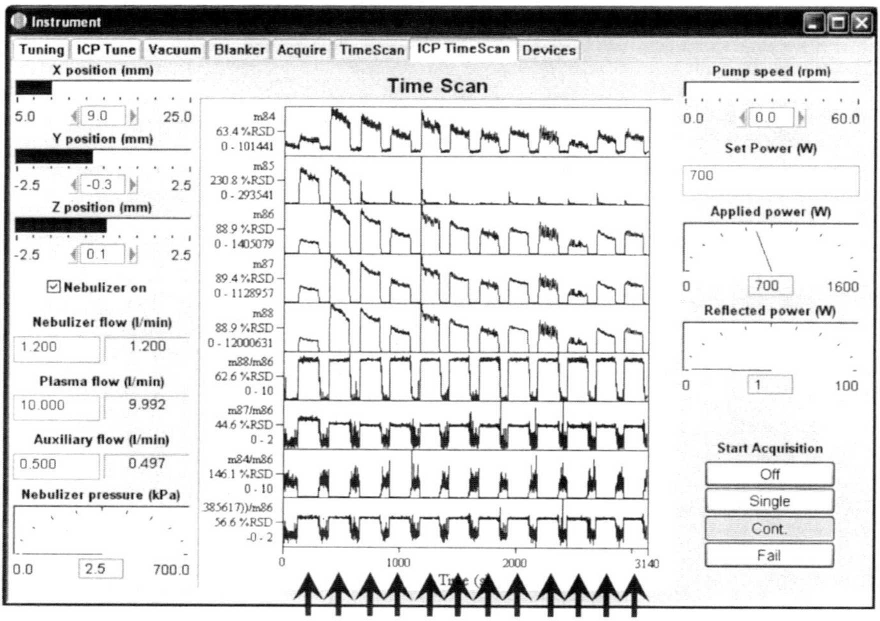

Figure 3. Timescan packets for dental enamel Sr isotope sampling. Arrows indicate individual samples (standards and teeth).

We tested this data acquisition method by first ablating bracketing concentration standard glasses that had been determined from previous measurements. All the samples in this analysis had total strontium elemental concentrations between 95 and 350 parts per million strontium. We then collected 150–300 one second ion count packets for ^{84}Sr, ^{86}Sr, ^{87} Sr and ^{88}Sr, plus the instantaneous ratio of each isotope to ^{86}Sr (Figure 3).

After examining the calibration curve for mass bias correction between the bracketing standards, and culling measurements arising from instrument noise or regions of sample contamination, all remaining one second count packets were averaged and bias corrected. This 'on the fly' sampling method, made possible by the simultaneous counting of the entire mass spectrum, enabled us to quickly determine the region on the sample with the highest elemental concentration, so that isotope ratio precision and accuracy could be optimized. It also permited rapid identification of regions of matrix variability, because aberrant transient signals were easily identified as spikes of analyte signal intensity on the graphical display. TOF-ICP-MS technology is perfectly suited to laser ablation sample introduction, in that it makes best use of the transient signal that is volatilized by the laser beam and can be temporally matched to the scan and pulse rate of ablation for systematic data acquisition.

Results from Archaeological Samples

Pb Isotopes in Glass Beads

We tested this method for lead isotope ratio measurement using archaeological glass trade beads from central California. These beads were used as trade goods between fort traders and Native Americans all over the western United States (*19, 20*). Central Californian Indian groups had manufactured and used shell and stone beads for centuries prior to the arrival of Europeans and quickly adopted the brightly colored glass beads as prestige and trade items. Linking the characteristic compositions of the beads found in both historic European forts and Native American archaeological sites is one method to determine the level of interaction between these groups. Specific recipes are believed to have been used by European bead workshops, and the addition of isotopic compositional data is a potentially powerful way to tie the chemical data to specific manufacturing regions.

Multiple compositional groups recognized in the data are illustrated on a bivariate plot of log-transformed potassium and antimony (Figure 4). These groups are believed to represent different recipes and, consequently, different regions of manufacture. We compared the ^{207}Pb/^{206}Pb and ^{208}Pb/^{206}Pb ratios to

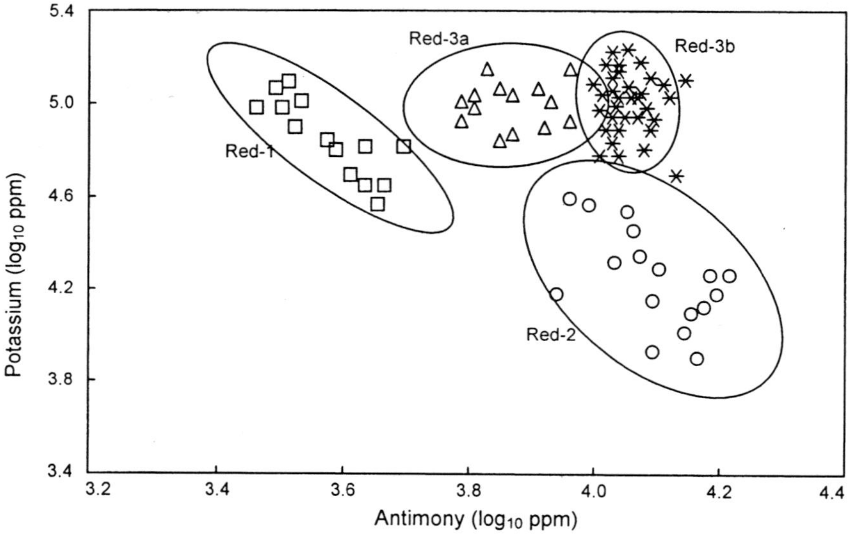

Figure 4. Provisional compositional groups among red beads determined on a plot of potassium and antimony.

see if similar group discrimination could be determined (Figure 5). Two groups, Red-1 and Red-2, are visible in this plot. The other groups defined elementally, Red-3a and Red-3b, do not show unique isotope ratio profiles, and the group identified as Red-3a shows an undesirable spread along both isotopic axes. Further analysis is required to understand whether this bead group is isotopically heterogeneous but of similar elemental recipe, or if the beads themselves are compositionally heterogeneous. Since the beads are considered to be aesthetically beautiful—as well as important—historical artifacts, large scale bulk analyses derived from crushing and vaporizing whole collections of beads is not an option for compositional determination and sourcing. For this reason, LA-TOF-ICP-MS may be the most appropriate tool for rapid elemental and isotopic characterization within this class of artifacts.

Sr Isotopes in Human Dental Enamel

As part of an ongoing research project to document the dietary and health status of the prehistoric population of Easter Island, dental samples were analyzed to determine our ability to discern marine versus terrestrial dietary groups on this small island (*12*). As an extension of a previous project to determine archaeological signatures of migration using total biologically

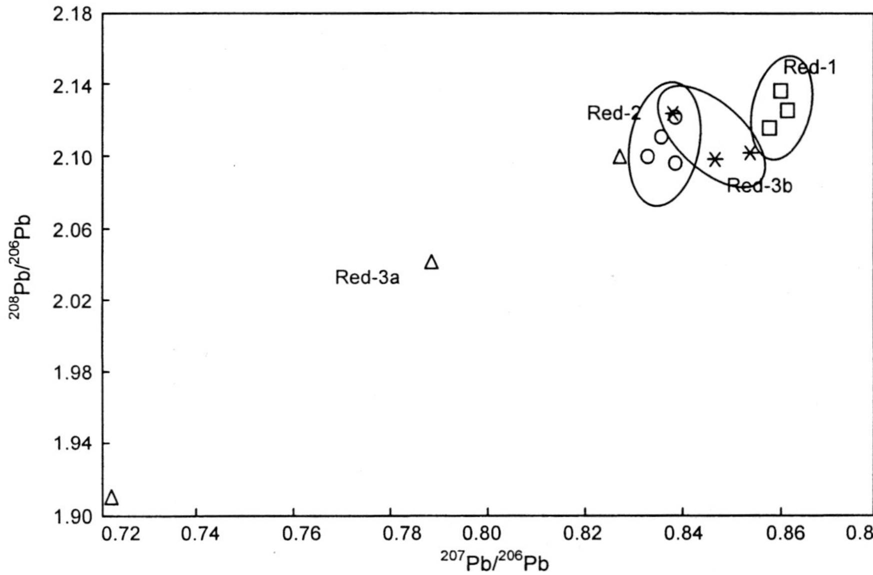

Figure 5. Possible group identification determined from a plot of $^{207}Pb/^{206}Pb$ and $^{208}Pb/^{206}Pb$.

incorporated strontium (*21*), we analyzed $^{87}Sr/^{86}Sr$ in a sample of skeletal remains representing coastal and inland populations. This preliminary effort attempted to generate $^{87}Sr/^{86}Sr$ ratios with precisions commensurate with data generated by TIMS. Following the methods outlined above, several hundred one second 'snapshot' measures of $^{87}Sr/^{86}Sr$ were collected in timescan mode, as well as the more traditional short interval, time-averaged integration method, and precision estimates calculated. Both results were compared to typical data created by TIMS, to assess the magnitude of precision differences between instrumental methods.

Precision estimates of our replicate measurements are shown below, along with average $^{87/86}Sr$ (Table II). Compared with precision estimates for published TIMS data and our standard laser ablation time of flight protocol (comprised of an average of five, five second integrations) we can see that our timescan protocol reduces the standard error of replicate measures by around 80%. This standard error is still about three times the average reported standard error of the TIMS measurements, but demonstrates the increased precision achievable with increased sampling.

It is anticipated that more experimentation using longer counting intervals will reduce the spread of replicate measures even more. However, it is not

Table II. Precision Estimates for $^{87/86}Sr$, by Instrument and Technique.

LA-TOF-ICP-MS (replicate integrations)	LA-TOF-ICP-MS (time scan)	TIMS (Univ. Wisconsin)
0.71034 +/- 0.0211	0.71034 +/- 0.0027	0.70843 +/- 0.0008
0.71034 +/- 0.0105	0.71034 +/- 0.0015	0.70788 +/- 0.0008
0.71034 +/- 0.0059	0.70824 +/- 0.0019	0.71259 +/- 0.0008
0.71034 +/- 0.0030	0.71191 +/- 0.0064	0.70856 +/- 0.0007
0.70414 +/- 0.0120	0.70474 +/- 0.0021	0.70821 +/- 0.0007
0.70919 +/- 0.0122	0.70856 +/- 0.0029	0.71869 +/- 0.0008
0.70109 +/- 0.0120	0.71240 +/- 0.0018	0.70872 +/- 0.0007
	0.71183 +/- 0.0018	0.70848 +/- 0.0007
	0.71468 +/- 0.0021	0.72279 +/- 0.0008
	0.71280 +/- 0.0018	0.73342 +/- 0.0007
	0.71208 +/- 0.0015	0.70867 +/- 0.0007

known whether discrete sampling techniques will equal the precision of bulk-averaged multi-collector mass spectrometer measurements like TIMS without some effort to average a heterogeneous sample prior to introduction to the ICP-MS torch. Presently, we believe the remaining noise in our strontium isotope determinations is due to: 1) inherent instability in the TOF measurements, presumably related to Poisson-distributed counting statistics and, 2) micro-spatial heterogeneity within the material being ablated. The latter may be explained by dietary variation during enamel mineralization, or micro-structural diagenetic variation (*22*). Both hypotheses impact the conclusions offered here, and as such deserve further exploration to assess the magnitude of their effect on our measurements. Additional research on samples measured by laser ablation time of flight ICP-MS, with validation by TIMS or other traditional high precision technologies will help determine whether LA-TOF-ICP-MS can address archaeological questions with the required degree of precision and accuracy to make verifiable statements about prehistoric migration and diet of animals and people.

Conclusion

While we think we have arrived at the right combination of sample introduction and system and ion counting parameters, it remains an empirical challenge to produce reliable isotope measurements using LA-TOF-ICP-MS. Adjusting for an appropriate instrumental count rate is key to approaching accurate measures, and averaging many multiples of replicates is effective at reducing the spread of the data points to achieve a precision that may be sufficient to assign samples to particular source groups or to identify migrants within populations. Unfortunately, the data required to prove the technology sufficient for the current application is essentially a moving target, because questions about the archaeological record are continually being asked at smaller scales, and answers that may render a meaningful interpretation of human prehistory today may not do so tomorrow. We are continually striving to compare more samples from different locales with the expectation that we can make meaningful distinctions between samples in most cases. We will continue to analyze artifacts previously characterized by TIMS and multi-collector ICP-MS to amass a database of isotope variability to compare our conclusions and validate our methods and results.

References

1. Speakman, R. J.; Neff, H. In *Laser Ablation-ICP-MS in Archaeological Research*; Speakman, R. J.; Neff, H., Eds.; University of New Mexico Press: Albuquerque, NM, 2005; pp 1–14.
2. Bentley, R. A.; Price, T. D.; Stephan, E. *J. Archaeol. Sci.* **2004**, *31*, 365–375.
3. Montgomery, J.; Evans, J. A.; Neighbour, T. *J. Geol. Soc. London* **2003**, *160*, 649–653.
4. Habicht-Mauche, J. A.; Glenn, S. T.; Schmidt, M. P.; Franks, R.; Milford, H.; Flegal, A. R. *J. Archaeol. Sci.* **2002**, *29*, 1043–1053.
5. Baker, J.; Stos, S.; Waight, T. *Archaeometry* **2006**, *48*, 45–56.
6. Becker, J. S. *J. Anal. At. Spectrom.* **2002**, *17*, 1172–1185.
7. Jackson, S. E.; Günther, D. *J. Anal. At. Spectrom.* **2003**, *18*, 205–212.
8. Willie, S.; Mester, Z.; Sturgeon, R. E. *J. Anal. At. Spectrom.* **2005**, *20*, 1358–1364.
9. Liu, C. Y.; Mao, X. L.; González, J.; Russo, R. E. *J. Anal. At. Spectrom.* **2005**, *20*, 200–203.
10. Tunheng, A.; Hirata, T. *J. Anal. At. Spectrom.* **2004**, *19*, 932–934.
11. Yang, L.; Sturgeon, R. E. *J. Anal. At. Spectrom.* **2003**, *18*, 1452–1457.
12. Dudgeon, J. V.; Neff, H. *Society for American Archaeology* 2005, Salt Lake City, UT.
13. Sturgeon, R. E.; Lam, J. W. H.; Saint, A. *J. Anal. At. Spectrom.* **2000**, *15*, 607–616.
14. Günther, D.; Hattendorf, B. In *Laser-Ablation-ICPMS in the Earth Sciences: Principles and Applications*; Sylvester, P., Ed.; Short Course Series; Mineralogical Assocation of Canada: St. Johns, Newfoundland, 2001; Vol. 29, pp 83–91.

15. Burgoyne, T. W.; Hieftje, G. M.; Hites, R. A. *J. Am. Soc. Mass Spectrom.* **1997**, *8*, 307–318.
16. Balsanek, W. J. Ph.D. thesis, University of Texas, Austin, TX, 2005.
17. Davidson, J.; Tepley III, F.; Palacz, Z.; Meffan-Main, S. *Earth Planet. Sci. Lett.* **2001**, *184*, 427.
18. Curzon, M. E. J. In *Trace Elements and Dental Disease*; Curzon, M. E. J.; Cutress, T. W., Eds.; John Wright/PSG Inc.: Littleton, MA, 1983; pp 1–9.
19. Spector, J. D. *Hist. Archaeol.* **1976**, *10*, 17–27.
20. Lightfoot, K. G.; Wake, T. A.; Schiff, A. M. *J. Field Archaeol.* **1993**, *20*, 159–175
21. Cucina, A.; Neff, H.; Blos, V. T. In *Laser Ablation-ICP-MS in Archaeological Research*; Speakman, R. J.; Neff, H., Eds.; University of New Mexico Press: Albuquerque, NM, 2005; pp 187–197.
22. Montgomery, J. Ph.D. thesis, University of Bradford, Bradford, West Yorkshire, UK, 2002.

Chapter 17

Lead Isotope Analysis of Roman Carthage Curse Tablets

Sheldon Skaggs

Department of Geology, University of Georgia, Athens, GA 30602

Lead curse tablets from Roman Carthage contain variable amounts of very small metallic inclusions. Electron microprobe analysis confirmed these metallic inclusions were bronze, brass, and a Sn-Sb alloy. This was interpreted as possible evidence of lead metal recycling. Six samples were chosen to represent a range of tablets containing the minimum to the maximum number of inclusions. Thermal ionization mass spectrometry of the Pb isotopes in the curse tablets appear to define a mixing line, with the tablets containing the least number of inclusions plotting closest to the Tunisian lead ore isotope ratios.

Introduction

Abundant evidence exists, in the form of stamped ingots, for the importation of lead metal into North Africa during Roman times (*1*). Currently, researchers believe that local North African ores were a minor constituent in the overall use and trade of lead in the region (*2*). Generally, Roman use of lead ores is argued to have been the result of serial exploitation; they worked only the richest mines available at any given time, and moved on when richer areas were conquered (*1*). However, there is a limited amount of evidence for the exploitation of local lead resources in the general vicinity of Carthage, located near the modern city of Tunis, Tunisia (*2*). Roman ceramic oil lamp fragments dating from 100 to 400

A.D. were found in old mines, and sizable slag deposits suggest at least some level of lead exploitation during Roman times (*2*). By the careful selection of Roman lead artifacts likely to have been manufactured in Carthage, and the use of lead isotope analysis, it may be possible to confirm the use of Tunisian lead ores during the Roman period. In this preliminary study, analysis of 22 curse tablets by electron microprobe analysis (EMPA) and six tablets by thermal ionization mass spectrometry (TIMS) suggest that these artifacts might be used to better understand Roman lead use and trade.

History of Carthage

The Phoenicians founded Carthage in the 9th or 8th centuries B.C. Punic Carthage became the capital of what historians call the Carthaginian Empire. Successful struggles with various Greek cities throughout the 5th and 4th centuries B.C. ultimately left Carthage as the dominant power of the western Mediterranean. A dispute over territory in Sicily in the year 264 B.C., led to a series of three wars between Carthage and the growing state of Rome. Carthage lost Sicily and later Spain to Rome as a result of the first two wars. The final war ended with the destruction of Carthage in 146 B.C. No further settlement was officially allowed at the site until Augustus established a veterans' colony late in the 1st century B.C. By the end of the 2nd century A.D., Carthage had become the third largest city in the Roman Empire and prospered as a center of trade. The Vandals conquered North Africa in 439 A.D. and occupied Carthage; it was recaptured by the Byzantines in 533 A.D. and then fell to the Arabs in 697 A.D. This study considers only artifacts of Roman Carthage from the 1st century B.C. to the 6th century A.D.

Background

Roman *defixiones*, or curse tablets, have been excavated at Carthage since the late 19th century. They are uncommon finds, and only 1000–1500 have been discovered in North Africa (*3, 4*). Curse tablets are a physical manifestation of a magical process meant to entice supernatural forces to affect their target. The curse is typically accomplished through the binding of the target, and the Latin word for curse tablets, *defixiones*, comes from defigere which means "to bind". Curse tablets were usually deposited in graves, wells, or in a location close to the victim (*3*). Curse tablets were usually made of thin sheets of lead, and are found rolled or folded with inscriptions on the inside. An unrolled tablet is shown in Figure 1. Those from Carthage are all written in Greek or Latin (*4*). The primary interest of archaeologists and curators has been to open these tablets, document

Figure 1. Example of a typical curse tablet before (A) and after (B) it is unrolled. Inscriptions can be seen on the left side and figures on the right side of the unrolled tablet.

the inscriptions, and interpret them (*3, 4*). Since the tablets are essentially pure lead, analysis of the isotopes could be useful in determining the original ore source location.

Lead isotope analysis is a well established archaeological method. Briefly, the amounts of lead isotopes 206, 207, and 208 in a lead ore are determined by the radioactive decay of the uranium isotopes 238, 235 and thorium isotope 232 initially deposited in the ore (*5*). Therefore, lead isotope abundance is controlled by the time elapsed since the ore was isolated from its uranium and thorium source. The lead isotope 204 is not radiogenic, denoting the stable abundance of this isotope from the initial deposition of the ore deposit (*5*). However, ore emplacement can be a very complex geological process. Multiple pulses of ore rich fluids can affect the abundances of lead isotopes, potentially giving each lead deposit a unique lead isotope fingerprint (*5*). Additionally, there is no guarantee that all areas of a lead ore deposit will have the same isotope fingerprint. Because of the high mass of lead compared to the difference in mass between lead isotopes, the ratios do not significantly change by fractionation during smelting and smithing of lead ores into lead metal (*5*). Typically, the ratios of $^{208}Pb/^{206}Pb$, $^{207}Pb/^{206}Pb$, and $^{206}Pb/^{204}Pb$ in an artifact are compared to the same ratios in the lead ore (*6*).

Curse tablets have some attributes favorable for this kind of lead isotope study. The curse tablets are essentially pure lead metal artifacts. Additionally, the size of the tablets would not require more lead than typically available in a single lead ingot. Therefore it is likely that tablet construction did not necessitate the practice of combining lead from different geographical sources. The mixing of geographically distinct ores would result in lead isotope values falling somewhere between those sources used to make the tablet. It is possible that the lead was obtained from the reuse of other lead objects, such as pipes and roofing material. The isotope values would then reflect the ore from which the original artifact(s) were made.

Additionally, the tablets are not likely to have been manufactured outside of Carthage. One reason being that the practice of magic, including making curse tablets, was illegal. It is unlikely that anyone would risk harsh punishment to import an item that could easily be manufactured by local means. Furthermore, the short time frame in which the curse tablet was to be effective suggests a limited geographical distance between manufacturing location and the place of use. Many of the translated curses were meant to affect legal proceedings, individual chariot races, or other competitive events (*3, 4*). Such time limitations would likely restrict the distance between the place of curse tablet manufacture and the location of the event.

Despite the advantages that curse tablets offer for such an analysis, there are a few problems that must be considered when analyzing these artifacts. First, only a very small portion of the curse tablet can be sampled and then used for analysis (*7*). Typical samples collected were only a few grams from the edge of the tablet. This limits the ability to ensure the tablets are homogenous. Furthermore, the samples were usually badly corroded (*7*). While this should not have an affect on the lead isotope values, it could change other trace element abundances.

Second, many of the tablets currently held in museum collections have no well-documented provenance. Those under study here are all from the collection of the National Museum of Carthage. They include several examples that were stratigraphically excavated by the University of Georgia excavation of the Roman Circus in the 1980s. The other tablets owned by the museum have been in their collection since at least the 1920s. The examples in this study fall into three categories: 1) those excavated by the University of Georgia, identified by a number prefix; 2) those catalogued and published by Audollent (*4*), identified by the prefix AUD; and 3) those others legally owned by the Museum from Carthage but whose exact find spots in the city are no longer known, identified by the prefix P. Only the University of Georgia and Audollent tablets have been opened, read and documented. Some of the tablets with the prefix P are opened, and may be tablets read and published by Audollent, but are no longer connected to that documentation (*7*). Most curse tablets are difficult to date, but the University of Georgia examples were excavated from Roman strata. All other tablets from Carthage are generally presumed by scholars to be Roman in date (*3*).

Previous Research

In 1991, Lisa Pintozzi at the University of Georgia performed lead isotope analysis by TIMS on eleven curse tablets stratigraphically excavated directly from the Circus in Carthage during the 1982–1990 seasons (*7*). She concluded that the sources for curse tablet lead ranged from Greece to Portugal (*7*). Her most interesting results suggested the source of lead for one tablet came from Ekaterinburg, Urals, and lead from two tablets originated in Pribram, Czech Republic (*7*). Additionally, Pintozzi collected samples of 94 curse tablets stored at the National Museum of Carthage in Carthage, Tunisia (the AUD and P tablets). These additional samples were not used, and have been stored at the University of Georgia ever since.

In the 15 years since Pintozzi's study, great improvements have been made in lead isotope studies. In particular, the instruments and techniques have greatly improved. Additionally, the foundation of a standardized archaeometry database (*8*) has greatly increased the required instrument precision required for acceptable sourcing of artifacts using the ore source data. As interesting as Pintozzi's results are, her method was not as precise as current TIMS analysis.

Research Design

The overall goal of this research is to understand Roman lead use and trade in Roman Carthage. Ideally, every available tablet would be analyzed by TIMS. Because of limited funding, a method of classification was needed to decide which tablets to analyze. Comparing the element maps of the different tablets suggested that EMPA might be a useful classification tool. The EMPA sends a beam of electrons onto the sample surface. The electrons from the beam excite the electrons of the sample into higher orbitals. When these excited electrons fall back to a lower orbital, they emit X-rays of a certain energy and wavelength characteristic of a particular element. Rastering the electron beam across the sample generates a map of the location and approximate concentrations for elements present within the sample.

A subsample of 22 tablets was selected for EMPA analyses to reflect a cross section of the University of Georgia, Audollent, and other curse tablets available. Given the rarity of these artifacts and the preliminary nature of this study, no tablet was rejected based on a lack of provenance.

The aim of this preliminary study was to repeat one of Pintozzi's analyses, along with five new samples using updated technology and a more comprehensive lead isotope database. Samples were chosen based on the observed patterns of inclusions from the EMPA analysis. Therefore, tablet 90.3–260 and five other tablets from the 22 analyzed by EMPA were then analyzed by TIMS. Beyond instrumentation, the key difference between this study and Pintozzi's is the inclusion of a purification procedure on the lead before TIMS analysis. Without this step, impurities can potentially cause beam instability and, more importantly, variable fractionation, leading to less precise results (*9*). TIMS results were compared to the EMPA classification to test the utility of the EMPA method. If possible, general conclusions about the lead ore sources would be drawn from the TIMS data.

TIMS operates by atomizing and ionizing a sample suspended in silica gel on a filament. The ionized atoms are then accelerated and travel through a magnetic field. Adjustments can be made in the acceleration or magnetic field to separate the ions according to the mass.

These same six samples were also analyszed by inductively coupled plasma–mass spectrometry (ICP-MS) to determine if the same element patterns detected in the EMPA analysis could be observed with a less labor intensive method. ICP-MS works in a similar manner as TIMS, expect the sample is dissolved and injected into a plasma flame in order to atomize and ionize the sample. Isotopes ratios can be determined in this manner, but generally the precision is less than with TIMS.

The heavy corrosion layer on tablet P6 was analyzed using X-ray diffraction (XRD) in order to identify the products that had formed. The corrosion on P6 contained all the different types of corrosion observed on other tablets to date. Corrosion products were considered a possible reason for variation in the ICP-MS results, but not the TIMS results. XRD analysis bombards the sample with X-rays of one wavelength and determines the distance between atoms and atomic units in the sample. Diffraction of the X-rays occurs at different angles and intensities depending on the minerals in the sample. The diffractograms are compared to known standards to determine what minerals are present in the sample.

The ore source field database for Tunisian lead ores is extremely limited. Ron Farquahar and Vanda Vitali have given permission to publish ten ore analyses which have previously only been represented in graphical form (*10*). Two additional Tunisian ore samples are included in the database, one from Sidi Amor and the other from Bou Grine. The mine locations, numbered 1–10, plus Sidi Amor and Bou Grine, are shown in Figure 2. Ideally, multiple samples would be taken from each ore location to ensure the deposit is homogenous. This was not possible due to the limited resources and exploratory nature of this study.

Procedures

EMPA

When possible, two subsamples weighing 20-100 mg from each curse tablet were mounted in a 25 mm epoxide resin disk. The surface was removed until

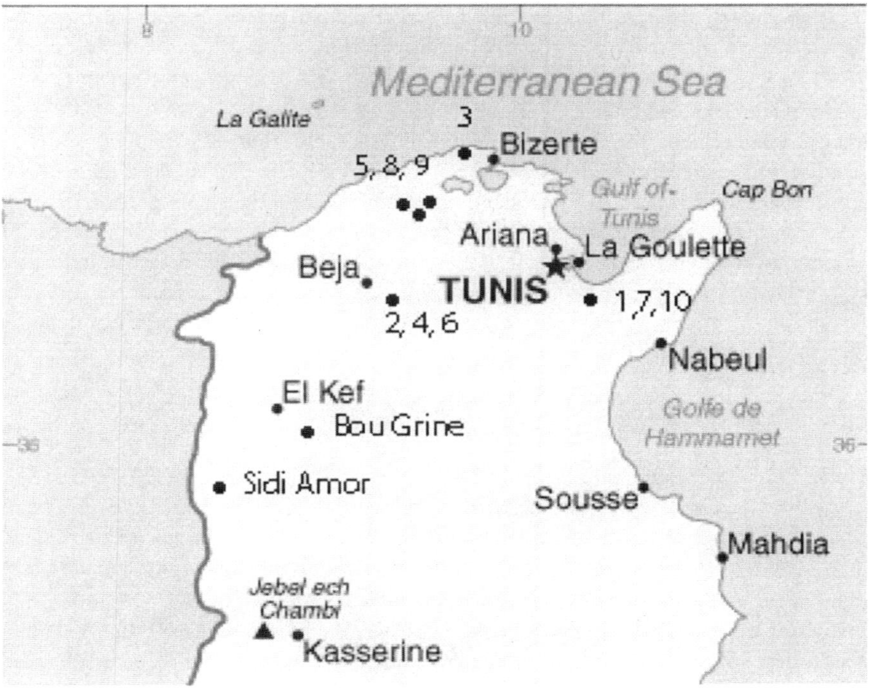

Figure 2. Northern Tunisian ore sampling locations. Map modified from reference 11.

significant interior lead metal was exposed using a diamond polishing wheel. A final polish to 0.25 micron was obtained using an oil-based diamond suspension and a MiniMet 1000 polisher. In addition to the curse tablets, one modern piece of lead (lead fishing weight pounded flat) and two ore samples from Tunisia were also mounted. The samples were coated with a thin layer of carbon to make the surface conductive. Each sample was then examined with the JEOL 8600 Super probe electron microprobe analyzer at the Department of Geology, University of Georgia. Operating conditions were generally 15 kV accelerating voltage, 15 mA beam current and 1 μm spot size.

An X-ray map of the elements As, Sb, Cu, and Ag, was created for each sample at 500X magnification with a wave dispersive spectrometer (WDS) using Geller dPICT 32 software. The X-ray map was created from 24 individual maps combined for a total mosaic area size of 0.5 mm by 1.0 mm. For WDS analysis, an indium arsenic compound was used as an arsenic standard. Otherwise pure element standards were used. A TAP crystal was used for detecting As, and Cu (position 145.465, bias 1666), while a PET crystal was used for Sb and Ag.

It was sometimes desirable to determine if the elements identified in phases distinct from the lead matrix by X-ray mapping were in a pure form or alloyed with undetected elements. This helped to ensure that the elements mapped were in comparable forms. A tablet containing only pure copper inclusions might not be appropriately classified in the same category as one with only copper alloy inclusions. An energy dispersive spectrometer (EDS) was used for this type of qualitative element detection. The EDS used was a PGT Avalon Multichannel system with a beryllium window and SiLi detector using Excalibur software. Generally, up to eight EDS analyses were performed on different regions in an X-ray map to determine if alloys were present.

TIMS

After selecting six samples that covered the range of EMPA results, 1 mg from each sample and the two Tunisian ores were prepared by the whole rock Pb clean up and analysis techniques used at the Geochronology and Isotope Geochemistry laboratory at the University of North Carolina, Chapel Hill. Each sample was dissolved in a mixture of 1 mL of 14 N HNO_3 and 1 mL of 6 N HCl and then heated for 8 hours. The liquid was then heated to dryness, and 10 ml of 6 N HCl added to the dried residue. A 100 µL aliquot was removed and heated to dryness. Next, 550 µL of 1.1 N HBr was added to the dry metal bead. Each solution was run through a column of Eichrom Anion Resin AG1x8 twice using 1.1 N HBr to rinse and 6 N HCl to elute. The elutions collected between runs were heated to dryness and dissolved again in 500 µL of 1.1 N HBr. Each final elution had 20 µL of dilute H_3PO_4 added to it before being heated to dryness. The now purified samples were loaded onto a 99.98% pure Re filament with silica gel and heated slowly to a 2.0 amp voltage. Finally, the filaments were loaded on the instrument.

TIMS analysis was performed on a fully automated VG Sector 54 mass spectrometer with eight adjustable faraday cups and a Daly ion-counting photomultiplier system. Analysis was performed in static mode. Each sample was analyzed 50 times to ensure acceptable precision. The TIMS analysis was standardized by use of the NIST SRM981 common lead standard. Multiple analyses of the SRM981 standard were used to determine a fractionation correction of 0.12% per amu and an overall error 0.06% per amu. Errors between runs of the same sample were below 0.01% per amu. This level of precision is comparable to the archaeometry database for lead isotopes (*8*).

ICP-MS

A 9 mL aliquot from each TIMS sample solution was submitted to the University of Georgia, Laboratory for Environmental Analysis, for inductively coupled plasma-mass spectrometry analysis (ICP-MS). A Perkin-Elmer Elan 6000 ICP-MS with quadrapole chamber mass detector system was used to analyze the solution for Ag, As, Cu, Sb, Sn, Pb, and Zn. Insufficient sample remained for further analysis or replicate samples. However, all appropriate blanks, dilutions, and standards were run.

XRD

A scraping from the surface of one heavily weathered curse tablet was placed on a zero background plate and analyzed using the University of Georgia, Geology Department Scintag XDS 2000 X-Ray Diffractometer. The source was run at 40 kV and 35 mA with a Co target (X-ray wavelength 1.789 Å). Analysis angle was from 10 – 80 degrees with a step size of 0.01 degrees and an eight-second count time. The lead fishing weight mentioned earlier was used as a lead metal standard.

Results

EMPA

The EMPA X-ray maps for As, Sb, Cu, and Ag revealed that the curse tablets differed in the numbers and types of metal inclusions present. Comparing Figures 3 and 4 shows the differences among typical curse tablets. The X-ray map was surveyed for areas of darkness clustered tightly together, which represent a concentration of the detectable element. These areas were then circled. In Figure 3, only the X-ray map for Cu shows any concentration of the element above the background noise level. The amount of Cu was qualitatively determined to be a minor amount. In Figure 4, all the X-ray maps show areas of elemental concentration above the background noise level. These were qualitatively determined to be major amounts of Cu and Sb and minor amounts of As and Ag in the sample. The reproducibility of the EMPA results were tested by mapping more than one location on two samples that were large enough to allow X-ray mapping of non-overlapping areas. In each case the ordinal level values of both areas were approximately the same. Additionally, pieces of one tablet which was originally

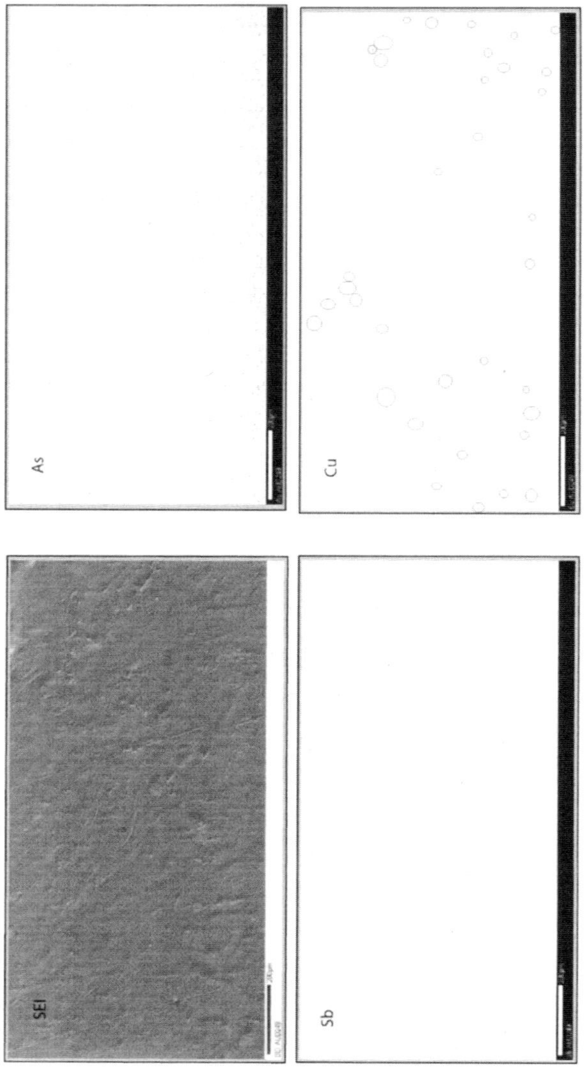
SEI
As
Sb
Cu

Curse Tablet AUD 249
Scale Bar = 200 µm
X-Ray Map color inverted

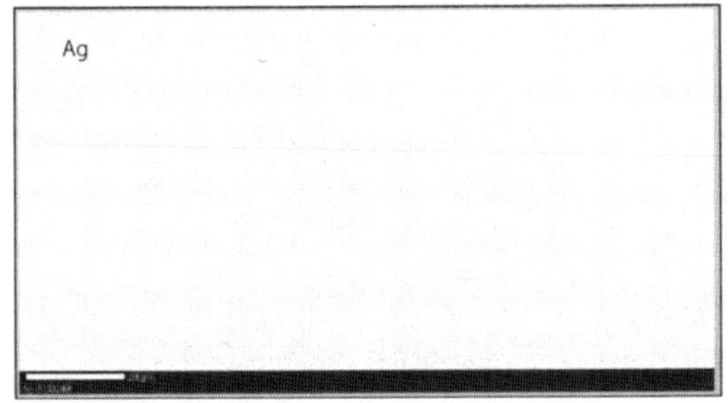

Figure 3. X-ray map for As, Sb, Cu, and Ag on curse tablet AUD-249. Regions with concentrations of elements above background level are circled.

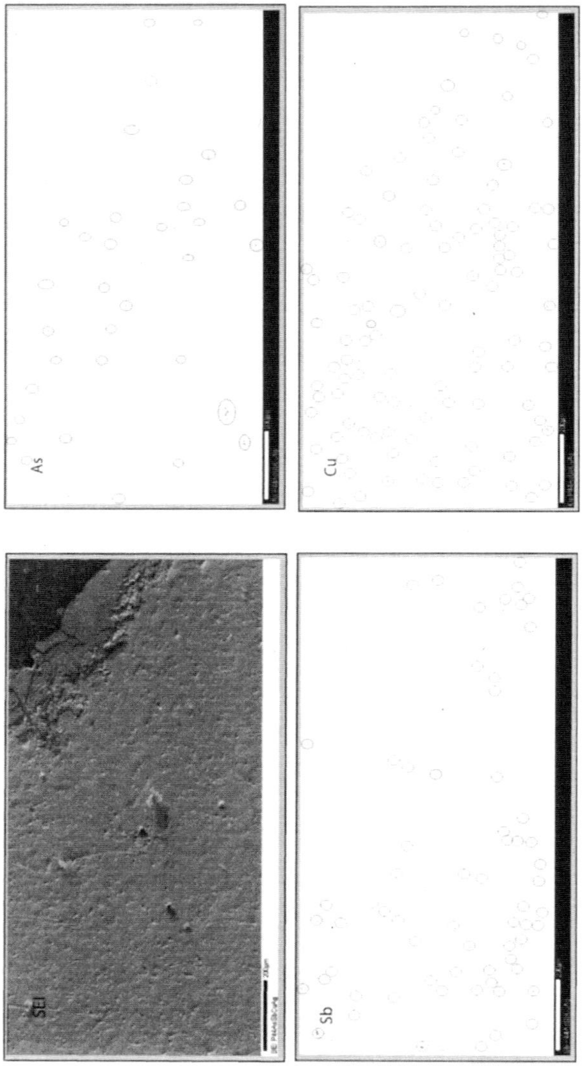
SEI
As
Sb
Cu

Curse Tablet P44
Scale Bar = 200 µm
X-Ray Map color inverted

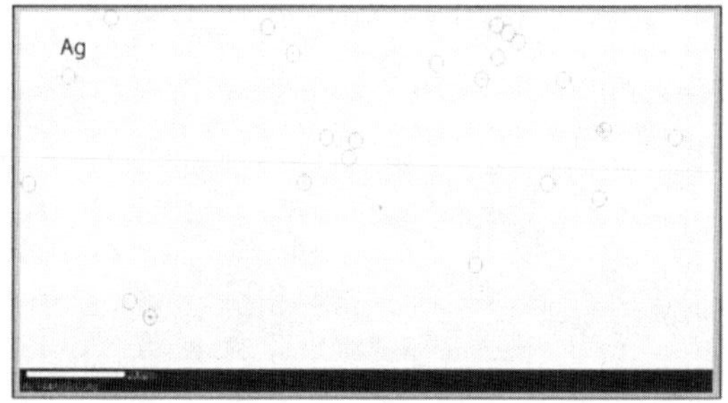

Figure 4. X-ray map for As, Sb, Cu, and Ag on curse tablet P44. Regions with concentrations of elements above background level are circled.

sampled twice (P15A and P15B) were X-ray mapped as two different samples. The ordinal level data for the two pieces were also approximately the same. The data for all tablets tested to date are compiled in Table I. From this ordinal data, six tablets representing the range of inclusion contents were chosen for TIMS analysis. Tablets 90.3–260 and AUD-259 were selected to represent samples with the most inclusions. Because tablet 90.3–260 was also analyzed by Pintozzi, it serves as a test of the utility of her results. Tablets P4 and P54 were selected to represent samples with a moderate number of inclusions. Finally, tablets AUD-249 and P6 were selected to represent samples with the lowest number of inclusions.

EDS results from a number of tablets demonstrated that the Cu inclusions are generally alloys with chemical compositions of bronze and brass. These inclusions also contained variable amounts of lead and arsenic. The inclusions were small enough to make it difficult to determine if the lead was from the lead matrix, or from the inclusion. Tablet AUD-259 was found to have a Sn-Sb alloy. Figure 5 shows the spectra of the largest of these Sn-Sb alloy inclusions. The two elements overlap somewhat on the EDS spectra but a double peak from the strongest lines of each element can be seen. WDS analysis revealed that the alloy is approximately 54% Sn, 43% Sb, 1.5% As, and 1.5% Pb. Silver inclusions are generally pure silver, according to the EDS spectra.

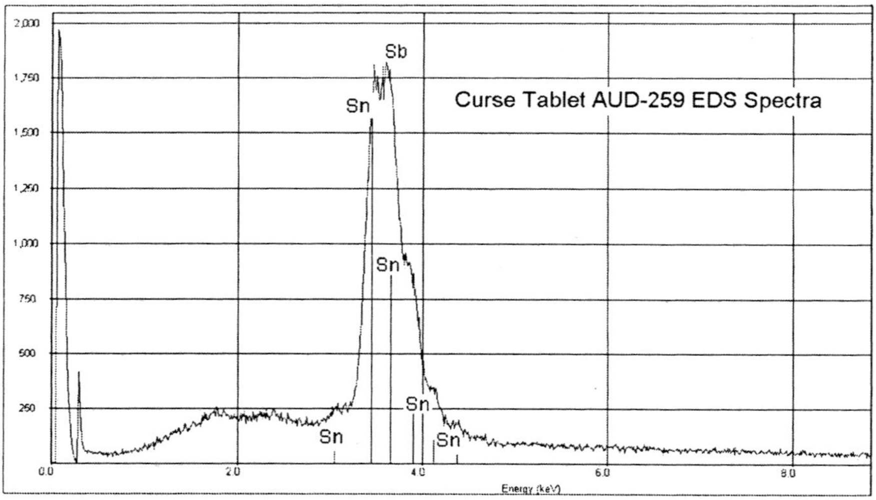

Figure 5. EDS spectra of curse tablet AUD-259 with Sn peaks and largest Sb peak labeled.

Table I. Results of mosaic X-ray mapping of Curse Tablets for As, Sb, Cu, and Ag.

Curse Tablet ID	*As*	*Sb*	*Cu*	*Ag*	*TIMS Sample*
90.3–206	X	Trace	Major	Minor	
90.3–260	Minor	Major	Major	Major	Most
90.3–588	X	Minor	Minor	Minor	
AUD-220	Minor	X	Major	Minor	
AUD-249	X	X	Minor	X	Least
AUD-256 Clear	X	X	Major	X	
AUD-256 Cloudy	X	X	Minor	X	
AUD-257	Minor	Minor	Major	Trace	
AUD-259	Minor	Major	Trace	Minor	Most
P4	Trace	Trace	Minor	Minor	Moderate
P6	X	X	Trace	X	Least
P7	Minor	Trace	Trace	Trace	
P15A	Trace	Minor	Major	Minor	
P15B	Trace	Minor	Minor	Minor	
P17	Minor	Minor	Major	Minor	
P18	Minor	Minor	Major	Trace	
P20	Trace	Trace	Major	Minor	
P28	X	X	Major	Minor	
P30	Trace	X	Major	Trace	
P44	Major	Minor	Major	Minor	
P45	Trace	X	Trace	Trace	
P54	Minor	Minor	Major	Minor	Moderate
Tun (gal/spal)	X	X	X	X	Ore
Sidi (galena)	X	X	X	X	Ore
Modern Lead	X	Trace	X	X	

Scale is from none of the element (X), to trace, minor, and major amounts. TIMS Sample column shows the samples chosen for TIMS analysis and the overall classification from least to most inclusions.

TIMS

In order to build a Tunisian lead ore isotope database, unpublished results from Farquahar and Vitali were used. These results are combined with the Sidi Amor and Bou Grine samples in Table II. There is only one sample from each mineral deposit. Northern Tunisia is broken up into three major geologic regions, namely the Nappa zone, the Dome zone, and the Shelf zone (*12*). Within these zones there are four primary types of Pb-Zn deposits (*12*). In fact, there are more than 45 areas where significant Pb-Zn deposits are known to exist (*12*). Clearly a dozen samples are insufficient to properly characterize such complex geologic formations. Although the lead isotope values appear to form 3–5 distinct groups of ores, this will almost certainly change and expand as more samples are analyzed.

The isotope ratios for the six analyzed curse tablets plus the results from Pintozzi's analysis of curse tablet 90.3–260 are summarized in Table III. These results will supersede those of Pintozzi in all future work due to the imprecision of her $^{206}Pb/^{204}Pb$ ratios. An estimate of her error in this ratio is approximately 0.6% per amu based on her four analysis of the SRM981 standard. This precision is not within the error limits of the current database. This is in no way a criticism of Pintozzi's analytical ability; it is just that the instrument and methods used at that time do not meet the requirements of a modern lead isotope study.

In Figure 6, the curse tablet lead isotope ratios create an interesting pattern, with those tablets identified by EMPA analysis to contain the least number of inclusions plotting between two areas of Tunisian lead ores. The curse tablets identified as containing the greatest number of inclusions are those tablets plotting farthest away from the Tunisian lead ores. Additionally, the tablets appear to define a mixing line of ratios between two or more sources.

While the curse tablets AUD-249 and P6 plot closest to the Tunisian ore sources, they are more isotopically similar to ore from the Rhodope Mountains in Bulgaria (*13*) as shown in Figure 7. Although this does not necessarily indicate that the lead originated in Bulgaria, it is a possibility. This region of lead isotope ratio space also overlaps with some Spanish and English ores. Tablets 90.3–260, AUD-259, P4, and P54 are unlikely to contain lead from the Tunisian ores. Of sources in the western Mediterranean, the tablets are the closest to ores from Sardinia (*13*). Due to the lack of correspondence between tablets and sources, the recycling of the metal in the artifacts is likely. A previous study found evidence of such recycling in the extraction of silver from bronze slags found in Carthage (*14*). Finally, silver inclusions were found in all of the tablets suspected of being recycled. This suggests the lead was used in a liquation or cupellation process.

Table II. Tunisian Lead Ore Isotope Database.

Map Location #	Location	Lab #	$^{208}Pb/^{206}Pb$	$^{207}Pb/^{206}Pb$	$^{206}Pb/^{204}Pb$
1	Ressas	961	2.0495 ± 0.0025	0.8277 ± 0.0004	18.968 ± 0.021
2	Pepperino	962	2.0691 ± 0.0025	0.8331 ± 0.0004	18.820 ± 0.017
3	Bechater	963	2.0685 ± 0.0025	0.8340 ± 0.0005	18.796 ± 0.016
4	Djebel Hallouf	964	2.0680 ± 0.0025	0.8330 ± 0.0004	18.815 ± 0.023
5	Djebel Semene	965	2.0663 ± 0.0029	0.8336 ± 0.0015	18.790 ± 0.034
6	Sidi Bou Aouane	966B	2.0672 ± 0.0025	0.8327 ± 0.0004	18.809 ± 0.019
7	Slata	967	2.0735 ± 0.0025	0.8371 ± 0.0004	18.724 ± 0.016
8	Safsaf	968	2.0664 ± 0.0025	0.8329 ± 0.0004	18.825 ± 0.019
9	Bazina	969	2.0690 ± 0.0025	0.8343 ± 0.0004	18.801 ± 0.018
10	Zriba	970	2.0540 ± 0.0026	0.8305 ± 0.0004	18.876 ± 0.019
Bou Grine	Bou Grine	Tun	2.0724 ± 0.0025	0.8364 ± 0.0005	18.731 ± 0.022
Sidi Amor	Sidi Amor	Sidi	2.0740 ± 0.0025	0.8369 ± 0.0005	18.737 ± 0.022

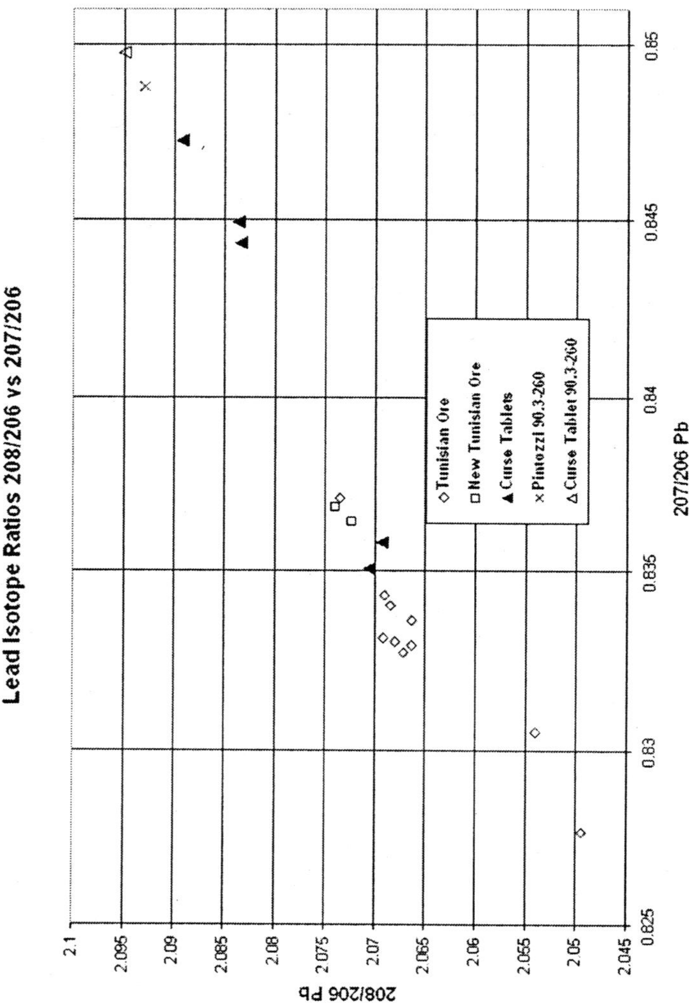
Lead Isotope Ratios 208/206 vs 207/206
208/206 Pb
207/206 Pb
Tunisian Ore
New Tunisian Ore
Curse Tablets
Pintozzi 90.3-260
Curse Tablet 90.3-260
2.1
2.095
2.09
2.085
2.08
2.075
2.07
2.065
2.06
2.055
2.05
2.045
0.825
0.83
0.835
0.84
0.845
0.85

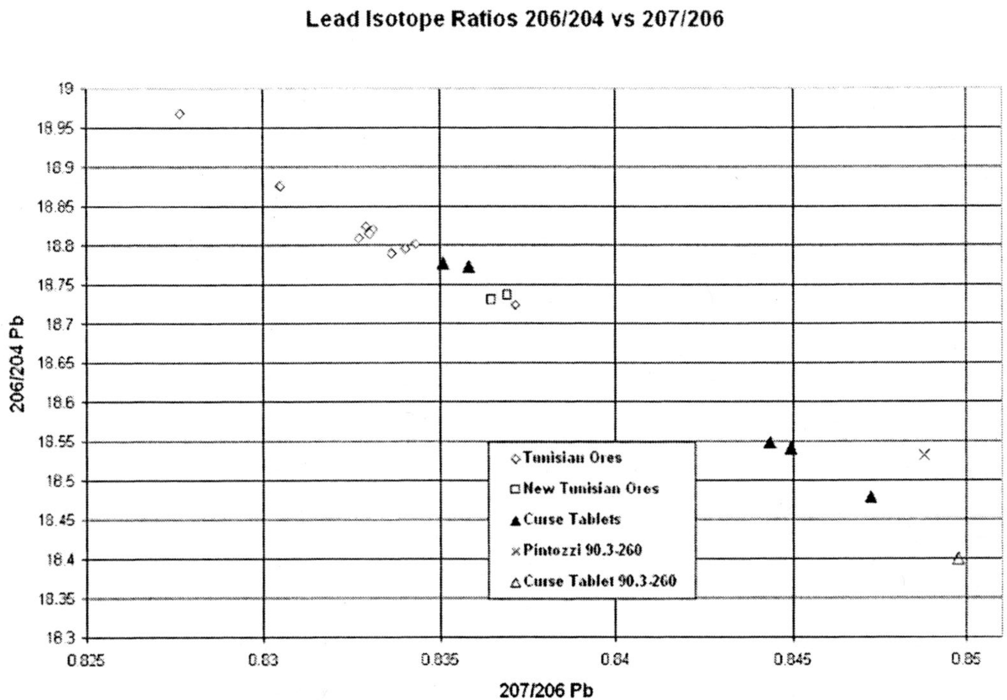

Figure 6. Lead Isotope ratios for curse tablets and Tunisian ores.

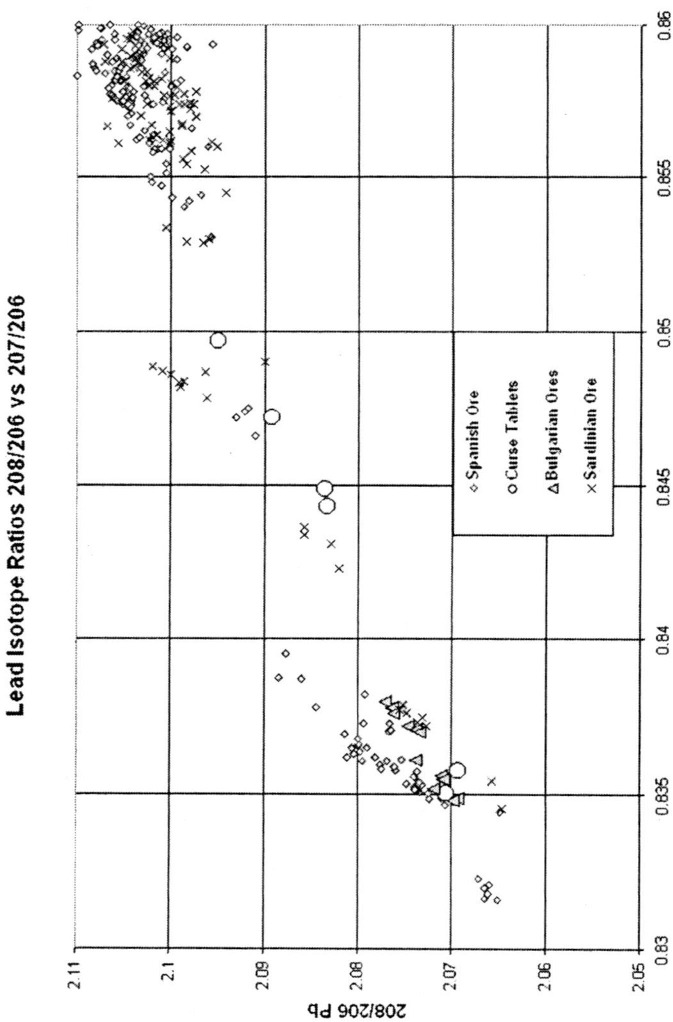
Lead Isotope Ratios 208/206 vs 207/206
208/206 Pb
207/206 Pb
Spanish Ore
Curse Tablets
Bulgarian Ores
Sardinian Ore
2.11
2.1
2.09
2.08
2.07
2.06
2.05
0.83
0.835
0.84
0.845
0.85
0.855
0.86

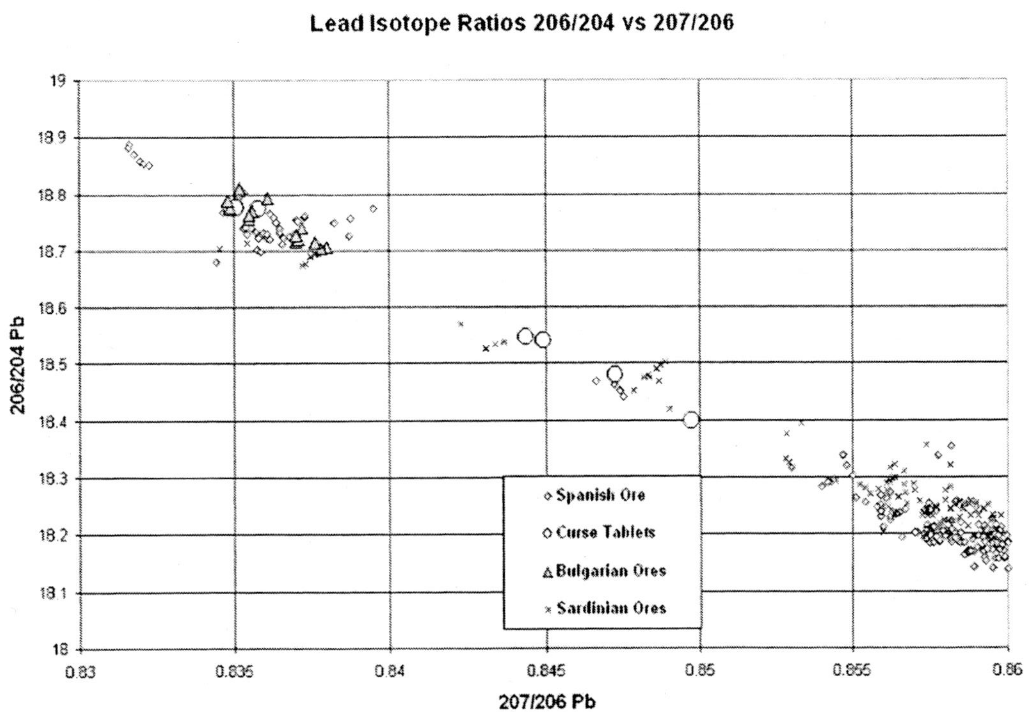

Figure 7. Lead isotope ratios for curse tablets and other ore sources.

Table III. Curse Tablet Lead Isotope Ratios.

Name	Material	$^{208}Pb/^{206}Pb$	$^{207}Pb/^{206}Pb$	$^{206}Pb/^{204}Pb$
P6	Tablet	2.0692	0.8358	18.773
AUD 249	Tablet	2.0705	0.8351	18.777
P4	Tablet	2.0833	0.8444	18.548
P54	Tablet	2.0836	0.8449	18.541
AUD 259	Tablet	2.0892	0.8472	18.479
90.3–260	Tablet	2.0949	0.8497	18.400
90.3–260	Pintozzi Tablet	2.0930	0.8488	18.532

Note: Error limits are the same as those reported for ores in Table II.

XRD and ICP-MS

Trace element abundances in the curse tablets were also investigated to see if there was a correlation between these abundances and the EMPA inclusion data. It is certainly possible that trace amounts of Cu, Sn, As, Ag, and Sb were alloyed during the smelting process of minerals such as galena, jarosite, or tetrahedrite. The last two minerals are particularly well known in Spanish and English lead deposits (*15*). This was an attempt to characterize the samples rapidly and more easily using ICP-MS. As shown in Table IV, the results were inconclusive. This is possibly due to the highly corroded nature of the samples hindering accurate analysis. Another problem results from partitioning of the trace elements between metal and slag during the smelting process.

XRD results of the surface scrapings positively identified both litharge (PbO) and cerussite ($PbCO_3$). Differential weathering of the trace elements into similar oxides and carbonates would certainly change the apparent concentrations. Considering the small amount of lead available for analysis, the relatively non-destructive EMPA is likely the better technique.

Conclusions

Preliminary results based on these six curse tablets appear to validate the use of EMPA analysis to identify and classify curse tablets for further TIMS

Table IV. Tunisian Ore and Curse Tablet ICP-MS Results.

Sample Name		Cu (ppb)	Zn (ppb)	As (ppb)	Ag (ppb)	Sn (ppb)	Sb (ppb)	Pb (ppb)
sample - H20	1	0.25	10.50	0.20	-0.05	0.00	0.02	0.08
sample AcidBlank	3	4.11	4.64	212.75	0.62	1.44	0.43	6.38
sample - sidi	2	1.11	2.26	510.75	13.20	0.06	16.80	46600.00
sample - tun	4	1.09	658.30	526.75	0.53	0.10	0.45	4053.33
sample P4	5	14.38	2.50	474.75	8.67	17.20	5.74	23133.33
sample P54	6	60.98	4.41	443.75	14.85	60.99	8.76	82783.63
sample P6	7	20.68	9.58	588.75	1.96	10.17	1.47	41232.00
sample AUD-249	8	46.38	0.97	452.75	17.16	2.85	8.76	108666.67
sample AUD-259	9	13.88	2.28	489.75	4.75	25.75	6.45	31633.30
sample 90.3-260	10	44.58	2.73	461.75	13.53	6.84	40.80	71110.30

analysis. It would be beneficial to design a more objective way to determine the number of inclusions in each tablet. This may not be possible, however, due to the way the software records the X-ray map data. These preliminary results cannot be used to determine if Tunisian or Bulgarian lead ore was used in Carthage during Roman times, although the results suggest it is possible. Clearly four of the six tablets were manufactured with lead from sources outside of Tunisia. Differences noted in the EMPA analysis are attributed to lead recycling during Roman times.

Future work will be necessary to complete EMPA analysis for all of the curse tablets. It may be possible to refine the current classification system of least to most inclusions to include considerations of which elements and/or alloys are present in the tablets. A protocol will then be written and used to choose 24 additional samples for TIMS analysis. The goal is to clarify the sources of lead used for curse tablet production. In conclusion, future work in the expansion of the limited Tunisian lead ore database will focus on resolving the current problem regarding the overlap of important ore sources.

Acknowledgments

Funding for all analyses was provided by the Wheeler-Watts grant from the University of Georgia, Department of Geology. Funding for chemicals, materials, and TIMS analyses was provided by grants from the Southeastern Section of GSA, and Sigma Xi. Chris Fleisher, and Dr. P. Schroeder made possible the analyses by EMPA and XRD at the University of Georgia. Dr. D. Coleman made possible the TIMS analyses at the University of North Carolina, Chapel Hill.

Ore samples were provided by the Smithsonian Institution National Museum of Natural History, Department of Mineral Science; Division of Mineralogy, and Breakwater Resources Limited. Dr. N. Norman, Department of Classics, University of Georgia provided the curse tablet samples.

I would like to thank Dr. E. Garrison and the rest of my committee for their valuable comments and revisions. Additionally I would like to thank Dr. S. Stos for her invaluable help, and Drs. R. Farquhar and V. Vitali for the Tunisian ore data. Finally, I thank the reviewers of my draft paper, and the ACS for helping this manuscript to publication.

References

1. Boulakia, J. D. C. *Am. J. Archaeol.* **1972**, *7*, 139–144.
2. Farquhar, R. M.; Vitali, V. *MASCA Research Papers in Science and Archaeology* **1989**, *6*, 39–45.

3. Jordan, D. R. In *The Circus and a Byzantine Cemetery in Carthage*; Humphrey, J. H., Ed.; University of Michigan: Ann Arbor, MI, 1988, pp 117–134.
4. Audollent, A. *Defixionum Tabellae Quofquot Innotuerunt Tam In Graecis Orientis Quam Occidentis Praeter Atticus*; Minerva GmbH: Frankfort/Main, 1967; pp 287–426.
5. Dickin, A. P. *Radiogenic Isotope Geology*; Cambridge University Press: Cambridge, UK, 2002; pp 104–132.
6. Gale, N. H. In *Thera and the Aegean World*; Doumas, C., Ed.; Thera and the Aegean World: London, UK, 1978; pp 529–545.
7. Pintozzi, L. A. M.A. thesis, University of Georgia, Athens, GA, 1991.
8. Stos-Gale, Z.; Gale, N. H.; Houghton, G.; Speakman, R. *Archaeometry* **1995**, 37, 407–415.
9. Horwitz, E. P.; Dietz, M. L.; Rhoads, S.; Felinto, C.; Gale, N. H.; Houghton, J. *Anal. Chim. Acta*. **1994**, *292*, 263–273.
10. Wolf, S.; Stos, S.; Mason, R.; Tite, M. S. *Archaeometry* **2003**, *45*, 405–420.
11. *CIA - The World Fact Book*; URL web address: www.cia.gov/cia/publications/factbook/geos/ts
12. Weisser, D.; Muller, D. *Natural Resources and Development* **1993**, *37*, 82–97.
13. Stos, S. University of Surry, Guildford. *Unpublished.*
14. Lyle, N. M. M.S. thesis, University of Georgia, Athens, GA, 2002.
15. Ixer, R. A.; Pattrick, R. A. D. In *Mining and Metal Production through the Ages*; Craddock, P.; Lang, J., Ed.; The British Museum Press: London, UK, 2003; pp 9–20.

Chapter 18

Laser Ablation–Inductively Coupled Plasma–Mass Spectrometry Analysis of Ancient Copper Alloy Artifacts

Laure Dussubieux

Department of Anthropology, Field Museum of Natural History, Chicago, IL 60605

The elemental composition of copper-based archaeological artifacts and museum objects provides information that can address questions related to manufacturing technologies and metal circulation, and also may be helpful for authenticity verification. As the range of the materials investigated using laser ablation-inductively coupled plasma-mass spectrometry (LA-ICP-MS) is growing wider due to technological improvements, we tested the performances of this analytical technique to determine copper alloy compositions. Standardization, detection limits and reliability of the method will be discussed. The application of our LA-ICP-MS protocol to the study of Matisse bronze sculptures will be presented.

Introduction

Laser ablation-inductively coupled plasma-mass spectrometry (LA-ICP-MS) is still a recent analytical technique that appeared in the middle of the 1980s and has yielded an increasing interest ever since *(1)*. At the beginning of the 1990s, applications to determine the composition of archaeological or ancient artifacts made of inorganic materials were developed (*2*). The wide range of elements detected with LA-ICP-MS, its low detection limits (in the range of the ppm or below) and the minimum damage caused to the artifacts make this technique particularly suitable to undertake provenance studies, to trace trade exchanges or to have a better understanding of ancient technologies. This technique does have some limitations: the material to be analyzed must be homogeneous and matrix-matched standard reference materials (SRMs) must be available. Natural and synthetic glasses were the first widely investigated materials using this technique since they meet the requirement for homogeneity and appropriate SRMs are easily accessible (*3, 4, 5, 6*). Metals are more problematic due to the existence of different phases or precipitates in some alloys. Nevertheless, analytical protocols have been developed and promising results presented for the compositional analysis of ancient gold (*7, 8*), silver (*9*) and iron alloys (*10*).

Until now, little attention has been given to the analysis of ancient copper alloys with LA-ICP-MS. This type of material is usually analyzed with fast or instrumental neutron activation analysis (FNAA or INAA), particle induced X-ray emission (PIXE), X-ray fluorescence (XRF), inductively coupled plasma-atomic emission spectrometry or inductively coupled plasma-atomic absorption spectrometry (ICP-AES or ICP-AAS). Some of these techniques are destructive and involve extensive sample preparation, some measure only surface compositions, and some require access to a cyclotron or a reactor. LA-ICP-MS is not affected by any of these inconveniences. We propose here an analytical protocol for copper alloys using LA-ICP-MS and present its application to the study of Matisse bronze sculptures.

Instrumentation and Analytical Protocol

Instrumentation and Parameters of Analysis

The analyses were carried out at the Field Museum of Natural History in Chicago, IL. The instrumentation is a Varian inductively coupled plasma-mass spectrometer (ICP-MS) equivalent to the actual Varian 810 instrument. A New Wave UP213 laser is connected to the ICP-MS for direct introduction of solid samples.

The Varian ICP-MS is a quadrupole mass spectrometer that takes advantage of a new technology. Instead of traveling linearly through the instrument, the ion beam is bent 90° by a series of mirrors and lenses before entering the quadrupole. This new design increases the sensitivity of the instrument 200 times compared to conventional quadrupole ICP-MS instrument without compromising on the analyte interferences or instrumental background signal (*11*).

The parameters of the ICP-MS are optimized to ensure a stable signal with a maximum intensity over the full range of element masses and to minimize oxides and doubly ionized species formation (XO^+/X^+ and $X^{++}/X^+ < 1$ to 2 %). For that purpose the argon flow rate, the RF power, the torch position, the lenses, the mirrors and the detector voltages are adjusted using an auto-optimization procedure.

Tests on copper SRMs show that a stable signal is obtained when each isotope is measured using the peak jumping mode and one point per peak, with a dwell time of 25,000 µs. The quadrupole mass spectrometer scans three times the mass range per replicate and accumulates 10 replicates for a total acquisition time of about 1 minute. For this application, 22 isotopes were selected (Table I).

Table I. Measured isotopes

^{9}Be	^{53}Cr	^{65}Cu	^{118}Sn
^{24}Mg	^{55}Mn	^{66}Zn	^{121}Sb
^{27}Al	^{57}Fe	^{75}As	^{125}Te
^{29}Si	^{59}Co	^{78}Se	$^{206, 207, 208}Pb$
^{31}P	^{60}Ni	^{107}Ag	^{209}Bi

The New Wave UP213 laser operates at a wavelength of 213 nm. The sample chamber is a cylinder with a diameter of 6 cm and a height of 5 cm. Hence, the sizes of the samples are limited. A CCD camera connected to a computer allows for the visualization of the surface of the sample. Helium is used as a gas carrier at a flow rate of 0.50 l/min. Stability and sensitivity requirements for the signal are met when the laser operates at 70% of its maximum energy (0.2 mJ) and at a pulse frequency of 15 Hz. The single point analysis mode is selected with a laser beam diameter of 55 µm. When a non-corroded surface is ablated, a 20–second pre-ablation time is set to be sure that possible surface contamination does not affect the results of the analysis and to eliminate the transient part of the signal. When the surface of the artifact is obviously corroded, two ablations are performed, at the same location. The laser beam diameter is set to 65 µm for the first ablation and 55 µm for the second one. The laser beam is focused at the bottom of the first crater before starting the second ablation. Only the signal acquired during the second ablation is assumed to be representative of the non-corroded copper alloy. The average of four measurements corrected from the blank is considered for the calculation of the concentrations.

Standardization

To improve reproducibility of measurements, internal standardization is required to correct possible instrument drift or changes in the ablation efficiency. The isotope ^{65}Cu was selected as an internal standard because it is present in every sample at relatively high concentrations allowing for accurate measurement. Figure 1 shows signals after normalization.

Quantitative results are obtained by comparing the signal intensity measured for a given element in a sample to the signal intensity for the same element in a SRM with certified concentrations. To prevent matrix effects, the compositions of SRMs must be as close as possible to those of the samples. We selected seven different SRMs with the largest number of elements and the widest range of concentrations as possible. Two SRMs are manufactured by the Centre de Développement des Industries de Mise en Forme des Matériaux in France (SRMs B10 and B12) and two others by the Bureau of Analysed Samples Ltd in England (SRMs 51.13–4 and 71.32–4). These four standards are bronzes which contain copper, tin, lead, zinc, nickel, phosphorus, iron, silica, manganese, arsenic, antimony, bismuth, aluminum, chromium and silver with concentrations ranging from about 100 ppm to a few percent. More elements were added to the previous list by using three SRMs from the National Institute of Standards and Technology: selenium and tellurium from SRM 500, beryllium from SRM C1123 and cadmium, magnesium and selenium from SRM 1275. The concentrations of the elements present in the samples were calculated assuming that the sum of their concentrations, in weight percent, is 100% (*3*).

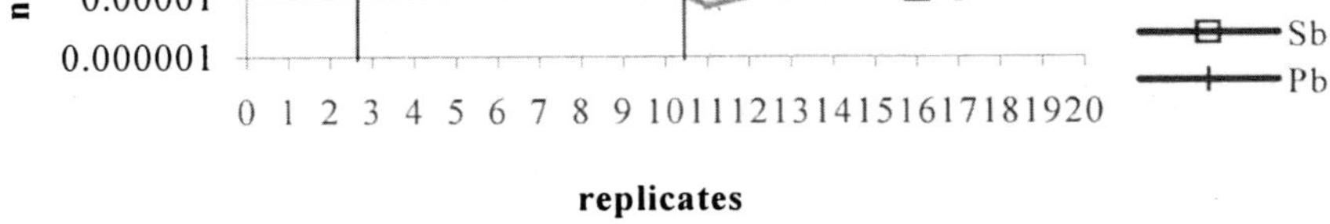

Figure 1. Signals of some elements normalized on the ^{65}Cu signal.

Evaluation of the Reliability of Our Analytical Protocol

To evaluate the performance of our method, the response linearity of the instrument was tested. We also determined the detection limits, reproducibility and accuracy.

Linearity

As the concentrations of some elements in the samples may vary over several orders of magnitude, it is important to verify that the response of the instrument is linear over a large range of concentrations. For this purpose, all SRMs were analyzed and calibration curves were traced with the accumulated data. The response of the instrument is all the more linear as the R-square value or correlation coefficient of the calibration curve for a given element is close to one. For each element tested, the number of standards included in the calibration curve, the range of concentrations, the slope and the correlation coefficient of the calibration curves are reported in Table II. No calibration curve could be generated for elements that are present in only one SRM. The linearity of the instrumental response is satisfied for all elements.

Table II. Characteristics of the calibration curves

Element	*Number of standards*	*Range of concentrations*	*Slope*	*Correlation coefficient*
Al	4	0.02 – 7.30 %	4.0266	1
P	3	0.02 – 0.53 %	0.6378	1
Mn	5	0.05 – 0.90 %	9.133	0.9934
Fe	6	0.04 – 1.81 %	0.2492	0.9988
Ni	6	0.01 – 2.6 %	0.9386	0.9996
Zn	6	0.01 – 6.52 %	1.9395	1
As	4	0.01 – 0.25 %	1.7052	0.9989
Sn	5	0.01 – 9.57 %	3.0571	0.9988
Sb	4	0.01 – 1.14 %	3.4331	0.9962
Pb	4	0.01 – 4.43 %	18.718	1
Co	3	0.0001 – 2.3 %	4.5457	1
Ag	3	0.009 – 0.034 %	5.5728	0.9886
Cr	2	0.001 – 0.05 %	0.12241	0.998

The Detection Limits

The detection limits are calculated as three times the standard deviation obtained from the measurement of ten blanks. Ideally, we should have measured these detection limits from multiple measurements of a pure copper material to take into account the contribution of the copper matrix to the background, but we did not have such a material. This may imply a slight underestimation of the detection limits for some elements. The detection limits range from about 1 ppb for bismuth to 2 ppm for iron (Figure 2).

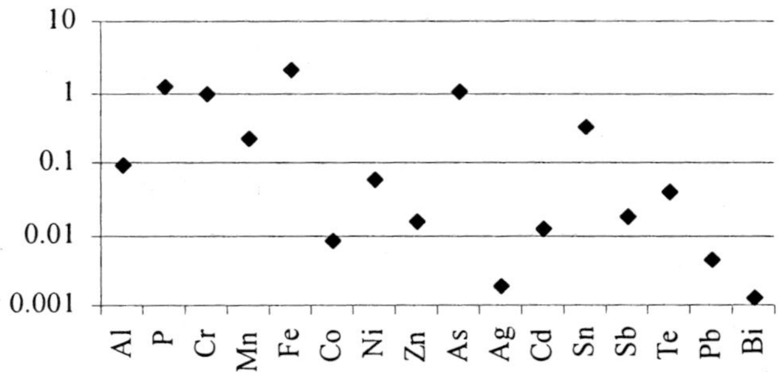

Figure 2. Detection limits in ppm.

Reproducibility

Reproducibility is calculated as the relative standard deviation obtained from 10 measurements on the SRMs B10, B12, 51.13–4 and 71.32–4 over several weeks. Reproducibility for the majority of the elements, in most SRMs, is better than 20 % and usually around 10 % (Figure 3). Phosphorus is a noticeable exception. The reproducibility of the measurements is better for phosphorus since the concentrations are higher. This observation applies to other elements such as arsenic. In SRM B10, arsenic has the lowest concentration and the poorest reproducibility. For some elements at low concentration, the dispersion of the results is more important.

Accuracy

Accuracy is the relative deviation between the certified concentrations and the average concentrations corresponding to 10 measurements of the SRMs B10,

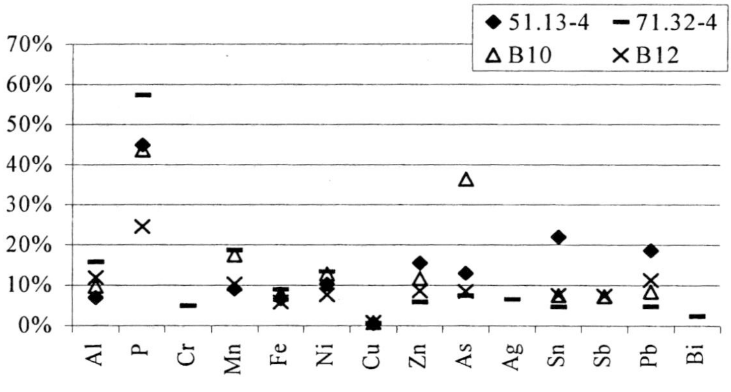

Figure 3. Relative standard deviations calculated from 10 measurements for four SRMs.

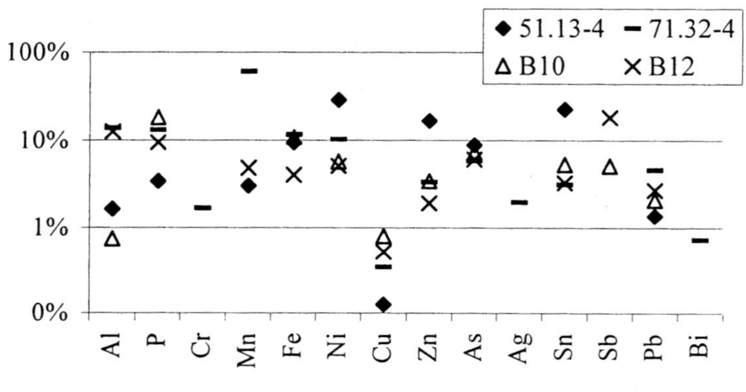

Figure 4. Relative deviation between the certified concentrations and the average concentrations measured by LA-ICP-MS for four SRMs.

B12, 51.13–4 and 71.32–4 by LA-ICP-MS over several weeks. The deviations between certified and LA-ICP-MS concentrations are about 10% or less for the majority of the elements and for most of the SRMs (Figure 4).

Application of LA-ICP-MS to the Study of Matisse Bronze Sculptures

The Baltimore Museum of Art hosts the Cone collection including many works by Matisse. Bronze sculptures by Matisse were cast using different methods (lost wax and sand cast) and in different foundries. Ann Boulton of the Baltimore Museum of Art initiated a project aimed at determining whether or not different compositions of Matisse bronze sculptures could be correlated to different manufacturing techniques or locations. The project started while the author was a post-doctoral fellow at the Smithsonian Center for Materials Research and Education, now Museum Conservation Institute, and involved other analytical techniques (*12*). The results presented here were obtained at the Field Museum of Natural History.

We investigated eight Matisse bronze samples that were either small pieces of metal or shavings saved from the drilling of mounting bolts. All sculptures were made using lost wax casting except for The Serf, which was sand cast (Table III).

Table III. List of the sculptures studied using LA-ICP-MS and ICP-MS

Acquisition number	*Title*	*Cast date*	*Foundry*
1950.439	Venus in a Shell I	1931	Claude Valsuani
1950.437	Reclining Nude III	1929	Claude Valsuani
1950.429	Reclining Nude I (Aurore)	1930	Claude Valsuani
1950.93	The Serpentine	1930	Claude Valsuani
1950.436	Large Seated Nude	1930	Claude Valsuani
1950.435	Crouching Venus	1930	Claude Valsuani
1950.423	Madeleine I	1925	Claude Valsuani
1950.422	The Serf	1908	Bingen-Costenoble

Samples were analyzed using both LA-ICP-MS and ICP-MS of solutions to assess the impact on the results of the sampling by laser ablation. For ICP-MS analysis, less than 1 mg of material was dissolved in double distilled nitric acid. SRMs B10, B12, 51.13–4 and 71.32–4 were prepared the same way, to obtain

Table IV. Characteristics of the calibration curves used for ICP-MS

Element	*Number of standards*	*Range of concentrations*	*Slope*	*Correlation coefficient*
Zn	4	0.34–6.52%	0.5805	0.9987
Sn	4	0.27–9.57%	4.1276	0.9774
Pb	4	0.05–0.90%	16.246	0.9958
As	4	0.01–0.25%	0.7189	0.9923
Ag	1	0.034%	7.5103	N/A
Bi	1	0.05%	15.097	N/A

matrix-matched standard solutions. The isotope ^{65}Cu was used as an internal standard. Table IV provides the characteristics of the calibration curves for zinc, tin, lead, arsenic, silver and bismuth.

Using the same calculation method for both LA-ICP-MS and ICP-MS, the concentrations of the elements in the samples were calculated assuming that their sum, in weight percent, is 100%.

Figure 5 compares the results obtained for the Matisse bronze samples with LA-ICP-MS and ICP-MS for six elements: zinc, tin, lead, arsenic, bismuth and silver. The best correlations between ICP-MS and LA-ICP-MS occur for silver, arsenic and tin. For zinc, lead and bismuth, the results by ICP-MS and LA-ICP-MS concur quite well for some samples. For the other samples, the values are either over or underestimated. Immiscibility of lead in copper, even at low concentrations, could explain the non-agreement for some samples between the ICP-MS and the LA-ICP-MS results (*13*). However, zinc and bismuth are miscible in copper at the concentrations encountered in the Matisse bronze sculptures.

The RSDs for each element of interest, calculated from the normalized signal intensities collected from the four ablations performed on each Matisse bronze sample range, on average, from 7% for zinc to 23% for bismuth. These same RSDs are much lower when considering SRMs: they range from 3% for zinc to 8% for bismuth. The difference in zinc and bismuth concentrations measured with ICP-MS and LA-ICP-MS seem to result from a slight heterogeneous distribution of the elements in the bronze. Under some alloying conditions, dendritic structures can form in metal (*14*). As a discrete sampling technique, laser ablation is sensitive to microstructural variations in the substrate. Performing a metallographic study of the samples may help to determine the source of the variations between ICP-MS and LA-ICP-MS results.

The main goal of this study was to identify different copper alloy compositions and relate them to different foundries or casting techniques. When considering the tin and the zinc contents in the sculptures, three different groups

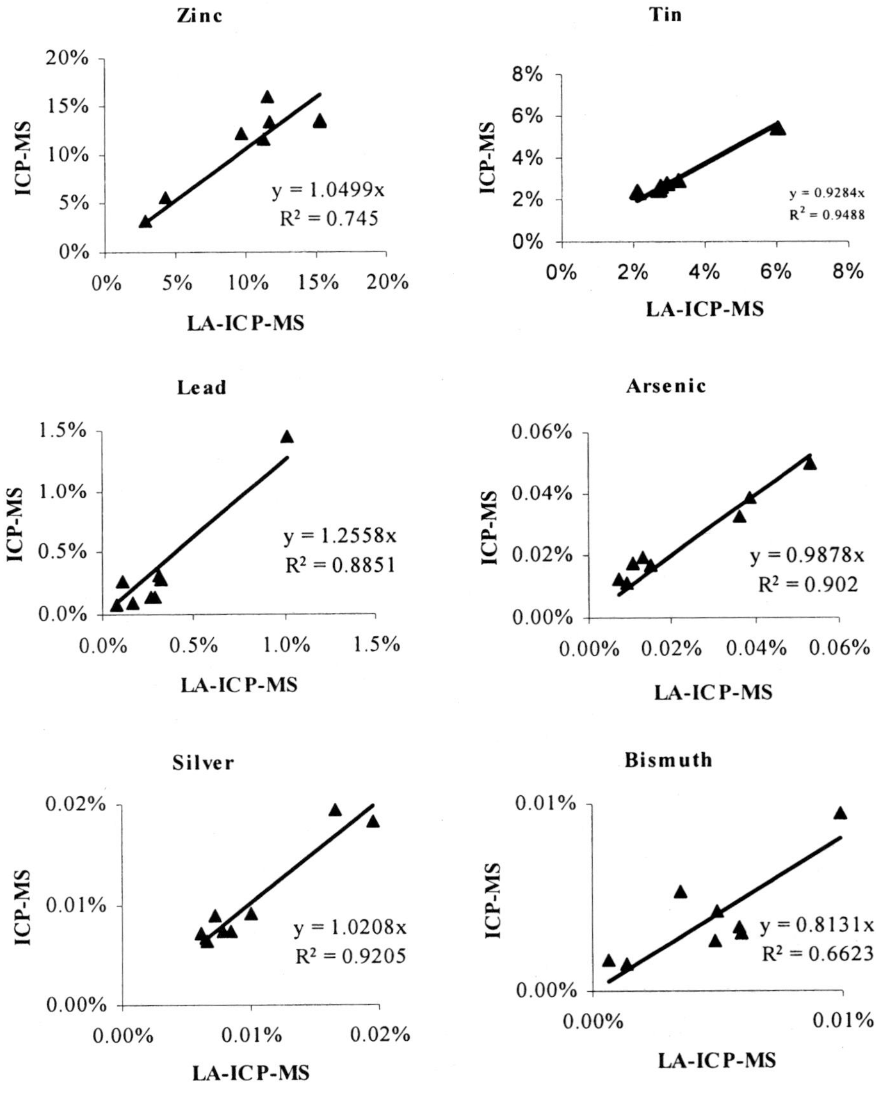

Figure 5. Comparison of concentrations (by weight percent) between LA-ICP-MS and ICP-MS.

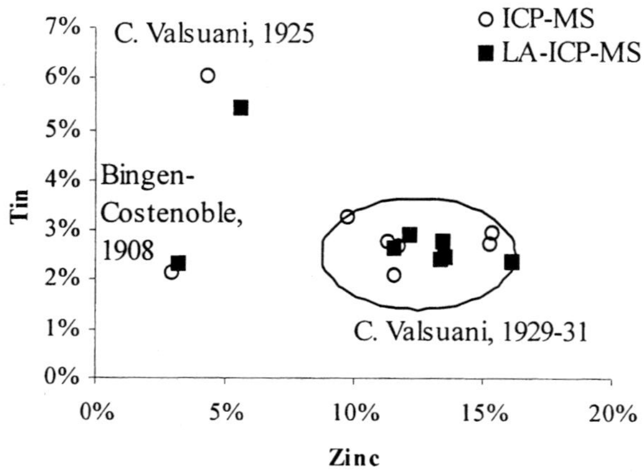

Figure 6. Zinc-tin binary diagram comparing concentrations measured by ICP-MS and LA-ICP-MS.

are evident, regardless of the analytical technique employed: LA-ICP-MS or ICP-MS (Figure 6).

The first group is comprised of 6 samples corresponding to sculptures produced using the lost wax technique from 1929 to 1931 by the C. Valsuani foundry. A second group (one sculpture, The Madeleine I) contains less zinc and more tin. It was cast by the same foundry (C. Valsuani) and with the same technique in 1925. The last group contains the sculpture The Serf. This sculpture has the lowest concentrations of tin and zinc. It was cast by the Bingen-Costenoble foundry, using the sand cast technique.

The conclusions, derived from the results obtained by LA-ICP-MS and ICP-MS are identical. This preliminary study shows that our parameters of analysis and more especially our parameters of ablation for LA-ICP-MS were found suitable to study Matisse bronze sculptures. These results are promising and further study should hopefully demonstrate the usefulness of LA-ICP-MS for the characterization of other types of copper alloys.

Conclusion

The analysis of copper alloys using LA-ICP-MS can be successfully undertaken, but is complicated by the scarcity of appropriate solid standard reference materials containing the proper suite of elements of interest and the heterogeneity of the ancient or archaeological materials investigated. The first problem was overcome by multipying the number of SRMs to increase the range of elements and concentrations. The comparison of ICP-MS and LA-ICP-MS results showed that laser ablation sampling can be sensitive to microstructural variations in the copper alloy but this did not affect the results obtained by LA-ICP-MS in such a way that it leads to erroneous conclusions. If bulk analysis is the most appropriate elemental analytical approach for ancient copper alloys, we believe we have demonstrated, at least for the case of the Matisse bronze sculptures, that LA-ICP-MS analysis is a suitable technique to address questions related to bronze sculpture production.

Acknowledgments

I am grateful to Ann Boulton from the Baltimore Museum of Art for the samples from the Matisse bronze sculptures and to Jia-Sun Tsang from the Museum Conservation Institute for having involved me in the Matisse bronze sculpture project.

The LA-ICP-MS laboratory at the Field Museum of Natural History was built with funds from the National Science Foundation (grant No. 0320903), an anonymous donation, and the Anthropology Alliance.

References

1. Günther, D.; Horn, I.; Hattendorf, B. *Fresenius J. Anal. Chem.* **2000**, *368*, 4–14.
2. Gratuze, B.; Giovagnoli, A.; Barrandon, J. N.; Telouk, P.; Imbert, J. L. *Rev. Archéom.* **1993**, *17*, 89–104.
3. Gratuze, B. *J. Archaeol. Sci.* **1999**, *26*, 869–881.
4. Gratuze, B.; Blet-Lemarquand, M.; Barrandon, J.-N. *J. Radioanal. Nucl. Chem.* **2001**, *247*, 645–656.
5. Robertshaw, P.; Glascock, M. D.; Wood, M.; Popelka, R. S. *J. African Archaeology* **2003**, *1*, 139–146
6. Neff, H. *J. Archaeol. Sci.* **2003**, *20*, 21–35.
7. Gondonneau, A.; Guerra, M. F.; Cowell, M.R. In *32nd Symposium of Archaeometry 2000, México City,* 2001, CD-ROM.
8. Dussubieux, L.; Van Zelst, L. *Appl. Phys. A, Mater. Sci. Process.* **2004**, *79*, 353–356.
9. Devos, W.; Moor, C. *J. Anal. At. Spectrom.* **1999**, *14*, 621–626.
10. Devos, W.; Senn-Luder, M.; Moor, C.; Salter, C. *Fresenius J. Anal. Chem.* **2000**, *366*, 873–880.
11. Elliott, S.; Knowles, M.; Kalinitchenko, I. *Spectroscopy* **2004**, *19*, 1.
12. Dussubieux, L.; Pinchin, S. E.; Tsang, J.-S.; Tumosa, C. S. In *14th Triennial*

Meeting The Hague, ICOM Committee for Conservation, September 2005, Preprints Volume II, 2005; pp 766–773.

13. Turhan, H.; Aksoy, M.; Kuzucu, V.; Yildirim, M. M. *J. Mater. Process. Technol.* **2001**, *114*, 207–211.
14. Henderson, J. *The Science and Archaeology of Materials*. London: Routledge, 2000; p 219.

Chapter 19

Laser Ablation–Inductively Coupled Plasma–Mass Spectrometry Analysis Applied to the Characterization of Peruvian Wari Ceramics

Laure Dussubieux[1], Mark Golitko[1,2], Patrick Ryan Williams[1], and Robert J. Speakman[3]

[1]**Department of Anthropology, Field Museum of Natural History, Chicago, IL 60605**
[2]**Department of Anthropology, University of Illinois at Chicago, Chicago, IL 60607**
[3]**Research Reactor Center, University of Missouri, Columbia, MO 65211**

Trace elemental analysis of ancient ceramics has been proven a very useful tool for tracing the circulation of this material. Instrumental neutron activation analysis (INAA) was for years the analytical technique of choice to measure the composition of ceramics because of the large number of elements it could determine and its good sensitivity. Lately, a few publications have shown that laser ablation-inductively coupled plasma-mass spectrometry (LA-ICP-MS) could provide similar results as INAA more quickly and at lower cost. A protocol has been developed to determine 51 elements using LA-ICP-MS and tested it on Wari period ceramics previously analyzed using INAA. We show how INAA and LA-ICP-MS analysis lead to the same conclusion in terms of sample groupings.

Trace element studies of ceramics have been undertaken for the purpose of locating source regions for archaeological materials since the 1960s. While a number of techniques have been used for this purpose, by far the most common and most effective has been instrumental neutron activation analysis (INAA), largely due to its excellent sensitivity, precision, accuracy and the large number of elements it can measure simultaneously.

Inductively coupled plasma-mass spectrometry (ICP-MS) has seen increasing use in trace element studies of ceramics in the last two decades. While early work using weak acid extraction proved problematic (*1*), recent work by Kennet *et al.* using microwave digestion (*2*) and by Larson *et al.* using laser ablation (*3*) have yielded considerable advances in the application of this technique.

The comparison by Larson *et al.* of results for the same dataset using both forms of ICP-MS and INAA in particular showed excellent correspondence between chemical patterning established by INAA and LA-ICP-MS, despite the fact that one technique measures bulk composition, and the other is a point sampling technique that typically measures only clay matrix in a ceramic sample.

We report here on results obtained using LA-ICP-MS on a set of Wari period ceramics, from sites in the Wari heartland previously analyzed by INAA at University of Missouri Research Reactor (MURR). These ceramic samples are good candidates for a study comparing INAA and LA-ICP-MS analysis because of their very fine paste and very little visible temper. We assumed that the contribution of the temper to the bulk composition obtained with INAA would therefore be negligible. Our results indicate a good match between patterning in the data set as measured by INAA, and as measured by LA-ICP-MS, further demonstrating the utility in some cases of this technique for analysis of archaeological ceramics.

The Wari Ceramics

Between 500 and 600 AD, the first expansive state of the central highlands of Peru emerged in the Ayacucho Basin. This state, known as Wari after its capital city located in the same region, quickly established far flung colonies covering much of the mountainous region of modern day Peru. Ongoing research in the heartland cities of Wari and Conchopata by Isbell, Cook, Ochatoma, Cabrera and others has yielded new insights into economic production of the early imperial state (*4*). Meanwhile, research in other parts of the Wari realm, including the southern frontier (*5, 6*) has focused on the integration of economic production in the far flung provinces. This research is part of a much larger study on Wari ceramic production and exchange in the heartland and in the provinces of the earliest expansive Andean state.

The initial sample consisted of 29 specimens excavated and selected by William Isbell and Anita Cook from investigations at the site of Conchopata from 1998–2002 and from the site of Wari in 1977. Samples were analysed via INAA at the University of Missouri Research Reactor (MURR) as part of a study on Wari ceramics by P. Williams in collaboration with the MURR team. The data reflect several distinct production locales, even within the local heartland region of the Wari state. These characterizations of Wari ceramic production are the subject of several publications in preparation by the research collaborators.

Instrumentation and Analytical Protocol

Instrumentation and Parameters of Analysis

The ICP-MS analyses were carried out at the Field Museum of Natural History in Chicago, IL. The instrumentation utilized was a Varian inductively coupled plasma-mass spectrometer (ICP-MS) equivalent to the actual Varian 810 instrument. A New Wave UP213 laser was used for direct introduction of solid samples.

The Varian ICP-MS is a quadrupole mass spectrometer that takes advantage of a new technology. Instead of traveling linearly through the instrument, the ion beam is bent 90° by a series of mirrors and lenses before entering the quadrupole. This new design increases the sensitivity of the instrument 200 times compared to conventional quadrupole ICP-MS without compromising on the analyte interferences or instrumental background signal (*7*).

The parameters of the ICP-MS were optimized to ensure a stable signal with a maximum intensity over the full range of masses of the elements and to minimize oxides and double ionized species formation (XO^+/X^+ and $X^{++}/X^+ < 1$ to 2 %). For that purpose the argon flows, the RF power, the torch position, the lenses, the mirrors and the detector voltages were adjusted using an auto-optimization procedure.

Tests on glass standard reference materials (SRM) from the National Institute of Standards and Technology (NIST) show that a stable signal is obtained when each isotope is measured using the peak jumping mode and one point per peak with a dwell time of 18,000 μs. The quadrupole mass spectrometer scans the entire mass range three times per replicate and accumulates 9 replicates for a total acquisition time of about 1 minute. For this application, fifty-three isotopes were selected (Table I).

The New Wave UP213 laser operates at a wavelength of 213 nm. The sample chamber is a cylinder with a diameter of 6 cm and a height of 5 cm,

Table I. Selected isotopes

^{9}Be	^{53}Cr	^{107}Ag	^{159}Tb
^{11}B	^{55}Mn	^{115}In	^{163}Dy
^{23}Na	^{57}Fe	^{118}Sn	^{165}Ho
^{24}Mg	^{59}Co	^{121}Sb	^{166}Er
^{27}Al	^{60}Ni	^{133}Cs	^{169}Tm
^{29}Si	^{65}Cu	^{137}Ba	^{172}Yb
^{31}P	^{66}Zn	^{139}La	^{175}Lu
^{35}Cl	^{75}As	^{140}Ce	^{178}Hf
^{39}K	^{85}Rb	^{141}Pr	$^{206, 207, 208}$Pb
^{44}Ca	^{88}Sr	^{146}Nd	^{209}Bi
^{45}Sc	^{89}Y	^{147}Sm	^{232}Th
^{49}Ti	^{90}Zr	^{153}Eu	^{238}U
^{51}V	^{93}Nb	^{157}Gd	

hence the size of introduced samples is limited. A CCD camera connected to a computer allows for the visualization of the surface of the sample. Helium is used as a gas carrier at a flow rate of 0.50 l/min. The signal intensities are stable when the laser operates at 70% of its maximum energy (0.2 mJ) and at a pulse frequency of 15 Hz.

For most of the materials investigated, the parameters of the laser ablation not only are chosen to insure the best sensitivity of the method and the best reproducibility of the measurements, but also to avoid any visible damage on the sample. This last point is not really critical as far as ceramics are concerned since we work on sherds. Ceramic is a particularly heterogeneous material and it is necessary to ablate a volume of material large enough so as to be representative to the whole object. No sample preparation is required aside from some light cleaning. The clay is targeted and temper avoided as much as possible. A pre-ablation time of 20 seconds is used to ensure that potential surface contamination does not affect the results of the analysis and to eliminate the transient part of the signal. Ten single point ablations are performed on each sample with a laser beam diameter of 100 μm. The average of these ten measurements corrected from the blank is used for calculating elemental concentrations.

Standardizations

To be able to produce quantitative data, an internal standard and external standards are required. Internal standardization corrects for possible instrument drift or changes in the efficiency of the ablation and thus improves the reproducibility of measurements. The element selected as an internal standard must be present in every sample at relatively high concentrations allowing for accurate measurement. The isotope ^{29}Si is the most logical choice for silica-based materials, such as ceramics.

Quantitative data are obtained by comparing the signal intensity measured for a given element in a sample to the signal intensity for the same element in a standard reference material (SRM) with certified concentrations. To eliminate matrix effects, the compositions of these SRMs have to be as close as possible to that of the samples. No ceramic standard matching the composition of ancient ceramic is available. We instead use NIST glass SRM 610 and 612 and NIST clay SRM 679 to calculate major and minor element concentrations in the samples assuming that the sum of these concentrations, in weight percent of oxide, is 100% according to the method proposed by Gratuze (*8*). SRM 679 is a clay powder that was shaped into a small pellet and fired at 900°C for 1 hour. Only SRM 610, using concentrations from Pearce *et al.* (*9*), was used for the calculation of trace element concentrations.

Evaluation of the Reliability of our Analytical Protocol

To evaluate the performance of our method, we have defined its detection limits, repeatability and accuracy.

The Detection Limits

The detection limits are calculated as three times the standard deviation obtained from the measurement of ten blanks. Ideally, we should have measured these detection limits from multiple measurements of a pure ceramic material to take into account the contribution of the ceramic matrix to the background but such a material does not exist. The detection limits range from less than 1 ppb to 2 ppm for copper (Figure 1). As the instrument is also used for copper analysis, it is likely that the high detection limit for this element is due to some contamination of the instrument.

Repeatability

Repeatability was calculated as the relative standard deviation (RSD) obtained from 10 consecutive analyses of the same material (Figure 2). The repeatability depends on the instrument and on the analyzed material. For SRM 612, considered as a homogeneous material, the RSDs range from 1 to 3% for

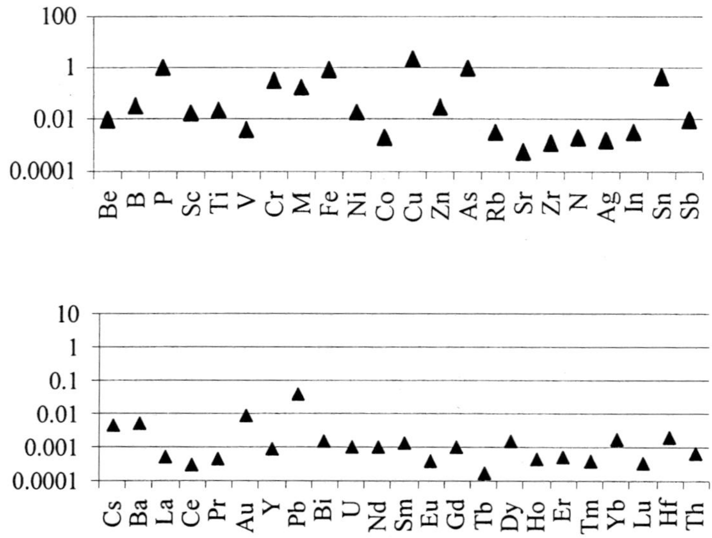

Figure 1. Detection limits in ppm.

most of the elements. For SRM 679 and Ohio Red clay, a very homogeneous clay analyzed multiple times by INAA, the RSDs on ten successive measurements are generally about 10 to 20%. As both materials, glass and clay, were analyzed by the same instrument and with the same operating conditions, this more important dispersion of the results for Ohio red clay reflects the more heterogeneous nature of the clay material compared to the glass. This dispersion of the results is slightly greater for archaeological samples (PRW219, 220, 221): the RSDs on ten measurements range from 10 to 30% depending on the sample and on the element. This result is not surprising. Even if we tried to avoid sampling temper, its presence in the form of very small particles may affect the results. Nevertheless, the figures obtained for the repeatability of Ohio Red clay and of the archaeological samples are quite close meaning that if the temper affects the results its influence is not very important in this case.

Reproducibility

Reproducibility is calculated as the RSD obtained from 37 measurements on Ohio Red clay, over a five-month period. These RSDs are less than 24% for all elements but chlorine. Chlorine has relatively high ionization energy and its measurement is not very reliable when present at low concentration. Only ten out of fifty-one elements have RSDs higher than 20% (Figure 3). Reproducibility

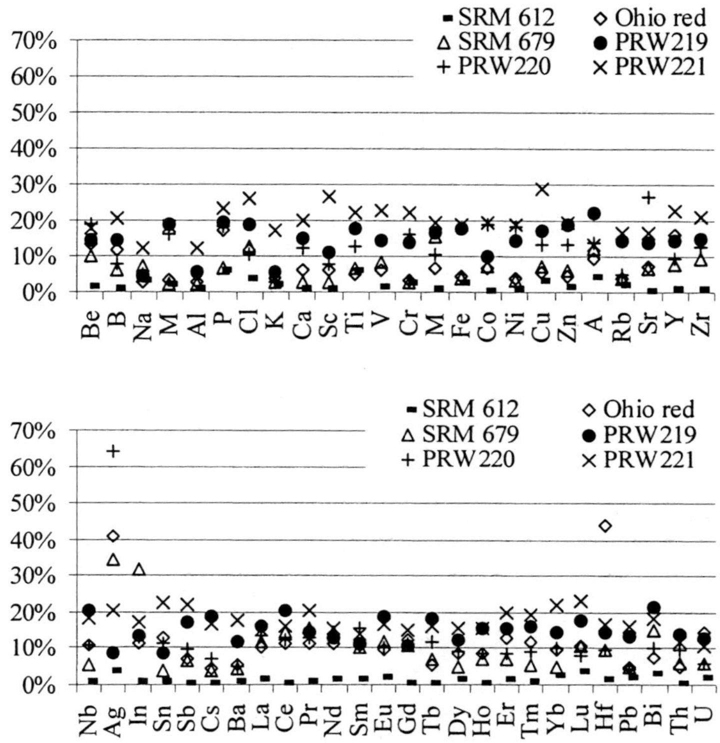

Figure 2. Repeatability for SRM612, 679, Ohio Red clay and samples PRW219, PRW220 and PRW221.

depends on the repeatability but also on errors introduced by changes in the instrument response over time. Another factor that could have affected the reproducibility in this case would have been the non-representativeness of our sampling technique when applied to heterogeneous materials. Repeatability and reproducibility are close. This shows that the instrument produces a stable response over time and that the average compositions calculated from 10 ablations does not vary from one day to another. These results validate the choice of our parameters of ablation.

Accuracy

Accuracy is the relative deviation between the concentrations obtained by INAA on Ohio Red clay and the concentrations measured with LA-ICP-MS.

Twenty-two elements are certified by INAA measurement. As measured by LA-ICP-MS, six elements exhibit accuracy better than 10% and seven better than 20%. Uranium and cesium exhibit more important differences between INAA and LA-ICP-MS results (Figure 3). Using a glass SRM instead of a better matched matrix standard to determine trace element concentrations may have affected our determination of the concentration of certain elements.

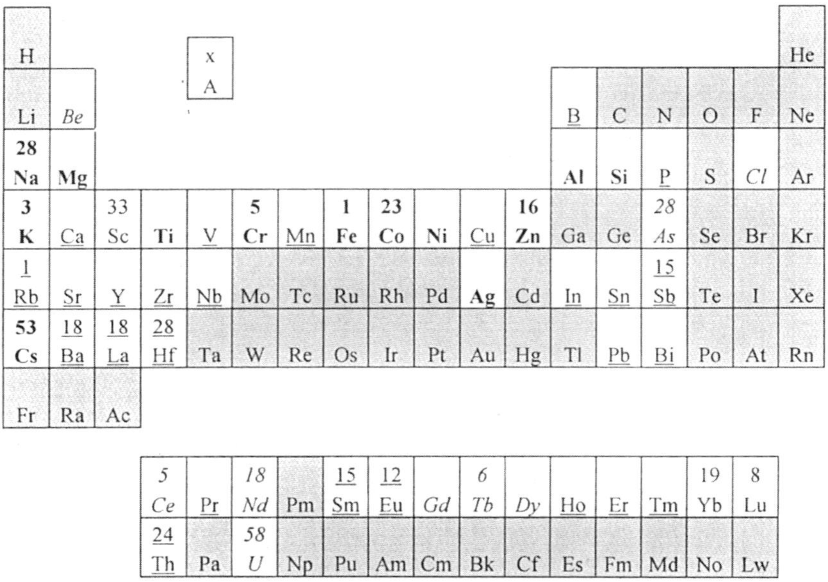

Figure 3. Reproducibility and accuracy. x is the accuracy as the relative deviation between the INAA and the LA-ICP-MS results, in %. A is the reproducibility. When A is bold, reproducibility is better than 10%, when A is underlined, reproducibility ranges from 10 to 20% and when A is in italic, the reproducibility is more than 20%. Elements in grey were not measured.

Results

Analysis of the original 29 Wari samples revealed three easily distinguishable chemical groupings, labeled Wari 1, 2, and 3, with two samples identified as outliers (labeled Mejia A and Wari-Unas). What remained of this material (20 samples) was subsequently returned to Williams at the Field Museum. We show here a principal components biplot of the INAA data with only 18 samples plotted (Mejia A and Wari-Unas have been excluded for purposes of clarity), showing the three main analytical groupings (Figure 4).

These were calculated using 16 elements out of the 32 measured that best differentiate between the three main chemical groupings: La, Sm, Yb, Ce, Eu, Hf, Rb, Sc, Tb, Th, Zn, Zr, Al, Dy, Mn, and Nd.

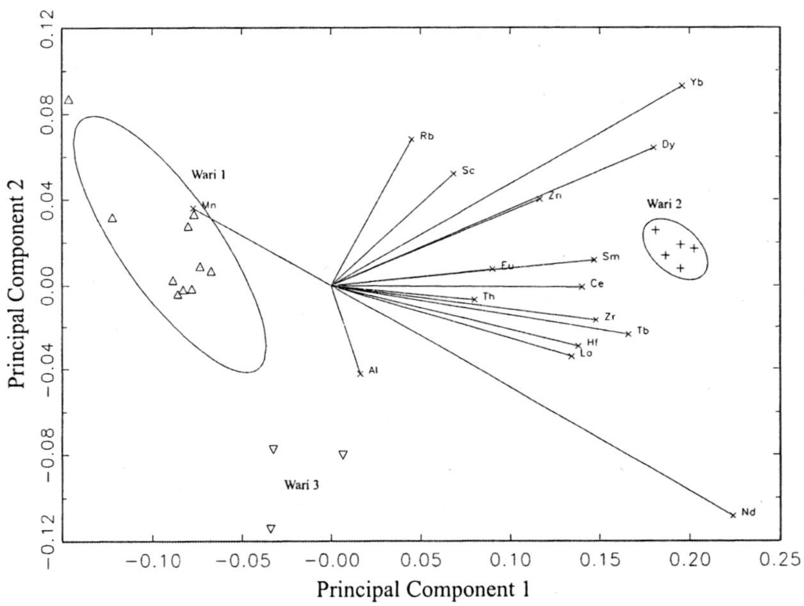

Figure 4. Principal Components 2 vs. 1 (INAA) with component loadings by element. Ellipses represent 90% confidence limits for group membership. PC1 summarizes 79.7% of the total variability in the utilized data, and PC2 10.7%.

These twenty samples were re-analyzed using LA-ICP-MS as described above. Statistical analysis was performed using SPSS v. 12.0 and Gauss Runtime statistical routines developed by MURR. Visual inspection of bivariate plots demonstrated that the same 16 elements produced best group separation in the LA-ICP-MS data as did in the INAA data. A principal component analysis was consequently carried out using these elements (Figure 5). The three major groupings are readily evident in the LA-ICP-MS data set as well (Mejia A and Wari-Unas samples are clear outliers, as in the INAA data, though not shown here), particularly for the first principal component analysis. Component loadings for each element show a broadly similar pattern across the two datasets, Al and Rb, for instance, provide the best separation of Wari 3 from the other two groupings, while Wari 2 has higher concentrations of Yb, Dy, Tb, Zn, Ce, Sm, Th, Eu, La, Nd, Zr, and Hf than Wari 1, and lower concentrations of Mn. For the most part, elemental correlations are similar, with La Nd, Zr, and Hf, Th and Eu, and Yb, Dy and Zn forming three groups of closely correlated elements in both datasets. There are however, some differences, particularly regarding Ce and Tb, which are closely correlated with La, Nd, Zr, and Hf in the INAA data, but are more closely correlated with Yb, Dy, and Zn in the LA-ICP-MS dataset. Whether this can be explained by the chemical contribution of temper will require further testing.

As a further means of comparing the datasets produced by the two techniques, group means, standard deviation from mean, and percent relative standard deviation from mean were calculated for each measured element as a way of quantifying the amount of dispersion present in each analytical group (Table IIa and b).

It should be stressed that this is not comparing the ability of LA-ICP-MS to correctly determine bulk composition as measured by INAA; as we have measured only clay matrix and excluded any inclusions in the paste, measured values for each element will of course differ somewhat. We are merely trying to quantify the ability of LA-ICP-MS to reproduce the broader structure of the data as revealed by INAA. While the values are in many cases quite close, this is likely a reflection of the homogenous fine paste from which these particular

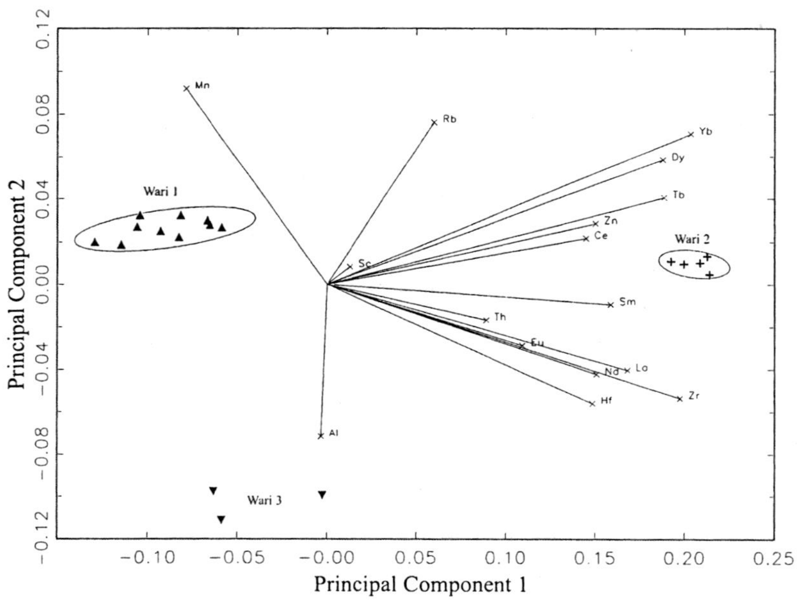

Figure 5. Principal Components 2 vs. 1 (LA-ICP-MS); 16 best discriminatory elements. PC1 summarizes 85.7% and PC2 10.5% of the total variability in the utilized data.

Table IIa. Comparison of mean values (ppm) by group for LA-ICP-MS and INAA by element, and RSD from group mean

	Wari 1				*Wari 2*			
	Mean (ppm)		*RSD*		*Mean (ppm)*		*RSD*	
	INAA	*LA-ICP-MS*	*INAA*	*LA-ICP-MS*	*INAA*	*LA-ICP-MS*	*INAA*	*LA-ICP-MS*
As	12.5	9.9	**23%**	35%	18.6	17.8	**13%**	22%
La	25.6	26.3	**5%**	10%	52.4	64.4	**2%**	7%
Lu	0.32	0.31	**11%**	13%	0.77	0.93	**4%**	7%
Nd	28.0	22.9	24%	**10%**	70.1	50.7	6%	**6%**
Sm	5.05	4.78	**4%**	10%	10.9	10.7	**2%**	4%
U	6.9	7.9	**9%**	11%	7.8	8.3	**12%**	15%
Yb	2.0	2.1	11%	**11%**	5.4	5.6	**2%**	7%
Ce	56.3	59.6	10%	**8%**	115.6	123.3	**4%**	9%
Co	8.6	11.2	**7%**	15%	9.6	10.3	**14%**	27%
Cr	25.1	23.7	9%	**8%**	22.0	21.9	**9%**	9%
Cs	10.3	13.6	**5%**	14%	13.6	17.6	**6%**	10%
Eu	0.90	0.98	**5%**	13%	1.4	1.7	**3%**	5%
Fe	23926	16170	**7%**	10%	20518	14539	6%	**5%**
Hf	4.2	4.1	**2%**	11%	8.8	9.1	**4%**	5%
Ni	20.9*	21.0	--	**12%**	0.00	18.1	**0%**	16%
Rb	170.4	189.9	**3%**	12%	209.3	239.1	**3%**	6%
Sb	1.0	1.2	**8%**	19%	0.9	1.2	7%	**5%**
Sc	8.9	16.1	**7%**	13%	12.3	16.8	**4%**	8%
Sr	178.1	168.0	31%	**28%**	0.00	100.3	--	**10%**
Tb	0.68	0.68	9%	**8%**	1.6	1.7	7%	**5%**
Th	16.1	18.6	**2%**	6%	24.6	29.5	**2%**	3%
Zn	77.0	93.3	**7%**	17%	138.2	191.2	**7%**	8%
Zr	111.9	105.3	17%	**7%**	242.6	301.3	8%	**4%**
Al	78131	82024	**3%**	5%	86406	84735	**2%**	3%
Ba	445.6	358.5	**32%**	32%	550.4	481.7	29%	**20%**
Dy	3.5	4.0	**6%**	9%	8.6	9.8	**3%**	4%
K	42384	40582	**6%**	11%	40320	35691	**3%**	5%
Mn	737.8	635.1	17%	**7%**	484.1	397.5	18%	**6%**
Ti	2368	2372	18%	**12%**	2678	3241	**12%**	15%
V	55.1	76.6	17%	**11%**	52.4	70.5	18%	**14%**
Ca	18121	12320	**34%**	36%	12638	9561	83%	**65%**

NOTE: Values in **bold** represent lower RSD value, * based on one sample with measurable levels of Ni

Table IIb. Comparison of mean values (ppm) by group for LA-ICP-MS and INAA by element, and RSD from group mean

	Wari 3			
	Mean (ppm)		*RSD*	
	INAA	*LA-ICP-MS*	*INAA*	*LA-ICP-MS*
As	9.6	10.5	**39%**	46%
La	41.6	38.3	14%	**7%**
Lu	0.27	0.23	13%	**12%**
Nd	44.0	33.5	7%	**6%**
Sm	5.8	5.8	6%	**5%**
U	5.8	6.5	13%	**5%**
Yb	1.5	1.6	12%	12%
Ce	74.8	59.3	32%	**17%**
Co	5.1	5.4	**3%**	19%
Cr	27.7	29.7	**11%**	21%
Cs	7.3	10.0	**1%**	4%
Eu	1.1	1.3	**6%**	14%
Fe	22584	16436	**11%**	12%
Hf	6.3	6.6	**4%**	6%
Ni	0.00	20.2	--	21%
Rb	113.2	125.9	**3%**	11%
Sb	0.60	0.81	**7%**	23%
Sc	6.9	15.5	11%	**8%**
Sr	240.3	193.2	13%	**9%**
Tb	0.58	0.62	**8%**	10%
Th	19.3	22.2	11%	11%
Zn	69.1	89.7	**9%**	28%
Zr	153.8	173.3	**5%**	10%
Al	108357	127544	6%	**6%**
Ba	715.9	570.5	**42%**	63%
Dy	2.7	3.3	**4%**	13%
K	33617	30992	**10%**	31%
Mn	531.8	335.4	22%	**6%**
Ti	2107	2253	31%	**11%**
V	36.1	55.0	**21%**	22%
Ca	8308	6005	**13%**	14%

NOTE: Values in **bold** represent lower RSD value.

ceramics were manufactured, and in other datasets it is expected that these values would be more divergent, unless the effects of temper were taken into account. For most elements, the two techniques produce comparable degrees of within group dispersion, though for ten elements (As, Co, Cs, Eu, Hf, Rb, Th, Zn, Dy, K), INAA measurement resulted in lower RSD from mean consistently across all three groupings, while LA-ICP-MS measurement resulted in lower RSD from mean for three elements (Ni, Mn, Sr) across all three groupings.

In a bi-variate plot of Neodymium and Hafnium concentrations (Figure 6), this is particularly evident for the Wari 1 grouping, where the INAA data is elongated relative to Nd, whereas the LA-ICP-MS data is elongated along the axis representing Hf concentrations. It should be noted however, that these comparisons are based on a small number of fairly homogenous samples, and the comparison of larger data sets needs to be undertaken before drawing any definitive conclusions regarding the relative performance of the two techniques in this regard.

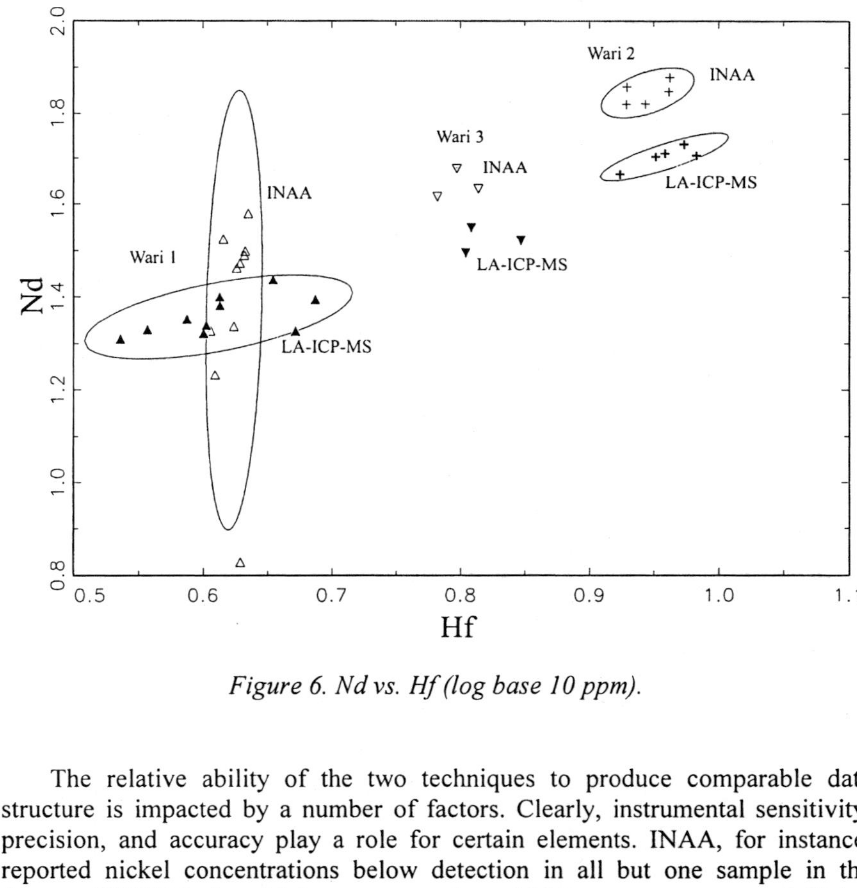

Figure 6. Nd vs. Hf (log base 10 ppm).

The relative ability of the two techniques to produce comparable data structure is impacted by a number of factors. Clearly, instrumental sensitivity, precision, and accuracy play a role for certain elements. INAA, for instance, reported nickel concentrations below detection in all but one sample in the dataset, PRW209, for which a concentration of 20.9 ppm was measured. Using LA-ICP-MS, a concentration of 20.8 ppm was measured for this same sample. However, this technique was also able to detect measurable concentrations of Ni in all other samples, ranging from 13 to 23 ppm. For other elements, high background noise to signal ratios limit the sensitivity, precision and accuracy of measurement by LA-ICP-MS. This is particularly a problem for two elements not measured by INAA: Cl and P. As mentioned before, the fact that LA-ICP-MS samples only a small amount of material relative to the total matrix present in a sample also likely introduces an amount of random deviation from the true total composition. The degree to which this causes dispersion within analytical groupings is at present unknown, but future work is planned to further test the validity of our methodology. While variability between measurements on a single sample was on the order of 10–20% for the samples reported on here (Figure 2), it is likely that for other datasets, there might be a larger degree of intra-sample variability, which might introduce a consequent larger amount of error into averaged bulk composition.

It can also be observed in the plot of Nd vs. Hf that for Wari 2 and 3, we measured consistently lower Nd concentrations by LA-ICP-MS as compared to measurements on the same samples via INAA. This can be observed in Tables IIa and IIb for a number of other elements. There are two possible explanations for this. This may result from our method of standardization relative to the method employed at MURR. Neodymium concentration measures low for Ohio Red clay as the certified value is 51.3 ppm, while we measure 39.0 ppm. However, the effects of temper on the INAA data relative to the values measured by LA-ICP-MS cannot be ruled out either. The presence of temper often acts to either increase or dilute concentrations of trace elements in a uniform way for discrete groups of ceramic wares, and could thus explain the overall higher Nd concentration for the INAA measurements relative to the corresponding ICP-MS values. In this case, the presence of inclusions not sampled by LA-ICP-MS seems to have had little effect on the overall three-group structure present in the data. However, in some cases, groupings that appear chemically homogenous when only matrix is measured are differentiable into separate chemical groupings when the chemical contribution of the temper is considered, as shown for instance by Larson *et al.* (*3*). Future work is planned to measure the composition of tempering materials found in the Conchopata Wari ceramics, as well as ceramics with a higher proportion of inclusions. The effects of this temper can then be mathematically introduced into the clay compositional data we have obtained via LA-ICP-MS as described by Neff *et al.* (*10*). At present, our results suggest that LA-ICP-MS has utility for measuring the composition of fine paste ceramics and producing results that are archaeologically useful, but the potential effects of heterogeneity in samples due to matrix variability or the presence of inclusions must be evaluated for each dataset analyzed.

Conclusion

The method to determine ceramic composition developed at the Field Museum of Natural History has produced encouraging results. Further study will be necessary to assess the contribution of tempering material in the bulk composition of the ceramics analyzed here, as well as in other groups of samples to be analyzed in the future. This will have to be done on a case-by-case basis, as the nature of temper can vary. The analysis of Wari ceramic samples using LA-ICP-MS and INAA give similar results, showing that LA-ICP-MS can be a promising alternative for ceramic provenance studies.

Acknowledgments

This project was made possible by the recent acquisition of a LA-ICP-MS at the Field Museum of Natural History. The LA-ICP-MS laboratory was built with funds from the National Science Foundation (grant No. 0320903), an anonymous donation, and the Anthropology Alliance.

References

1. Triadan, D.; Neff, H.; Glascock, M. D. *J. Archaeol. Sci.* **1997**, *24*, 997–1002.
2. Kennett, D. J.; Sakai, S.; Neff, H.; Gosset, R.; Larson, D. O. *J. Archaeol. Sci.* **2002**, *29*, 443–455.
3. Larson, D. O.; Sakai, S.; Neff, H. In *Laser-Ablation-ICP-MS in Archaeological Research*; Speakman, R. J.; Neff, H., Eds.; University of New Mexico Press: Albuquerque, 2005; pp 95–104.
4. William, I. H.; Cook, A. G. In *Andean Archaeology II: Art, Landscape, and Society*; Silverman, H.; Isbell, W. H., Eds.; Kluwer Academic, New York, 2002; pp 249–307.
5. Moseley, M.; Nash, D.; Williams, P.; DeFrance, S.; Miranda, A.; Ruales, M. *Perú Proceedings of the National Academy of Sciences* **2005**, *102*, 17264–17271
6. Williams, P. R. *Lat. Am. Antiq.* **2001**, *12,* 67–83
7. Elliott, S.; Knowles, M.; Kalinitchenko, I. A. *Spectroscopy* **2004**, *19*, 1.
8. Gratuze, B. *J. Archaeol. Sci.* **1999**, *26*, 869–881
9. Pearce, N. J. G.; Perkins, W. T.; Westgate, J. A.; Gorton, M. T., Jackson, S. E.; Neal, C. R.; Chenery, S. P. *Geostandards Newsletter* **1997**, *XXI*, 114–115.
10. Neff, H.; Bishop, R. L.; Sayre, E. V. *J. Archaeol. Sci.* **1988**, *15*, 159–172.

Chapter 20

Characterization of Building Materials from the Brick Chapel at Historic St. Mary's City

Ruth Ann Armitage[1], Leah Minc[2], Silas Hurry[3], and Melissa Doolin[1]

[1]Department of Chemistry, Eastern Michigan University, Ypsilanti, MI 48197
[2]Oregon State University Radiation Center, Oregon State University, Corvallis, OR 97331
[3]Archaeology Research Laboratory, St. Mary's City, MD 20686

Instrumental neutron activation analysis (INAA) was used to compare bricks from the 17th-century Chapel at St. Mary's City, Maryland to clay sources representing a range of geologic contexts within the area, particularly those near archaeological evidence of brick making. Clays from the transition slope near the purported brickyard at St. Barbara's reflect the same quartz sand dilution pattern observed in the Chapel bricks, and display a trace-element signature associated with heavy mineral sands (i.e., high in Zr and Ti). This paper reports the most recent results from our ongoing investigation toward understanding the technology of the Chapel bricks and their relationship to other bricks and source materials from St. Mary's City.

St. Mary's City, the first capital of the Maryland Colony, was founded in 1634 on the banks of what is now the St. Mary's River (Figure 1). The English settlers shared the original site with the local Native American people, but the city slowly evolved into an Italian Baroque-style designed town. The buildings of St. Mary's City incorporated brick to varying degrees (*1*). St. John's, where

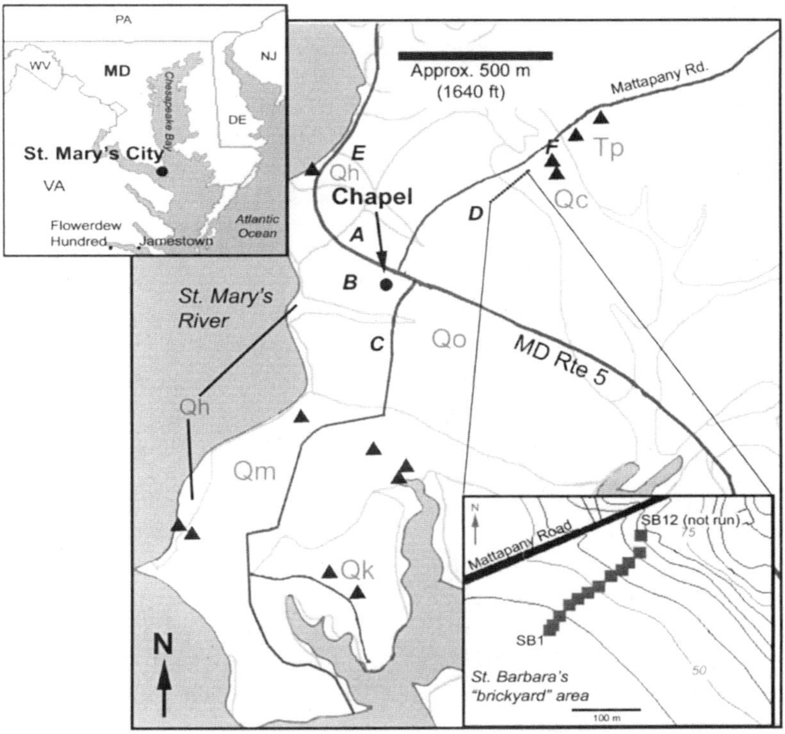

Figure 1. Map of St. Mary's City, Maryland area (adapted with permission from Reference 13; copyright 1989, Maryland Geological Survey). Inset shows elevation changes along Mattapan Road, up the St. Mary's Hill, with transect sampling sites marked by boxes. Triangles on larger map indicate most recent sampling locations. Sample sites A-F were examined by Armitage et al. (14).

construction of Phase I dates to 1638, had partially brick chimneys. Van Sweringen's Ordinary (a kind of 17th century inn or tavern) was brick veneered, and the Priest's House was mainly a wood frame building with some brick elements. The State House, built in 1674, and St. Peter's, also dating to the 1670's, were wholly brick buildings. Perhaps the most impressive brick structure in this small city was the brick Chapel. The exact dates of the Chapel's construction are not known; the historic record contains no information about the building of the Chapel, but indicates that a resident of St. Mary's City was charged with breaking a window in the Chapel in 1670 (*2*). As the Chapel was the first brick Catholic church in the English colonies in North America, it may have been politically expedient to not keep records of its construction.

The capital of Maryland moved from St. Mary's City to Annapolis in 1696, and in 1704, the Chapel was ordered locked shut and never again used as a house of worship (*3*). Many of the buildings at the old capital were torn down soon after, and the construction materials utilized elsewhere. Of the major brick constructions at St. Mary's, only the State House remained until the 1840s, when the bricks were reused in the nearby Trinity Church (*4*). Bricks from the Chapel superstructure were likely reused at the Jesuit manor house at St. Inigoes, a few kilometers from the center of the city. Today, the site of St. Mary's City is a historical and archaeological park, with a museum and several reconstructed buildings.

Excavations at the Chapel site in 1938 and again in the 1980s uncovered the foundations of that building (*5, 6*). The Chapel would have been a formidable structure for the time: the foundations, approximately 1 m thick and penetrating the ground nearly 1.5 m, indicate walls almost 8 m high. Based on contemporaneous European examples of Jesuit architecture, the Chapel might have appeared as in the artist's rendering in Figure 2, in the shape of a Latin cross.

The Chapel is currently being reconstructed on top of the original 17th century foundations. This new Chapel will serve as an architectural counterbalance for the 1930s reproduction of the 1674 State House. Additionally, it will provide the museum with a new interpretive possibility to better tell the important story of Maryland's early practice of religious toleration. Reproduction bricks for the building were made using clays extracted from an artifact-free zone within the city boundaries. However, the actual source of the materials – clay and possibly sand – used to make the historic bricks was unknown. Bricks from the Chapel have a unique appearance, as they are characterized by red to purple-black amorphous inclusions. This appearance has lead to speculation that the bricks might have been imported, as were many of the goods to this tobacco-producing colony. The phrase "English brick," often used to describe Colonial-era bricks leads to confusion: were English bricks brought from England, or were they made locally using standard English measurements and methods?

Bricks were certainly made in the Colonial Tidewater region. Kilns dating to the 17th century have been excavated in the Jamestown, VA area, and later 18th century kilns near Williamsburg, VA are also known (*7, 8*). Firing bricks in permanent kilns may have been the exception, however. Clamps – temporary firing sites built of green bricks to be fired – were more likely used by the

Figure 2. Artist's rendering of the probable appearance of the 1667 Chapel at St. Mary's City.

itinerant brick makers who plied their Old World training throughout the new colonies (*9*). At St. Mary's City, no archaeological evidence has yet been discovered for a brick kiln. Excavations at the Mill Field by Forman (*10*) uncovered a pit filled with waster mullion bricks that he interpreted at the time as evidence of brick making in the area. Surface surveys in the St. Barbara's field also found over- and under-fired brick wasters (*11*). Neither of these sites has since had significant subsurface investigation. Comparing the composition and mineralogy of the locally available clays to the bricks used at St. Mary's City should help to clarify if the bricks were made using those materials, or if they were imported from elsewhere. Previous studies (*12*) also used INAA and petrographic analysis to study a suite of building materials from St. Mary's City and sites in Virginia. This work incorporates this existing data with new INAA results on additional source materials from the St. Mary's Hill to further clarify the relationship between the Chapel bricks and locally available source materials.

Materials

Brick and tile samples from buildings including the Chapel at St. Mary's City and sites in Virginia were prepared and analyzed by INAA as part of an earlier study (*12*). These data have been combined with new results from further geologic sampling at the site.

The geology of St. Mary's County is quite diverse. Six geologic formations lie within the area surrounding Historic St. Mary's City (Figure 1), all of which bear clays of varying quality. The terrain of Southern Maryland is made up of uplands (elevations typically >100 feet) and lowlands (elevations typically 0 -50 feet), with a distinct, intervening slope marking the transition (*13*). Surface deposits in the area belong to five Quaternary formations: the Kent Island (Qk), the Maryland Point (Qm), the Omar (Qo), and the Chicamuxen Church (Qc) Formations (Figure 1) consist of sediments reworked from adjacent highlands and deposited in shallow estuarine environments during the middle to late Pleistocene (*14*). The Tertiary Park Hall Formation (Tp) is located at the highest point of the upland terrace and contains sediment derived directly from the Piedmont and Appalachian Mountains (*14*), and therefore is expected to be a significant source of heavy mineral sands containing elevated levels of Zr, Hf, and Ti (*13*).

The transition slope between upland and lowland is located near the St. Barbara's "brickyard" area; locally, the slope is referred to as "St. Mary's Hill." The local clay deposits were initially sampled (*12*) based on archaeological evidence – or lack thereof – of brick making activity. Subsequently, samples for this work have been selected to represent the full range of geological deposits as mapped within the St. Mary's City area. Previous work (*12*) showed that the material from the St. Mary's Hill, within the Qc deposit on the St. Mary's Hill, showed two distinct strata. Of these, the lower stratum bore significant similarities to the chapel bricks chemically, mineralogically, and visually, and yet statistically did not show a significant probability of membership to the compositional group defined by the Chapel bricks. To investigate the possible link between the archaeological surface survey at the St. Barbara's field and the composition of clays at the top of the St. Mary's Hill, a transect (Fig. 1, inset) was sampled along the elevation of the hill. Twelve samples were collected. Sample #12 was almost all sand, and thus was not included for bulk analysis as it was not clayey enough to form a brick.

Methods

Clay samples were rehydrated with deionized water, formed into small tiles (1.5 x 1.5 x 4 cm), dried at 100 °C overnight, and fired in an oxidizing environment to 800 °C. Brick samples were analyzed from four 17th century houses (St. John's, St. Peter's, the Priest's House, and Van Sweringen's ordinary), the 1674 State House, and the brick Chapel. All brick and experimental tiles were cleaned by abrading away the outer surfaces, rinsed in DI water, allowed to air dry, and then crushed for encapsulation for INAA.

In total, 90 samples, including those from a previous study (*12*) were prepared for INAA. Approximately 200 mg of crushed brick was encapsulated and submitted for irradiation; analyses were initiated at the University of Michigan's Ford Nuclear Reactor and completed at Oregon State University's Radiation Center. In each case, the samples were exposed to two separate irradiations and a total of four separate counts of resulting gamma activity, following protocols established for the multi-element analysis of archaeological materials (*15–20*). Details of the experimental conditions are provided elsewhere (*12*). Element concentrations were determined for 31 elements using three replicates of NIST SRM-1633A as comparison standards. Determination of Ca used NIST SRM-688. Interlaboratory calibrations, based on the New Ohio Red Clay check standard, indicated that differences in analytic results between labs were not significant.

To directly visualize all of the relationships between sample groups based on all of the elements measured by INAA, principal component analysis (PCA) is typically employed to reduce the dimensionality of the data. PCA is a mathematical procedure that transforms the original variables, based on their degree of inter-correlation or covariance, into a fewer number of principal components, which are simply linear combinations of the original variables that maximize the amount of covariance accounted for and which are themselves uncorrelated (*21*). The resulting principal components represent the major dimensions of variability within the data set, and can be interpreted from the total structure coefficients, or the degree of correlation between the original variables and the new composite ones. A plot comparing the primary principal components will thus illustrate the overall differences in elemental compositions among groups of samples.

Results

Petrographic analyses of the Chapel bricks and the clays carried out in the previous study indicated large and variable amounts of quartz sand (SiO_2) in the bricks (*12*). Neither Si nor O can be easily measured by INAA, so sand effectively dilutes the contribution of the clay component in the composition of the bulk sample. This is observed as a reduction in the concentrations of elements associated with clay. Bivariate plots of aluminum (representative of the clay fraction of the bulk composition) and each of the other elements were used to examine the effect of dilution. For the local clays from the Qo, Qm, and Qk deposits, most elements covaried positively with Al, a pattern reflecting their association with the clay fraction and relative dilution by varying amounts of quartz sand. The Chapel bricks, in contrast, show a distinctive pattern (Figure 3) in which elements characteristic of heavy minerals (Zr and Hf from zircons, and Ti from rutile) negatively covary with Al, suggesting their association, not with the clay fraction, but with the sand fraction. This observed negative correlation is

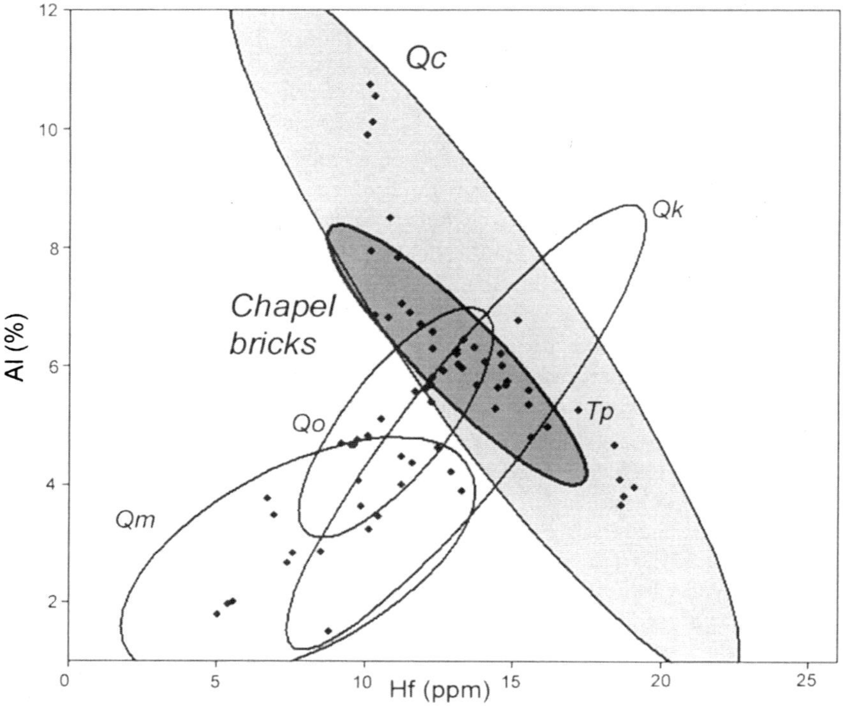

Figure 3. Bivariate plot of Hf and Al showing the negative correlation between most of the locally available clays and the Chapel bricks. Note that the correlation for the Chicamuxen Church Formation clays (Qc) is similarly negative. Ellipses indicate 95% confidence limits.

also found in the Qc clays, and suggests that these materials share a heavy mineral sand component not naturally found in other local clays, such as the Qo or Qm deposits. Petrographic analysis of the new clay samples is needed to confirm this mineralogically. However, the Zr content may be in the silt fraction and would thus be difficult to observe optically.

The transect samples showed significant variations in elemental concentrations (Figure 4). Manganese in particular was highest at base of slope (40–45 ft elevation) and decreased with increasing elevation. The heavy mineral sand elements (Hf and Zr) are highest near the top (>60 ft elevation) of the St. Mary's Hill. Local differences in depositional history, as well as in situ weathering of sediments, may explain these clear changes in composition over a linear distance of less than 200 m and an elevation change of only about 20 feet (6 m).

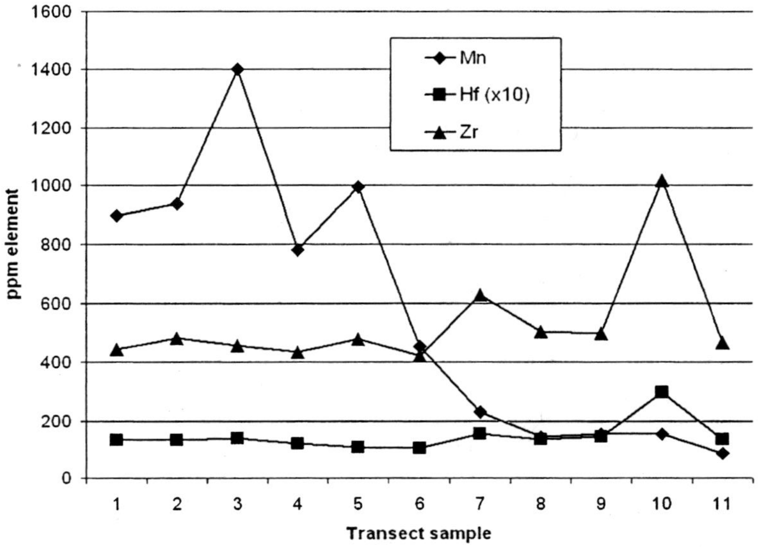

Figure 4. Elemental profiles for transect up the St. Mary's Hill; Hf values are shown x10 for scale. Samples 7–9 show the most similarity to the Chapel bricks.

In the trace-element data, the first principal component accounts for over 50% of the variance. Aluminum and most other elements correlate positively with the first principal component, a pattern consistent with simple dilution (*22,23*) – in this case, by quartz sand temper. In contrast, the second principal component (accounting for an additional 15% of the variance) represents the heavy mineral sand component (Ti, Hf, Zr), which negatively covaries with cobalt, manganese, antimony, and arsenic. The Qo and Qm clays from the lowlands are broadly similar in composition (Figure 5). The Qc deposit differs significantly, i.e., the low PC2 scores indicate high concentrations of the characteristic of heavy mineral sands (Ti, Hf, Zr). The Qk and Tp samples span range of composition, but are represented by only 2 samples each.

The question is whether or not the Chapel bricks were made using the locally available materials. Based on the principal components plot shown in Figure 6, it would appear that the Chapel bricks belong to the group defined by the Chicamuxen Church Formation (Qc) clays. Multivariate probabilities show this to be more complicated. Mahalanobis distance statistic calculations using the first four principal components show that only one of the Kent Island Formation samples (Qk) and sample #8 from the transect have significant probabilities ($P > 0.07$) of belonging to the compositional group defined by the Chapel bricks. For contrast, the brick from St. Peter's (1670's) also groups with Chapel bricks ($P = 0.43$), indicating that these building materials share a clay source.

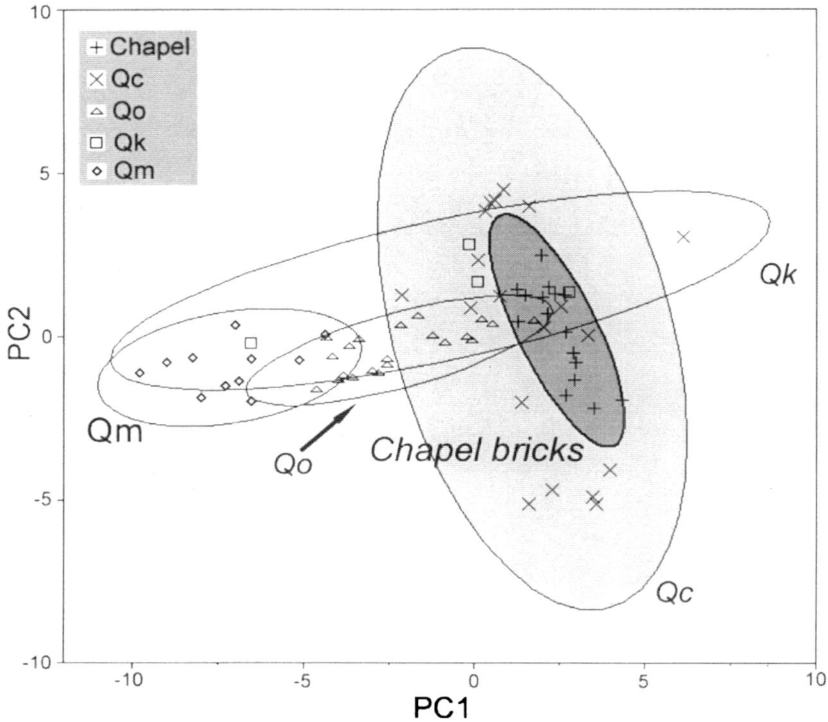

Figure 5. Principal component analysis showing relationships between Chapel bricks and local clays. Ellipses indicate 95% confidence limits.

To relate the elemental compositions to the archaeology, a selected sample of the data must be examined more closely. There are two areas with archaeological evidence of brick making: the Mill Field (Figure 1, A) and St. Barbara's brickyard (D) at the base of St. Mary's Hill. Although the surface scatter of waster bricks at St. Barbara's was observed on the flat area, clays could easily have been removed from the adjacent hillside, left to weather over the winter (and thereby changing the concentrations of mobile cations such as Na and K, and possibly Mn), and made into bricks on that lowland flat area. A smaller compositional group defined by only the clays from the hillside (excluding those from the flat) again shows significant similarities to the Chapel bricks (Figure 6). Although, Chapel bricks are distinctly lower in Na content, consistent with a higher degree of weathering. The number of clay samples from St. Mary's Hill is too small (N=13) to provide a well-characterized reference group. But preliminary comparisons indicate that 9 out of 21 Chapel bricks show

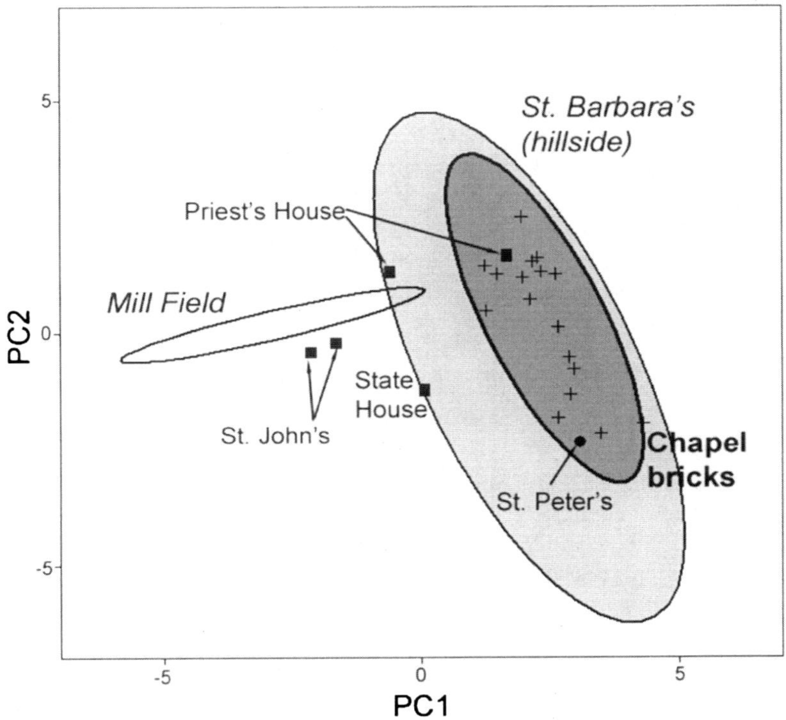

Figure 6. Principal components analysis showing the relationship between the areas with archaeological evidence of brick making activity and bricks from 17th-century St. Mary's City. Ellipses indicate 95% confidence limits.

low to moderate multivariate probabilities of group membership in this subset of Qc clays, based on PC scores. Bricks from other historical buildings at St. Mary's City show similar affiliations to the hillside clays, but those from Virginia do not. Before any conclusions can be drawn, subsurface archaeological fieldwork, even test pits, would aid in better understanding the possibilities of making bricks in this area.

Conclusions

It appears that the clays from the hill in the St. Barbara's field may have been the source materials for the Chapel bricks. The St. Barbara's clays reflect the same dilution pattern observed in the Chapel bricks, and display the trace-element signature associated with heavy mineral sands. The ratio of heavy mineral sands changes with slight differences in elevation; this may indicate that the 17th century brick makers were mixing the clays from the hillside. This is consistent with the appearance of the Chapel bricks as well.

Because of the geologic and compositional complexity present in this part of Southern Maryland, many more samples (especially more transects of the St. Mary's Hill) would aid in better understanding the observed trace element distributions. Archaeological excavation in the St. Barbara's brickyard area will be important in determining if this might have been the site where the Chapel bricks were made.

Acknowledgements

Research funding was provided by DOE University Reactor Sharing Program and the Seller's Research Fund. Nan Richards and Carré Zalma, former EMU students, also participated in this project. Special thanks to James Reger, Maryland Geological Survey, and the anonymous reviewers for their helpful comments, revisions, and suggestions in preparing the manuscript.

References

1. Leone, M. P.; Hurry, S. *Hist. Archaeol.* **1998**, *32*, 34–62.
2. *Archives of Maryland*; Proceedings of the Provincial Court 1666–1670; Vol. 57, p 610.
3. *Archives of Maryland*; Proceedings and Acts of the General Assembly September 5, 1704–April 19, 1706; Vol. 26, p 46.
4. Hammett, R. C. *History of St. Mary's County, Maryland*; St. Mary's County Bicentennial Commission: Ridge, MD, 1977; p 176.
5. Miller, H. M.; Riordan, T. B.; Hurry, S. D. *Maryland Hist. Mag.* **2004**, *99*, 351–373.
6. Riordan, T. B., *Dig A Grave Both Wide and Deep: An Archaeological Investigation of Mortuary Practice in the 17th-Century Cemetery at St. Mary's City, Maryland*; Historic St. Mary's City Archaeology Series no. 3; Historic St. Mary's City Commission: St. Mary's City, MD, 2004.
7. Harrington, J. C., *Virginia Magazine of History and Biography* **1950,** *58*, 16–39.
8. Metz, J. In *Historical Archaeology, Identity Formation, and the Interpretation of Ethnicity*; Franklin, M.; Fesler, G., Eds.; Colonial Williamsburg Foundation: Williamsburg, VA, 1999; pp 11–30.

9. O'Connor, R. P., Ph.D. thesis, University of Pennsylvania, Philadelphia, PA, 1987.
10. Forman, H. C. *Maryland Archeology: J. Archeol. Soc. Maryland* **1986**, *22,* 1–19.
11. Riordan, T. B. *The Mill Field at St. Mary's City: An Intensive Surface Collection of a Portion of the 17th Century Townlands*; Historic St. Mary's City, unpublished work, 1989.
12. Armitage, R. A.; Minc, L.; Hill, D. V.; Hurry, S. D. *J. Archaeol. Sci.* **2006,** *33*, 615–627.
13. Reger, J. Maryland Geological Survey. Personal communication, 2003.
14. McCartan, L. *Geologic Map of St. Mary's County*; State of Maryland, Department of Natural Resources, Maryland Geological Survey: Baltimore, MD, 1989.
15. Arnold, D. E.; Neff, H.; Bishop, R. L. *Am. Anthropol.* **1991**, *93*, 70–90.
16. Bishop, R. L.; Blackman, M. J. *Acc. Chem. Res.* **2002**, *35*, 603–610.
17. Bishop, R. L.; Neff, H. In *Archaeological Chemistry IV*; Allen, R. O., Ed.; Advances in Chemistry Series No. 220; American Chemical Society: Washington, DC, 1989; pp 57–86.
18. Glascock, M. D.; Neff, H. *Meas. Sci. Technol.* **2003**, *14*, 1516–1526.
19. *Chemical Characterization of Ceramic Pastes in Archaeology*; Neff, H., Ed.; Prehistory Press: Madison, WI, 1992.
20. Neff, H.; Glascock, M. D. *J. Radioanal. Nucl. Chem. Art.* **1995**, *196*, 275–285.
21. Dunteman, G. H. *Principal Components Analysis*; Quantitative Applications in the Social Sciences, Vol. 69; SAGE Publications: Newbury Park, CA, 1989.
22. Neff, H.; Bishop, R. L.; Sayre, E. V. *J. Archaeol. Sci.* **1988**, *15*, 159–172.
23. Neff, H.; Bishop, R. L.; Sayre, E. V. *J. Archaeol. Sci.* **1989**, *16*, 57–69.

Chapter 21

Characterization of 15th–16th Century Majolica Pottery Found on the Canary Islands

Javier Garcia Iñañez[1], Jaume Buxeda i Garrigós[1], Robert J. Speakman[2], Michael D. Glascock[2], and Elena Sosa Suárez[3]

ERAUB (Equip de Recerca Arqueomètrica de la Universitat de Barcelona), Dpt. Prehistòria, H.Antiga i Arqueologia, Facultat Geografia i Història, Universitat de Barcelona, c/ Montalegre, 6–8, 08001 Barcelona, Catalonia, Spain
[2]Research Reactor Center, University of Missouri, Columbia, MO 65211
[3]Canarias Cultural y Arqueológica, S. L., Pancho Guerra, 22, 3–B, 35014 Las Palmas de Gran Canaria, Canary Islands, Spain

To assess the provenance of majolica pottery found in the Canary Islands, a sample of 55 sherds was obtained from two sites on Gran Canaria Island: La Cueva Pintada (Gáldar) and El Antiguo Convento de San Francisco (Las Palmas de Gran Canaria). The pottery was studied by X-ray fluorescence, instrumental neutron activation analysis, and X-ray diffraction. The results show one group that matches a reference group from Seville, an assessment that supports the historical record. However, the data also reveal samples whose provenance corresponds to other production centers on the Iberian Peninsula, such as Manises, Barcelona, and, possibly, an unknown Portuguese center. Moreover, it is possible that Italian and Dutch pottery have been identified thereby providing a complexity factor to the historical accounts.

Introduction

The purpose of this paper is to present a preliminary chemical and mineralogical assessment of majolica pottery found on Gran Canaria sites dating from the 15th–16th centuries as evidence of the important trade that occurred between Spain and the newly discovered Americas. This study forms part of an on-going research project focused on the archaeometric characterization of Spain's primary pottery production centers (*1–5*), and forms part of the Ph.D. research of one of the authors (J. Garcia-Iñañez).

Majolica is an earthenware pottery characterized by a creamy light-buff-colored paste and an opaque white tin-lead glaze coating the entire outer surface of the vessel. However, the most outstanding feature of the majolica pottery perhaps lies in its decoration, being the subject of numerous studies by art historians. Majolica decorations are produced from metallic oxides that always are applied on top of the tin glaze.

Majolica opaque-white glaze is obtained from the addition of opacifying particles of tin oxide (SnO_2), and from quartz and feldspar inclusions contained within the glazing material. Majolica glaze is essentially sand (SiO_2) and lead. The lead acts as a flux that is necessary to decrease the temperature required for melting SiO_2. The tin-lead glaze and the bubbles produced by the firing process absorb, scatter, and/or reflect incident light, turning what is essentially a transparent glaze into a white one. Due to this opacity, majolica decoration is always applied to the outer surface of the glaze (*3, 6, 7*).

The conquest of the Americas by Spaniards began in 1492. However, this process would not have been possible without conquest of the Canary Islands by Spanish troops just a few years earlier. In that sense, although the Canary conquest started at the end of the 14th century (with the Spanish arrival at Lanzarote), the archipelago was not totally occupied by the Castilian Kingdom until 1494 (with the conquest of Tenerife), two years after the first voyage of Columbus to the Caribbean. The Canary Islands represented the key stopping point for Spanish access to the Americas. Indeed, between the 15th and 18th centuries, the Canary Islands were a required stopover for Spanish trade with the Americas. This was due to the necessity of taking the advantage of the ocean currents and Trade Winds that blow directly to the Caribbean from the eastern Atlantic. Moreover, the Canary Islands signified, for all the Spanish ships, the last geographic region controlled by the Spanish before starting their long open-sea voyage. Additionally, the Canary Islands were a forced stop for legal reasons in which the Castilian Kingdom established a rigid control over the trade of items to the New World. In order to supervise the traffic of goods, a bureau of trade was established in 1503, called *Casa de Contratación*, only eleven years after discovery of the Americas. The *Casa de Contratación* had, for more than 200 years (1503–1717), its primary office in the city of Seville, an inland fluvial port protected from pirate attacks. For that reason, as the commercial trade between the Americas and Europe increased, the port of Seville emerged as Spain's most important trading port because it was the departure point and the final destination for most of the ships that traveled between Spain and the Americas. As a direct result of this organization, Seville became one of the most important production centers for majolica and other types of pottery shipped to the Americas.

Goals and Sampling

The main objective of the present work is to study the provenance of majolica pottery found on important Gran Canaria Island archaeological sites that have been subject of archaeological research. Furthermore, by means of archaeometric characterization, the role played by the majolica produced in Seville that was documented historically as an import into the Canary Islands, is also assessed. Majolica from different production centers also is considered, thereby adding a complexity factor to the Canary trade relationships panorama in a historical period when, in theory, a very tight control existed over trade with the Americas.

La Cueva Pintada

La Cueva Pintada is one of the most important Prehispanic settlements on Gran Canaria Island, and is located near the present-day city of Gáldar (Figure 1). The name of this site comes from an exceptional Prehispanic cave, painted with geometrical motifs that were first documented in 1862. The native settlement, of Agaldar, had a continuous occupation from the 6th to the 16th centuries, until it was abandoned after the Spanish conquest. Agaldar was the capital of the main indigenous chiefdom on the island, and after the conquest it was one of the first Spanish settlements. From the 18th century until 1970, the area where the site is located was used for agriculture. Archaeological and conservation work (*8, 9*) was initiated in 1970.

San Francisco Convent

The ancient San Francisco convent at Las Palmas de Gran Canaria on Gran Canaria Island was built at the end of the 15th century (Figure 1). It is one of the most important historical sites on the island because it represents the earliest founding of the Franciscan order on the island during the conquest of the Canary Islands. The convent remained active until 1835, when the Franciscans were forced to leave after the lands were confiscated by the Spanish government (i.e., Mendizabal's confiscation). After being used by the Spanish army, the building was abandoned during the mid-20th century. In 1992, the archaeological section of the Museo Canario de Las Palmas de Gran Canaria initiated an archaeological excavation of the San Francisco convent remains. A large quantity of majolica pottery, not only Spanish, but also of Italian and Portuguese origin was recovered.

Pottery recovered from the archaeological site of the San Francisco convent was classified, on the basis of stylistic attributes, into three main categories: (1) cuerda seca, a hispano-mudejar unglazed pottery, probably produced in Southern

Figure 1. Map of the main sites cited in the text. 1: Delft; 2: Liguria; 3: Barcelona; 4: Vilafranca del Penedès; 5: Lleida; 6: Muel; 7: Villafeliche; 8: Reus; 9: Teruel; 10: Paterna; 11: Manises; 12: Talavera de la Reina; 13: Puente del Arzobispo; 14: Seville; 15: Cueva Pintada; 16: San Francisco's convent.

Spain during 15th and first half of the 16th centuries; (2) Spanish and Italian 16th century majolica—primarily white and blue-on-white, and (3) Portuguese faience dating to the 17th century (*10*).

Sampling and Methods

All samples analyzed from La Cueva Pintada and the San Francisco convent were selected based their chronology and typology. From the site of La Cueva Pintada, 18 majolica and glazed pottery sherds, dating from the late 15th and 16th centuries were selected (Table I); 37 sherds from the San Francisco convent, mostly majolica, dating from the late 15th to 17th centuries also were obtained (Table II). Most of the analyzed majolica are plain white or exhibit blue patterns over an opaque white background. Nevertheless, a few of the specimens show different decorations, such as polychrome or green-on-white motifs. A few of the sherds from La Cueva Pintada exhibit a transparent lead glaze coating which is green or honey colored.

Forty-four of the 55 samples were chemically characterized by X-ray fluorescence (XRF). Thirty-seven of these samples were also characterized mineralogically by X-ray diffraction (XRD). Finally, 11 out of the 44 samples analyzed by XRD (4 specimens from La Cueva Pintada site and 7 from the San Francisco's convent) were analyzed by instrumental neutron activation analysis (INAA). Additionally, 11 samples from San Francisco convent, which had an insufficient weight to be characterized by XRF, were analyzed by INAA as well. XRD analyses are now being conducted for the 18 samples not yet characterized.

Analytical Procedure and Data Analysis

In the present study, 12 g from each sherd was collected. Glazes and other surfaces were mechanically removed to minimize contamination from glaze and other external compounds. Following this process, specimens were powdered and homogenized in a Spex Mixer (Mod. 8000) tungsten carbide cell mill for 12 minutes. Powdered specimens were transferred to clean polyethylene vials for transportation to the laboratory.

The chemical compositions of the samples analyzed by XRF[1] were determined using a Phillips PW 2400 spectrometer with a Rh excitation source. Samples were prepared for analysis using two different methods. Duplicates of glassy pills made by fusing 0.3 g of dry powder with 5.7 g of $LiBO_4$ to determine the major and minor elements with the exception of Na_2O. Trace elements and Na_2O were determined from 5 g pills of powdered dry sample. The quantification of the concentrations was obtained using a calibration line based on 60 international geological standards. The elements determined were Fe_2O_3 (as total Fe), Al_2O_3, MnO, P_2O_5, TiO_2, MgO, CaO, Na_2O, K_2O, SiO_2, Ba, Rb, Mo, Th, Nb, Pb, Zr, Y, Sr, Sn, Ce, Co, Ga, V, Zn, W, Cu and Ni. The loss on ignition (LOI) was determined by firing 0.3 g of dried powder at 950°C for three hours. An extended description of the XRF analytical procedures, including the details of accuracy and precision, has been published elsewhere (*11*).

Chemical analyses by INAA were conducted at the Archaeometry Laboratory at the Missouri University Research Reactor Center (MURR). Aliquots of sample were oven-dried at 100 °C for 24 h. Amounts of approximately 150 mg were weighed into small polyvials used for short irradiations. At the same time, 200 mg of each sample was weighed into high-purity quartz vials used for long irradiations. Along with the majolica samples, reference standards of SRM-1633a (coal fly ash) and SRM-688 (basalt rock) were similarly prepared, as well as quality control samples of SRM-278 (obsidian rock) and Ohio Red Clay (standards treated as unknowns).

---

[1] The XRF and XRD analyses were conducted by the Scientific-Technical Services laboratory at the University of Barcelona.

Table I. Summary of Samples from La Cueva Pintada Site

Sample ID	*Description*	*XRF*	*XRD*	*INAA*	*Classification*
MJ0236	Lusterware	x	x	x	Manises
MJ0237	Sevillan White	x	x	x	Seville
MJ0238	Blue-on-white	x	x		Seville
MJ0239	Green	x	x		Ungrouped
MJ0240	Blue tile	x	x		Seville
MJ0241	Sevillan White	x	x	x	Seville
MJ0242	Blue-on-white	x	x		Seville
MJ0243	Blue-on-white	x	x		Seville
MJ0244	Honey	x	x		Ungrouped
MJ0245	Honey	x	x		Seville
MJ0246	Blue-on-white	x	x		Seville
MJ0247	Blue	x	x		Seville
MJ0248	Honey	x	x		Seville
MJ0249	Green	x	x		Ungrouped
MJ0250	Green	x	x		Seville
MJ0251	Honey	x	x		Ungrouped
MJ0252	Honey	x	x		Ungrouped
MJ0253	Green	x	x	x	Seville

At MURR, INAA of pottery consists of two irradiations and three gamma counts. Short irradiations involve a pair of samples being transported through a pneumatic tube system into the reactor core for a 5 s neutron irradiation at a flux of 8 X 10^{13} n $cm^{-2}{\cdot}s^{-1}$. Following a 25-min decay, the samples are counted for 720 s using a high-resolution germanium detector. This count yields data for nine short-life elements: Al, Ba, Ca, Dy, K, Mn, Na, Ti and V. For the long irradiation, bundles of 50 or 100 of the encapsulated quartz vials are irradiated for 24 h at a flux of 5 X 10^{13} n $cm^{-2}{\cdot}s^{-1}$. After the long irradiation, samples are permitted to decay for seven days, and then are counted for 2000 s (known as "mid count") on a high-resolution germanium detector coupled to an automatic sample changer. This mid count yields data for seven medium half-life elements: As, La, Lu, Nd, Sm, U, and Yb. After an additional two-week decay, a second count of 10,000 s is carried out on each sample. This measurement allows quantification of 17 long-life elements: Ce, Co, Cr, Cs, Eu, Fe, Hf, Ni, Rb, Sb, Sc, Sr, Ta, Tb, Th, Zn, and Zr (*12*).

The mineralogical composition of all the samples included in the study was determined by XRD, using the same powdered sample prepared for XRF analysis. Measurements were performed using a PANalytical X'Pert PRO alpha1 powder diffractometer (radius = 240 mm) using the Cu Kα radiation (λ = 1.5418 Å), with a working power of 45 kV – 40 mA. The incident beam was passed through a 0.04 radians Soller slit, and the diffracted beam passed through a second slit. Moreover, the diffracted beam was Ni filtered. An X'Celerator

Table II. Summary of Samples from San Francisco Convent Site

Simple ID	*Description*	*XRF*	*XRD*	*INAA*	*Classification*
MJ0254	Lusterware	x	x		Manises
MJ0255	Lusterware	x	x		Seville
MJ0257	Columbia Simple	x	x		Seville
MJ0258	Columbia Simple	x	x	x	Seville
MJ0259	Columbia Simple	x	x		Seville
MJ0260	Columbia Simple	x	x		Ungrouped
MJ0261	Sevillan White	x	x		Seville
MJ0262	Sevillan White	x	x	x	Seville
MJ0263	Sevillan White	x	x		Seville
MJ0264	Sevillan White	x	x		Seville
MJ0265	Sevillan White	x	x		Seville
MJ0266	Sevillan White	x	x		Seville
MJ0268	Isabela Polychrome	x	x	x	Seville
MJ0269	Delft series			x	Ungrouped
MJ0270	Ligurian blue-on-white	x	x	x	Ligurian
MJ0271	Ligurian blue-on-white	x	x		Ligurian
MJ0272	Ligurian blue-on-white	x	x	x	Ligurian
MJ0273	Ligurian blue-on-white	x	x		Ligurian
MJ0274	Ligurian blue-on-white	x	x		Ligurian
MJ0275	Portuguese blue-on-white	x	x	x	Portuguese
MJ0276	Portuguese blue-on-white	x			Portuguese
MJ0277	Portuguese blue-on-white	x			Portuguese
MJ0278	Italian Polychrome	x			Ligurian
MJ0280	Ichtuknee blue-on-blue	x			Seville
MJ0281	Ichtuknee blue-on-blue	x			Seville
MJ0282	Ichtuknee blue-on-blue	x			Seville
MJ0284	Sgraffito	x		x	Ungrouped
MJ0285	Lusterware			x	BCN-SC
MJ0286	Lusterware			x	Seville
MJ0287	Columbia Simple			x	Seville
MJ0288	Sevillan White			x	Seville
MJ0289	Delft series			x	Seville
MJ0290	Ligurian blue-on-white			x	Ligurian
MJ0291	Ligurian blue-on-white			x	Ligurian
MJ0292	Portuguese blue-on-white			x	Portuguese
MJ0293	Catalan blue-on-white			x	BCN-SC
MJ0294	Green and Manganese			x	Ungrouped

Detector, with active length = 2.122 °, was used. θ/2θ scans were reorded from 4 to 70 ° 2θ (step size=0.017°; time=50 s per step). The evaluation of crystalline phases was carried out using the DIFFRACT/AT program by Siemens, which includes the Joint Committee of Powder Diffraction Standards (JCPDS) data bank.

Data Treatment

Although sample preparation was conducted under great care to minimize the analytical error, the potential for contamination does exist. As a precaution, tungsten and cobalt were removed from consideration during the statistical analysis of data because samples were powdered using a tungsten carbide cell grinder in which W is a major component and Co is a minor component.

Additionally, the technology of majolica pottery production is characterized by Sn and high Pb in their glazes, causing analytical problems because of their diffusion into the clay matrix body. Therefore, Sn and Pb concentrations have not been considered because of the possibility of contamination from the glaze. Moreover, chemical results have shown that Pb content largely exceeds the upper XRF regression limit (928 ppm) (Figure 2). In XRF, unusually high Pb results in interferences with nearby peaks. Thus, trace elements such as Ga and Y, and Th and Rb, might have high error. Consequently, these elements were not considered in the analysis of the XRF data. Moreover, Mo, as determined by XRF, is always below the lower detection limit and cannot accurately be determined. Finally, the element Ni was below the detection limits in most samples analyzed by INAA and was subsequently removed from consideration.

For statistical analysis, the D-compositional chemical data were transformed into log-ratios following Aitchison's and Buxeda's recommendations, by taking the natural logarithm of the ratio of all determined components to one component selected as the divisor according to the equation

$$\mathbf{x} \in S^d \rightarrow \mathbf{y} = \ln\left(\frac{\mathbf{x}_{-D}}{x_D}\right) \in R^d,$$

where $\mathbf{x}_{-D}=(x_1,\ldots,x_d)$ (*13, 14*). The variation matrix was used as an exploratory tool to identify sources of compositional variability, to quantify this variability, and to identify the appropriate element to be used as a divisor in the log-ratio transformation (*15*). Canary Island specimens were compared against several reference groups already established for the primary productions centers on the Iberian Peninsula. To date, 425 samples (Table III) have been chemically analyzed and reference groups have been defined for Barcelona, Reus, Vilafranca del Penedès, Lleida, Paterna, Manises, Muel, Teruel, Villafeliche, Talavera de la Reina, Puente del Arzobispo and Seville (Figure 1). All 425 samples have been also mineralogically characterized by XRD. However, the analyses for 15 samples from Lleida, and 57 samples from Seville, corresponding to the kiln sites of Pureza (attributed to Niculoso Pisano), Valladares, and Plaza de Armas, are still in process.

Results and Discussion

XRF Data

The variation matrix generated for the XRF data exhibits a total variation (vt) of 2.034 which can be considered high (*15*). The components responsible for most of this variability are P_2O_5, Cu, Na_2O, Sr, K_2O, CaO, MgO, Ni and Ba ($vt/\tau_{.i}$<0.6). On the other hand, the component that introduces the least amount of variation is Fe_2O_3 ($vt/\tau_{.i}$=0.91). Consequently, Fe was selected as divisor for the logratio transformation. Figure 4 shows a dendrogram resulting from the cluster analysis performed on the following subcomposition: Al_2O_3, MnO, TiO_2, MgO, CaO, Na_2O, K_2O, SiO_2, Ba, Nb, Zn, Sr, Ce, V, Zn, Ni and Cr. As explained above, Fe_2O_3 was used as a divisor in the logratio transformation. The analysis was performed using the squared euclidean distance and the centroid agglomerative algorithm from S-Plus 2000 (*16*).

Figure 2. Histogram of Pb concentrations determined by XRF. The arrow indicates the upper regression limit (928ppm)

An examination of the dendrogram (Figure 3) suggests that most of the Canary Island samples are of Sevillan origin (26 samples). In addition, two clusters that do not correspond to any of the known Spanish production centers were identified. The first unknown cluster groups five possible Ligurian samples with one Italian polychrome majolica. The second unknown cluster is comprised of the three samples classified as Portuguese. Moreover, two of the nine

Table III. Summary of Reference Samples Analyzed by XRF and INAA.

Production Centers	*XRF*	*XRF & INAA*	*INAA*	*Total*
Barcelona	72	27		99
Lleida	15		15	30
Manises	13	7		20
Muel	9	22		31
Paterna	14	6		20
Puente	9	7		16
Reus	11	11		22
Seville	71	6		77
Talavera	14	6		20
Terol	13	21		34
Vilafranca	23	12		35
Villafeliche	10	11		21
Total	274	136	15	425
Gran Canaria	33	11	11	55
Total	307	147	26	480

remaining samples match with the Manises reference group. The remaining seven samples are unclassified. A second cluster analysis based on only the Canary Island samples and the Sevillian and Manises reference groups is shown in Figure 4. This dendrogram clearly shows the structure of the dataset in four main clusters (groups 1, 2, 12 and 14), and the unassigned samples (Figure 4, Table IV).

Group 1 is distinct from the others based on higher Ni and Cr concentrations (Figure 5). The relatively high Ni and Cr content may be characteristic of ophiolitic zones, such as the Ligurian region and other parts of northern Italy (*17*). These results are reinforced by the archaeological classification as either Ligurian or Italian production for these samples.

Group 2 is composed of three samples recovered from the San Francisco convent identified as Portuguese majolica. This classification combined with the chemical differences warrants the placement of these samples in a different group despite the fact that this is not a particularly homogeneous group. In any event, Group 2 samples do not match with any of the characterized production centers from the Castilian and Aragonese Kingdoms thereby supporting the possibility of a Portuguese origin for these samples.

It should be pointed out that the primary differences between Manises and Seville, based on XRF, are differences between the Na_2O and K_2O

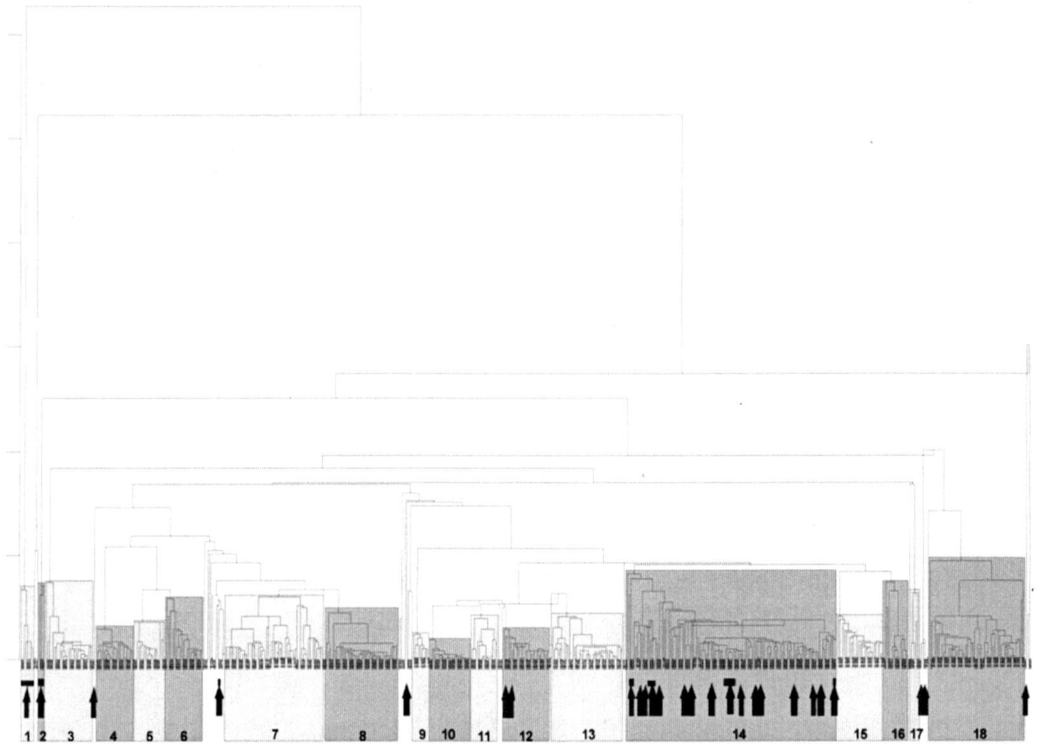

Figure 3. Dendrogram on the 454 individuals analyzed by XRF. Arrows point to individuals from Gran Canaria. 1: possible Ligurian group; 2: possible Portuguese group; 3: Villafeliche; 4: Paterna; 5: Teruel 1; 6: Teruel 2; 7: BCN-DR/PI; 8: VdP; 9: BCN-SC; 10: Muel 2; 11: Muel 1; 12: Manises; 13: Talavera-Puente; 14: Seville; 15: Reus; 16: Lleida 1; 18: BCN-DR/PI altered.

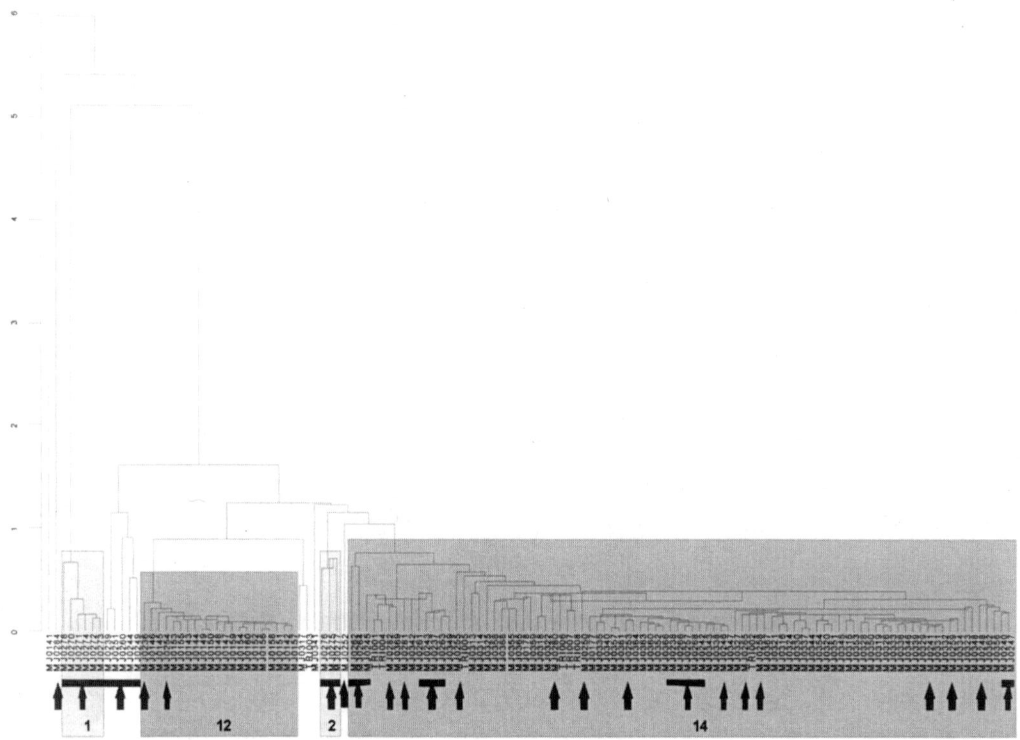

Figure 4. Dendrogram on 131 individuals analyzed by XRF. Arrows point to individuals from Gran Canaria. 1: possible Ligurian group; 2: possible Portuguese group; 12: Manises; 14: Seville

Table IV. Element Concentrations in the Ceramic Sample from Gran Canaria Analyzed by XRF.

XRF	*Seville (n=90)*		*Manises (n=22)*		*Ligurian (n=6)*		*Portugal (n=3)*		*MJ0284*	*MJ0260*
	mean	*s.d.*	*mean*	*s.d.*	*mean*	*s.d.*	*mean*	*s.d.*	*(Sgrafitto)*	
Fe_2O_3 (%)	4.88	0.35	4.78	0.20	5.85	0.09	4.24	0.09	8.08	5.59
Al_2O_3 (%)	12.72	0.98	14.51	0.67	13.21	0.41	12.13	0.23	18.96	15.26
MnO (%)	0.08	0.01	0.06	0.01	0.07	0.01	0.04	0.01	0.14	0.05
P_2O_5 (%)	0.25	0.07	0.10	0.03	0.21	0.06	0.25	0.01	0.19	0.19
TiO2 (%)	0.68	0.04	0.62	0.03	0.66	0.02	0.89	0.03	0.91	0.83
MgO (%)	3.24	0.35	2.93	0.29	4.59	0.48	2.24	0.05	2.97	2.68
CaO (%)	21.81	2.53	23.99	2.05	19.93	1.46	28.27	4.36	7.29	8.50
Na_2O (%)	1.03	0.28	0.42	0.06	1.30	0.10	1.23	0.33	0.85	0.97
K_2O (%)	1.99	0.55	3.40	0.18	1.67	0.26	1.30	0.32	3.04	2.97
SiO_2 (%)	53.18	2.23	49.06	1.60	52.34	1.29	49.30	3.87	57.41	62.83
Ba (ppm)	342	43	333	21	254	22	209	35	449	319
Nb (ppm)	17	1	17	1	15	0	23	1	23	20
Zr (ppm)	198	22	179	8	152	11	192	7	164	246
Sr (ppm)	438	60	331	32	651	83	326	65	242	343
Ce (ppm)	52	9	55	7	40	9	41	5	65	64
V (ppm)	71	10	67	10	80	7	57	3	127	103
Zn (ppm)	86	9	71	3	118	13	73	5	148	108
Cu (ppm)	83	72	94	52	73	55	40	14	74	43
Ni (ppm)	34	6	35	4	226	21	33	5	120	54
Cr (ppm)	78	9	64	4	232	25	72	5	137	96

concentrations. Unfortunately, these elements are strongly affected by a common alteration and contamination process that affects high-fired and overfired calcareous pottery. Such a process seems to imply a first phase of alteration of the glassy phase that provokes the leaching of potassium, followed by a second phase of analcime crystallization. The crystallization of this Na-zeolite fixes sodium from the environment into the sherd. The result can be a significant loss of potassium together with an enrichment of sodium (*14, 18, 19*), even in relatively modern ceramics, such as 17th century majolica (*1, 2*). In the present study, XRD analyses indicate that a large number of samples exhibit analcime, and therefore it would seem justified to disregard Na_2O and K_2O concentrations. Even so, their discriminant role does not support such a decision. A bivariate plot of $\ln(Na_2O/Fe_2O_3)$ versus $\ln(K_2O/Fe_2O_3)$ shows that the unaffected samples from Manises and Seville are clearly separated (Figure 6). None of the samples from Manises have undergone such an alteration and contamination process, as opposed to the majority of the samples from Seville (Figure 7). The attribution of the two Canary Island samples to the Manises group is therefore unambiguous.

The largest group (Group 14) comprises most of the samples from the reference group of Seville. In addition, it contains a majority of the samples from the two Canary Island sites.

Finally, five of the seven unassigned samples seem to be displaced from the Seville group because of lower CaO and Sr concentrations. However, this fact does not enable us to establish a relationship between these samples and a Sevillian provenance. For the two remaining unassigned samples (Table IV), no explanation is suggested for their provenance. Although sample MJ0260 is typologically classified as "Columbian simple", sample MJ0284 is an example of *sgrafitto* pottery and exhibits Ni and Cr contents higher than the norm, but not as high as the Ligurian group (Figure 5).

INAA Data

The INAA data were examined in a manner similar to that described for the XRF data. A variation matrix generated for the INAA data show a total variation (vt) of 2.245 that also can be considered high (*15*). The components responsible for most of this variability are Na, Sr, Sb, As, Cs, Ca, Rb, Ba, Cr and K ($vt/\tau_{.i}<0.5$). In contrast, La is the component that introduces lowest variability ($vt/\tau_{.i}=0.93$), and was subsequently selected as divisor in the logratio transformation. Cluster analysis was then performed, with S-Plus 2000 (*16*), by employing the squared euclidean distance and the centroid agglomerative algorithm on the following subcomposition: Lu, Sm, Yb, Al, Ba, Ca, Dy, Mn, Ce, Cr, Eu, Fe, Hf, Sc, Sr, Ta and Th. As previously stated, La was used as the divisor in this transformation. The results (Figure 8) suggest an agreement exists between the XRF and INAA results. Canary Island samples analyzed by both

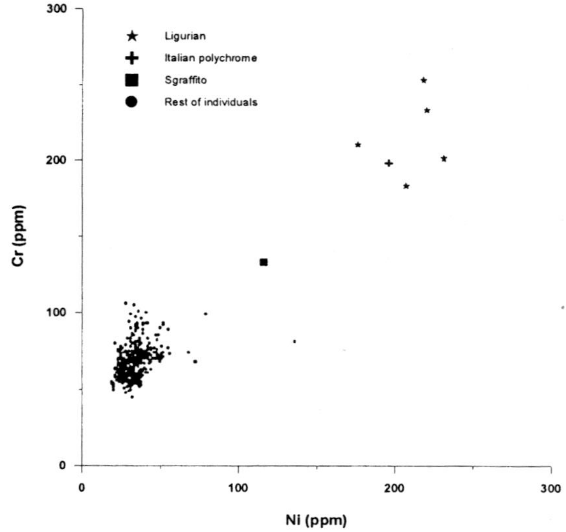

Figure 5. Bivariate plot of raw XRF Ni and Cr contents

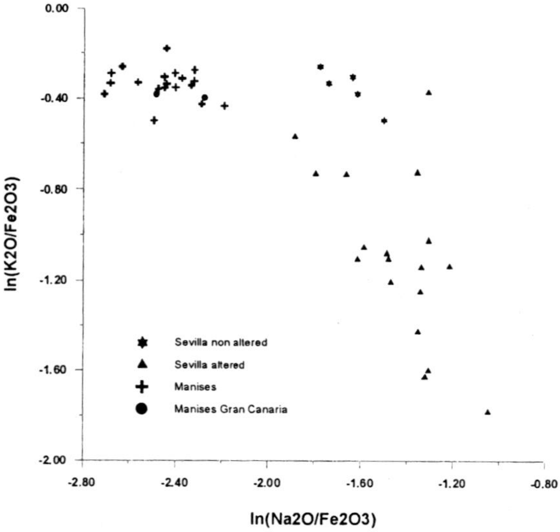

Figure 6. Bivariate plot of $ln(Na_2O/Fe_2O_3)$ versus $ln(K_2O/Fe_2O_3)$

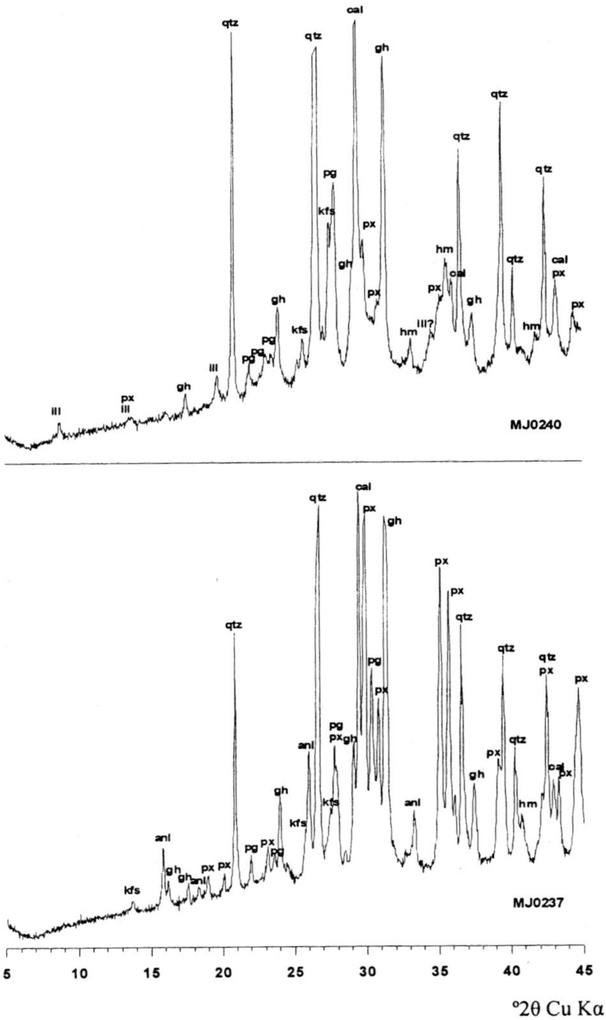

Figure 7. Upper: Diffractogram of the sample MJ0240, representative of the unaffected Sevillian pottery. Lower: Diffractogram of the sample MJ0237, representative of the affected Sevillian pottery. anl: analcime; cal: calcite; gh: gehlenite; hm: hematite; ill: illite-muscovite; kfs: alkaline feldspar; pg: plagioclase; px: pyroxene; qtz: quartz

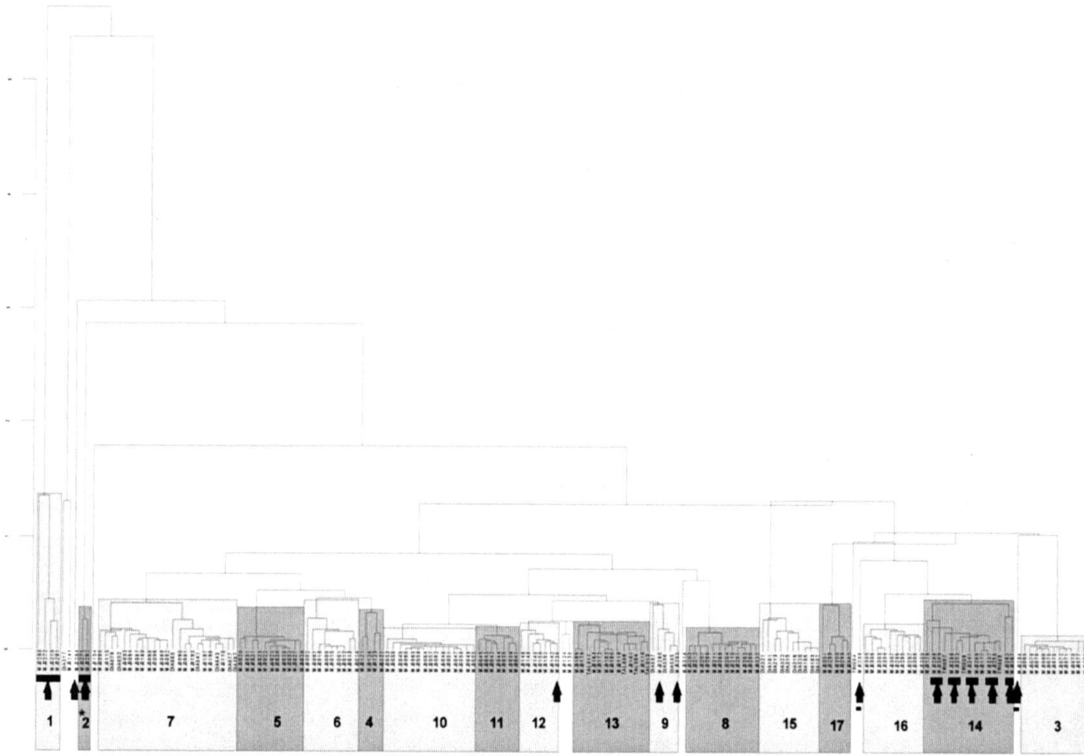

Figure 8. Dendrogram on the 173 individuals analyzed by INAA. Arrows point to individuals from Gran Canaria. 1: possible Ligurian group; 2: possible Portuguese group; 3: Villafeliche; 4: Paterna; 5: Teruel 1; 6: Teruel 2; 7: BCN-DR/PI; 8: VdP; 9: BCN-SC; 10: Muel 2; 11: Muel 1; 12: Manises; 13: Talavera-Puente; 14: Seville; 15: Reus; 16: Lleida 1; 17: Lleida 1

techniques are assigned to the same groups (Table V). One main difference, however, is the unification of the group BCN-DR/PI— previously split by XRF in two different groups (Groups 7 and 18) (Figure 3), according to the extension of the alteration and contamination process that affected sodium and potassium contents. Given that these elements are not considered here, the division is not apparent. Even so, Manises and Seville are clearly differentiated, contrary to the XRF data.

The possible Ligurian group includes four samples, two of which were previously assigned by XRF to that group, but the two additional samples had not previously been analyzed by XRF. These four specimens were archaeologically classified as Ligurian productions, and they exhibit typically high concentrations of Ni and Cr.

The possible Portuguese group is comprised of 2 samples, one new and one previously analyzed by XRF. As was the case previously, both samples were typologically classified as Portuguese products. Based on the INAA data, a clear-cut distinction exists between these samples and the reference groups established for the Castilian and Aragonese Kingdoms.

The Manises reference group includes one sample previously classified as a Manises production by XRF. It is important to note that Na and K were not used in this classification, but a clear separation exists between the Manises and Seville groups, reinforcing the chemical differences between the groups.

The reference group for Seville is comprised of six samples previously classified as Sevillan productions by XRF and four new samples that were not previously analyzed. It is essential to point out that one of the latter sherds is typologically classified as a Delft product.

The remaining samples classified in a previously identified reference group include two samples, not analyzed by XRF, that belong to the BCN-SC reference group from Barcelona. In fact, one of the samples had been classified as Catalan blue-on-white. Until now, the study of majolica production at Barcelona has revealed the existence of a very large group, labeled BCN-DR, because it is related to the Drassanes kiln dump associated with the Barcelona potter's guild (*1*). This group is related to a second reference group, BCN-PI that does not exhibit a clear-cut chemical difference. This large Barcelona reference group represents the greatest number of the majolica samples analyzed from Barcelona, and it constitutes the only established group that matches the majolica from the consumption centers (*3*). In addition to these groups, a third reference group with a very low number of samples has also been identified at Barcelona. This latter group, BCN-SC, seems to be a minor constituent of Barcelona productions. Therefore, it is surprising that the only Barcelona products identified in this study do not belong to the BCN-DR/PI groups, but to the BCN-SC group, being the first identification out of Barcelona.

The final three samples from Gran Canaria analyzed by INAA remain unassigned (Table V). The three samples can be considered as different from one

Table V. Average Concentrations of Elements in the Ceramic Sample from Gran Canaria Analyzed by INAA.

INAA	*Ligurian (n=4)*		*Portugal (n=2)*		*Manises (n=7)*		*BCN-SC (n=5)*		*Seville (n=15)*		*MJ0284*	*MJ0269*	*MJ0294*
	mean	*s.d.*	*mean*	*s.d.*	*mean*	*s.d.*	*mean*	*s.d.*	*mean*	*s.d.*	*(Sgrafitto)*	*(Delft)*	
As (ppm)	15.71	7.38	32.59	10.16	19.22	1.74	33.16	13.70	12.93	3.86	5.59	12.48	5.64
La (ppm)	26.72	1.37	26.70	2.15	35.86	0.74	36.24	2.26	31.12	1.44	39.97	23.62	34.72
Lu (ppm)	0.34	0.02	0.30	0.01	0.36	0.03	0.37	0.03	0.32	0.04	0.36	0.26	0.37
Nd (ppm)	23.88	0.88	24.80	2.10	31.10	1.17	30.75	0.84	27.34	1.94	34.48	21.19	27.77
Sm (ppm)	5.04	0.13	5.05	0.27	6.34	0.15	6.33	0.39	5.58	0.28	6.42	4.17	5.98
U (ppm)	2.41	0.27	2.86	0.25	3.26	0.38	2.63	0.57	2.34	0.28	2.34	1.62	2.98
Yb (ppm)	2.22	0.10	1.90	0.27	2.49	0.06	2.69	0.21	2.42	0.23	2.47	1.65	2.46
Ce (ppm)	55.19	2.85	55.75	4.25	69.86	1.93	72.34	4.48	62.01	3.30	83.81	49.14	66.39
Cr (ppm)	300.86	28.09	80.44	4.44	64.84	2.61	65.83	6.61	75.08	9.28	159.17	74.31	81.59
Cs (ppm)	3.65	1.28	3.26	0.07	9.55	1.11	6.86	1.80	5.28	0.73	8.46	5.62	9.94
Eu (ppm)	1.03	0.04	1.07	0.09	1.20	0.03	1.22	0.09	1.15	0.07	1.44	0.85	1.18
Fe (%)	4.04	0.24	2.77	0.08	3.21	0.16	3.23	0.31	3.12	0.26	5.65	2.31	3.73
Hf (ppm)	4.15	0.51	4.76	0.56	4.77	0.24	5.00	0.24	5.63	0.62	4.26	4.39	4.67
Rb (ppm)	60.24	33.27	50.33	12.64	130.56	8.78	104.82	30.25	71.52	16.74	151.88	38.82	141.29
Sb (ppm)	2.00	0.59	0.64	0.04	1.30	0.18	5.03	5.95	1.32	0.47	0.81	0.55	1.23
Sc (ppm)	14.36	0.87	10.24	0.82	11.41	0.38	12.43	0.93	11.25	0.77	19.65	8.58	14.15
Sr (ppm)	675.80	180.17	359.44	15.92	350.30	44.39	246.99	54.81	497.76	73.24	179.79	530.99	684.96
Ta (ppm)	0.91	0.14	1.40	0.12	1.19	0.07	1.13	0.18	0.95	0.08	1.63	0.79	1.34
Tb (ppm)	0.77	0.03	0.64	0.05	0.84	0.06	0.83	0.07	0.76	0.07	0.82	0.55	0.82
Th (ppm)	8.69	0.55	8.08	0.82	11.49	0.36	12.10	0.71	9.31	0.47	12.64	7.51	12.31
Zn (ppm)	107.93	6.32	66.38	5.59	62.96	2.24	86.75	13.45	81.89	36.03	141.04	62.26	111.58
Zr (ppm)	120.05	9.50	145.97	11.39	132.42	8.15	121.74	19.08	152.46	26.59	115.56	137.27	116.90
Al (%)	6.66	0.29	5.76	0.42	7.04	0.36	6.91	0.43	5.85	0.41	9.15	5.02	8.53
Ba (ppm)	238.58	63.12	159.77	0.31	352.60	20.74	345.00	38.21	305.12	38.32	293.19	148.22	304.53
Ca (%)	12.50	3.17	19.93	0.80	15.57	1.02	16.71	1.29	13.97	1.42	5.19	16.23	14.54
Dy (ppm)	3.79	0.25	3.33	0.05	4.44	0.26	4.42	0.25	4.07	0.21	3.95	3.01	4.55
K (%)	1.32	0.49	0.96	0.35	2.45	0.18	1.98	0.48	1.41	0.39	2.18	0.91	2.21
Mn (ppm)	486.05	92.13	245.67	48.92	501.86	76.11	542.80	76.43	592.82	140.48	1104.80	319.91	224.66
Na (%)	0.94	0.14	0.74	0.24	0.26	0.03	0.41	0.12	0.69	0.19	0.52	0.72	0.29
Ti (%)	0.35	0.02	0.45	0.02	0.34	0.01	0.34	0.02	0.34	0.04	0.47	0.23	0.37
V (ppm)	103.99	7.48	57.67	1.19	93.16	20.51	93.05	14.25	88.12	13.52	132.46	76.11	90.24

another, rather than as comprising a new group. Labeled with an asterisk in Figure 8, sample MJ0284, previously analyzed by XRF, is the only *sgrafitto* ware in this study. According to XRF and INAA results, it is not related to any other materials. Sample MJ0269, labeled with a circle in Figure 8, was archaeologically classified as a Delft type. The fact that it did not cluster with any of the reference groups suggests, in the absence of other data, an actual Dutch provenance. Finally, sample MJ0294, labeled with a square in Figure 8, is the only green and black majolica in this study. Although it has not been assigned to any previously known group, its chemical composition is consistent with the Lleida 2 reference group. However, the Mn content, which is around 1/3 of the Lleida 2 reference group, clearly precludes its assignment to this group.

In terms of types of pottery, the results show that lusterware recovered at Gran Canaria had origins in Manises, Barcelona, and Seville. In the case of Manises, lusterware is the only type found in Gran Canaria. This seems to be reasonable, given that the most important production from Manises is lusterware. And, it was the most luxurious ceramic ware until the beginning of the 17th century (*20*). Regarding the Barcelona (BCN-SC) pottery, the occurrence of lusterware also seems to be logical given that this is the only known type from Barcelona. Surprisingly, this lusterware sample is grouped with a Catalan blue-on-white vessel. Therefore, the results from Gran Canaria are not just the first identification of BCN-SC products out of Barcelona, but also provide the first example of a non-lusterware that has been identified as a product from Barcelona's BCN-SC group. Finally, two lusterwares can be attributed to the Sevillan production.

The Sevillian production center is by far the most represented group in Gran Canaria. Moreover, it is very heterogeneous in terms of types of pottery documented. In addition to lusterware, Sevillan white (n = 9), Columbia simple (n = 4), Blue-on-white (n = 4), Ichtuknee blue-on-blue (n = 3), and Isabela polychrome (n = 1) are the other majolica types identified. In addition, one blue tile also has a Sevillan origin. A special case is raised by one sample, MJ0289, classified as Delft.

In fact, as mentioned above that sample MJ0269 was archaeologically classified as a Delft product. The fact that chemically this sample does not cluster to any defined group seems to support an actual Dutch origin at Delft, even though we do not have any reference groups for Delft products. On the contrary, sample MJ0289, which chemically corresponds to Seville reference group, might be a Delft imitation. The existence of such imitations has been hypothesized on historical and archaeological grounds (*21*) and has now been confirmed archaeometrically.

A different problem related to Sevillian pottery is that several specimens are not considered to be true majolica, but rather a lead-glazed pottery. A few of these samples have been considered in the present study, five of which are clearly of a Sevillian origin (2 honey samples, 2 green samples and 1 blue sample). Moreover, as discussed above, XRF indicate that five of the 7 ungrouped samples could be Sevillan products, but contain significantly lower concentrations of CaO and Sr. All five were lead-glazed (3 honey samples and 2 green samples). All this, clearly suggests that Seville was also a production center, and that it played a significant role in this trade network.

Regarding the Columbian sample type, all of them can be classified as a Sevillan as it has been explained above, and the only exception is the sample MJ0260 whose provenance cannot be inferred.

The blue-on-white type is present, as stated above, at the BCN-SC and Seville groups, but also in the Portuguese (4 samples) and Ligurian groups (7 samples). Hence, this type of majolica exhibits the highest dispersion concerning its origin among the pottery from Gran Canaria. The Italian group also includes one polychrome specimen.

Finally, the *sgrafitto* sample could not be assigned to one of the compositional groups. However, it is noted that the Fe content is the highest for this sherd among the studied materials. This feature, together with a relatively medium Ca content, provides a characteristic red color for the clay body. This color, which would be undesirable in majolica, is a desired effect for *sgrafitto* pottery (*22*).

Conclusions

The origin of the American trade monopoly can be attributed to the establishment of the *Casa de la Contratación* in Seville in 1503. This organization directly controlled all trade of goods that were shipped to the Americas from the Castilian Kingdom. Consequently, between the 15th to 18th centuries Sevillan potters produced the majority of the majolica exported to the Americas (*23*). For that reason, Sevillan pottery has a high occurrence in the archaeological records of the Canary Islands and in sites in the Americas. The present study confirms that most of the ceramics from the sites of La Cueva Pintada and the San Francisco's convent are in agreement with a Sevillan origin. Nevertheless, Seville was not the only production center that was exporting pottery shipped to the Americas.

The historical sources, especially numerous notarial and commercial manuscripts generated by the *Casa de la Contratación* and preserved in the *Archivo de Indias,* tell us about the existence of a strong control in this trade network. In the same sense, the repetitive pleas by the Canary Island merchants to increase their market share with America also point to such a situation. The occurrence of pottery of different origins might then be explained by several possibilities that do not exclude each other. On the one hand, despite the very tight control of the Spanish monopoly, a vigorous smuggling activity could be plausible based on the archaeological and archaeometric studies. It has to be highlighted that, from the Iberian Peninsula (except Portugal), only Castilian merchants had license to the trade with America, therefore no one else could legally provide pottery from other sources. On the other hand, objects of non-Sevillan origin could have been included in cargos by Castilian merchants themselves, when the potters from Seville were unable to provide a comparable product (*21, 24*).

The present study suggests the arrival of ceramics and possibly other goods, from different places around the Mediterranean, such as Italy, Manises, Barcelona, Portugal, and possibly Holland. The extent of the participation of all these different productions centers, and their impact and the variety of products involved in this process remains to be determined. However, this study shows that the trade network, the most important one at that time, was far more complex than previously assumed. The exact mechanisms by which the different pottery enters these flows, whether it is licit or illicit, is not easily understood, but the work must be extended in order to obtain a more comprehensive picture. The Canary Island's role in this network becomes, then, a central subject of research.

Acknowledgments

The work presented here forms part of the project "*Identification, Recovery and Improvement of ancient Mediterranean ceramic manufacturing technologies for the reproduction of ceramic artifacts of archaeological value* (CERAMED)", from the Research Directorate General, European Commission, European Community (ICA-3–CT-2002–10018). Javier Garcia Iñañez is also indebted to the support of *Departament d'Universitats, Recerca i Societat de la Informació de la Generalitat de Catalunya i Fons Social Europeu*. Authors are also indebted to the *Museu de la Ceràmica de Barcelona*, Museo Arqueológico de Sevilla, Museo Canario (Las Palmas de Gran Canaria) and Parque Arqueológico de la Cueva Pintada (Gáldar, Gran Canaria) for their collaboration providing the samples. This work was also supported in part by National Science Foundation grant #0504015.

References

1. Buxeda i Garrigós, J.; Madrid i Fernández, M.; Gurt i Esparraguera, J. M. In *La ceràmica catalana del segle XVII trobada a la plaça Gran (Mataró)*, Cerdà, J. A., Ed.; Associació Catalana de Ceràmica Decorada i Terrissa: Barcelona, 2001; pp 155 - 170.
2. Buxeda i Garrigós, J.; Madrid i Fernández, M.; Garcia Iñañez, J. *Estudio de caracterización arqueométrica de las producciones de cerámica vidriada de Talavera*; FBG301873; Equip de Recerca Arqueomètrica de la Universidad de Barcelona: Barcelona, 2003.
3. Garcia Iñañez, J.; Buxeda i Garrigós, J.; Madrid Fernández, M.; Gurt i Esparraguera, J. M.; Cerdà i Mellado, J. A. In *Archaeometric characterization of Middle Age and Renaissance tin lead glazed pottery from Barcelona*, EMAC ' 05, Proceedings of the 8th European Meeting on Ancient Ceramics Lyon, 26th-29th October, 2005; Lyon, in press.
4. Garcia Iñañez, J.; Schwedt, A.; Madrid i Fernández, M.; Buxeda i Garrigós, J.; Gurt i Esparraguera, J. M. In *Caracterización arqueométrica de los principales centros*

productores catalanes de cerámica mayólica de los siglos XVI y XVII, Actas del 6° Congreso Ibérico de Arqueometría (Girona 16-19 November 2005), in press; Molera, J., Ed. (Girona 16-19 November 2005).

5. Garcia Iñañez, J.; Buxeda i Garrigós, J.; Glascock, M. D.; Speakman, R. J. In *Archaeometric characterization of Renaissance tin lead glazed pottery from Talavera de la Reina, Puente del Arzobispo and Seville (Spain)*, 36th International Symposium on Archaeometry (ISA 2006), Quebec (CA), forthcoming; Quebec (CA).
6. Molera, J.; Pradell, T.; Salvado, N.; Vendrell-Saz, M. *J. Am. Cer. Soc.* **1999**, 82, 2871-2875.
7. Schwedt, A.; Buxeda i Garrigós, J.; Madrid Fernández, M. In *Study of Technological Differences in Tin Lead Glazes*, paper presented at EMAC ' 05, Proceedings of the 8th European Meeting on Ancient Ceramics Lyon, 26th-29th October, 2005; Lyon.
8. Morales Padrón, F., *Canarias: crónicas de su conquista.* Museo Canario Ayuntamiento de Las Palmas de Gran Canaria: Las Palmas de Gran Canaria, 1978.
9. Sáez Sagasti, J. I.; Rodríguez Santana, C. G.; Onrubia Pintado, J., *La Cueva Pintada: un nuevo concepto de museo.* Servicio de Patrimonio Histórico, Cabildo de Gran Canaria: Las Palmas de Gran Canaria, 2003.
10. Sosa Suárez, E. In *Las cerámicas del antiguo convento de San Francisco de Las Palmas: un modelo cronológico para el estudio de los yacimientos del Archipiélago Canario*, XV Coloquio de Historia Canario-Americana Las Palmas de Gran Canaria, 2002; Ediciones del Cabildo Insular de Gran Canaria: Las Palmas de Gran Canaria, pp 1999 - 2021.
11. Hein, A.; Tsolakidou, A.; Iliopoulos, I.; Mommsen, H.; Buxeda i Garrigós, J.; Montana, G.; Kilikoglou, V. *Analyst* **2002**, 127, (4), 542-553.
12. Glascock, M. D. In *Chemical Characterization of Ceramic Pastes in Archaeology*, Neff, H., Ed.; Prehistory Press: Madison, Wisconsin, 1992; Vol. 7, pp 11 - 25.
13. Aitchison, J., *The statistical analysis of compositional data.* Chapman and Hall: London - New York, 1986.
14. Buxeda i Garrigós, J. *J. Archaeol. Sci.* **1999**, 26, 295-313.
15. Buxeda i Garrigós, J.; Kilikoglou, V. In *Patterns and Process, a festschrift in honor of Dr. Edward V. Sayre*, Van Zelst, L., Ed.; Smithsonian Center for Materials Research and Education: Suitland, Maryland, 2003; pp 185 - 198.
16. MathSoft *S-PLUS 2000. User's Guide*, Data Analysis Products Division, MathSoft: Seattle, 1999.
17. Maggetti, M. *Archaeometry* **2005**, 47, 389-401.
18. Buxeda i Garrigós, J.; Mommsen, H.; Tsolakidou, A. *Archaeometry* **2002**, 44, 187-198.
19. Schwedt, A.; Mommsen, H.; Zacharias, N.; Buxeda i Garrigós, J. *Archaeometry* **2006**, 48, 237-251.
20. Soler, M. P. In *Cerámica española*, Sánchez-Pacheco, T., Ed. Espasa Calpe, S.A.: Madrid, 1997; Vol. 42, pp 135 - 178.
21. Pleguezuelo, A.; Sánchez Cortegana, J. M. In *Transferències i comerç de ceràmica a l'Europa Mediterrània: Actes de les XV Jornades d'Estudis Històrics Locals*, Institut d'Estudis Baleàrics: Palma de Mallorca, 1997; pp 333 - 363.
22. Lazzarini, L.; Calogero, S.; Burriesci, N.; Petrera, M. *Archaeometry* **1980**, 22, 57 - 68.
23. Sánchez Cortegana, J. M., *El Oficio de ollero en Sevilla en el siglo XVI.* Diputación Provincial de Sevilla: Sevilla, 1994; Vol. 65, p 165.
24. Pleguezuelo, A. In *Cerámica y Cultura. The Story of Spanish Mexican Mayólica*, Farwell, R.; Pierce, D.; Pleguezuelo, A., Eds.; University of New Mexico Press: Albuquerque, 2003; pp 24 - 47.

Chapter 22

Intraregional Provenancing of Philistine Pottery from Israel

David Ben-Shlomo

The Institute of Archaeology, The Hebrew University, Mt. Scopus, Jerusalem 91905, Israel

Chemical characterization of pottery is often used to identify imported pottery by comparison to locally-made reference materials. However, identifying intra-regional production centers is less straightforward. In this study Iron Age Philistine pottery from Israel was analyzed by ICP-AES and ICP-MS to identify local production centers and trade patterns between closely located Philistine city sites on the southern coast of Israel (Philistia). Decorated Philistine pottery together with reference materials were sampled from the four excavated Philistine sites as well as from other regional sites. In addition, all of the samples were analyzed by thin section petrography. The results showed that all pottery produced in the region of Philistia had a relatively similar profile. However, in most cases it was possible to differentiate between a coastal-Philistia profile and an inner-Philistia profile. The results of thin section petrography proved to be useful in many cases in which chemical provenancing was inconclusive.

Chemical characterization of ancient pottery is often used to identify imported pottery when compared to locally made reference materials. This approach, based on the assumption of a typical chemical fingerprint per clay source, can clearly distinguish between pottery produced in different geographical regions. However, identifying intra-regional production centers is less straightforward. It is not obvious that one can differentiate between closely-located production centers, especially in a region which is geologically homogeneous. In this study Iron Age Philistine pottery from Israel was analyzed by inductively coupled plasma-atomic emission spectroscopy (ICP-AES) and inductively coupled plasma-mass spectrometry (ICP-MS) to identify local production centers and trade patterns between closely located Philistine sites on the southern coast of Israel (Philistia, Figure 1). An additional technique used was thin section petrography.

Philistine pottery was one of the main components of the new material culture brought to the southern coastal region of Israel by immigrants from the Aegean region and Cyprus during the 12th century B.C. (*1*, *2*, *3*). This pottery is most common in the main Philistine cities (Figure 1): Ashdod, Ashkelon, Ekron and Gath, and appears to a lesser extent in other regional sites and rarely in sites outside Philistia. The Philistine pottery can be divided into the earliest Monochrome style, the Bichrome style (both appear in the Iron Age I, 12th and the 11th centuries B.C.) and the Iron Age II (10th-9th centuries B.C.) ware of red-slipped, late Philistine pottery (also termed Late Philistine Decorated Ware [henceforth LPDW] or 'Ashdod Ware')(*4*). The early stages of this pottery show high resemblance in form, decoration and technology to contemporary Mycenaean wares from the Aegean region and Cyprus (*2*, *3*). Later, there is a decline in the Aegean character of the Philistine material culture.

Previous NAA and petrographic studies showed the Iron I Philistine pottery to be locally made in Philistia (*3*, *5*, *6*). Moreover, several of the wares indicated specified technological aspects of clay selection and treatment; this was especially shown by Killebrew in her petrographic and stylistic study of Monochrome pottery from Ekron (*3*, *7*). Thus, the Philistine pottery represents, at least in the early stages, pottery made in Philistia by potters with some Aegean technological skills.

Several questions remain open including the regional patterns in the manufacture and trade of this pottery. Could the comparison of Philistine pottery from all four excavated Philistine cities result in the identification of more specified production centers of this pottery? The later Iron II Philistine pottery, which has been minimally studied in the past, could be similarly studied to compare it with the earlier wares. Other minor questions included the possible identification of a potter's workshop at Ashdod Area G (*8*). These questions were examined by a combination of several disciplines and methods including both chemical and petrographic analyses. It was expected that the combination of

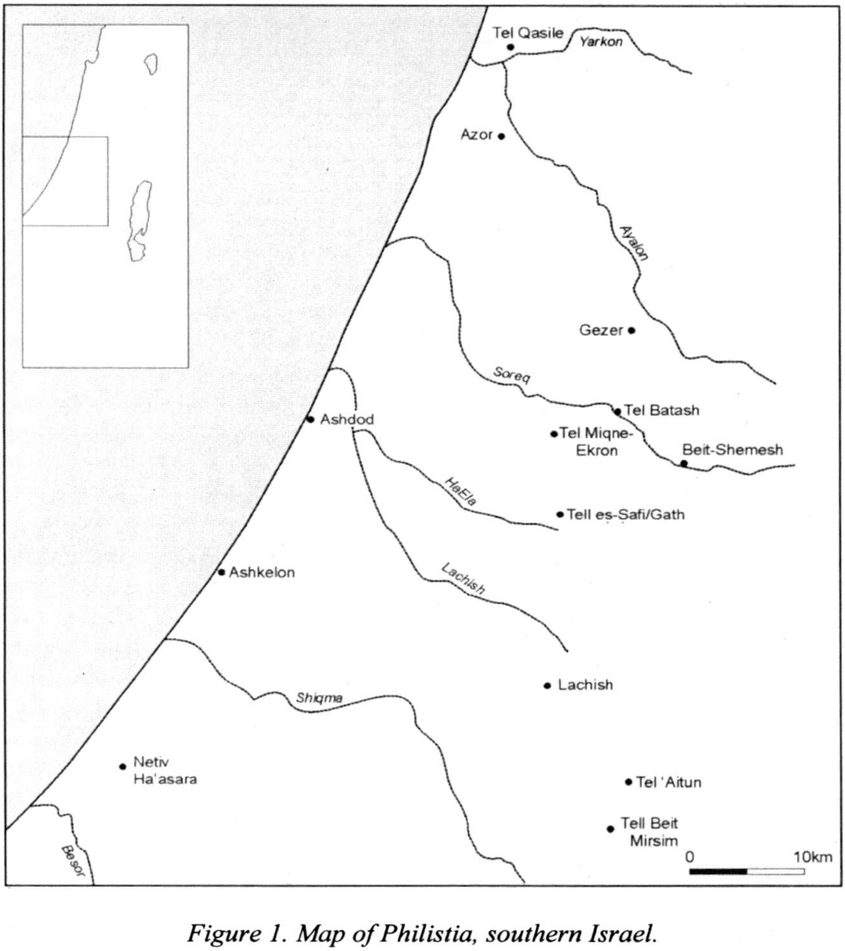

Figure 1. Map of Philistia, southern Israel.

the two techniques would give more meaningful results, as indicated by other provenance studies (*9*, *10*, *11*, *12*, *13*).

Visual examination showed that the earliest Philistine Monochrome had more variability in fabric. A distinct subware was defined as fine Monochrome; it was made of a much lighter colored and highly levigated clay. This subware was most common at Ekron, rarer at Ashdod, and absent at Ashkelon. The Bichrome pottery was more homogenous in appearance and had a wider geographical distribution; the LPDW pottery was abundant only at Ashdod and Gath and less common in the relevant assemblages.

Method

The archaeometric analyses included 225 samples chemically analyzed by ICP and thin section petrography and an additional 100 samples were analyzed only by petrography. The samples originated from 25 different sites, although the majority are from the Philistine cities (Ashdod, Ekron, Gath and Ashkelon). About 30% of the samples were considered to be reference material (e.g., clay samples, vessels from kilns, coarse ware and undecorated, common types) (Figure 2).

The chemical analyses by ICP-MS and ICP-AES were undertaken at the Earth Sciences Department of the University of Bristol, and was made possible by the European Commission Program for Access to Research Infrastructures (Contract No. HPRI-1999–CT-00008). The powdered samples were dried for 12 hours in 110 °C. A 200 ±3 mg sample (or rarely 100 mg if too small) was dissolved in an acid cocktail of 5 ml of HF 40%, 5 ml of HNO_3 (concentration 2.5%), and 1 ml of $HClO_4$. This mixture was left in covered beakers for at least 12 hours and then heated at 100 °C for 2 hours; the solution was then dried at 230 °C. Finally, the sample was leached in distilled water with 1% of HNO_3. The solution was then mixed with HNO_3 and internal standards (10 ppb in solution of Bi, In, Re, and Ru, which are very rare elements in soils) in a glass flask, to 100 ml (a dilution of 1:500) (*4*, *14*). Somewhat similar procedures were adopted in several other cases in which ICP has been used for chemical fingerprinting of ceramics and other materials (*15*, *16*, *17*, *18*). The analytic equipment used for the analyses included a VG PQ2 (Plasma Quad) ICP-MS and a YJ Ultima II ICP-AES The calibration of the ICP-MS was made using five synthetic standards (the cocktail of elements prepared according to the elements and required concentrations) and in ICP-AES according to both synthetic and international rock standards. It was found that rock standards give more accurate results, as they can reflect spectral matrix effects as well (resulting from having many spectral lines in the background from the sample); therefore, these were used for calibration when possible. Elements obtained by ICP-AES were primarily major

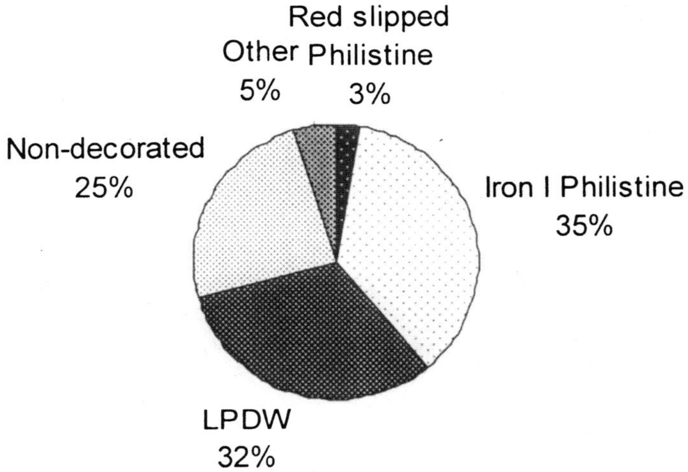

Figure 2. Breakout of chemical and petrographic sampling according to pottery ware groups (total=325; LPDW=Late Philistine Decorated Ware).

and minor elements, including Na, Mg, Al, P, K, Ca, Ti, Mn, and Fe. And, in most samples six trace elements were also obtained (e.g., V, Cr, Co, Ni, Cu, Zn, and Sr). Elements measured by ICP-MS include the heavier trace elements and rare earth elements (e.g., Sc, V, Cr, Co, Ni, Cu, Zn, Rb, Sr, Y, Cs, Ba, Hf, Ta, La, Ce, Pr, Nd, Sm, Eu, Gd, Tb, Dy, Ho, Er, Tm, Yb, Lu, Th, and U).

Large-scale pottery provenance studies employing ICP are uncommon. Therefore, most ICP laboratories have not yet developed a definite procedure for such a study including sample preparation, choice of standards for calibration, choice of elements etc. Here there was an attempt to obtain elements that are accurately measured by NAA as well and use well-known international rock standards, so comparison with results of other labs could be made in the future.

The compositional data was analyzed by multivariate statistics using 24 well-acquired elements (i.e., measured in all samples), free of contamination and dilution effects. These included: Al, Ti, V, Cr, Mn, Fe, Co, Y, La, Ce, Pr, Nd, Sm, Eu, Gd, Tb, Dy, Ho, Er, Tm, Yb, Lu, Hf and Ta. The log-transformed data with various treatments of the raw data, was analyzed by hierarchical cluster analysis, discriminant analysis and principal component analysis.

Petrographic analysis was undertaken for almost all 225 samples and an additional 100 samples from the same pottery groups. The thin section slides were examined through a petrographic polarizing microscope (Nikon Labophot-Pol and Zeiss [for photography] models were used).

Results

Chemical Analysis

When data for all 225 samples were examined using multivariate statistics, a major group of 138 samples including most of the reference material from Philistia was identified (Group I, Figure 3). This group had a spread of around 15% from the mean value in most elements. Another group of 43 samples had a high and highly variable calcium value. A group of 26 samples had a slightly different composition than the major group. Eighteen samples are unassigned. These initial results indicate that the vast majority of all Philistine ware were produced in Philistia. The high-calcium group (Group II, Figure 3) includes most of the fine Philistine Monochrome vessels and the reference samples from Ekron, implying that these vessels had distinct technological features and were manufactured at Ekron. Inspection of log-transformed data resulted in the identification of two sub-groups within this group (Groups 4A and 4B). By using best relative fit values, i.e., obtaining individual dilution factors per each sample, this group is more compact and distinct (Figures 5–6).

The 'best relative fit' examines the possibility of a dilution factor between a given sample (xi) and a group's profile (c_i) as explained by Mommsen (*19*) where:

$$f=\frac{1}{m}\sum_{i=1}^{m}\frac{c_i}{x_i}.$$

A low spread of f (σ_f) indicates a good chemical match and a low dilution factor (a value near 1 means there is no dilution) (*20*). Thus, resolving of dilution effects or factors and chemical matching can be combined. The elemental values in this procedure can be weighted according to errors (and group spread) as well. This data treatment technique is advantageous especially when one of the chemical groups has a clear variable dilutant included in it. However, this technique should also be used with caution as it may obscure some of the more subtle differences between the groups. Note, that in this case (Figure 6) the grouping with best relative fit appears to be somewhat sharper.

The general compositional homogeneity of the pottery from the main Group I can be interpreted in several ways: (1) All of this pottery was produced in a single production center; (2) The group may represent several regional production centers, but the composition of the clay used in the different centers is similar to the degree of the source compositional variability and results in a single chemical profile; (3) The group represents several production centers with closely related, but different, clay profiles; the analytical method, however, is not sensitive enough to distinguish between them; (4) There are different clay

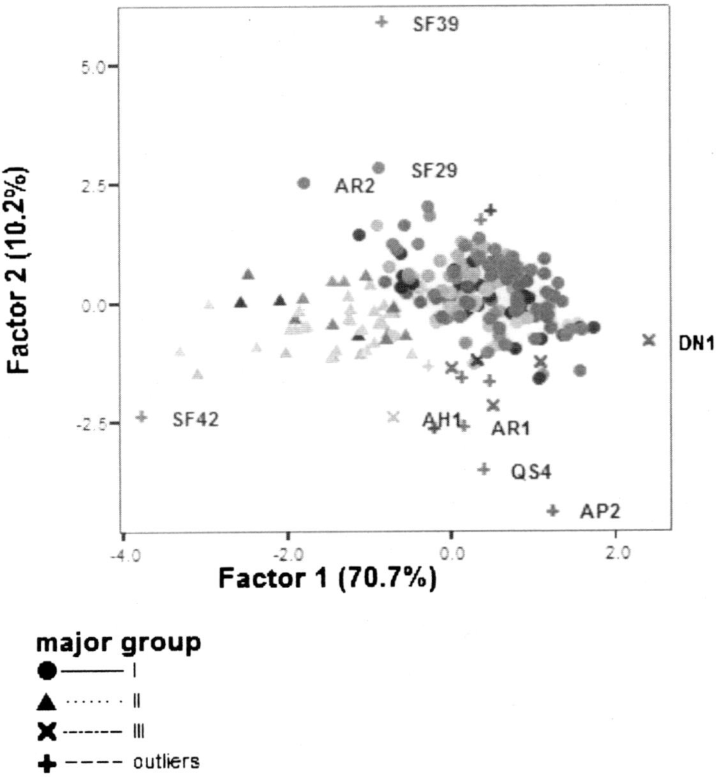

Figure 3. Scatter plot of principal component analysis including all 225 sample showing main chemical groups.

profiles within the main group, but their differences are obscured by dilution effects or larger measurement errors for some of the elements.

All of these possibilities must be considered. Option 1 seems highly unlikely from an archaeological perspective, given that the main group includes pottery from several different major sites including typological groups of decorated and undecorated pottery and a time span of about 400–500 years. It is not reasonable to expect that all these sites produced all their pottery throughout most of the Iron Age at the same workshop (especially since we have archaeological records for at least two workshops, at Ashdod and at Ekron). Option 2 cannot be ruled out considering the geological homogeneity of Philistia, and the fact that several rivers carry clay from the inland to the coast. It is, however, a rather pessimistic conclusion that impedes any further research. Option 3, relating to the precision

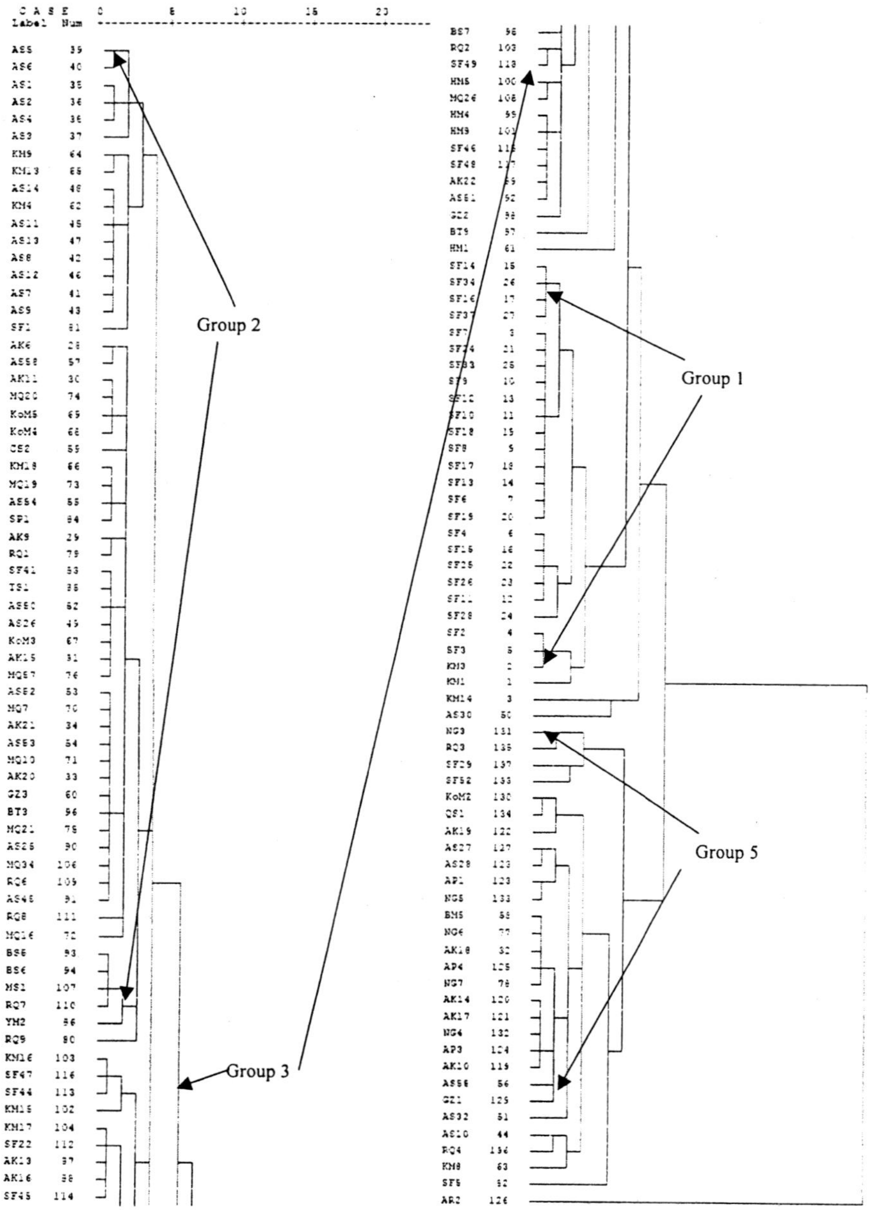

Figure 4. Dendrogram showing sub-division of main chemical group according to cluster analysis (Euclidean distance; Ward's method).

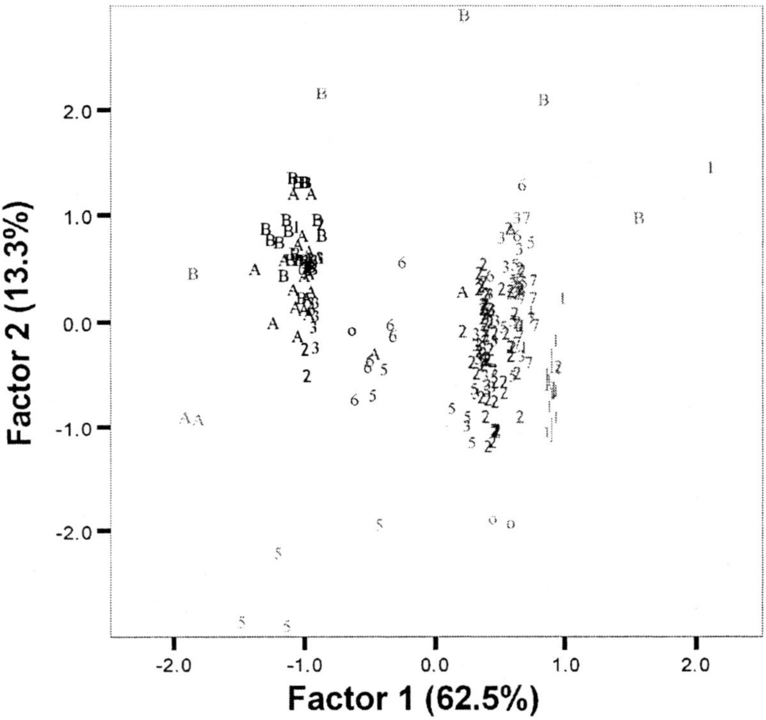

Figure 5. Scatter plot showing principal component analysis for all 225 samples after 'best-relative-fit' factors were applied (symbols indicate chemical subgroups; 'o'=outliers/loners; 'A'= Subgroup 4A; 'B'= Subgroup 4B).

of the analytical method does not seem likely as the relative precision (according to the quality control standards and repeat samples) of all elements used for grouping was 5–6% at the most, which is below the common variance within most chemical profiles of pottery. Therefore, Option 4 was the working hypothesis, and the groupings were made according to the sub-groups seen in the multi-variate statistics, using the petrographic results as a control.

Cluster analysis, employing Euclidean distance, yielded four subgroups (Figure 4, termed here as chemical Groups 1, 2, 3, and 5). These groups were more compact in their composition, and most elements had a spread of 5–10% from the mean values (see Table I). Although compositional differences in many elements between the subgroups is not large as in Fe, Al, La or other elements, one can see distinct differences in more specific elements. For example, samarium and tantalum differentiate Groups 1, 2, 3 and 5. Discriminant analysis

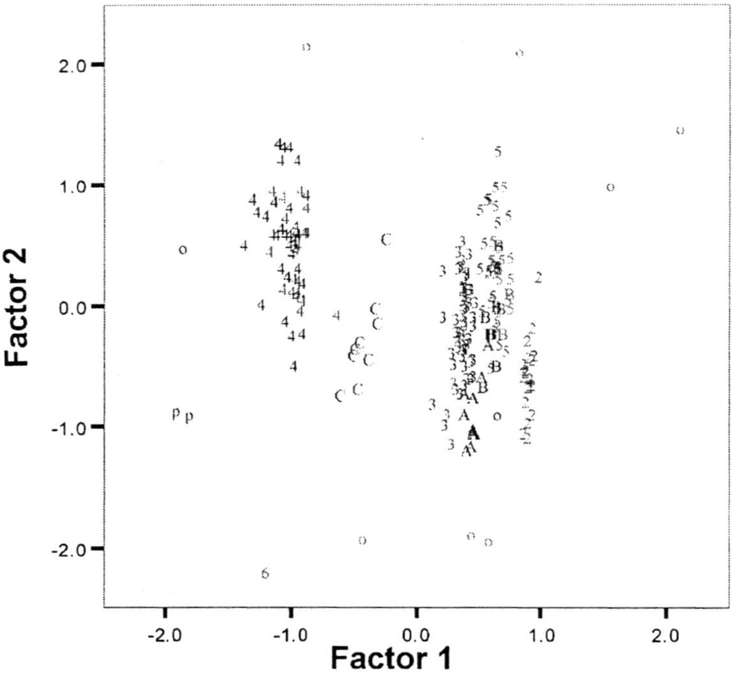

Figure 6. Scatter plot showing principal component analysis for all 225 samples after 'best-relative-fit' factors were applied; symbol shape indicates best-relative-fit groups ('A'= 1a; 'B'=1b; 'C'=3a; 'o'=outliers/loners; 'p'=pairs).

according to the subgroups also indicates a viable difference between Groups 1 and 2 as well (Figure 7); Groups 3 and 5 are less distinctive.

Group 1 includes 27 samples which cluster together. The group is relatively chemically compact and based on 21 of the measured elements (including Sc and Th not measured in other groups) varying under 10% and six elements under 20% (mostly under 15%). The group's average has a moderate concentration of Ca (5.17% with a 43% spread) and relatively high Hf (4.64 ppm) and Ta. This profile seems to represent clay sources from the inner plains, most probably from the site of Gath or its vicinity; this is according to the reference material from the site. The group includes most of Gath reference group of plain common pottery, most of the LPDW vessels from Gath.

Group 2 includes 59 samples that cluster together. The group is not as compact chemically as Group 1, though still consistently compact. Eighteen elements vary under 10%, ten under 20% and only two over 20%. The Ca values are quite similar to Group 1 (5.88% with 41% spread), the Al and Fe (as most

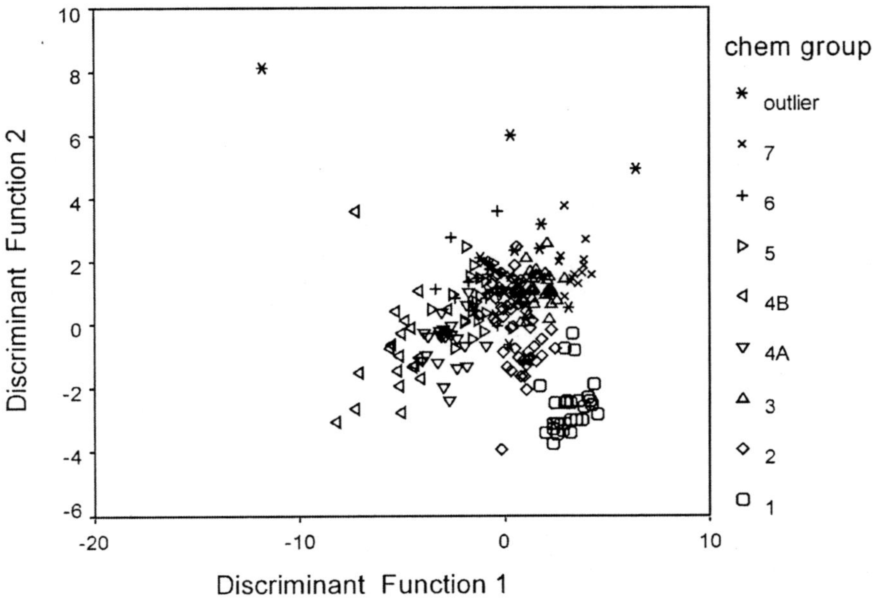

Figure 7. Scatter plot of two major discriminant functions of data set according to chemical grouping.

other elements) are slightly lower at 5.4% and 3.88% respectively; although all the elements are in a lower concentration of 5–15% compared to Group 1, this is not due to a constant dilution factor. Moreover, Hf is considerably lower (2.86 ppm) as is Ta, while Co and Y are practically identical in the two groups. Chemical Group 2 is relatively homogenous considering the petrographic groups represented: thirty-three samples belong to petrographic Group A (see below).

An attempt was made to compare the current ICP results to earlier NAA results when possible. Eight vessels from Ashdod previously analyzed by NAA were analyzed by ICP; all belong to chemical Sub-Group 2 (the elemental results were obtained with the permission and assistance of Prof. Frank Asaro). These vessels were mostly assigned to Group 1a of Perlman and Asaro (*21*) also indicating an Ashdod or coastal provenance. The fact that all samples previously provenanced by NAA to Ashdod belong to chemical Group 2 further strengthens its identification with a production center at Ashdod.

Group 3 clusters relatively close to Group 1 (see overlap in dendogram, Figure 4). Generally, the average values for most elements in Group 3 are either slightly higher than in Group 1 or lie in between the values of Group 1 and Group 2. However, Ti, Co, Cr, Mn and Y are considerably higher than in Groups

Table I. Means and standard deviation, as percentage of mean, of chemical Groups 1–5

	Group 1		*Group 2*		*Group 3*		*Group 4A*		*Group 4B*		*Group 5*	
Element (ppm)	*Mean (27)*	*CV%*	*Mean (59)*	*CV%*	*Mean (32)*	*CV%*	*Mean (23)*	*CV%*	*Mean (20)*	*CV%*	*Mean (20)*	*CV%*
Al(%)	6.01	5.8	5.50	5.5	5.85	5.8	4.12	6.2	3.33	10.0	5.06	9.4
Fe(%)	4.55	7.2	3.91	7.4	4.18	7.5	2.86	7.0	2.25	9.3	3.55	11.6
Ca(%)	5.17	43.4	5.87	39.1	5.93	30.1	13.09	29.5	18.08	27.9	4.44	37.1
Ti(%)	0.64	8.0	0.60	8.8	0.65	10.8	0.43	9.9	0.34	14.4	0.59	12.0
K(%)	1.14	16.5	1.29	16.3	1.37	21.3	1.33	22.2	1.37	20.5	1.40	21.9
Na(%)	0.65	18.4	0.61	11.4	0.68	24.9	0.45	20.1	0.35	24.7	0.59	14.3
Co	18.36	8.2	18.39	11.6	20.77	10.2	12.32	12.5	9.30	21.7	17.87	12.3
Cr	90.21	8.3	89.33	12.9	99.80	8.7	76.17	7.8	67.93	14.1	86.50	14.4
Mn	741	8.7	729	15.6	814	10.4	507	13.4	401	23.5	662	10.8
Sr	318.6	19.3	296.8	24.9	314.1	18.8	412.2	20.0	434.8	27.6	272.4	17.3
V	118.6	10.1	103.3	20.3	101.5	8.5	70.6	11.8	63.9	14.7	85.6	10.9
La	30.6	5.9	27.9	6.7	31.0	6.9	24.4	7.5	21.2	8.2	24.1	8.1
Ce	63.7	4.9	57.8	5.5	64.0	6.7	46.4	7.5	39.2	7.5	52.6	7.1
Pr	7.28	6.1	6.77	4.9	7.54	5.0	5.55	6.8	4.89	6.9	6.03	7.7
Nd	29.35	6.1	26.80	4.5	29.66	4.3	21.69	6.1	19.10	7.5	23.16	8.3
Eu	1.44	5.8	1.33	6.1	1.46	6.2	1.13	9.01	0.94	9.55	1.12	10.3
Sm	6.27	5.8	5.57	5.4	6.11	5.1	4.58	6.87	3.97	9.38	4.78	8.5
Tb	0.78	7.7	0.70	5.8	0.76	6.1	0.58	6.79	0.51	9.14	0.58	9.9
Gd	5.14	6.5	4.75	5.1	5.24	5.1	3.97	4.87	3.47	8.81	3.97	9.7
Dy	4.36	8.1	3.98	5.6	4.42	4.9	3.46	5.75	3.01	8.54	3.31	9.6
Ho	0.85	8.5	0.75	5.3	0.83	6.2	0.63	7.49	0.57	9.28	0.63	9.3
Er	2.34	10.0	2.08	5.7	2.29	6.8	1.80	6.98	1.61	9.43	1.72	8.4
Tm	0.33	8.5	0.31	7.8	0.36	11.5	0.27	10.00	0.26	10.41	0.29	12.0
Yb	2.10	9.9	1.90	5.1	2.13	7.7	1.71	7.04	1.52	9.51	1.64	8.1
Lu	0.31	9.1	0.28	6.9	0.31	9.7	0.24	7.99	0.22	10.38	0.24	8.5
Y	20.74	14.9	21.03	13.9	24.81	10.3	21.50	8.10	19.12	12.04	18.52	7.5
Hf	4.64	8.8	2.85	15.1	2.88	8.8	2.21	8.82	1.70	15.40	2.41	15.5
Ta	1.85	11.4	1.16	11.1	1.34	10.1	0.84	13.66	0.63	21.99	1.04	26.3

1 and 2, while V and Hf are low, similar to Group 2. The Ca value is slightly higher at 5.93%. Group 5 is not as compact as the other groups. The Ca is slightly lower than Groups 1–3 at 4.4%; fourteen elements vary below 10%, eleven under 20%, K is high at 1.4% and varies 21.9% and Ta varies 26%. Most elements are about 10% lower than in Group 2, and Mn, V, Eu, Gd and Dy are even lower (20–30% lower than Group 2); Hf is also even lower than Group 2 at 2.4 ppm. According to the petrographic analysis, several members of this group were made of clay with more *loess* soil (see below, Petrographic Group B, Table III). Therefore, a provenance of southern Philistia—the vicinity of Ashkelon (or a hypothetical provenance of Gaza) may be suggested. Interestingly, several Philistine vessels from northern Philistia (Aphek and Qasile) belong to this group as well.

Using 'best relative fit' on the raw data produced a slightly different clustering, with the new subgroups (denoted by symbol shapes on Figure 6), naturally, being more compact. The main difference between the regular groups (denoted by symbol shape in Figure 5) and the 'best relative fit' groups is that Group 5 becomes more distinct, while Groups 6 and 7 seem to collapse into the main group. Groups 1 and 2 remain distinct and Groups 4A and 4B together are clearly separated as one unit. It should be stressed that these sub-groupings are tentative, indicating potential groups, and are used as a starting point for further research and analysis.

According to the reference material and the petrographic features of the respective samples, two groups were tentatively assigned to the coastal region of Philistia: Group 2 (representing Ashdod) and possibly Group 5 (maybe representing Ashkelon). The two other groups, 1 and 3 were assigned to the inner plains of Philistia, probably originating from the region of Gath. As noted, Groups 4A-4B were assigned to Ekron. This seems to be the limit of the chemical fingerprinting of this region at this stage. The coastal or inland cities themselves: that is Ashdod and Ashkelon or Ekron and Gath, which are only nine km apart, cannot be distinguished by chemical fingerprinting.

The samples taken from the alleged workshop at Ashdod, Area G, were not more homogenous in their composition than the general group of samples from the site. Thus, the identification of the locus as a workshop could not be substantiated.

Chemical Subgroups 6 and 7 possibly represent clay from southern Israel, though not enough reference material from the sites was analyzed to achieve better provenancing. Of the other samples several group together (see Figure 3); these are mostly vessels from northern Israel resembling Philistine pottery; the petrographic analysis shows these vessels to be produced in the north as well. Other outliers may represent various clay recipes not fitting any group but still local to Philistia, or imported vessels.

Petrographic Analysis

Petrographic analysis was undertaken for nearly all 225 chemically analyzed samples in addition to about 100 samples from the same pottery groups. It was anticipated that thin section petrographic analysis would complement the chemical analysis and assist in defining the intra-regional production centers. The region under investigation, Philistia, is relatively small, and most distances between sites are less than 20 km. The geology of this region illustrates several aspects: the coastal strip of about 10–15 km is relatively homogenous and covered with brown soil, sand dunes and kurkar ridges. The southern part of this coast yields more *loess* type soil, which originates in Nilotic alluvial sediments coming from the southeastern Mediterranean. The eastern part of Philistia is called the Inner Plains or Inner Philistia. Its western part lies still on the same brown soil but its eastern part shows a change in the geological formation exposing carbonatic Eocene formations with chalk and limestone. Several rivers, some adjacent to the Philistine sites, carry clays from the inner plains to the coastal area making the situation more complicated, though the coastal soils should still be more sandy with less calcareous inclusions. The question is whether different ceramic fingerprints can be identified within the same geological formations.

Most of the samples were represented by five petrographic groups with several subgroups therein (see Table II). These fabrics probably represent clay which could be found in the region of Philistia. The largest group was of a fabric representing a porous, optically inactive clay matrix with a high quantity of quartz inclusions (Group A; Figure 8:A-B). This fabric can be described as a non-calcareous fabric, usually with a non-active, relatively dark matrix. This clay is quite porous with usually 20–30% voids, with a single to double-spacing. The voids are sometimes aligned in a laminated fashion testifying to some organic matter used in the clay; the voids sometimes represent decomposed calcite as well (with visible calcite margins). The silty component is of coarse silt. The predominant component of the inclusions is quartz, consisting in most cases of 15–25% of the slide. The quartz inclusions sometimes show poly-crystalline texture with cracking (due, possibly, to high temperature firing). Other inclusions are much more rare and include few angular limestone or chalk fragments in fine sand size (kurkar chunks are very rare), ferrous minerals (rounded shape) and mica (usually sub-angular) both in medium silt to fine sand sizes. A few feldspar inclusions (usually angular up to 100 microns) and other heavy minerals as hornblende and zircon also occur but very rarely and mostly in worn conditions.

This fabric, probably deriving from brown soils, could be further subdivided into two subgroups, according to the texture of the quartz inclusions and the frequency of various calcareous inclusions as limestone and chalk fragments. Subgroup A1 (Figure 8:A) has a bimodal texture of quartz comprising of a coarse silt-very fine sand (30–80 microns) size and angular element and a fine-

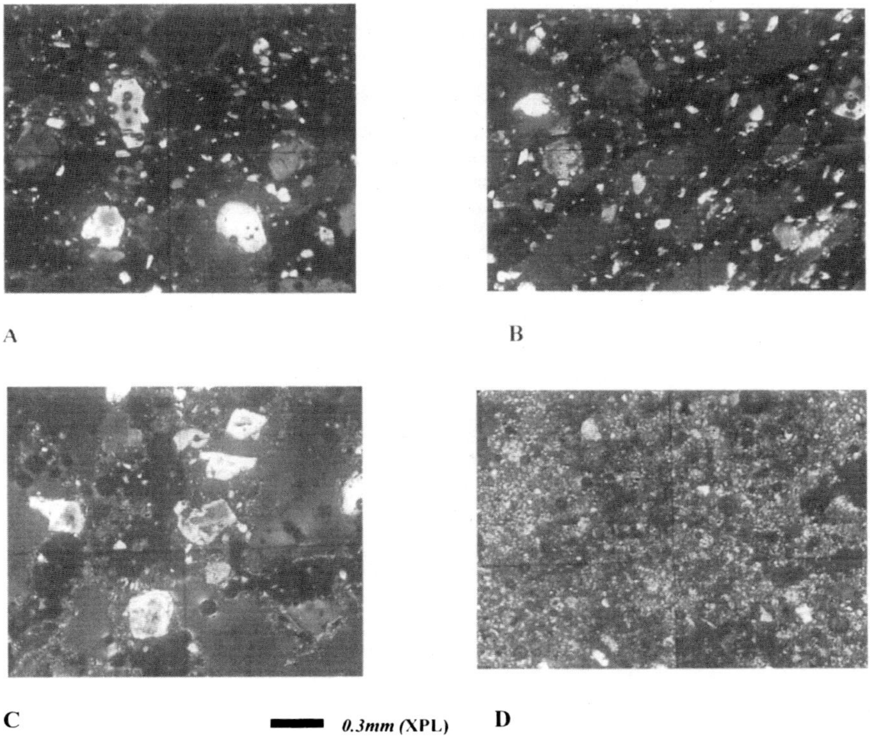

Figure 8. Thin sections of pottery sherds (XPL): A. Petrographic Group A1; B. Petrographic Group A2; C. Petrographic Group B; D. Petrographic Group C1.

medium sand (120–350 microns) component, usually with rounded or sub-rounded shape; hardly any calcareous inclusions occur. This subgroup was tentatively assigned to a coastal origin on both petrographic analysis and chemical analysis of the reference material; somewhat similar petrographic groups were defined at 7th century BCE Ashkelon (*22*). Subgroup A2 (Figure 8:B) did not show this bimodal texture and had quite more calcareous inclusions. Similarly, Subgroup A2 was identified as having an inland provenance. Another group, Group D, also represents a similar soil, with a higher well-sorted component of angular coarse silt quartz component. This group could not be geographically traced within Philistia. Petrographic Group B (Figure 8:C) represents clay made of *loess* soil, identified by its silty carbonatic matrix; this

Table II. Characteristics of the main petrographic groups

Group	*No. of samples*	*Soil*	*Matrix*	*Main inclusions*	*Proposed provenance*
A1	75	Dark brown (quartzite)	Inactive, moderately silty	Bimodal quartz (coastal sand)	Coastal
A2	38	Dark brown (quartzite)	Inactive/slightly active, moderately silty	Moderately sorted quartz, calcareous	Inner plains
B	29	Loess	Calcareous, silty	Quartz, feldspar, heavy minerals	Southern Israel
C1	33	Loess/grumusol?/ rendzina	Calcareous, fine, silty, compact	Calcareous, low quartz, ferrous	Inner plains
C2	8	Loess/rendzina	Calcareous, fine, silty, compact	Foraminifers (chalk)	Inner plains
D	17	Dark brown/hamra?	Inactive	Sorted silty angular quartz	Southern Israel
E	19	Brown/ Terra Rossa?	Inactive/slightly active, reddish	Quartz, calcareous, ferrous	Southern Israel

soil is present in southern coast of Israel, although appearing in more inland locations as well (*22, 23*). The bimodal quartz, however, suggests a coastal origin.

Another petrographic group (C) is quite different than other groups (Figure 8:D). The matrix is usually active, particles are double to open spaced, the fine silt component is moderate to very high, and the voids are lower than previous groups, at 5–15% in most cases. A distinct characteristic of the inclusions of this petrographic group is the relatively low quantities of quartz, rarely above 10% of the slide area. The quartz inclusions are poorly to moderately sorted, very fine to fine sand size (50–100 microns) and in variable shapes, usually angular to sub rounded. The calcareous inclusions become more dominant, usually this component is 5–10% of the slide, but in several cases 20%. This includes moderately sorted limestone/calcareous concentrations fine sand (60–120 microns) fragments, of sub-rounded shapes; few larger particles (up to 500 micron) also occur. Fine to medium sand chalk inclusions of rounded shapes are also common in the samples. In addition foraminifers appear in various quantities, from several inclusions up to 15% of the slide area. Most foraminifers are rounded and are fine sand in size.

This fabric, overlapping the high calcium chemical group, represents a well-levigated clay possibly originating from the Eocene Maresha and/or Adulam formations which have outcrops less that two km from both Ekron and Gath. This could be some sort of a mixture of soils occurring in the border zone between the coastal plains and the southern Shephelah: brown/dark brown soil, loess and pale rendzina. The fabric is similar to Killebrew's Miqne-A1 fabric (*7*, *24*), though the exact clay source was not identified. Note that the calcareous inclusions are imbedded in the clay matrix and are not intentionally added. The fifth petrographic group, Group E, was less distinctive and relatively similar to Group A. It had a more reddish color of the clay, possibly representing a Terra Rossa soil source. It should be stressed that in many cases the differentiation between the groups or at least some of them was difficult, leaving a significant portion of the samples as petrographically indecisive.

The samples analyzed from an Iron Age II kiln site of Kfar Menahem near Gath posed a few questions concerning some of the reference material. This site was excavated during 2001 in a salvage excavation of the Israel Antiquities Authority, directed by Ygal Israel, and unearthed a series of four well-planned rectangular pottery kilns dated to the 8th century BCE (*14*, *25*). The samples show a relatively variable chemical and petrographic profile; thus, it was not clear which profile should be used as the reference material for this production center. This illustrates that the use of vessels from kiln sites as reference material should also be made with some criticism. An explanation for this phenomenon could be that in a production center there would be more variance in clay sources as several recipes are used and experimented with regularly. A similar phenomenon was reported for pottery kilns at Late Minoan Kommos, Crete (*26*) and in modern workshops in northwestern Spain (*27*). Alternatively, this site was not used in its final stage as a pottery workshop. Further analysis of the material from the Kfar Menahem kiln site may aid in resolving this issue.

Notwithstanding these difficulties, the results of thin section petrography proved to be useful in many cases in which the chemical provenancing was inconclusive. Generally, if the petrographic conclusions are plotted against chemical groupings, about 90% of the decisively designated samples agree (Figure 9, Table III). Petrographic Groups A1 and B correspond to chemical Groups 2 and 5, respectively; petrographic Group C to corresponds chemical Group 4 and petrographic Group A2 corresponds to chemical Groups 1 and 3.

Discussion and Conclusions

Provenance studies of pottery from various regions around the globe indicate that identifying distinct chemical or compositional profiles of pottery production centers, which are located in a relative proximity to each other, is not simple (*9*, *11*, *12*, *22*, *23*, *26-30*). While the success of identifying such profiles

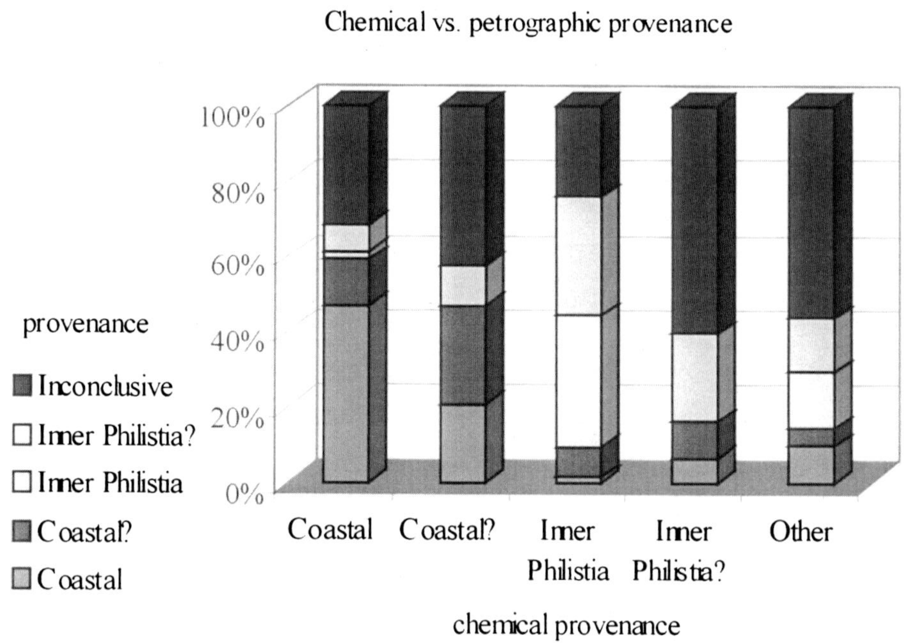

Figure 9. Comparison between the chemical groups and the proposed provenance according to petrography.

Table III. Samples from the petrographic groups according to major sites, typological groups and chemical groups

Petrographic group	*A1*	*A2*	*B*	*C1*	*C2*	*D*	*E*	*Other*	*Inconclusive*
Ashdod	30	0	6	5	2	7	0	1	8
Ashkelon	12	0	3	1	0	0	0	1	5
Tel Miqne-Ekron	5	9	4	24	5	1	0	2	10
Tell es-Safi/Gath	4	14	3	3	0	4	5	14	9
Monochrome fine	0	0	0	22	5	0	0	1	5
Monochrome gray/red	11	3	1	3	1	8	0	0	10
Philistine Bichrome	5	1	3	2	3	3	1	6	9
Late Philistine	30	11	14	0	0	2	6	7	21
Chemical Group 1	1	10	2	0	0	1	2	6	2
Chemical Group 2	27	3	7	1	0	5	3	1	10
Chemical Group 3	2	4	3	0	0	6	1	4	11
Chemical Group 4A/4B	2	2	3	18	4	0	0	2	10
Chemical Group 5	4	1	5	0	0	1	2	2	4
Proposed provenance	Coastal Philistia	Inner plains	Souther n Israel	Inner plain s	Inner plain s	Souther n Israel	Souther n Israel	?	Southern Israel?

was variable in these studies, as in many others, certain issues could be noted. The chemical profiles may differ to a range not far from the level of experimental errors, and when the number of samples is increased the picture may become even blurrier. Obtaining as many elements as possible as well as a combination of chemical and petrographic analysis is desirable to improve the identification and definition of the clay sources. In general, chemical profiles or compositional pottery groups can be classified either hierarchically—according to geographical parameters: that is a wider profile of a larger geographical region and a 'sub-profile' of a sub-region; or continuously—according to technological parameters: a paste may be diluted by temper or mixed with other clay in any amount, creating a continuous range of compositional profiles. Eventually, chemical, petrographic, and archaeological reasoning should be brought together to achieve a suggestion for fine provenancing. Selection of good reference materials, combining both kiln wasters and large homogenous groups of common pottery from the sites, is more effective. The inter-calibration of chemical results from various laboratories and techniques is vital as well. Measuring of absolute elemental compositions rather than relative ones, achieved by using the same international rock standards for calibration in all laboratories, may improve the final outcome. Thus, the ability to achieve intra-regional provenancing of pottery is dependent on both the geological and pedological intra-regional differences and the resolution and comprehensiveness of the archaeometric analysis.

The archaeological conclusions of this provenance study indicate that during the Iron IA period a distinct sub-type of Philistine Monochrome pottery, defined here as fine ware, was produced in Ekron. It seems that the workshop at Ekron specialized in this sub-ware, which has the highest resemblance to Mycenaean pottery, and exported it to the other Philistine sites, especially to Ashdod. During the Iron IB period, there was less demand for the fine Monochrome pottery made from the special clay recipe, which was hard to procure and to produce, and therefore, its manufacture was discontinued. The other Monochrome pottery fabrics and the Bichrome wares were made of a more common clay recipe, possibly produced by traditional Canaanite potters. The forms and decorations were more influenced by local pottery traditions, resulting in the development of the Philistine Bichrome pottery. Each Philistine city could provide for its own need for this type of pottery, but occasionally there was some natural trade and movement of vessels between these cities, which were strongly linked to each other. It should be noted that there is no evidence that the Philistine Monochrome pottery production, even at Tel Miqne, was produced on a large scale—factory or large-scale industry, i.e., a larger scale than the one existing in other Iron I pottery workshops. The kilns at Tel Miqne are small and few in number, and although some resemblance to Aegean kilns and workshops is suggested, this is not sufficiently clear in this stage. The standardization of the pottery is relatively minimal and various sub-wares were all produced in the same place. Influences of a higher mode of production was suggested by Killebrew (*7*), relating to the large scale industry of Mycenaean IIIB in the Aegean; an example of such an industry is illustrated by compositional analysis (*31*). While it is not certain that the Mycenaean pottery itself was the result of 'large-scale' production (although it shows high quality and standardization), the production of Philistine Iron I pottery, particularly the Monochrome ware, clearly cannot be viewed as a product of large-scale production, especially, as noted above, due to its low level of standardization.

It is assumed that the Philistine pottery production in Iron I Philistia was conducted solely by Philistine potters, immigrants from the Aegean or Cyprus, who brought with them the technology of Mycenaean pottery production (*1*, *2*, *3*, *7*). While there is no direct evidence on the ethnicity of the Iron I potters in Philistia, this assumption seems likely. However, the option that traditional Canaanite potters living in the Philistine cities also produced some Philistine pottery cannot be altogether dismissed. One may suggest that Philistine wares made of the more common clay (not the fine sub-ware), could have been made by Canaanite potters. The use of calcareous clay for non-Aegean pottery groups at Tel Miqne may indicate that either Philistine potters produced Canaanite pottery as well, or that the Philistine potters influenced Canaanite potters. It could be also suggested that in the initial phase Philistine potters produced their own decorated pottery and the Philistine population obtained other pottery forms from the Canaanite workshops. Later on, the differences between the pottery groups were obscured and workshops included both Canaanite and Philistine potters working together. This could have also resulted from inter-marriages between the populations. The Philistine pottery does not include all classes of vessels in any case, so a certain dependency on Canaanite workshops or traditions always existed for the Philistine population.

The Philistine Bichrome pottery shows much more influence of the local pottery traditions. Its manufacture from regular clays also could have made it easier and cheaper to produce. The mixture of various styles, Aegean, Cypriote, Canaanite and Egyptian would have appealed more to the local non-Philistine population. The potters in Philistia, possibly a second generation to the immigrants, conformed more to the local pottery tradition by this stage (*32*). At the same time, rising demand for decorated Philistine pottery could have induced the Philistine (or Canaanite) potters to combine local pottery traditions in the pottery they produced and marketed. Thus, the Philistine Bichrome pottery was exported to non-Philistine sites, as it became a popular decorated tableware. In the final stages of the Iron I there is a reduction in the production of Philistine pottery, and the degenerated style and red-slipped treatment is introduced. In the Iron IIA, Philistine and Aegean-related forms almost disappear. The LPDW replaces this pottery in Philistia but in much smaller quantities. This pottery illustrates specific decorative techniques but with a much more limited repertoire of decorative motifs. In this period there seems to be a somewhat opposite trend in the trade of Philistine pottery. The Late Philistine pottery seemed to have been produced in Philistia with possibly two production centers (Figure 7): Ashdod (chemical Group 2) and Gath (chemical Groups 1 and 3). Previous analysis of some of these vessels by NAA produced the same results concerning Ashdod (*21*). There are no imports of Late Philistine pottery made in inland Philistia at the coastal Philistine cities, while several Late Philistine vessels from Gath and from Ekron are imported from the coast. Nevertheless, each Philistine city produced this pottery locally as well. Sites other than the four Philistine cities usually imported this pottery from either coastal or inner Philistia production centers. The distribution of LPDW probably illustrates the general decrease in demand for decorated pottery. The relatively smaller proportion of this ware even in Philistia may imply it had a different symbolic meaning than the Iron I Philistine ware. One may suggest that the Philistine identity in this stage, partly acculturated with local elements, may have started to develop new ethnic identity markers.

The intra-regional trade patterns of Philistine pottery should, thus, be viewed in the perspective of other archaeological and historical evidence. During the Iron IA Tel Miqne-Ekron was the most rapidly growing city illustrating the strongest characteristics of an Aegean-Philistine culture (*3*, *14*). Later, during the Iron IB and Iron IIA Ashdod and Tell es-Safi/Gath become stronger; a process reaching its peak in the 9th and 8th centuries BC (*14*); meanwhile Ekron was reduced to a small city (*14*). Thus, in this period, components of material culture with a certain ethnic value, i.e., the decorated Philistine pottery, were produced in, and exported from Ashdod and Gath. During the close of the Iron Age, the 7th century BC, Gath is diminished and Ashdod is significantly reduced as well; meanwhile, Ekron became again a large and strong city. Albeit, at this stage decorated Philistine pottery was probably not being produced any more. While the ethnic and political identity of the Philistines was still vibrant, the expression of this identity through pottery (and in most other aspects of material culture) was no longer essential for the Philistines. They conformed to the general Levantine tendency of the period, producing more standardized, non-decorated pottery. Therefore, the patterns of intra-regional trade in Philistine pottery during the Iron Age I and II fit well with the archaeological and historical picture, keeping in mind that this pottery is an important ethnical marker of the material culture in Philistia.

References

1. Dothan, T. *The Philistines and Their Material Culture*; Israel Exploration Society: Jerusalem, 1982.
2. Killebrew, A. E. In *Mediterranean Peoples in Transition. Thirteenth to Early Tenth Centuries B.C.E.*; Gitin, S.; Mazar, A.; Stern, E., Eds.; Israel Exploration Society: Jerusalem, 1998; pp 379–405.
3. Dothan, T.; Zukerman, A. *Bull. Am. Sch. Oriental Res.* **2004**, *333*, 1–54.
4. Ben-Shlomo, D.; Shai, I.; Maeir, A. M. *Bull. Am. Sch. Oriental Res.* **2004**, 335, 1–36.
5. Asaro, F.; Dothan, M.; Perlman, I. *Archaeometry* **1971**, *13*, 169–175.
6. Gunneweg, J.; Perlman, I.; Dothan, T.; Gitin, S. *Bull. Am. Sch. Oriental Res.* **1986**, *264*, 3–16.
7. Killebrew, A. E. Ph.D. thesis, Hebrew University, Jerusalem, 1999.
8. Dothan, M.; Porath, Y. *Ashdod V.* 'Atiqot 23. Jerusalem: Israel Antiquities Authority, 1993, p 54, pls. 12, 13:1–2, 14:1.
9. Adan-Bayewitz, D.; Wieder, M. *J. Field Archaeol.* 1992, *19*, 189–205.
10. Stoltman, J.B.; Burton, J.H.; Haas, J. In *Chemical Characterization of Ceramic Pastes in Archaeology*; Neff, H., Ed.; Madison WI: Prehistory Press, 1992; pp 85–92.
11. Buxeda I Garrigós, J.; Kilikoglou, V.; Day, P. M. *Archaeometry* **2001**, *43*, 349–371.
12. Hein, A.; Tsolakidou, A.; Mommsen, H. *Archaeometry* **2002**; *44*, 177–186.

13. Montana, G.; Mommsen, H.; Iliopoulos, I.; Schwedt, A.; Denaro, M. *Archaeometry* **2003**, *45*, 375–389.
14. Ben-Shlomo, D. *Decorated Philistine Pottery: An Archaeological and Archaeometric Study*, B.A.R. International Series; Archaeopress: Oxford, 2006.
15. Hart, F. A.; Storey, J. M. V.; Adams, S. J.; Symonds, R. P.; Walsh, J. N. *J Archaeol. Sci.* **1987**, *14*, 577–598.
16. Beith, M.; Shirav, M.; Halicz, L.; Bogosh, R. *Geological Survey of Israel Report*; GSI/15/88, 1988.
17. Porat, N.; Yellin, J.; Heller-Kallai, L.; Halicz, L. *Geoarchaeol.* **1991**, *6*, 133–149.
18. Young, S. M. M.; Budd, P.; Haggerty, R.; Pollard, A. M. *Archaeometry* **1997**, *39*, 379–392.
19. Mommsen, H.; Hein, A.; Ittameier, D.; Maran, J.; Dakoronia, Ph., In *Archaeometry Issues in Greek Prehistory and Antiquity*; Bassiakos, E.; Aloupi, Y.; Facorellis, Z., Eds.; Hellenic Society of Archaeometry: Athens, 2001; pp 343–354.
20. Mommsen, H. *Archaeometry* **1988**, *30*, 50.
21. Perlman, I.; Asaro, F. In *Ashdod IV. Excavation of Area M*; Dothan, M.; Porath, Y., Eds.; 'Atiqot 15. Jerusalem, 1982; pp 70–90.
22. Master, D. M. *Bull. Am. Sch. Oriental Res.* **2003**, *330*, 47–64.
23. Goren, Y.; Finkelstein, I.; Na'aman, N. *Inscribed in Clay: Provenance Study of the Amarna Tablets and Other Ancient Near Eastern Texts*; Tel Aviv Monograph Series No. 23. Institute of Archaeology: Tel Aviv, 2004; pp 18–19, 295.
24. Wieder, M.; Gvirtzman, G. *Catena* **1999**, *35*, 219–237.
25. Israel, Y.; Nachshoni, P.; pers. comm.
26. Buxeda i Garrigós, J.; Kilikoglou, V.; Day, P. M. *Archaeometry* **2001**, *43*, 366–369.
27. Buxeda i Garrigós, J.; Cau Ontiveros, M. A.; Kilikoglou, V. *Archaeometry* **2003**, *45*, 15–16.
28. Adan-Bayewitz, D. *Common Pottery in Roman Galilee: A Study in Local Trade*. Bar-Ilan University Press: Ramat-Gan, 1993.
29. Arnold, D. E.; Neff, H.; Bishop, R. L. *Am. Anthropol.* **1991**, *93*, 70–90.
30. Arnold, D. E.; Neff, H.; Glascock, M. D. *Archaeometry* **2000**, *42*, 301–316.
31. Buxeda i Garrigós, J.; Jones, R. E.; Kilikoglou, V.; Levi, S. T.; Maniatis, Y.; Mitchell, J. J.; Vagnettil, L.; Wardle, K. A.; Andreou, S. *Archaeometry* **2003**, *45*, 263–284.
32. Wood, B. G. The Sociology of Pottery in Ancient Palestine. The Ceramic Industry and the Diffusion of Ceramic Style in the Bronze and Iron Ages. JSOT/ASOR Monographs 4. Sheffield Academic Press: Worcester, 1990.

Chapter 23

The Technology of Mesopotamian Ceramic Glazes

David V. Hill[1], Robert J. Speakman[2], Michael D. Glascock[2], and Hector Neff[3]

[1]Archaeological Research and Technology, Inc., Albuquerque, NM 87102
[2]Research Reactor Center, University of Missouri, Columbia, MO 65211
[3]Department of Anthropology and Institute for Integrated Research in Materials, Environments, and Societies, California State University, Long Beach, CA 90840

Colored glazes were used to decorate ceramics in Mesopotamia for about 3500 years. The earliest glazes were produced using an alkaline-based flux. Beginning in the eighth century A.D., glazes containing a lead-based flux appear. Through the use of various analytical techniques, the technological history of ceramic glazes in Mesopotamia and the appearance of lead-based ceramic glazes can be reconstructed.

Introduction

Ceramics produced using a white-colored, tin-opacified, lead-based glaze represent a technical and artistic tradition known throughout the Middle East, Europe, and the western hemisphere under various terms such as majolica, delft, or faïence. It was the addition of lead as a glaze flux in Mesopotamian ceramics that enabled an entire series of technological and decorative changes including the use of tin oxide to create a shiny white background that could then be further decorated. The following study will focus on the appearance of lead-based ceramic glazes in Mesopotamia. An understanding of the process by which lead-based glazes appear in Mesopotamia will contribute to our understanding of how technological innovations occur.

Glazes first appear in Mesopotamia on quartz beads dating to the Third Millennium B.C. (*1*). Glazed ceramic vessels first appear in the archaeological record about five hundred years later (*2, 3*). The ceramic glazes were produced by combining plant ashes with sand or crushed quartz and a colorant, most commonly copper or iron and were used to produce vessels of a single color (*4, 5*). Beginning about 750 B.C., ceramics were produced that were decorated using one or more colors on a white background composed of lead to which tin oxide had been added as an opacifier. Through compositional analysis of ceramic glazes spanning the timing of the development of lead-based glazes, it is possible to reconstruct the technological history of pottery glazes in Mesopotamia. Isotopic characterization of lead present in glazes provides evidence for the origins of lead-based glazes. By combining the analysis of glaze compositions with the characterization of ceramic pastes, it is possible to place the development of lead-glazes within a spatial framework.

The Technologies of Mesopotamian Ceramic Glazes

Ceramic glazes have three basic components: silica, a flux used to lower the melting point of the silica, and a colorant. In Mesopotamia, sources of silica include crushed and sifted mineral quartz or quartz-rich sands. Two different types of fluxes are used. One is based on the use of the ashes from alkali-rich plants and the other on lead oxide. Colorants were derived from metallic oxides, the sources of which include native ores and debris from metal-crafting (*2, 6, 7*).

Alkaline and lead-based glazes have different firing properties. In Mesopotamia, alkaline-based glazes are prepared by combining alkali-rich plant ashes and quartz and firing them together. The resulting material is ground to a powder and combined with colorants for use. Alkaline-based glazes present a glassy appearance and mature at around 1150 degrees centigrade (*8*). Alkaline-based glazes are also “stiff” and can be applied over painted designs and fired without diffusing the underlying image. The Under-Glaze painted ceramics of twelfth to seventeenth century A.D. Iran were produced using alkaline-based glazes (*6*). The stiffness of alkaline-based glazes also means that they are prone to crazing (i.e., the formation of cracks in the ceramic glaze) due to differences in the shrinkage rates of the glaze relative to the clay body (*8*).

Lead-based glazes are produced by combining lead oxide with silica and colorant without the pre-fritting stage required by alkaline-based glazes. Depending on the amount of silica present, lead glazes mature between about 850 and 1050 degrees centigrade (*8*). Lead-glazes present a smooth glassy surface and are more "fluid" thus avoiding the crazing that often occurs with the use of alkaline-based glazes (*9*).

The relationship between the flux and the colorant greatly affects the appearance of the glaze. Glaze colorants consist of a solution containing transition metal ions. The outer electrons in the transition metal ions absorb all but certain wavelengths of light (*10, 11*). An alkaline-based flux containing copper will result in a blue-colored glaze. When combined with a flux composed of 30 percent lead, copper will produce a green-colored glaze (*11, 12*).

Previous chemical analysis of Mesopotamian pottery glazes indicated that prior to the eighth century A.D. pottery glazes were produced using only alkaline-based fluxes (*4, 5, 13, 14, 15, 16*). Although vessels decorated with an alkaline-based glaze continued to be produced, ceramics with lead-based glazes appear suddenly in the late eighth century.

Trace amounts of lead have been reported previously in monochrome-glazed ceramics dating to the Parthian and Sasanian Periods (*4, 5, 12*). The presence of these trace amounts of lead have been attributed to differences in the sources of pottery, accidental inclusion, or the smooth appearance of the glaze (4).

Compositional Analysis of Ceramic Glazes

To examine changes in the technology and sources of glazed ceramics from Mesopotamia, a sample of ceramics was analyzed spanning the time prior to the development of lead-based glazes until after the production of lead-glazed ceramics became an integral part of Middle Eastern ceramic technology. Glazed ceramics dating between the Parthian Period and the seventeenth century A.D. recovered from the Deh Luran Plain located in southwest Iran were selected for analysis (*17, 18, 19, 20*). An additional nine sherds of monochrome, blue-colored glaze from Siraf, located in Iran on the Persian Gulf and a *sgraffito* sherd collected from Yemen were also analyzed for comparison with similar ceramics from the Deh Luran assemblages (*22, 23*). A total of 183 sherds were analyzed.

The compositions of the ceramic glazes were examined using laser-ablation inductively coupled-mass spectroscopy (LA-ICP-MS) at the University of Missouri Research Reactor (MURR). LA-ICP-MS is a surface analysis technique requiring little sample preparation. Because of its small spot size, areas of weathered glaze could be avoided during analysis (*24*).

Three different flux compositions were identified through LA-ICP-MS analysis and are shown in Figure 1. The points along the left side of the ternary plot represent ceramic glazes with an alkaline-based flux. Glazes with a 30 percent or greater lead and representing true lead-based glazes plot on the lower right of the figure. Lead-based glazes are found on dark-green-colored and polychrome glazes dating to the late 700s and later.

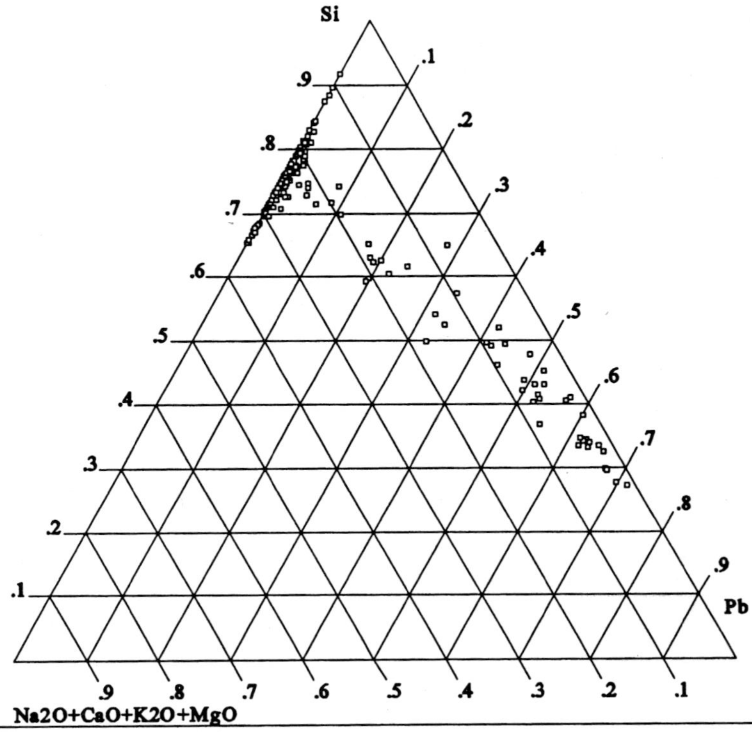

Figure 1. Plot of glaze flux compositions derived from LA-ICP-MS.

Of the 80 samples of blue-colored glaze analyzed, only two were classified as alkaline-based glazes. With the exception of two lead-glazed sherds, the other blue-glazed ceramics were produced using a low-lead, alkaline-based glaze. Beginning with the Parthian period small amounts of lead were added to alkaline-based glazes colored by copper. Low-lead, alkaline-based glazes continued to be produced during the Sasanian and Islamic periods. While rich blue-colored ceramic glazes can be produced using an alkaline-based glaze, the resulting vessels are subject to the common problems associated with alkaline-based glazes: solubility in water and uneven rates of shrinkage between the ceramic body resulting in the crazing of the glaze (*12, 25*). In one set of experiments with copper-based, turquoise-blue-colored glazes, at least 13% lead could be added to an alkaline-based glaze without affecting the blue color of the fired vessel (*12*). The results of the LA-ICP-MS analysis of the ceramic glaze samples indicate that up to 25% lead can be present in a blue-colored alkaline glaze, before the glaze takes on a greenish hue. The addition of lead to an

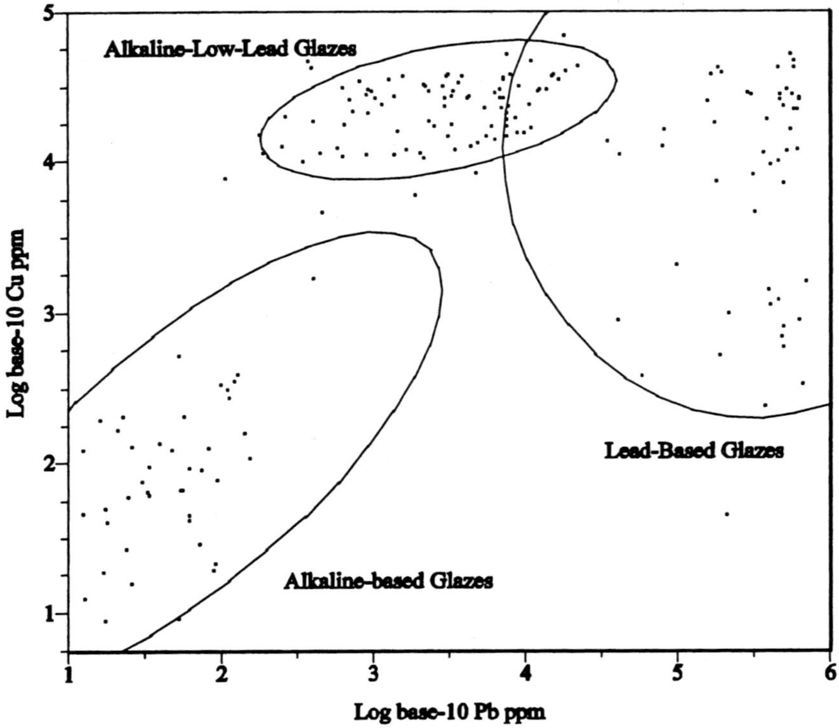

Figure 2. Plot of the relationship of copper to lead in the glazed ceramic sample. The ellipses represent 90% confidence intervals.

alkaline-based glaze also causes the blue color to become more intense and increases the reflectivity of the glaze. Adding a small amount of lead to alkaline-based ceramic glazes lowers the viscosity of the glaze, smoothes out glaze defects, and alleviates the mismatch between the glaze and the ceramic body that results in crazing (*8, 25, 26*). Contemporary industrial ceramicists add 0.5% to 2% lead to alkaline glazes to improve the appearance of the finished product (*27*). The presence of low amounts of lead in alkaline-based glazes indicates that potters located in Basra and elsewhere in southern Mesopotamian understood the beneficial properties of lead when added to their glazes. The use of lead thus became part of local craft practice. Low amounts of lead were most likely added to blue-colored glazes to enhance the appearance and assure a successful match of the glaze to the clay body. It was the familiarity with low-lead, alkaline-based glazes that most likely served as a precursor technology to the development of lead-based glazes.

Monochrome black, pea-green and white, or white glazes mottled with black were produced using primarily alkaline-based glazes. Only one example of each of these types was also produced using low-lead, alkaline-based glazes. Ceramics in this group date from the Parthian, Sasanian and Islamic Periods.

While post eighth-century ceramic types including the Splash glazes, *Sgraffito* and press-molded were produced using lead-based glazes, a few sherds were not. One of the *sgraffito* sherds was made using an alkaline-based glaze as was one of the Splash glazes.

Low-lead, alkaline-based glazes continued to be produced into the Islamic Period. The samples with exterior blue glazes and interior black-colored splash and blue glazed barbotine-decorated ceramics were produced between the eighth and tenth centuries A.D. (*17*). Low-lead, alkaline-based glazes were used to produce the four Under-glaze sherds, one sherd with black Splash-glaze decoration on white and one white glazed sherd with a painted black line. Under-glaze-painted ceramics were produced sometime after 1150 A.D. The low-lead, alkaline-based glaze composition was chosen by Islamic potters to avoid diffusion of the decoration during firing.

Lead Isotope Analysis of Ceramic Glazes

It is likely that the low percentages of lead observed in Mesopotamian ceramic glazes result from a technological choice made by potters, rather than an accidental inclusion. The argument for continuity of the use of lead from low-lead, alkaline-based glazes to lead-based glazes would be strengthened if it could be demonstrated that there was also continuity in the source of lead.

A sample of 63 sherds from the Deh Luran Plain, and Siraf, were selected for lead isotope analysis. The sixty-three sherds were selected based on the presence of 0.5% or greater amounts of lead in their glaze compositions as determined by LA-ICP-MS. The 0.5% cut-off allowed for the analysis of low-lead, alkaline-based and lead-based glazed ceramics. The 0.5% cut-off also allowed for sufficient lead to be present for lead isotope analysis using a time-of-flight ICP-MS (TOF-ICP-MS). The sixty-three samples were separated into two groups distinguished by age. The "Early" group included Sasanian and Early Islamic blue-glazed monochromes, Early Islamic period white-colored glazes with green and cobalt-blue Splash decoration, Press-molded, Lusterware, and cobalt-on-white. The "Late" group composed of ceramics types post-dating 850 A.D., contained dark green monochrome glazes, *sgraffito*, Splash glazes with a yellow-colored background, pseudo-calligraphy-decorated, and blue–colored Under-Glaze painted.

The TOF-ICP-MS used in the study is located at IIRMES (Institute for Integrative Research in Materials, Environments and Societies) located at California State University at Long Beach. The strength of the TOF-ICP-MS for isotope ratio analysis originates in the fact that different masses are detected simultaneously, by monitoring how long it takes them to reach a detector, the heavier ions taking longer than the lighter ions. This contrasts with quadrupole or magnetic-sector mass spectrometers, which are inherently sequential. The instrument used in this study was a GBC Optimass 8000. Signals were generated by laser ablation, using a New Wave UP213 system. Spot size, laser power, and other laser settings were varied in order to achieve signal intensities that would produce RSDs below 0.3 percent on the $^{207}Pb/^{206}Pb$ ratio (*28*).

Because there was no lead glass standard with known isotopic composition available for this study, isotope ratios were calibrated using samples of archaeological glass that had been analyzed previously by thermal ionization mass spectroscopy. These were obtained from Robert H. Brill of the Corning Glass Museum. These samples were run periodically during the analysis, and isotope ratios for the unknowns were corrected using the average corrections calibrated from the knowns. During the current analysis an attempt was made to keep the ^{208}Pb signal intensity between two million and eight million, as previous experiments (28) had shown that the relative standard deviations (RSDs) on the $^{207}Pb/^{206}Pb$ were optimized and relatively stable in that range. Recent experiments reported by Dudgeon et al. (this volume), however, have yielded further improvements in precision for all isotope ratios and in techniques for calibrating the ratios. The results presented here are reasonably precise and accurate for the $^{207}Pb/^{206}Pb$ ratio, but the $^{208}Pb/^{206}Pb$ and, especially, the $^{204}Pb/^{206}Pb$ ratio, are unreliable.

Results of the Lead Isotope Analysis

Despite the problems with the lead-isotope data outlined above, there is a suggestion that the lead used in Mesopotamian glazed pottery came from multiple sources. The basis for this suggestion is shown presented in Figures 3 and 4. The Early glazes show a smaller range of isotopic variation than the later glazes. Two of the four press-molded sherds have a greater $^{207}Pb/^{206}Pb$ ratio than did other members of the Early group of glazed ceramics and lie outside the 90% confidence interval for this group. Four blue-glazed sherds, one white-glazed, and one white-glazed sherd with yellow and green Splash were also located just outside the 90% confidence interval for the Early group and may also represent the use of different sources of lead. The blue-glazed sherds from Siraf plot within the 90% confidence interval of the Early group, suggesting a common source of lead for these sherds and the blue-glazed ceramics from Deh Luran.

The Late glazes are isotopically much more variable than the Early glazes, which may indicate an increase in the number of sources of lead used by Mesopotamian potters. The increased isotopic variability could indicate that

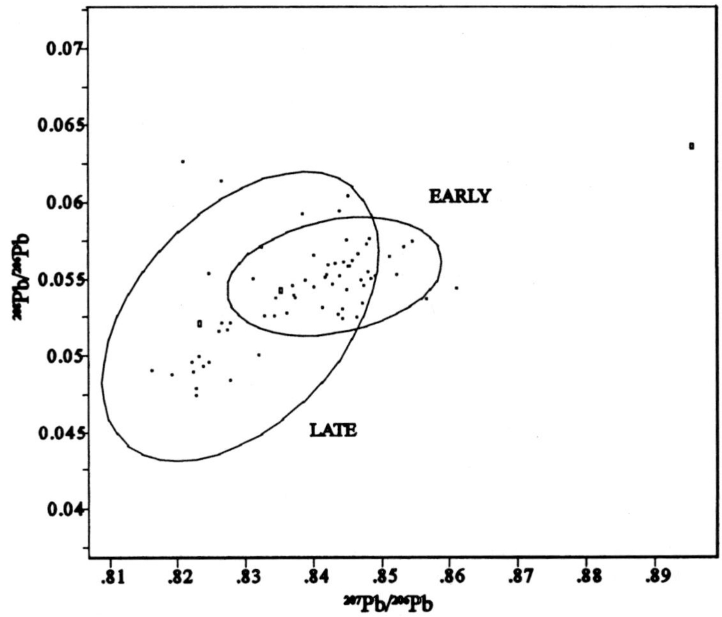

Figure 3. Plot of the ratios of $^{208}Pb/^{206}Pb$ and $^{207}Pb/^{206}Pb$. The ellipses represent 90% confidence intervals. The three open squares represent plots of the three standards.

either lead or lead oxide became a more widely traded commodity or that, as the technology of lead-based glazes spread from its original location, different sources of lead were exploited by the newly established lead-glazed pottery industries.

The ore bodies used by the Early group potters probably did not go out of use. Seven Late group sherds plot within the 90% confidence interval for the Early group. These include the two Under-glaze blue sherds in the sample, two dark-green-glazed sherds, and individual sherds of cobalt-blue-on-white, *sgraffito*, and brown-Splash-on-yellow glaze.

LA-TOF-ICP-MS may be a useful tool for distinguishing different sources of lead found in low-lead, alkaline-based and lead-based glazes from Mesopotamia. Even with the limitations of the present study, lead-isotope ratios determined by TOF-ICP-MS indicate that multiple ore sources were utilized by makers of Islamic lead-glazed pottery. With the further refinements of Dudgeon et al. (this volume), LA-TOF-ICP-MS promises to be a rapid, cost-effective approach to lead-isotope ratio analysis of glazed pottery.

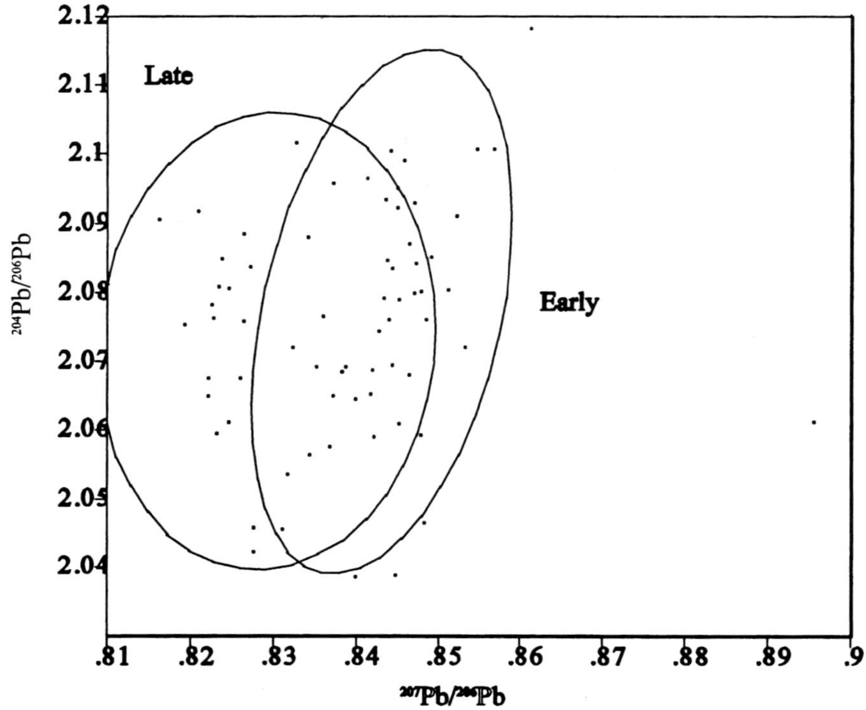

Figure 4. Plot of Ratios of $^{204}Pb/^{206}Pb - ^{207}Pb/^{206}Pb$. The ellipses represent 90% confidence intervals. The three open squares represent the values of the glass standards.

Compositional Analysis of Mesopotamian Glazed Ceramics

The following section focuses on the variability of the clay bodies or pastes of the glazed ceramics previously characterized by LA-ICP-MS and LA-TOF-ICP-MS. The purpose of this study is to compare the variation in the ceramic pastes with the different glaze decorative technologies through time. The variability in the ceramic pastes will be characterized through instrumental neutron activation analysis (INAA) and petrographic analysis.

Neutron activation of the ceramic glazes was conducted at MURR. In addition to the ceramics an unglazed sherd and a "crow's foot" recovered during the survey of the Deh Luran Plain was also included in the INAA sample. In Middle Eastern kilns, glazed bowls are fired in stacks (*6*). Crow's feet are unglazed clay triangular-shaped objects about 5 cm. across that are placed between glazed bowls during the process of firing so that the glaze will not cause the vessels stick to one another (*6, 7*). Bowls fired with crow's feet display a pattern of three evenly spaced breaks in the glaze where the crow's foot touched the glaze during firing. Such breaks were observed in the bottoms of bowls during the analysis of ceramic from the Deh Luran Plain. This object represents the only evidence of pottery production recovered during the Deh Luran project. An undecorated sherd was also included in the INAA sample as well for comparison with the composition of the glazed ceramics.

The samples for INAA were prepared by abrading the exterior surfaces with a tungsten-carbide drill to remove possible contamination from the glaze and other material adhering to the ceramics surface. The samples were then crushed into a fine powder in an agate mortar to homogenize the sample. An aliquot of each sample was weighed into high-purity quartz and polyethylene vials. As discussed by Glascock (*29*), INAA of pottery at MURR consists of two irradiations and three gamma ray counts on the quartz and polyvials. A short irradiation is carried out on the polyvials. The pneumatic tube system transports the samples to the reactor core where they are exposed to a neutron flux of 8 x 10^{13} neutrons cm^{-2} s^{-1} using five-second irradiations. Following irradiation, the sample decays for 25 minutes before beginning a 12–minute count with the HPGe detector to measure the short-lived elements Al, Ba, Dy, K, Mn, Na, Ti, and V. In order to measure elements with longer half-lives, an irradiation of 24 hours is also carried out on the samples that were sealed in quartz vials using a neutron flux of 5 x 10^{13} neutrons cm^{-2} s^{-1}. Seven days after the end of irradiation, the samples were counted for 1,800 seconds with an HPGe detector coupled to an automatic sample changer. The second count yields seven medium-lived elements: As, La, Lu, Nd, Sm, U, and Yb. After additional three-to-four week decay, a final count of 9,000 seconds was conducted to measure the long-lived elements: Ce, Co, Cr, Cs, Eu, Fe, Hf, Ni, Rb, Sb, Sc, Sr, Ta, Tb, Th, Zn, and Zr. The data from this study have been published previously (*17, 30*).

The resulting INAA data were analyzed using a suite of multivariate statistical procedures. The underlying objectives to facilitate identification of compositional groups. The first step in the analysis was to transform the abundance data into base-10 logarithms. This was done to reduce wide variations in the magnitudes of the different elements measured. Initially, the data were subjected to an average link cluster analysis (CA) to identify preliminary groups. The groups identified through CA were then evaluated using principal components analysis (PCA)—a pattern-recognition procedure used to give an idea of the subgroup structure of chemical compositional data. PCA calculates the orientations and lengths of axes of greatest variance in the data. These are found by eigenvector extraction. Element biplots were also used to highlight the compositional differences between the paste composition groups.

Results

Neutron Activation Analysis

The results of the principal components analysis are presented in Figure 5. Figure 6, presents a bivariate plot of cerium and iron serves to support the results of the clusters identified by the PCA. Based on the data presented in Table I, three compositional groups were identified though INAA. Group 1 contains primarily ceramics decorated using alkaline-based glazes. With one exception, the 16 black-colored glazed ceramics were produced using alkaline-based glazes. The mottled white and black sherds were all made using alkaline-based glazes and are also a component of Group 1. The single anomalous black-glazed sherd is a member of Group 2. Of the 10 pea-green-colored glazes, all but two are present in Group 1. The black, pea-green, and white mottled with black glazes were produced during the Parthian and Sasanian periods. Only two lead-based glazes are present in Group 1, a monochrome blue and a white glaze decorated using yellow and green-colored splash. A single pea-green sherd with green splash was also assigned to Group 1. Splash glazes originated during the Sasanian period and this may represent an example of such a piece.

Blue-colored glazes were more variable in terms of the relationship between the compositions of the ceramic paste to the type of glaze used in their decoration. Eleven blue-colored sherds with low-lead, alkaline-based glazes were assigned to paste Group 1. Two blue-glazed, lead-based sherds were also part of Group 1 and represent the only lead-based glazes in this group. The single unglazed sherd examined during the INAA study was also assigned to Group 1.

INAA compositional Group 2 is much more variable in the types present. Sixty-seven blue-glazed sherds with low-lead, alkaline-based glazes were present in Group 1. The nine blue-glazed sherds from Siraf, Iran were also assigned to Group 2 and indicate trade in this type of pottery. Except for the two blue-colored sherds mentioned previously and the *sgraffito* sherds, all of the lead-based glazed ceramics assigned to Group 2. With the exception of the pea-green glaze with green splash present in Group 1, all of the other Splash glazes are assigned to Group 2. Splash glazes were first produced in the late eight century A.D. and were thought to be glazed with lead (*31*). The presence of one Splash-glaze, alkaline-based sherd in Group 2 suggests an experimental specimen, as the other Splash glazes are decorated using either low-lead, alkaline-based or lead-based glazes. The four press-molded sherds and the single lusterware sherd, also produced during the late eight century were made using lead-glazes and were assigned to Group 2. One *sgraffito* sherd was also attributed to Group 2. Group 3 is composed exclusively of *Sgraffito* sherds. *Sgraffito* was produced from the late ninth into the eleventh century A.D. (*31*).

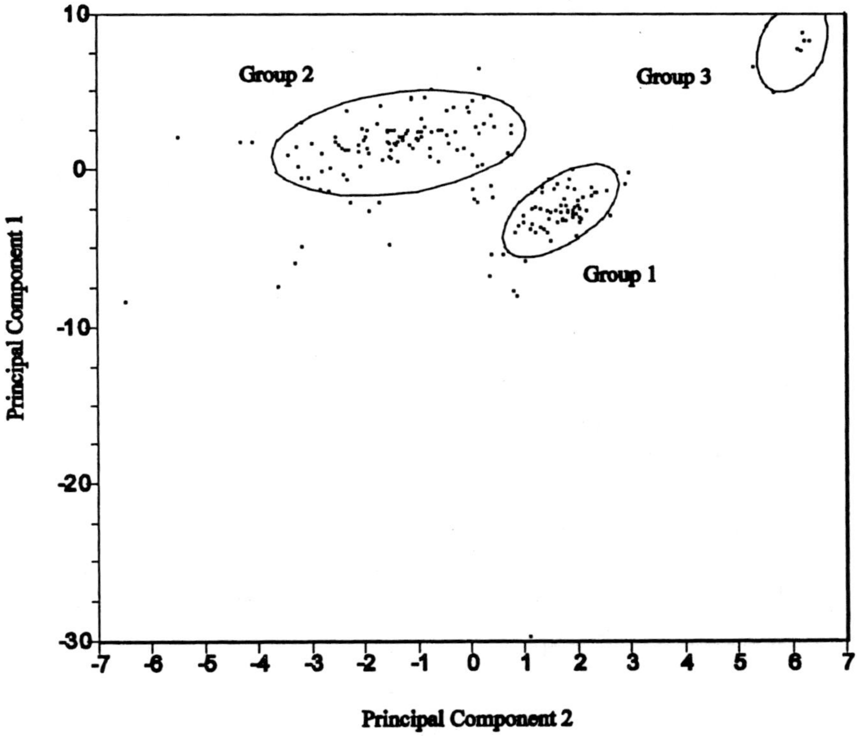

Figure 5. Principal domponents analysis of the ceramics pastes from Mesopotamia. The ellipses represent 90% confidence intervals.

The "crow's foot" could not be attributed to any of the paste compositional groups. Glazed bowls are known to be transported in stacks with the "Crow's Feet" still in place to prevent breakage (*32*). It is likely that the "crow's foot" does not represent the local production of glazed pottery, but is possible evidence of trade. However, it is still possible that the crow's foot could have been made from a different paste than the glazed pottery and be a local product.

Petrographic Analysis

Petrographic analysis was conducted of a sample of sherds from Groups 1, 2, and 3, of the INAA paste composition groups. The petrographic study was conducted for the purpose of comparing the current INAA analysis to previous petrographic studies of Early Islamic ceramics (*33*).

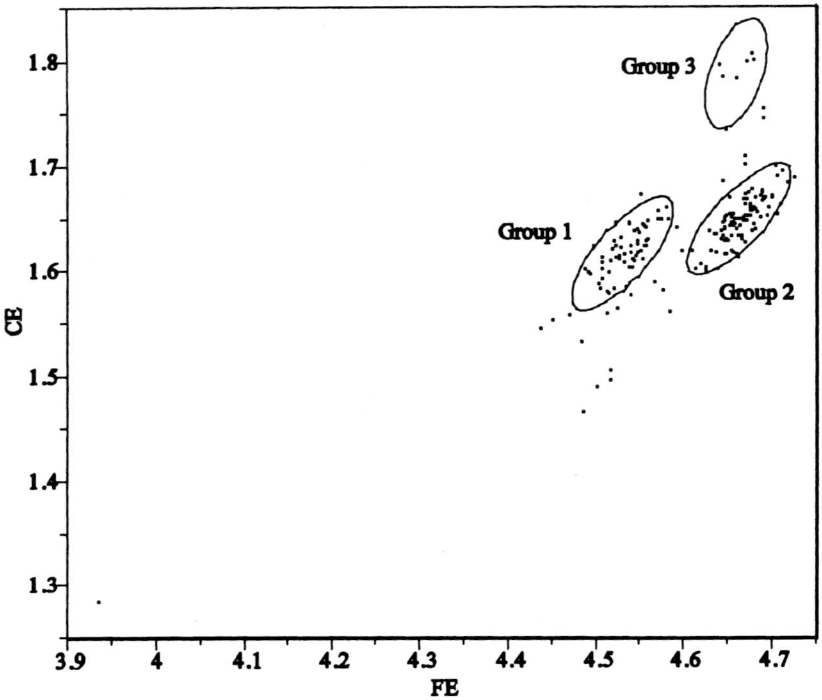

Figure 6. Bivariate plot of log-base 10 values for iron and cerium. The ellipses represent a 90% confidence interval for each group.

Eleven sherds assigned to the Deh Luran INAA Paste Group 1 were examined through petrographic analysis. Ceramics in the sample included glazed pottery from the Sasanian and early Islamic periods. A distinctive characteristic of these sherds was the presence of 5 percent or less of 0.065 to 0.5 mm rounded calcareous inclusions in the paste observed in the thin-sections. The calcareous inclusions have a micritic texture and occasionally contain isolated rounded grains of quartz or untwinned alkali feldspar. It is assumed that these rounded calcareous inclusions represent weathered and possibly redeposited soil carbonates, and as such represents a natural inclusion in the paste of the compositional Group 1 ceramics. The paste of Group 1 ranges from a medium to dark reddish brown color and contains between 1 to 5 percent 0.01 to 0.5mm sub-angular grains of quartz, untwinned alkali feldspar and brown biotite. Trace amounts of plagioclase, reddish brown volcanic tuff of aphanitic texture with or without secondary chalcedony, andesine plagioclase in a reddish aphanitic groundmass, and biotite schist are also present in the paste of Group 1 ceramics.

Table I. INAA data from MURR.

Element	*Deh Luran Group-1*			*Deh Luran Group-2*			*Deh Luran Group-3*		
As	7.8	±	1.3	13.8	±	8.3	21.7	±	9.6
La	20.4	±	1.1	22.1	±	1.2	31.2	±	0.7
Lu	0.28	±	0.02	0.31	±	0.03	0.40	±	0.01
Nd	18.1	±	1.5	19.7	±	1.7	26.2	±	1.3
Sm	3.88	±	0.17	4.26	±	0.22	5.62	±	0.11
U	2.05	±	0.41	1.81	±	0.29	3.14	±	0.38
Yb	1.83	±	0.12	2.06	±	0.14	2.67	±	0.14
Ce	41.4	±	2.4	44.7	±	2.5	62.6	±	1.2
Co	17.6	±	1.2	27.4	±	2.2	21.6	±	1.05
Cr	230.	±	28.	316.	±	74.	158.	±	5.
Cs	3.62	±	0.71	2.68	±	1.05	7.05	±	0.51
Eu	0.89	±	0.04	1.01	±	0.05	1.23	±	0.03
Fe (%)	3.41	±	0.21	4.60	±	0.28	4.60	±	0.16
Hf	3.35	±	0.18	3.39	±	0.32	4.06	±	0.10
Ni	117	±	20.	234.	±	34.	118.	±	18.
Rb	59.1	±	7.5	46.8	±	22.1	104.9	±	6.6
Sb	0.70	±	0.07	0.96	±	0.60	4.10	±	0.38
Sc	12.1	±	0.8	17.3	±	1.0	16.9	±	0.5
Sr	494.	±	124.	497.	±	113.	487.	±	153.
Ta	0.72	±	0.04	0.81	±	0.05	1.06	±	0.04
Tb	0.55	±	0.06	0.63	±	0.08	0.85	±	0.11
Th	6.01	±	0.39	6.44	±	0.38	10.4	±	0.3
Zn	73.8	±	6.4	87.8	±	8.4	127.0	±	8.2
Zr	93.	±	16.	84.	±	16.	120.	±	17.
Al (%)	5.39	±	0.25	6.34	±	0.40	7.99	±	0.25
Ba	368.	±	164.	466.	±	288.	694.	±	286.
Ca (%)	1.12	±	0.11	1.16	±	1.16	5.84	±	0.37
Dy	3.02	±	0.27	3.45	±	0.36	4.39	±	0.17
K (%)	1.67	±	0.26	1.27	±	0.30	2.99	±	0.12
Mn	765.	±	58.	960.	±	71.	882.	±	110.
Na (%)	0.72	±	0.10	1.17	±	0.39	0.93	±	0.05
Ti (%)	0.33	±	0.03	0.37	±	0.04	0.40	±	0.04
V	81.1	±	7.5	113.94	±	15.2	108.89	±	5.54

Individual grains of augite, chert, and basalt were observed in one or more of the sherds, but were not universal to the total sample.

The single study of the sources of sediment in the Tigris-Euphrates drainage network (*34*) can be used to suggest potential source areas for the production of INAA Paste Group 1 ceramics. The Tigris River in its upper part receives the majority of its water and sediment from the Greater and Lesser Zab Rivers. These rivers drain the Nappe Zone, a region composed of highly folded and faulted igneous and metamorphic rocks that would have served as sources for water-transported sediments (34). The presence of both igneous and metamorphic rock fragments in Group 1 indicates a source of the ceramics above Baghdad, where most of the sediments transported by the Tigris drop out as it becomes a meandering stream. Given that most of the Group 1 ceramics are of Sasanian age, it is possible that Group 1 glazed ceramics were produced in Northern or Central Iraq and may have been produced at or near the Sasanian capitol at Ctesiphon. Group 1 ceramics continued to be produced into the early Islamic period.

A previous petrographic study of Islamic ceramics from Siraf and Basra reports a similar set of inclusions, including 5 percent quartz, 2 percent each of biotite, felsic volcanics, and biotite schist, 1 percent untwinned alkali feldspar, plagioclase, and trace amounts of plagioclase, amphibole, muscovite, clinopyroxene and opaque inclusions designated the "Siraf 4" Petrofabric (*33*). The calcareous inclusions observed in the current study were apparently not present in Mason's Islamic material from Siraf or Basra. However, carbonates have been recognized in a small sample of Islamic pottery thought to originate in Hira (*33*). But the rest of the mineral suite recorded for the Hira ceramics differs considerably from that observed in the Group 1 pottery. Consequently, INAA Group 1 represents a newly recognized source of ceramic production, one with considerable time depth.

Ten sherds from Deh Luran in Group 2 were also examined through petrographic analysis. The glazed ceramics examined include Parthian, Sasanian, and Early Islamic period pottery. The paste of the sherds belonging to Group 2 is characterized by a light yellow or gray very fine, featureless paste. The paste contains 3 percent or less of silt-sized rounded grains of quartz with trace amounts of brown biotite and untwinned alkali feldspar. Trace amounts of medium to coarse sized rounded grains of quartz and untwinned alkali feldspar are also present in Group 2 ceramics. The results of the petrographic study of Group 2 accords well with the description of the "Basra Petrofabric" (*6, 33, 36*).

The Basra Petrofabric was identified in an extensive sample of early Islamic ceramics and kiln furniture from Basra. Some of the kiln furniture examined by Mason formed a part of another INAA study (*32*). The Basra Petrofabric, as identified in ceramics and kiln furniture, is characterized by a light yellow featureless groundmass containing 2 to 3 percent angular quartz, with a trace amount of amphibole and biotite, less than 0.05 mm. in size (*33*). Mason also reports a smaller population of 0.25 to 1.0 mm of trace to 10 percent quartz, trace to 3 percent of weathered untwinned alkali feldspar, and trace to 1% clear plagioclase (*33*).

Reanalysis of data from a previous INAA of early Islamic period ceramics using the same program of statistical manipulation applied to the MURR study results in Figure 7. Group 1 represents a wide range of early Islamic glazed ceramics including the same types examined during the MURR study including turquoise-blue glazes, Splash glazes on a white background, luster ware, cobalt-blue-on-white. Also included in Group 1 were four kiln rods used to support glazed vessels during firing (6), kiln wasters, and several unglazed ceramics. The kiln rods and wasters are believed to have been recovered from Basra (*32*). Group 2 represented by four sherds is composed of one turquoise-colored sherd and three white-glazed sherds recovered respectively from Iran, and Iraq. Group 3 is composed of four sherds of unglazed pottery believed to have been produced on the Arabian Peninsula (*32*).

Given the similarity of the petrographic data between the analysis of the Deh Luran Group 2, previously published descriptions of the Basra Petrofabric (33) and Halletts' (*32*) previous INAA study of Early Islamic period glazed ceramics recovered from across the Near East and southeast Asia, it is likely that these three studies are examining the same source of ceramic raw materials. One mineralogical study reports that the sediments at Basra are mainly clayey, a condition consistent with the low-energy deposition of sediments (34)

A previous petrographic study of ceramics dating to the Mesopotamian Early Dynastic period (ca. 2900 to 2400 B.C.) also reports a significant decrease in the overall amount and size of isolated mineral grains present in the ceramics between central and southern Iraq (*37*). Although these ceramics date from a much earlier period than that covered in this paper, the decrease in the size of sand grains present in the ceramic clays results from the slowing of the Tigris and Euphrates rivers as they approach the Persian Gulf was observed in the Group 2 ceramics and the ceramics from the earlier study. It is proposed that the resource zone and the source of Group 2 clay used by Islamic potters living in the vicinity of Basra represents continuity in the use of a local compositionally homogeneous clay source since at least the Parthian Period and that a significant portion of the glazed pottery found in Deh Luran came from Basra or elsewhere in southern Mesopotamia.

Given the limited number of sherds, only seven, that comprise the Deh Luran Group 3 composed exclusively of *sgraffito* sherds, only two sherds were examined. The paste of these two sherds in the MURR Group 3 has a dark-reddish-brown color. The two sherds contain roughly equal proportions of quartz, untwinned alkali feldspar, and brown biotite, which account for 5 percent of the ceramic paste. These mineral grains range in size from 0.065 to 0.5mm. Single grains of moderately well sorted (*38*) arkose or arkosic sandstone

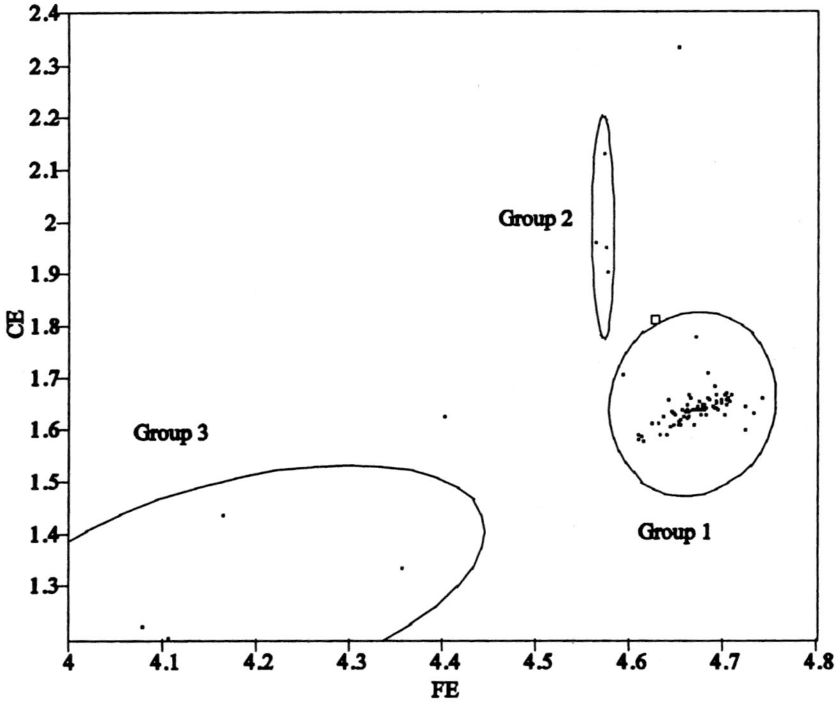

Figure 7. Bivariate plot of log-base 10 values of iron and cerium derived from a previous INAA study of early Islamic period glazed ceramics(32). The ellipses represent a 90% confidence interval for each group. Note that the data for Group 2 fall in the same range as the Fe and Ce values for Group 2 in Figure 6.

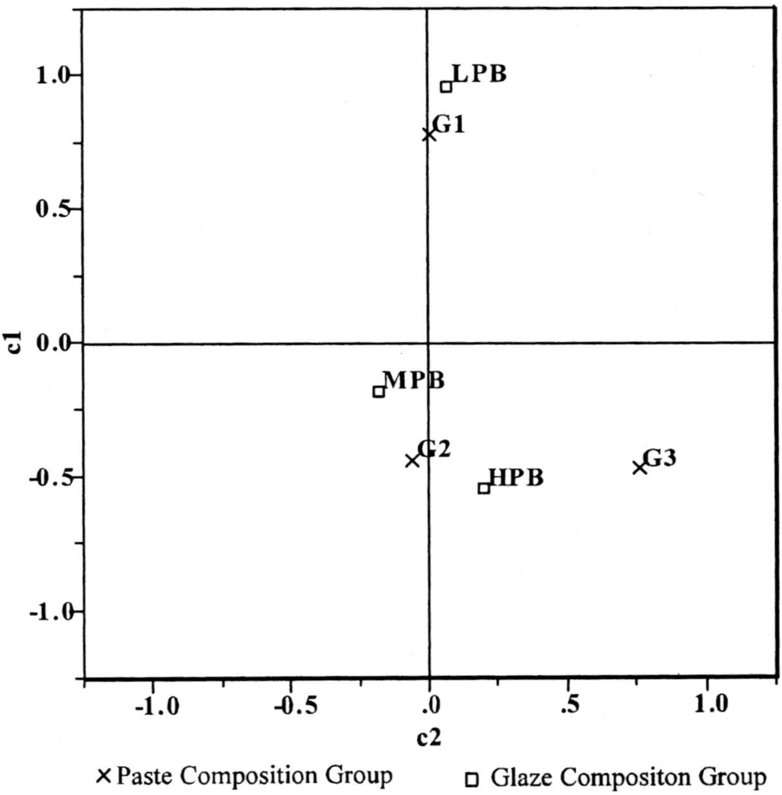

Figure 8. Correspondence analysis plot showing the relationship of the three ceramic paste groups and glaze composition groups. Alkaline-based ceramics glazes are more likely to be associated with INAA Group 1 (G1) ceramic pastes than with other glaze or paste composition groups. Alkaline, low-lead (MPB) and lead-based (HPB) ceramic glazes are more likey to be associated with INAA Group 2 (G2) and Group 3 (G3). With only one alkaline-based glaze in the Sgarffito sample (G3) this group is more strongly associated with the lead-based glaze composition group.

characterized by reddish brown cement occurs in both sherds. Splash-glaze *sgraffito* sherds with the above paste composition have not been previously identified. There are arkosic sandstones of Plio-Pleistocene age located along both the Euphrates and Tigris Rivers near present-day Baghdad (*34*). The *sgraffito* sherds that constitute Group 3, including a sherd from the Royal Ontario Museum's collection from Yemen, were probably produced in central Iraq.

Based on the INAA dataset and the petrographic study, at least two compositionally distinct sources provided the raw clays used for making the monochrome glazed pottery traded into the Deh Luran Plain beginning in the Parthian period and continuing into at least the twelfth century A.D. It is only with the appearance of *sgraffito*, apparently first at Basra and later elsewhere that a third compositionally distinct source of raw clay appears.

Discussion

The ceramic types subsumed within INAA compositional Groups 1 and 2 represent a continuity of the use of ceramic clay resources from the Parthian until the Islamic period. The paste used in Group 1 produced pottery with alkaline-based and low-lead, alkaline-based glazes during this time (Figure 9). Group 2 produced a little alkaline-based pottery. However, Group 2 served as the primary production source of blue-glazed pottery made using a low-lead, alkaline-based flux beginning in the Parthian period with monochrome colored glazes (Figure 10). The new decorative styles including Splash glazes, press-molded, and luster ware were produced using both low-lead, alkaline-based and lead-based glazes (Figure 11). It is suggested that the familiarity with potters, presumably living in southern Mesopotamia near modern-day Basra, had with the properties of lead when added to ceramic glazes led to their development of lead-based glazes, the earliest in Mesopotamia. The continuity in the source of lead found in ceramic glazes also serves as an indicator of continuity in ceramic traditions, as it is not until the Islamic Period when new sources of lead are incorporated into the ceramic technology. The relationship of Mesopotamian ceramic glaze compositions to the paste composition groups characterized through INAA and petrographic analysis are presented graphically in Figure 8.

Sgraffito, which appears to have been produced in more southern and more commonly at a third location different from the other two ceramic sources identified during this study. *Sgraffito* appears in its new location as a fully developed ceramic technology without local antecedents. The sudden appearance of *sgraffito* production represents the migration of potters, most likely from southern Mesopotamia since one *sgraffito* sherd was assigned to INAA Group 2.

Figure 9. This alkaline-based glaze with a managanese Splash-glaze decoration was assigned to INAA Group 1. This sherd was directly dated by Optically Stimulated Luminescence to 1230 +/- 330 A.D. (OxL-1353) (17).

Figure 10. Parthian and Sasanian ceramics. Upper left, Sasanian white alkaline-based glaze. Upper right, Parthian copper/iron alkaline-based glaze. Lower sherds, Sasanian copper-colored, low-lead, alkaline-based glazes.

Figure 11. Selected Islamic Period Ceramics. Upper left, Green splash on yellow glaze. Upper center, Green splash glaze on white glaze. Upper right, Sgraffito. Lower, blue-green splash on white glaze.

Table II. Historical periods and technological developments in Mesopotamian ceramic technology.

Historical Period	*Time Range*	*Technological and Stylistic Development*
Parthian	171B.C.- 244 A.D.	Low percentages of lead are added to alkaline-based glazes colored with copper in southern Mesopotamia. INAA Group 2 ceramic produced. Alkaline-based glazes are produced as well (INAA Group 1).
Sasanian	244 A.D.- 637 A.D.	Continued production of copper-colored, low-lead, alkaline-based glazes (INAA Group 2). Ceramics with alkaline-based glazes continue to be produced.
Early Islamic	ca. 750 A.D.	The first lead-based ceramic glazes are produced in Basra and likely elsewhere in southern Mesopotamia. New decorative treatments including tin-opacified glazes, Splash glazes, Luster painting and the use of cobalt as a glaze colorant appear. All of the Early Islamic period lead-glazed ceramics have INAA Group 2 paste compositions.
Later Islamic	Post-Ninth century A.D.	The first lead-glazed pottery, *sgraffito* is produced outside of southern Mesopotamia (INAA Group 3). Alkaline-based and low-lead, alkaline-based glazed pottery continues to be produced (INAA Group 1). INAA Group 2 ceramics continue to be produced.

By using a battery of analytical techniques it has been possible to outline the history of the development of lead-based ceramic glazes in Mesopotamia. The innovations in Mesopotamian ceramic technology are summarized in Table II. Future studies of the technology of Mesopotamian ceramic glaze technology should include additional compositional analysis of glazes and lead isotope analysis. Analysis of Iranian and western Turkish lead ores should be conducted to trace the sources of lead used in the lead-containing glaze recipes. Additional INAA studies should also be conducted to establish the variability within each of the paste compositional groups identified here.

Acknowledgements

The authors acknowledge the support for this project, in part, received from the National Science Foundation grants to the University of Missouri Research Reactor (grant no. 0504015) and California State University-Long Beach (grant no. 0228187).

References

1. Vandiver, P. B.; Fenn, M.; Holland, T. A. In *Materials Issues in Art and Archaeology III*; Vandiver, P. B.; Druzik; J. R.; Wheeler G. S.; Freestone I. C., Eds.; Materials Research Society: Pittsburgh, PA, 1992; Vol. 267, pp 519–525.
2. Moorey, P. R. S. *Ancient Mesopotamian Materials and Industries: The Archaeological Evidence*; Eisenbrauns: Winona Lake, IN, 1999.
3. Vandiver, P. B. In *Early Pyrotechnology: The Evolution of the First Fire-Using Industries*; Wertime, T.A.; Wertime, S. F., Eds.; Smithsonian Institution Press: Washington, DC, 1982; pp 73–92.
4. Hedges, R. E. M.; Moorey, P. R. S. *Archaeometry* **1975**, *17*, 25–43.
5. Hedges, R. E. M. *Archaeometry* **1976**, *18*, 209–211.
6. Keblow-Bernsted, A. M. *Early Islamic Pottery: Materials & Techniques*; Archaeotype Publications, Oxford, U.K. 2003.
7. Wulff, H. E. *The Traditional Crafts of Persia*; M.I.T. Press: Cambridge, MA, 1966.
8. Rhodes, D. *Clay and Glazes for the Potter*; Chilton Book Company: New York, NY, 1957.
9. Tite, M. S.; Freestone, I. C.; Mason, R.; Molera J.; Vendrell-Saz, M.; Wood, N. *Archaeometry* **1998**, *40*, 241–260.
10. Nassau, K. *The Physics and Chemistry of Color: The Fifteen Causes of Color*; John Wiley & Sons, Inc.: New York, NY, 1982.
11. Vandiver, P. B. *Scientific American* **1990**, *4*, 106–113.

12. Berdel, E., *Ceramic Industry* **1931**, *17*, 195–198.
13. McCarthy, B. E. Ph.D. thesis, The Johns Hopkins University, Baltimore, MD, 1996.
14. Debevoise, N. *Parthian Pottery from Seleucia on the Tigris*; University of Michigan Press: Ann Arbor, MI, 1934.
15. Toll. N. In *The Excavations at Dura-Europos*. Rostovtzeff, M. I.; Bellinger, A. R.; Brown, F. E.; Wells, C. B., Eds.; Yale University Press: New Haven, MA, 1943; Vol. IV, Part I.
16. Matson, F. R. In *The Excavations at Dura-Europos*; Rostovtzeff, M. I.; Bellinger, A. R.; Brown, F. E.; Wells, C. B., Eds.; Yale University Press: New Haven, MA, 1943; Vol. IV, Part I.
17. Hill, D. V. Ph.D. thesis, The University of Texas, Austin, TX, 2005.
18. Neely, J. A. *Iran* **1970**, *8*, 202–203.
19. Neely, J. A. In *Irrigations Impact on Society*, Downigh, T. E.; Gibson, M., Eds.; Anthropological Papers 25, University of Arizona Press: Tucson, AZ, 1974; pp 21–44.
20. Neely, J. A.; Wright, H. T. *Early Settlement and Irrigation on the Deh Luran Plain: Village and Early State Socities in Southwestern Iran;* University of Michigan Museum of Anthropology Technical Reports, No.26 The University of Michigan: Ann Arbor, MI, 1994.
21. Whitehouse, D. *Estrato da Annali dell'Instuto Orientale di Napoli,* Naples, IT, 1979; Vol 39.
22. Tampoe, M. *Maritime Trade between China and the West: An Archaeological Study of Ceramics from Siraf (Persian Gulf), 8th Century to 15th Century*. British Archaeological Reports International Series, Hadrian Books: Oxford, U.K., 1989; Vol. 555.
23. Cuik, C.; Keall, E. *Zabid Project Pottery Manual 1995: Pre-Islamic and Islamic Ceramcis from the Zabid Area, North Yemen*; British Archaeological Reports International Series, Hadrian Books: Oxford, U.K., 1996; Vol. 655.
24. Speakman, R. J.; Neff, H. *Laser Ablation ICP-MS: A New Frontier in Archaeological Characterization Studies*; University of New Mexico Press: Albuquerque, NM, 2005.
25. French, M. M. *J. Am. Ceramic Soc.* **1923**, *6*, 405–408.
26. Norton, F. H. *Ceramics for the Artist Potter*; Addison-Wesley Publishing Company: New York, NY, 1956.
27. Nordyke, J. S. *Lead in the World of Ceramics*; American Ceramic Society, Inc.: Columbus, OH, 1984.
28. Neff, H. *Analysis of Lead Isotopes in Glass Beads from Igbo-Ukwu*; Unpublished manuscript on file with the author, 2005; Department of Anthropology, California State University, Long Beach, CA.
29. Glascock, M. D. In *Chemical Characterization of Ceramic Pastes*; Neff, H., Ed.; Prehistory Press: Madison, WI, 1992; pp 11–26.

30. Hill, D. V.; Speakman R. J.; Glascock M. D. *Archaeometry* **2004**, *46*, 585–606.
31. Kennett, D. *Sasanian and Islamic Pottery from Ras al-Khaimah: Classification, Chronology, and Analysis of Trade in the Western Indian Ocean*; Society for Arabian Studies Monograph No.1 and British Archaeological Reports, International Series; Hadrian Books: Oxford, UK, 2004; Vol. 1248.
32. Hallett, J. R. Ph.D. thesis, Oxford University, UK, 1999.
33. Mason, R. B. Ph.D. thesis, Oxford University, UK, 1994.
34. Phillips, G. *J. Sedim. Petrol.* **1968**, *38*, 35–44.
35. Dunnington, H. V. In *Habitat for Oil*; Weeks, L. G., Ed.; American Association of Petrolium Geologists: Tulsa, OK, 1958; pp 1194–1251.
36. Mason, R. B.; Keall, E. J. *Iran* **1991**, *29*, 51–66.
37. Mynors, H. S. In *The Proceedings of the 22nd Symposium on Archaeometry.*; Aspinall, A.; Warren, S. E., Eds.; University of Bradford: UK, 1983; pp 377–387.
38. Folk, R. *Petrology of Sedimentary Rocks*; Hemphill Publishing Company, Austin, TX, 1980.

Chapter 24

Analysis of Historic Latter-day Saint Pottery Glazes by Laser Ablation–Inductively Coupled Plasma–Mass Spectrometry

Nicole C. Little[1], Timothy J. Scarlett[2], Robert J. Speakman[1], Michael D. Glascock[1], and Christopher W. Merritt[2]

[1]Research Reactor Center, University of Missouri, Columbia, MO 65211
[2]Michigan Technological University, Houghton, MI 49931

In recent years, laser-ablation (LA) systems coupled to state-of-the-art inductively-coupled-plasma mass-spectrometers (ICP-MS) have gained increased popularity in archaeology for providing chemical analyses of a variety of inorganic and organic matrices. Such analyses have enabled archaeologists to address questions concerning provenance, trade, and technology through the analysis of metals, lithics, ceramics, and other archaeological materials. One area of research that has proven particularly fruitful has been the analysis of glazes found on pottery. This paper demonstrates the feasibility of detecting different glaze recipes. We will discuss how compositional groups correspond to manufacturing locales of ethnohistorically known Latter-day Saints' potters in Utah during the late 19th and early 20th centuries.

During the late 19th century, European and Euroamerican immigrants established more than 45 potteries in more than 26 towns and cities throughout the Intermountain Region of North America. These potters immigrated with the Church of Jesus Christ of Latter-day Saints and brought their technological expertise into this arid region. Many of the individuals who continued to make pottery in their new settlements successfully extended their specialized technical knowledge from segmented industrial production systems to smaller-scale artisanal production. Although they attempted to manufacture high-quality, semi-vitreous, white earthenware, none were successful in sustaining production beyond earthenware or stoneware (*1, 2*).

Among the many aspects of potting with which immigrants struggled, creating glazes that matched their new clays was a significant challenge. Due to the increased porosity of earthenwares (20–25% per volume), glaze was essential to seal a pot's surface and thus prevent it from absorbing liquids during use. No specific reference to any glaze recipe survived from the nineteenth century in Utah. Potters traditionally use a wide variety of glaze materials to produce a particular end product, whether it is clear, colored, translucent, opaque, etc. These potters traditionally used a main glaze body of powdered lead oxide (PbO_2), galena (lead sulfide, PbS), or a mix of powdered lead, clay, and sand (*3*).

Although no known historical documents record a specific glaze formula, some records account for general procedures and practices. In the early days of colonization, lead was an expensive and rare commodity due to shipping costs. Many potters, therefore, experimented using locally available galena ore, which they pounded and ground to a fine powder (*4–6*). Lead was so expensive that potters scavenged tiny quantities from sources such as the small sheet of lead that came wrapped around tea imported from Japan. Tea lead was crumbled, put into a kettle over the fire, and stirred until it disintegrated into a fine powder (*6, 7*). At least one potter mined lead by sifting for spent bullets from the shooting ranges at Fort Duchesne (*5*).

By the last quarter of the nineteenth century, most potters ordered their lead from eastern retailers. Frederick Hansen of Brigham City, for example,

> procured the lead for glazing, in a crude state, from the East. He ground it up, placed it in a large urn or pan over the fire, and stirred it constantly with an iron ladle until it became liquid. It was then cooked until it formed a dry powder which, when cool, he rubbed over his wares (*8*).

While Hansen was running the Brigham City Co-Operative Pottery, the Co-Operative store ordered many kegs of red lead (lead oxide, Pb_3O_4) and white lead (carbonate of lead, $2PbCO_3{\cdot}Pb(OH)_2$). Some kegs of red lead were undoubtedly for consumption in the pottery shop, although much may also have been an ingredient for paint.

Once the heat rendered the ore or scrap lead clean of impurities, it was mixed with other ingredients. These ingredients included highly refined and powdered clay (feldspars preferably containing high alumina content) and silicates (ground flint, quartz sand, and/or perhaps even ground glass). Small quantities of a wide variety of other ingredients may also have included calcium carbonate or potash to increase the quantity of calcium or potassium in the glaze (*9, 10*). One potter supposedly combined three or four hundred pounds of lead with two parts clay and produced approximately one hundred gallons of glaze (*11*). Mrs. Burton's explanation that the lead to clay ratio in the glaze (1:2) may also be a slight exaggeration, since standard glaze mixtures include a lower ratio closer to 1:1 or even lower (*9*).

Precisely how each potter developed their glaze recipe and matched it to their clay body is an important question that remains to be resolved. Given that glaze recipes are not known and that different combinations of ingredients can produce visually similar effects, it has been impossible to determine if potters used individual formulas or if recipes were shared among the Latter-day Saint communities in the region. Nor can researchers determine if a particular glaze can be attributed to a particular potter or group of potters. This study is a first step in characterizing the potters' glazes and reconstructing their strategies. The study included 65 separate ceramic sherds, including some of the kiln sites where the pots were made and other samples from consumption sites throughout the region (Figure 1).

LA-ICP-MS

Inductively coupled plasma-mass spectrometry (ICP-MS) has been utilized as a bulk technique for the analysis of obsidian, chert and ceramic compositional analyses (*12–14*). However, due to the high level of spatial variation of ceramic materials, increased sample preparation is necessary with volatile acids coupled with microwave digestion (MD-ICP-MS) to properly represent the variability of ceramic assemblages (*15, 16*). Due to the increased sample preparation and exposure to volatile chemicals, researchers have continued to utilize neutron activation analysis (INAA) as the preferred method of chemical characterization of archaeological ceramics (*17*).

Nevertheless, ICP-MS has proven to be an integral analytical technique for the analysis of glaze recipes when coupled with laser ablation (LA-ICP-MS) (*12, 18–20*). Because the laser only ablates the surface of ceramics, data is generated for glazes and paints alone. This allows the researcher to focus on various glaze and paint colors on a single ceramic surface, which would not be possible using traditional bulk analytical techniques such as INAA and MD-ICP-MS.

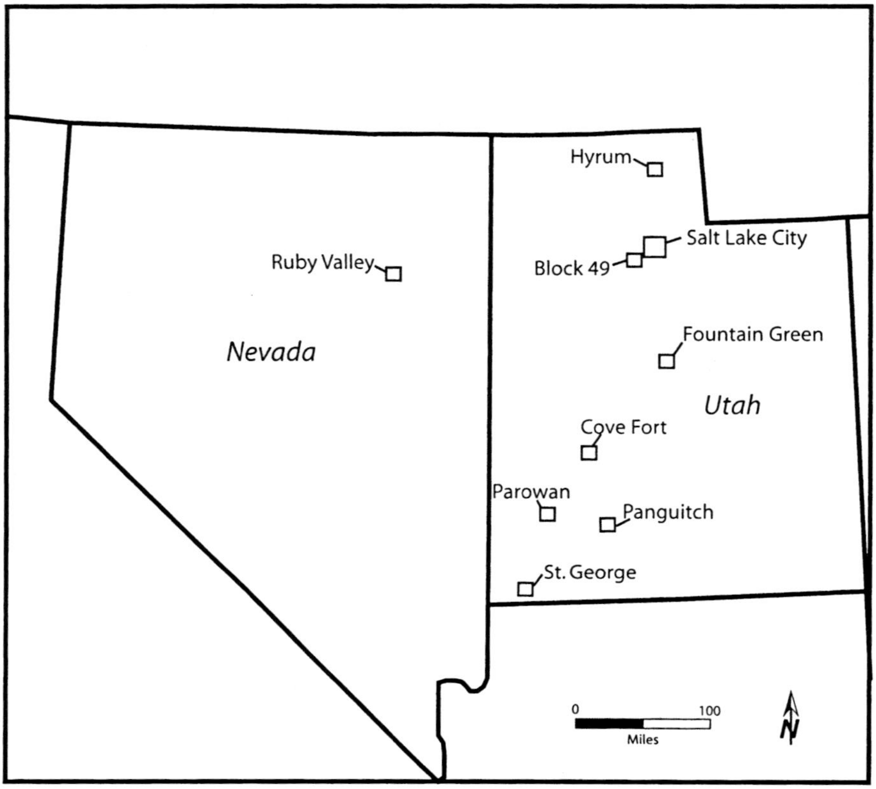

Figure 1. Map showing location of sampled sites in Utah and Nevada.

Additional surface techniques such as X-ray fluorescence (XRF) do not allow the researcher to target points of interest on the ceramic body.

A major advantage of LA-ICP-MS as a microprobe analytical technique is the ability to obtain data for virtually any element in the periodic table. LA-ICP-MS also can be used to quantify elements that are present in the low parts-per-million (ppm) to parts-per-trillion (ppt) range. In contrast, other surface techniques such as, SEM, XRF, and PIXE are limited by the number of elements detectable and have higher detection limits than ICP-MS.

In LA-ICP-MS, the sample is placed inside a sample holder or laser cell where ablation takes place. The ablated area varies in size depending on the sample matrix but is usually smaller than 1000 X 1000 μm and less than 30 μm deep. The ablated material is flushed from the laser cell using a 1.1–1.3 liter/minute flow of argon or an argon/helium-mixed carrier gas through Tygon tubing, and introduced into the ICP-MS torch where argon gas plasma capable of sustaining electron temperatures between 8000 and 10,000 K is used to ionize the injected sample. The resulting ions are then passed though a two stage interface designed to enable the transition of the ions from atmospheric pressure to the vacuum chamber of the ICP-MS system. Once inside the mass spectrometer, the ions are accelerated by high voltage and pass through a series of focusing lenses, an electrostatic analyzer, and finally a magnet. By varying the strength of the magnet, ions are separated according to mass/charge ratio and passed through a slit into the detector that records only a very small atomic mass range at a given time. By varying the instrument settings the entire mass range can be scanned within a short period of time.

The instrument used in the studies reported here is a high resolution VG Thermo Elemental Axiom magnetic-sector inductively coupled plasma mass spectrometer. The ICP-MS is coupled to a Merchantek Nd:YAG 213–nanometer wavelength laser ablation unit. The laser can be targeted on spots as small as five microns in diameter. The small spot size and the high sensitivity of magnetic-sector ICP-MS to a wide range of major, minor, and trace elements make LA-ICP-MS a very powerful microprobe. Moreover laser ablation is virtually non-destructive to most samples considering that the ablated areas are often indistinguishable with the naked eye (Figure 2).

Prior to data acquisition, samples were pre-ablated with the laser to remove possible surface contamination. Power settings for the laser were adjusted to prevent the laser from burning through the paint during analysis, ensuring that the material introduced to the ICP-MS was actually glaze and not the underlying clay matrix. NIST SRM 610 and 612 (glass wafers spiked with ~60 elements), Brill Glass Standards B, C, and D, and Glass Buttes obsidian were used as standards to calibrate data.

Data were calibrated using an approach adopted from Gratuze (*19, 21–23*) to produce oxide concentrations for the elements analyzed in each sample. In this approach, the signal (blank-subtracted counts) for each sample is standardized by calculating ratios of the counts to a single element (e.g., Si). A response coefficient obtained from multielement standards is then used to calculate the composition of each sample. The basic assumption of this approach is that the elements measured represent essentially all of the material, other than oxygen, that is ablated from the sample. Oxygen is then taken into consideration by converting the elemental signals to signals of their oxides. Some error may be introduced at this point for elements that occur in more than one oxidation state, particularly iron, which may be present as FeO as well as Fe_2O_3. Additionally, any water in the material is unaccounted for in the summation to 100%, as are some elements, such as chlorine, which may be present but were not measured. These missing measurements may contribute to a slight overestimation of the various measured oxides. Nonetheless, this approach yields linear responses across a wide range of concentrations (e.g., Figure 3)

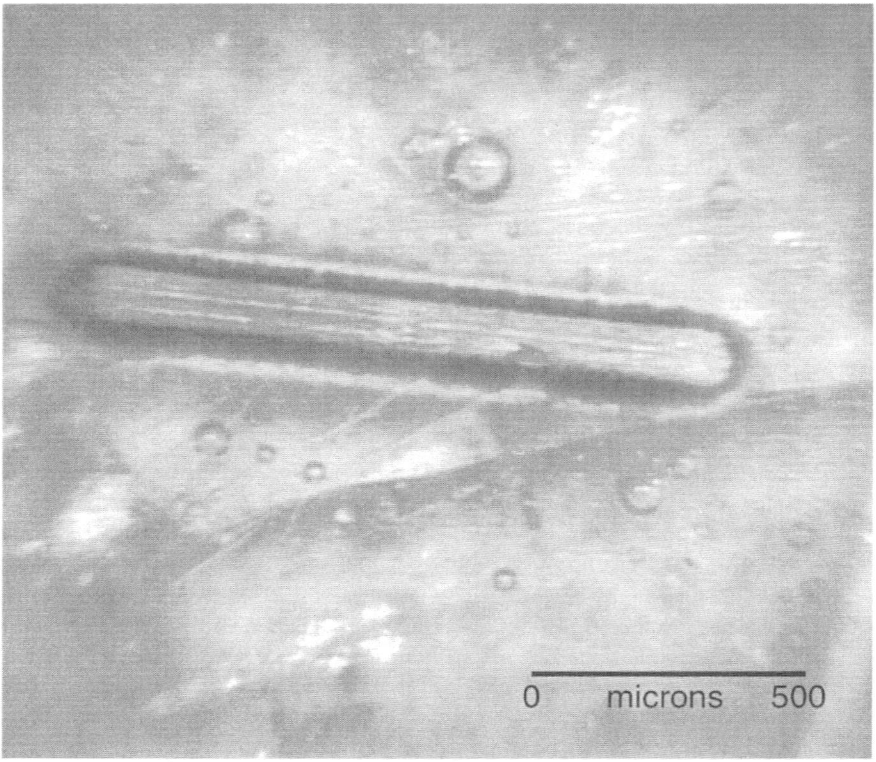

Figure 2. Glazed surface that has been ablated with the laser for ICP-MS analysis.

Results

Chemical analysis via LA-ICP-MS verified 9 distinct technological groups within the dataset of 75 glazes represented on 65 ceramic sherds (Figures 4–5). Glaze groups 1 and 2 (GL1 and GL2) are made up entirely of ceramics from the Ruby Valley Pony Express Station (five from GL1 and seven from GL2). Glaze group 3 (GL3) contains eight samples from Fountain Green, four samples from Salt Lake City, and three samples from Panguitch. Glaze group 4 (GL4) contains five samples from Parowan, two samples from Panquitch, and one from Salt Lake City. Glaze group 5 (GL5) contains two samples from Parowan and one from Salt Lake City. Glaze group 6 (GL6) contains eight samples from Salt

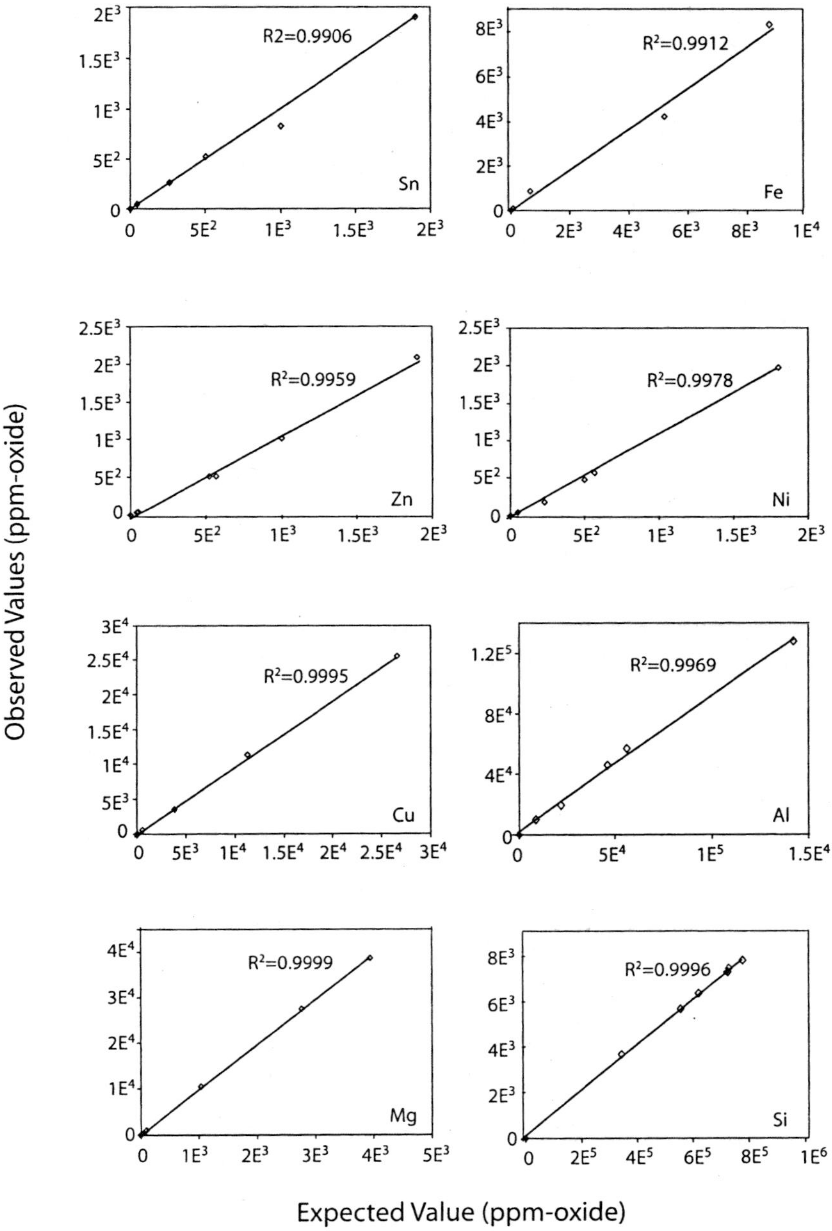

Figure 3. Plots of expected versus observed values obtained from LA-ICP-MS standards analyzed for this study.

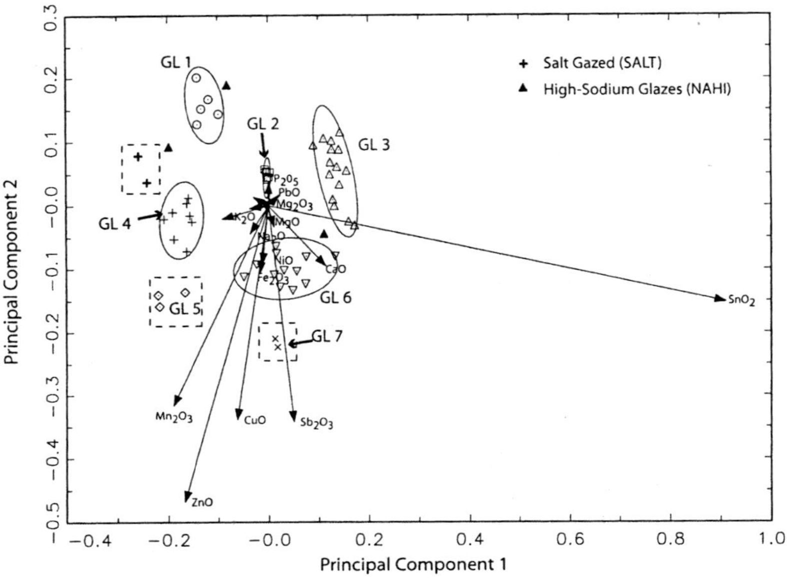

Figure 4. Projection of nine compositional groups by principal components 1 and 2. Ellipses represent 90% confidence interval for group membership.

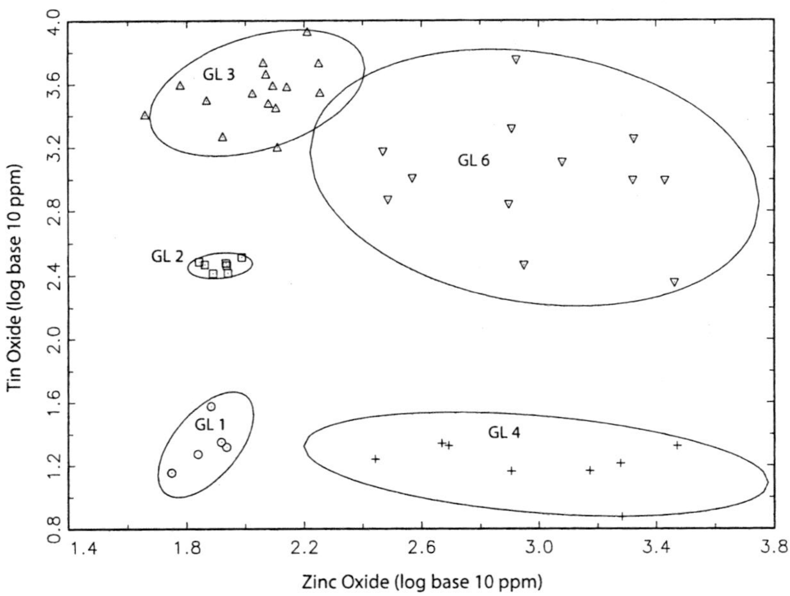

Figure 5. Bivariate plot of zinc oxide and tin oxide base-10 logged concentrations showing distribution of the five main compositional groups.

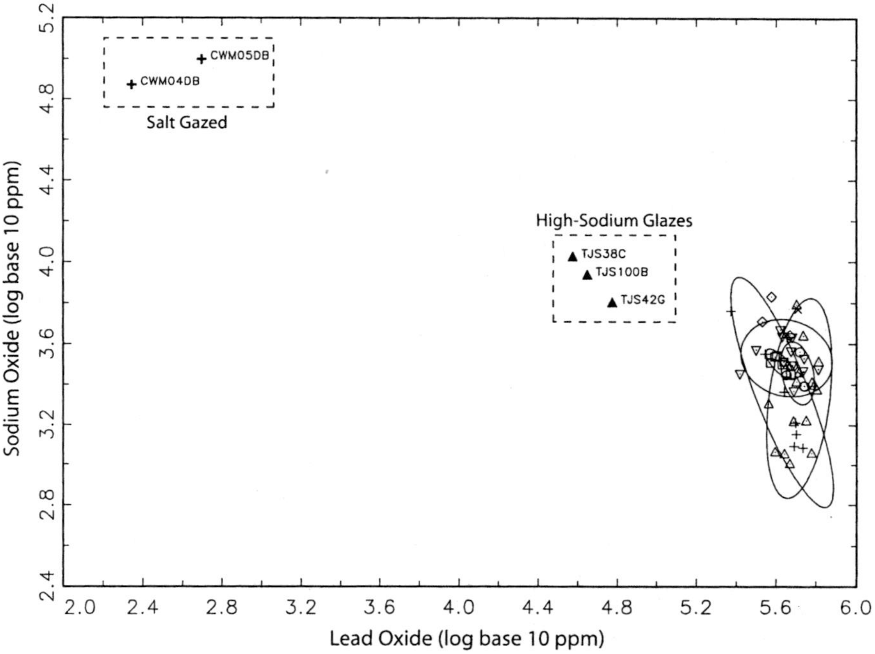

Figure 6. Bivariate plot of lead oxide and sodium oxide base-10 logged concentrations projecting high sodium and salt glazes against the remaining technological groups. Ellipses represent 90% confidence interval for group membership.

Lake City, and one each from Ruby Valley, St. George, Fountain Green, and Parowan. Glaze group 7 (GL7) contains two samples from Salt Lake City.

The two remaining glaze groups are chemically distinct from all other groups. Three samples (group NAHI) are particularly high in sodium and were sampled from the towns of Hyrum, Fountain Green, and Panguitch. The final glaze group (group SALT) is made up of two samples from Ruby Valley (Figure 6). The SALT group is named for its extremely high sodium and low lead content. Eighteen remaining samples are unassigned.

Discussion

Samples from the Ruby Valley Pony Express Station divide into four compositional groups, with yellow and green glazes belonging entirely to GL1, brown glazes making up GL2, and Ruby Valley's single red glaze sample belonging to GL6. Ruby Valley is also the origin of the two most distinct samples in the dataset (SALT). These samples are remarkably high in sodium (Na_2O) and extremely low in lead (PbO) content, suggesting these samples were salt-glazed (or glazed using powdered high-sodium glass) as a substitute for the more common lead-glazed variety. Records indicate the only workshop historically known to produce salt-glazed stoneware in the Utah territory was E. C. Henrichsen's Pottery in Provo, Utah (*24, 25*), suggesting salt-glazed pottery from the study area likely originated at this production facility. However, ceramics from Henrichsen's workshop have yet to be analyzed by INAA, and cannot be attributed as the production locale at this time.

More than half the samples from compositional group GL3 are from Fountain Green, with the remaining samples evenly distributed between Salt Lake City (n=4) and Panguitch (n=3). These data either suggest communication among potters from these towns in regard to the technology of glaze recipes or the distribution of finished wares from a single production locale. Because 80 percent of samples in group GL3 were translucent or yellow glazes, it is likely that technology for making these particular glazes was widely distributed throughout Utah. Compositional group GL4 is primarily made up of brown or reddish brown glazes from the town of Parowan. Although Parowan is represented in several compositional groups, the vast majority of glazes analyzed were a brown or dark brown/reddish color with few exceptions. This suggests the preference by potters in Parowan for darker/opaque glazes in their pottery manufacture.

Group GL5 consists of glazes that are exceptionally dark and opaque brown or red. This suggests a precise glaze formula was adapted to obtain a darker hue, consisting of elevated quantities of manganese (MnO) and copper (CuO). Group GL6 consists of a variety of colored glazes on ceramic samples primarily from Salt Lake City (approximately 67%). The data suggest this compositional group represents a locality of production instead of a recipe for a specific color. The two samples that make up group GL7 are both from Salt Lake City and are dark brown in color. The elemental composition of these samples reflects the earlier description of GL5 samples in that manganese (MnO) and copper (CuO) were elevated compared to more translucent glazes.

The high sodium (Na_2O) samples from group NAHI are neither uniform in their coloring nor their town of origin. However, due to these samples' elevated levels of potassium (K_2O), sodium (Na_2O), and magnesium (MgO), it is likely powdered glass was used as a main ingredient to the glazes for these samples. The use of glass for glazes was not uncommon in the manufacture of historic pottery and often resulted in chemical composition similar to those samples in the NAHI group.

Conclusion

The lack of glaze formulas available in the written records for historic Latter-day Saints potters has forced researchers to venture into compositional analyses in the form of LA-ICP-MS to determine the variability of glaze recipes throughout Utah. Quantitative analysis of 75 glaze colors from historic Latter-day Saints' pottery has yielded invaluable information regarding a variety of technological differences in glaze manufacture. For instance, to increase opacity and darkness in hue, Latter-day Saints potters seemed to add manganese (MnO) and copper (CuO) to their recipes.

Lead glazes seem to be the most popular technology used throughout Utah, though potters also seemed to have experimented with melted glass and salt in their glaze recipes. Although few examples were found in the analyzed ceramic assemblage, it is likely additional sampling and analysis of glazes from the region would yield further evidence of salt and melted glass technologies. In several instances, glaze recipes seem to have been restricted to a particular color at manufacturing facilities. One example being pottery manufactured for the Ruby Valley Pony Express Station, where chemically distinct glaze recipes produced characteristic brown (GL2) glazes.

Although small sample sizes preclude any strenuous statistical testing of compositional groups, the separation of groups in bivariate space as well as principle component analysis (PCA) suggests the validity of the defined groups. It is likely that additional sampling and analysis will provide further evidence to technological styles suggested in this study.

Acknowledgments

The authors would like to acknowledge the cooperation given to us by the staff and/or members of the history Department of the Church of Jesus Christ of Latter-Day Saints, the Church Museum of History and Art, This Is The Place Heritage Park, the Daughters of the Utah Pioneers, the Mariott Library at the University of Utah, the Utah State History Archive, and the State's Antiquities division. The LA-ICP-MS research was conducted at the MURR Archaeometry Laboratory supported by NSF grant No. 0504015.

References

1. Scarlett, T. J. Ph.D. thesis, University of Nevada-Reno, Reno, NV, 2002.
2. Scarlett, T. J.; Speakman, R.J.; Glascock, M.D. *Hist. Archaeol.* **2007**.

3. Rice, P. M. *Pottery Analysis: A Sourcebook*; University of Chicago: Chicago, IL, 1987.
4. Mortensen, E. N. *Pioneer Potters [Hyrum section]*; Daughters of the Utah Pioneers: Salt Lake City, UT, 1958; pp 476–479.
5. Nielson, E. C. *The Development of Pioneer Pottery in Utah*; Brigham Young University: Provo, UT, 1963.
6. Burt, O. W. *The Utah Magazine*; January, 1948; pp 27–30.
7. *Heart Throbs of the West Volume II*; Carter, K. B., Ed.; Daughters of the Utah Pioneers: Salt Lake City, UT, 1940.
8. *History of Box Elder County;* Forsgren, L. W., Ed.; Daughters of the Utah Pioneers: Box Elder County, UT, 1937.
9. Langenbeck, K. *The Chemistry of Pottery*; Chemical Publishing Company: Easton, PA, 1895.
10. Norton, F. H. *Ceramics for the Artist Potter*; Addison-Wesley Publishing Company, Inc.: Reading, MA, 1956.
11. Burton, D. K. *Settlements of Unitah County*; Unitah County Library: Vernal, UT, 1998.
12. Speakman, R. J.; Neff, H.; Glascock, M. D.; Higgins, B. J. In *Archaeological Chemistry: Materials, Methods, and Meaning*; Jakes, K. A., Ed.; ACS Symposium Series No. 831; American Chemical Society: Washington, DC, 2002; pp 48–63.
13. Larson, D. O.; Sakai, S.; Neff, H. In *Laser Ablation ICP-MS in Archaeological Research*, Speakman, R. J.; Neff, H., Eds.; University of New Mexico Press: Albuquerque, NM, 2005; pp 94–103.
14. James, W. D.; Dahlin, E. S.; Carson, D.L. *J. Radioanal. Nucl. Chem.* **2005,** *263*, 697–702.
15. Larrera, M. T.; Gomez-Pinilla, I.; Farinas, J. C. *J. Anal. Atom. Spectrom.* **1997**, *12*, 1324–1332.
16. Kennett, D. J.; Sakai, S.; Neff, H.; Gossett, R.; Larson, D. O. *J. Archaeol. Sci.* **2002**, *29*, 443–455.
17. Little, N. C.; Kosakowsky, L. J.; Speakman, R. J.; Glascock, M. D.; Lohse, J. S. *J. Radioanal. Nucl. Chem.* **2004**, *262*, 103–110.
18. Hill, D. V.; Speakman, R. J.; Glascock, M. D. *Archaeometry* **2004**, *46*, 585–606.
19. Neff, H. *J. Archaeol. Sci.* **2003**, *30*, 21–35.
20. Rodriguez-Alegria, E. Ph.D. thesis, University of Chicago, Chicago, IL, 2002.
21. Gratuze, B. *J. Archaeol. Sci.* **1999**, *26*, 869–882.
22. Gratuze, B.; Blet-Lemarquand, M.; Barrandon, J. N. *J. Radioanal. Nucl. Chem.* **2001**, *247*, 645–656.
23. Speakman, R. J.; Neff, H. In *Laser Ablation ICP-MS in Archaeological Research*; Speakman, R. J.; Neff, H., Eds.; University of New Mexico Press: Albuquerque, NM, 2005; pp 1–15.

24. Henrichsen, L. E.; Henrichsen. K. B. *The Henrichsen Pottery: A Historical Sketch of the Provo Pottery Company and its Proprietor/Operator E. C. Henrichsen*; Marriott Library, Special Collections Department, University of Utah: Salt Lake City, UT, 1984.
25. Henrichsen, K. B. *Utah Historical Quarterly* **1988**, *56*, 360–395.

Chapter 25

Fingerprinting Specular Hematite from Mines in Botswana, Southern Africa

Adam V. Kiehn[1], George A. Brook[2], Michael D. Glascock[3], Jonathan Z. Dake[3], Lawrence H. Robbins[4], Alec C. Campbell[5], and Michael L. Murphy[6]

Departments of [1]Geology and [2]Geography, University of Georgia, Athens, GA 30602
[3]Research Reactor Center, University of Missouri, Columbia, MO 65211
Department of Anthropology, Michigan State University, East Lansing, MI 48824
[5]P.O. Box 306, Crocodile Pools, Gaborone, Botswana
[6]Kalamazoo Valley Community College, Kalamazoo, MI 49003

Specular hematite, or specularite (Fe_2O_3), was a valued cosmetic in Southern Africa during the Late Stone Age (LSA) and Iron Age (IA), and there are ancient mines throughout the region. Early explorers reported that it was applied to the body and hair with animal fat and was traded extensively. We analyzed specularite samples from seven prehistoric mines in Botswana, including five in the Tsodilo Hills, and one each at Dikgatlampi and Sebilong to test the feasibility of geochemically fingerprinting sources. Most of the mines examined exploited specularite-rich schist or hydrothermal quartz veins. The heavy mineral fractions (density >2.9 g/cm^3) of 73 samples, separated by heavy liquid and consisting largely of specularite, were subjected to instrumental neutron activation analysis (INAA). Multivariate statistical analyses of the compositional data suggest that specularite can be fingerprinted to specific mines or groups of mines separated by tens to hundreds of kilometers. The method appears robust enough for determining the provenance of archaeological samples throughout the region and so could provide valuable information on past trade routes and patterns.

Archaeological occurrences and early historic accounts indicate that in Southern Africa specularite was heavily exploited and highly valued as a cosmetic from the Early Iron Age through the 19th century *(1, 2)*. It was prepared for use by grinding and mixing with grease then applied to the hair and body, giving the wearer a shimmering appearance *(2, and references therein)*.

Specular hematite or specularite (Fe_2O_3) is a mineral with steel gray to black color, metallic luster, tabular or platy crystals, and a specular or micaceous habit. It can occur in igneous, metamorphic, and sedimentary geologic settings, but the most desirable crystals, such as those sampled for this study, are most commonly found in hydrothermal vein deposits and metamorphosed hematite-rich sedimentary rocks. Specularite from these deposits is chemically and geologically similar to other pigments, such as red and yellow ochre, which should allow similar methods to be applied to these materials also.

Being able to provenance archaeological materials is crucial to understanding why people procured, processed and used them. In the case of specularite, very little has been done to provenance archaeological finds. In Australia, the isotopic and magnetic characteristics of sedimentary hematite have been used successfully to fingerprint sources *(3–6)*. However, specularite in Southern Africa has not been fingerprinted despite widespread archaeological occurrence. The ability to source specularite, which is common at archaeological sites, could provide answers to such contentious questions as how prehistoric groups in the Kalahari region interacted *(7–9)*. This study is a first attempt to geochemically characterize and fingerprint specularite deposits from mines in Botswana by applying multivariate discriminant analysis to chemical compositions determined by instrumental neutron activation analysis (INAA).

Objectives and Assumptions

The main objectives of the study are to determine if specularite from prehistoric mines can be distinguished by its geochemical signature, and at what scale this is possible (mine, local, regional). To answer these questions, specularite ore was collected in Botswana from individual mines at Sebilong and Dikgatlampi near Gaborone, and from five separate mines in the Tsodilo Hills. Samples from the Tsodilo Hills mines, which are at most a few kilometers apart, provide a good test of whether specularite can be sourced to individual mines at the local scale. As the Tsodilo Hill mines are hundreds of kilometers from the Sebilong and Dikgatlampi mines, ores from these mines provide a test of how well specularite can be sourced at the regional scale.

Another aim of this study is to determine which elements are most reliable and useful for fingerprinting specularite and other heavy minerals commonly associated with it. Preliminary analyses and previous literature *(5, 10, 11)* indicate that transition metals (TM) and rare earth elements (REE) have unique signatures associated with genetic and metamorphic processes. This study will test how well these two groups of elements can source specularite from different mines in Botswana by comparing the results of discriminant analysis for both a TM model and a TM+REE model *(10, 11)*. The final step, if this preliminary work is successful, is to develop a consistent and applicable methodology and begin building a database of sources that can be used in future archaeological provenance studies.

All provenance studies make implicit assumptions. According to the provenance postulate, the sourcing of materials requires that "there exist some qualitative or quantitative chemical or mineralogical difference between natural sources that exceeds the qualitative or quantitative variation within each source" *(12)*. We assume that this postulate applies to the specularite sources examined here and that the inter-source variation is greater than the analytical errors of the techniques employed. Provenance studies are also limited by the fact that an artifact cannot be definitely assigned to a single source until all possible sources have been analyzed. However, this study does not attempt to source artifacts but merely to determine if specularite ores from different mines or mining areas can be differentiated on the basis of geochemical characteristics.

Sampling Regions

Samples of specularite ore were collected in 1995 and 2005 from five prehistoric mines at the Tsodilo Hills and from single mines at Sebilong and Dikgatlampi in Botswana (Figure 1). The samples were generally taken from tailings piles at or near the mines and some of the most specularite-rich fragments collected. This was done as an attempt to take samples from the tailings discarded by ancient miners that would have been most similar to the material collected and actually used and eventually deposited in archaeological contexts and to avoid analyzing samples with specularite content incomparably lower than archaeological samples. A few hundred grams for each distinct bedrock type was obtained from each mine site to help determine similarities and differences between the mines, and several samples were taken from different areas of each of the larger mines. The result of the sampling was roughly 2–4 samples per mine at Tsodilo Hills and 8–12 for Dikgatlampi and Sebilong.

Tsodilo Hills

The Tsodilo Hills World Heritage Site is located in northwestern Botswana, in the Kalahari Desert, about 40 km from the Okavango River and is surrounded by relict linear sand dunes. This is one of the most interesting archaeological localities in southern Africa, in part because hills in the region are rare, the nearest to Tsodilo being about 200 km away. Evidence of human activity at the

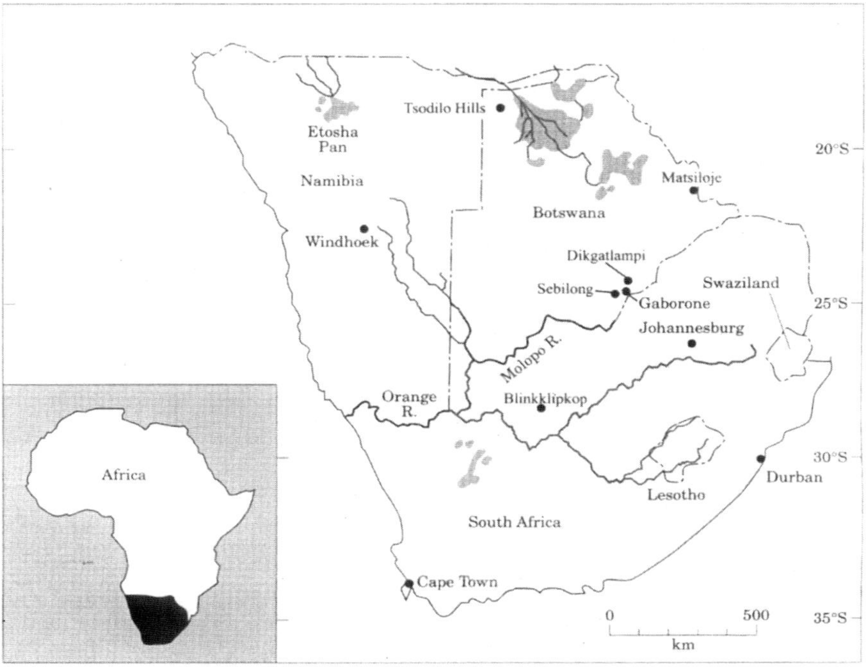

Figure 1. Map of Botswana showing the locations of prehistoric specularite mines examined in this study (modified from (13)).

Tsodilo Hills includes two Early Iron Age (EIA) villages, more than 4,000 Middle (MSA, c. 150k B.P. - 30k B.P.) and Late Stone Age (LSA, c. 30k B.P. - 1k B.P.) rock paintings in rock shelters and caves, and numerous mica schist and specularite mines *(7–9, 13–16)*. There are four named hills at Tsodilo: Male, Female, Child, and North *(7)*. Male Hill rises 410 m above the surrounding landscape, the Female Hill is 300 m high, and the other two hills are 40 m high or less. All are composed of compositionally mature Precambrian quartzites and schists that have been faulted and altered in many places. Despite this metamorphism, hematite concentrated on the original bedding planes still shows clear cross-bedding features as shown in Figure 2. Judging by the common occurrence of these features at the shallow mines, they may have signaled to prehistoric miners that significant specularite deposits were present. Evidence for occasional specularite mining at Tsodilo Hills dates to before 5300±160 ^{14}C yr B.P. at Rhino Cave *(13)*. Radiocarbon ages for soot deposits formed during fire-spalling indicate that there was intensive specularite mining in the period ca. A.D. 800–1025, which coincides with occupation of the EIA villages of Nqoma and Divuyu on the top of Female Hill.

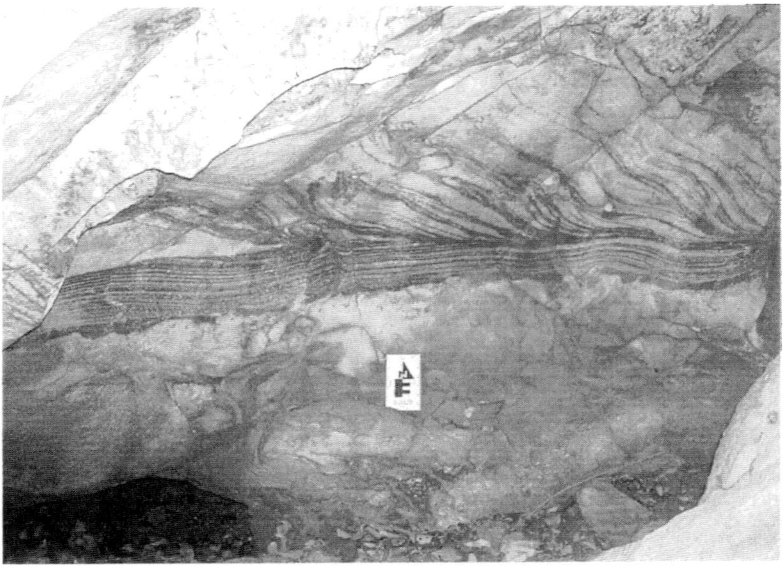

Figure 2. Hematite concentrated on paleo-crossbedding planes in Precambrian quartzite at a prehistoric mine in Tsodilo Hills.

The large scale of production from the mines at this time implies that specularite was being traded or distributed beyond local populations. More than 1000 tons of rock may have been removed from Upper Male Hill mine alone *(7)*. Ostrich-eggshell and ceramic containers containing specularite have been found in Swaziland and Namibia, respectively *(17, 18)*. The Namibian find has no known source of specularite within 100 km or more, demonstrating long distance transport *(19)*. Glass beads and marine shells found at the Nqoma, dating to around 900 A.D., are evidence that the people living at Tsodilo were involved in trade networks extending as far as the Indian Ocean *(8)*.

Sebilong and Dikgatlampi

Sebilong and Dikgatlampi are located in southeastern Botswana, approximately 40 km west-southwest and 60 km north-northwest of Gaborone, respectively. The topographies of both areas consist of plains and flat alluvial valleys interrupted by sharp hills and tablelands. The Sebilong mine is high on a cliff face at the edge of a plateau in the Sesitajwane Hills between Thamaga and Moshupa *(20)*. The plateau is composed of the Mannyelanong Formation of the early to mid-Proterozoic Waterberg Group. This formation consists mainly of cross-bedded red quartz sandstones and is brecciated. Large grained specularite is concentrated in and near quartz veins located in these altered zones *(21)*.

Fire-spalling techniques were used to extract the specularite, and extensive mining is evidenced by tailings that form a scree from the workings near the top of the 80 m high cliff to its base where slag from iron-smithing has been found *(20)*. The site has not been excavated but thermoluminescence dates place intensive workings at the 14th C. A.D. while historical accounts indicate that mining had ended by the 19th century, with evidence for iron smelting at the base of the cliff dating to the latter part of this period *(20)*. In the area around the village of Thamaga, there are several hills and ridges similar to the one exploited at Sebilong that may also have been mined for specularite in prehistoric times.

The Dikgatlampi workings are located in one of a series of low hills rising from a plain 8 km southeast of the village of Lentsweletau. There are roughly two dozen pits in a two-hectare area. The pits exploit specularite found in brecciated zones of the sandstone and conglomerate that form the hill. As at Sebilong, these rocks belong to the Mannyelanong Formation of the Waterberg Group *(20, 21)*. Excavations in one of the pits revealed pottery and metal tools thought to date to the 17th Century, but direct dates have not been obtained. Although there is no clear evidence of iron working or prehistoric settlement, earlier mining of the site cannot be ruled out *(20, 22)*.

Research Methods

Laboratory Methods

Hand samples, ranging from about 50–300 grams each, were crushed using a Plattner mortar then ground to a monomineralic grain size with an alumina mortar and pestle. The heavy mineral fraction was obtained by gravity separation in Tygon tubing filled with a sodium polytungstate solution with density greater than 2.9 g/cm^3. Once the samples had completely settled the tubing was clamped to separate the light and heavy fractions. Each fraction was rinsed several times with deionized water and the sodium polytungstate solution was recovered by vacuum filtration. The heavy mineral fraction was washed from the filter paper with ethanol or acetone and then allowed to dry at room temperature. The samples were inspected and described at 10–45x binocular magnification before INAA analysis. Generally, the remaining heavy mineral fraction weighed a few grams, but ranged from 0.5–15 grams depending on the size of the original hand sample and its relative hematite content. The goal of the heavy mineral separation was to concentrate the specular hematite in order to reduce spurious variation introduced by non-hematite minerals. We recognize that in trying to source archaeological samples they would have to be concentrated in the same way before analysis.

Instrumental neutron activation analysis was conducted at the University of Missouri Research Reactor (MURR). Samples of approximately 50–100 mg were subjected to long and short irradiations using the same methodology used on pottery and other materials with appropriate reference standards *(23)*. The analysis provided abundances of 33 elements for 72 samples.

It is important to note that the heavy mineral separation process only removes the majority of the light minerals but not all of them. Also, any heavy minerals other than hematite will remain in the sample. The main goal of the separation process was simply to concentrate the heavy minerals, especially hematite, and remove the light minerals so that our tailings samples more closely approximated the specularite-rich archaeological samples. Although the remaining light minerals also contribute to the elemental signature of the samples, the specific elements examined in this study are generally found at concentrations at least an order of magnitude higher in the hematite than in the light minerals *(10)*. Thus, we assume that the contribution of the light minerals to the geochemical signature is negligible for the elements of the subcomposition in consideration.

Data Analysis

Statistical Procedures

The INAA data were subjected to discriminant function analysis (DFA), which uses the variance of variables within predetermined groups of observations in the data set (e.g., individual mines) to develop linear functions of the variables that can best discriminate between the groups. The centroid of each group is calculated using the geometric means of the values with the discriminant functions as the axes. Each observation is then classified based on its Mahalanobis distance to the group centroids with the nearest group centroid as the most probable candidate for group membership. The accuracy of a model can be evaluated by examining the percentage of misclassified observations in this, the initial classification.

Cross-validation of the DFA model is conducted by casewise deletion, re-estimation of functions, and classification. In other words, for each observation in the data set, that observation is omitted and the discriminant functions are re-estimated using the full data set minus that observation. Then that observation is classified based on the re-estimated functions. The accuracy of the cross-validation can be used to evaluate the reliability of the DFA and the potential impact of group outliers. In essence, the cross-validation process is the same process used to determine provenance of an unsourced archaeological sample where the discriminant functions are developed independently of the sample and then used to determine its most likely source.

The actual analyses were conducted using SPSS 13.0. The results presented were obtained by entering all variables together and using the within groups covariance matrix options. The stepwise entering of variables was also conducted in parallel to each test because it allows the exclusion of variables if they do not significantly contribute to the model's variance and allows the evaluation of each element's utility for fingerprinting hematite at a given scale. However, the presented methodology was chosen because of a lack of clear reasons based in statistical theory for the alternative, because the alternatives had identical cross-validation results, and because the groups generated by simultaneous entering of variables had a nearly identical form but were slightly more discrete than by stepwise entering of variables.

Previous work using subcompositions, or selected groups of elements from the total composition of the rock, in geochemical data analysis has suggested that raw compositional data should not be examined using multivariate methods such as discriminant function analysis *(24)*. The reasons for this include the constrained nature of the data. Several transformations have also been suggested in the geochemical statistical literature, and for the purposes of this study the additive log-ratio transform (ALR) is used. ALR was one of the first transformations suggested by Aitchison (1986) and serves to unconstrain the data while preserving the elemental ratios regardless of the subcomposition or denominator chosen.

The ALR transformation process consists of taking the selected group of elements and dividing each element by an arbitrarily chosen denominator, in most cases one of the elements of the group that is present in but variable across all samples. The logarithm, natural or base 10 depending on personal preference, is then taken of each ratio. The group of elements used in this study consists of 11 transition and rare-earth elements and Fe was chosen for the denominator, which essentially normalizes each transformed subcomposition by its relative hematite content. Additionally, the small degree of analytical error associated with most INAA analyses does not significantly affect the log-ratios in this study because most elements in the subcomposition occur at levels several orders of magnitude lower than Fe.

Data Preparation

Data preparation began by excluding any possibly unreliable and irrelevant data from the set of 33 elements. The heavy mineral solution could have imparted excess sodium (Na) and was thus ignored. The mortar and pestle used could have contaminated the aluminum (Al). Based on the hardness of the material and relative contribution to the sample this is likely not a significant problem; however the role of Al was monitored closely during the data analysis. Previous literature on the geochemistry of specularite *(10, 11)* and preliminary analyses in this study suggested that the most likely and reliable elements for fingerprinting specularite sources would be transition metals and other ion species that could easily substitute for Fe^{3+} in the crystal lattice. The ratios of these elements would be affected by initial composition and subsequent petrogenic and metamorphic processes which would hopefully leave a unique signal for each source.

One drawback to this transformation is that a log-ratio cannot be taken for elements not present or below detection limits. Sophisticated multiple imputation techniques, simple replacement, and variable omission are strategies for dealing with these zeroes *(24)*. In this preliminary study, elements not present in all samples were simply omitted from the analysis as most of the relevant transition metals were present in all samples. Due to these restrictions on missing abundances, 20 elements (As, Ba, Ca, Ce, Cs, Dy, Eu, Hf, K, Lu, Nd, Ni, Rb, Sr, Ta, Tb, U, Yb, Zn, Zr) were excluded from the analyses.

Elements with ionic species commonly associated with quartz and mica minerals such as Na, Zr, and Hf are more likely to co-vary with those minerals rather than hematite. The ratio of these latter elements to Fe-substituting elements would be dependent on the relative amounts of hematite and other minerals in the sample rather than just the rock-forming processes. For this reason, elements that commonly substitute into quartz and other abundant minerals in the samples were excluded from the discriminant analysis. Some rare earth elements may also play a role in fingerprinting the sources, and their importance was tested by comparing discriminant analysis models with and without La, Sm, and Th.

Samples that were very low in hematite (<20 wt %), even after heavy mineral separation, should be excluded from or used with caution in the initial discriminant function analysis to avoid skewing the function with samples that may have a significant contribution to the geochemical signature from light minerals. This limitation only excluded one sample, a very specularite poor schist. Also, similar samples of such low specularite content are unlikely to be used in the described archaeological contexts.

Replicate samples were also excluded to avoid artificially increasing confidence in the classification functions. The subtraction of these seven replicate samples reduced the data set to 60 samples. For the interest of space and clarity, observations and elements excluded from DFA have been omitted from Tables III–V, but may be obtained from the primary author. After excluding limiting the data set to TMs and REEs common to hematite and without missing values, a subcomposition of 11 elements remained. This data set of 11 elements (Fe, Sc, Ti, V, Cr, Mn, Co, Sb, La, Sm, and Th) was transformed into 10 log-ratios for the remaining 60 non-replicate samples for discriminant analysis. The TM and REE model included all 10 log-ratios while the TM model excluded the log-ratios of Sb, La, Sm, and Th. Comparison of these two models tests whether the REEs can significantly contribute to the discriminating power of the TM model.

Results and Discussion

Promising results were obtained using discriminant analysis on the INAA geochemical data, the results of which are summarized in Tables III, IV, and V at the end of this paper. However, the analysis was not as successful as those conducted on some other materials including obsidian *(25)*. The three mining areas were successfully separated at the regional scale (Figures 3 and 4) but results were mixed at the local, mine-to-mine scale. The statistical analysis also showed that when REEs were added they contributed to the TM and REE model but the resulting model was not significantly better than the TM model.

The regional scale discriminant analysis was conducted using 60 samples grouped into three mining regions; 27 from Tsodilo Hills, 21 from Sebilong, and 12 from Dikgatlampi. Figure 3 shows the discriminant function plot for transition metals and rare earth elements (TM+REE). Discriminant analysis was successful in this case, with all 60 cases correctly classified by the initial functions and only misclassifying the one major outlier from the Tsodilo Hills group with the Dikgatlampi samples during cross-validation. The absolute values of the standardized canonical function coefficients, summarized in Table I, show the relative contribution of each element to the function. While only the results of the discriminant analysis using simultaneous entering are presented graphically here, the variables omitted by the stepwise method (Co, Sb, La) can be considered to contribute insignificantly at this scale. The coefficients of Function 1 indicate that Sc, Cr, Mn, and the REEs are useful for identifying Tsodilo Hills specularite from that from the other regions. The coefficients of Function 2 mainly indicate that Sc and Cr can be used to separate Dikgatlampi from Sebilong specularite.

In general, a simpler model is easier to implement and interpret, especially when using a log transform on geochemical data, because useful elements sometimes occur below detection limits and cannot be included in multivariate analysis without adjustment. Therefore, if the simpler TM model does not show a drastic change of form or loss of accuracy compared with the more complicated TM+REE model, it is the preferred option in most cases. In Figure 4 the horizontal axis of the TM model has been arbitrarily flipped, but otherwise the groups are nearly identical to the TM+REE model.

Table I also shows the same classification accuracy for both models. So, despite high standardized coefficients in the TM+REE model, the REEs are not as useful for fingerprinting these mining regions as are Sc, Mn, and Cr.

The local scale discriminant analysis was conducted using the same elements as in the regional analysis for 27 samples from Tsodilo Hills grouped

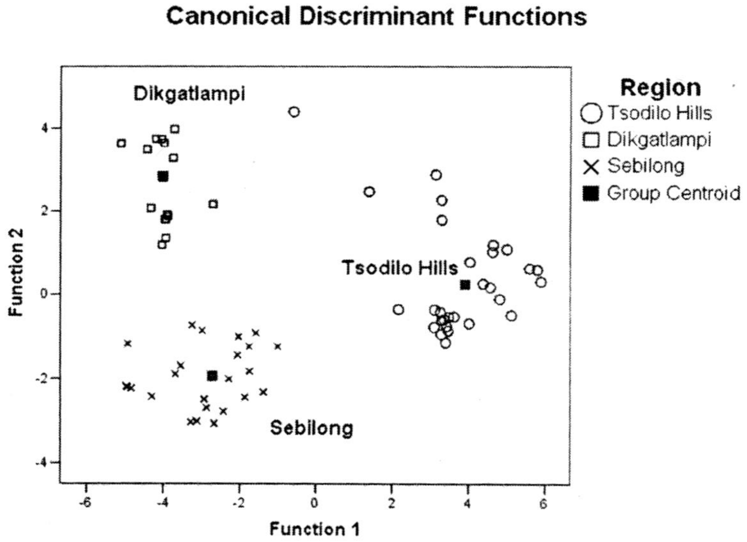

Figure 3. Regional discriminant function plot of TM and REE Model.

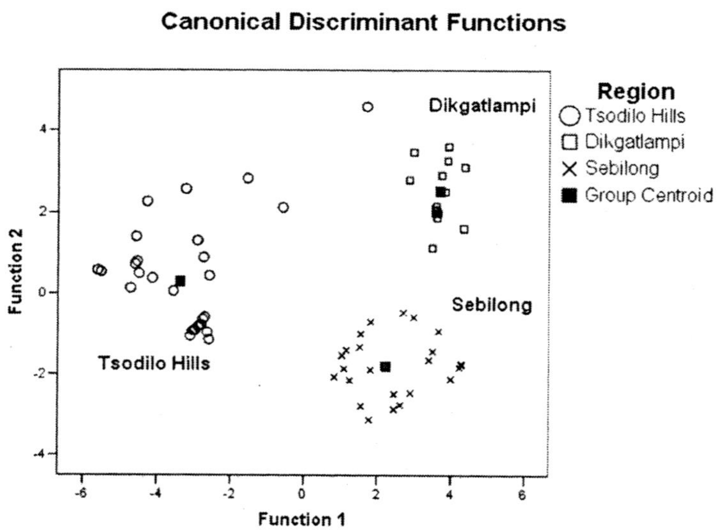

Figure 4. Regional discriminant function plot of TM model.

Table I. Regional Discriminant Analysis Summary

	TM and REE Model		*TM Model*	
Initial Misclassifications	0/60		0/60	
Cross-Validation Errors	1/60		1/60	
	Function		*Function*	
	1	*2*	*1*	*2*
alr(Sc)	-1.012	0.968	1.065	1.456
alr(Ti)	1.232	-0.153	-0.81	-0.294
alr(V)	-0.764	0.21	0.89	-0.068
alr(Cr)	-1.041	-0.889	0.72	-0.743
alr(Mn)	1.261	-0.163	-1.266	-0.186
alr(Co)	0.292 [a]	0.136 [a]	-0.329 [a]	0.146 [a]
alr(Sb)	-0.122 [a]	0.106 [a]	-	-
alr(La)	1.721 [a]	0.612 [a]	-	-
alr(Sm)	-2.531	0.286	-	-
alr(Th)	1.215	-0.395	-	-

[a]Denotes coefficient of variables omitted by stepwise discriminant function analysis.

into five individual prehistoric mines; 2 from Big Mine, 7 from Below Sex Rocks Mine, 5 from Lower Male Hill Mine, 11 from Upper Male Hill Mine, and 2 from Near Sex Rocks Mine. The very small sample sizes from several of the mines add to the concern of accuracy and reliability of the method at this scale. Hopefully this problem can be addressed by the sampling strategy of future work to add confidence to the conclusions.

Fingerprinting individual mines in a small area and in similar geologic formations, as with the dozens of mines at Tsodilo Hills, was not as successful as separating mining regions at some distance from one another. The results summarized in Table II show that specularite from some mines is geochemically too much alike to be distinguishable using either the TM or TM+REE models. In both models, the discriminant analysis misclassified two samples by the initial functions and five, or nearly one quarter, of the samples during cross-validation.

The misclassification of samples is apparent in Figure 5. Although some samples plotted in this figure are replicate samples that were not used in the calculation of the discriminant functions but only classified by them. The TM+REE model clearly separates three of the mines, however, the other two mines, Below Sex Rocks and Upper Male Hill, have a great deal of overlap, and it is in these fields where all of the misclassified samples occur. Interpretation of the standardized coefficients is rather complicated with so many variables, but indicates that La, Sm, Mn, and Cr are most useful for fingerprinting with this data set. Surprisingly, the stepwise analysis of this model excluded all of the REEs and Sc even though their coefficients suggested they have discriminating power.

Table II. Local Scale Discriminant Analysis Summary

	TM and REE Model		*TM Model*	
Initial Misclassifications	2/27		2/27	
Cross-Validation Errors	5/27		5/27	
	Function		*Function*	
	1	*2*	*1*	*2*
alr(Sc)	0.047 [a]	-0.305 [a]	-0.347 [a]	-0.62 [a]
alr(Ti)	3.575	1.616	1.379	1.078
alr(V)	-3.364	1.961	-3.223	1.294
alr(Cr)	5.165	1.844	1.088	0.654
alr(Mn)	8.418	2.86	3.92	0.588
alr(Co)	-4.004	-4.341	-0.253	-1.711
alr(Sb)	-3.306 [a]	-1.053 [a]	-	-
alr(La)	-8.043 [a]	-6.198 [a]	-	-
alr(Sm)	7.046 [a]	5.217 [a]	-	-
alr(Th)	0.869 [a]	1.095 [a]	-	-

[a]Denotes coefficient of variables omitted by stepwise discriminant function analysis.

The discriminant plot of the TM model for Tsodilo Hills (Figure 6) shows a small but important change from the TM and REE model. In the TM model, Big Mine no longer plots separately from the Below Sex Rocks and Upper Male Hill mines. This indicates that the REEs excluded from this model can play an important role in distinguishing Big Mine from the others. Despite this graphic change in form, the classification accuracy remains unchanged from the TM and REE model, so it appears that the REEs are only somewhat useful for fingerprinting mines at Tsodilo Hills but their contribution to the overall model is statistically insignificant.

The results from Tsodilo Hills are also further complicated by geologic differences and similarities between and within the mines. Whereas within the Sebilong and Dikgatlampi mines there is a rather homogeneous distribution of rock types and mineral assemblages, the mines at Tsodilo Hills have low metamorphic grade schist and quartzite rocks adjacent to hydrothermal vein deposits which all contain high quality specularite. In fact, much of the overlap found in Figures 5 and 6 is due to greater similarity among similar rock types located at different mines than among different rock types found within an

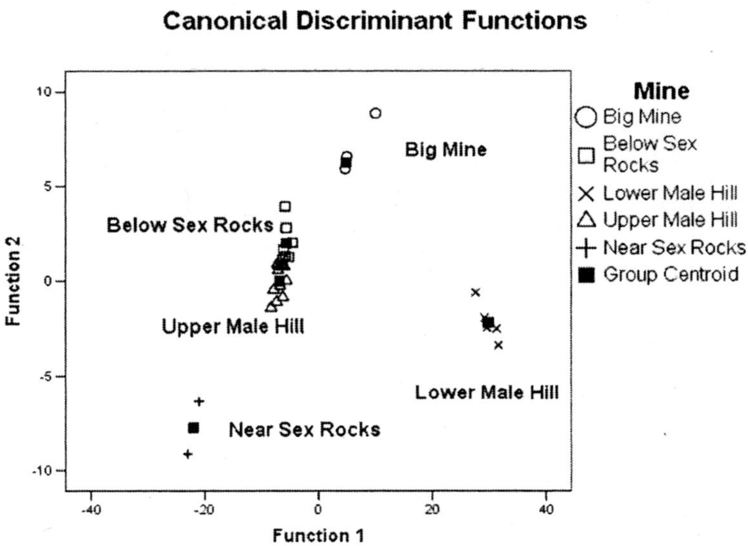

Figure 5. Local scale discriminant function plot of TM and REE Model.

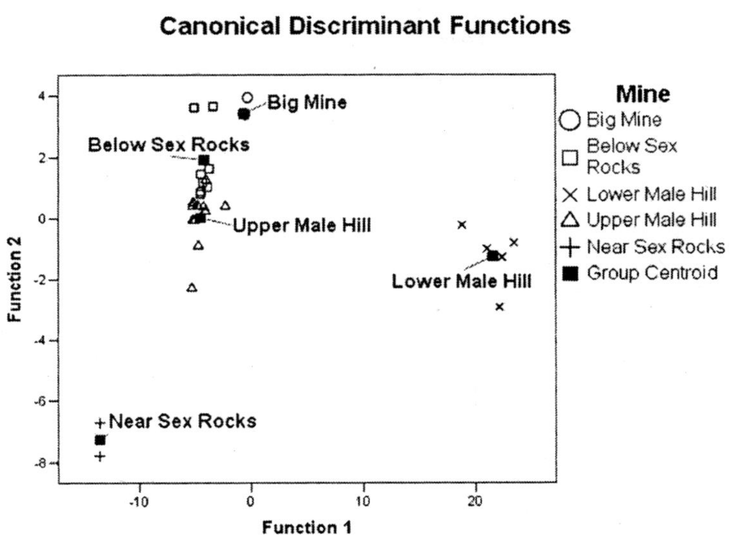

Figure 6. Local scale discriminant function plot of TM Model.

individual mine. In the case of Tsodilo Hills, specularite found in schist at the Upper Male Hill mine is more similar to that in schist at the Below Sex Rocks mine than it is to specularite from a hydrothermal vein deposit located within the same Upper Male Hill mine. Thus, unlike the regional scale, all of the assumptions required for accurate provenance are not completely valid at the local scale for Tsodilo Hills.

Unfortunately for future prospects of provenance at this scale, further discriminant analysis of the TM and REE model after subdividing the Upper Male Hill mine into two groups, one for schist (S) and one for hydrothermal vein (V) specularite, still shows the overlap visible in Figures 4 and 5. This is illustrated in Figure 7, where the Male Hill hydrothermal vein plots discretely from all other groups, but the Male Hill schist group continues to plot closely with the Below Sex Rocks mine samples. Thus, the problem is not simply a case of grouping by rock type or formation rather than mine or geographic unit, but rather a more complicated combination of processes recorded in the geochemical signature of the specularite.

The overlap effect due to rock types does not appear to be a problem at the larger regional scale, as the difference among rock types and formations between the far-flung mining regions easily dominate the geochemical signatures.

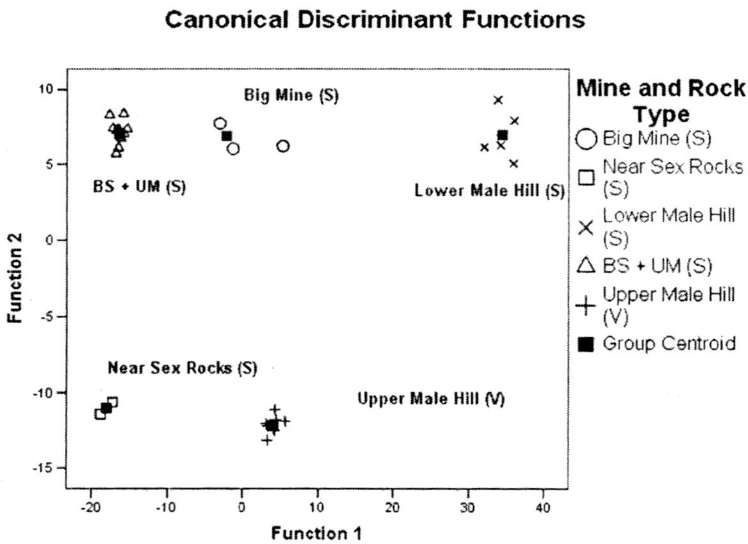

Figure 7. Local scale discriminant analysis plot of TM model after subdividing groups into schist (S) and hydrothermal vein (V) rock types.

Table III. Results of INAA on Heavy Mineral Fraction of Specularite Samples from Tsodilo Hills

Sample	*Mine*	*Rock ype*	*Fe*[a]	*Sc*	*Ti*	*V*	*Cr*	*Mn*	*Co*	*Sb*	*La*	*Sm*	*Th*
BMO030	Big Mine	Schist	384445	6.6	19882	295	448	473	24.8	4.7	42.4	4.9	27.9
BMO031	Big Mine	Schist	75816	4.6	7090	93	96	102	7.8	3.1	61.9	8.6	25.3
BSO009	Below Sex Rocks	Schist	398649	6.3	18390	217	358	308	24.6	6.5	41.7	6.4	26.6
BSO010	Below Sex Rocks	Schist	518539	8.3	24111	304	458	414	32.0	8.2	64.6	9.2	45.8
BSO023	Below Sex Rocks	Schist	582752	8.8	25304	314	525	455	36.1	8.1	34.8	6.1	29.7
BSO024	Below Sex Rocks	Schist	587780	8.9	26512	315	530	470	36.9	8.2	43.4	7.0	33.3
BSO025	Below Sex Rocks	Schist	536097	8.6	26200	296	497	422	32.7	8.8	60.6	9.2	39.5
BSO026	Below Sex Rocks	Schist	372062	6.0	24719	262	412	290	24.8	12.8	54.7	6.5	35.7
BSO027	Below Sex Rocks	Schist	463287	6.9	24533	349	482	371	26.9	8.5	45.5	5.2	23.4
LMO011	Lower Male Hill	Schist	149703	12.6	21739	81	273	519	33.4	17.2	54.8	7.9	13.0
LMO012	Lower Male Hill	Schist	225002	11.6	29063	93	396	819	51.1	23.0	41.2	6.5	15.4
LMO013	Lower Male Hill	Schist	405648	28.9	43029	119	758	1484	82.9	27.2	53.5	7.5	26.5
LMO020	Lower Male Hill	Schist	381625	14.1	50670	119	649	1359	81.2	40.4	122.6	16.9	31.7
LMO021	Lower Male Hill	Schist	331678	12.0	37119	123	554	1358	74.1	34.1	115.8	15.8	43.8
MHO001	Upper Male Hill	Schist	229288	6.6	7916	362	45	543	30.1	0.9	32.6	5.7	12.2
MHO002	Upper Male Hill	Vein	619028	11.1	32323	1276	54	1453	86.8	1.0	1.0	0.2	0.4
MHO003	Upper Male Hill	Schist	422094	7.2	20712	221	393	313	25.9	7.4	57.3	8.1	33.1
MHO004	Upper Male Hill	Schist	568112	8.9	22799	311	514	443	35.3	7.6	42.4	7.6	36.4
MHO005	Upper Male Hill	Vein	626277	11.5	33072	1146	29	1443	83.8	1.5	0.7	0.2	0.6
MHO006	Upper Male Hill	Schist	201870	12.2	13573	376	57	467	29.0	4.1	33.9	3.2	11.3
MHO016	Upper Male Hill	Schist	591650	8.3	22267	292	531	453	36.5	6.8	29.8	4.9	23.7
MHO017	Upper Male Hill	Vein	583941	8.9	24164	865	44	1313	72.0	1.3	0.8	0.2	0.5
MHO018	Upper Male Hill	Vein	579679	10.4	31272	1161	39	1492	85.7	1.2	3.4	0.7	0.6
MHO019	Upper Male Hill	Vein	521745	13.0	31391	925	227	939	82.6	9.6	1.4	0.5	2.2
MHO022	Upper Male Hill	Vein	619950	11.2	34530	1176	29	1470	83.1	1.5	0.7	0.2	0.3
NSO028	Near Sex Rocks	Schist	491514	6.8	4555	319	74	645	31.4	0.1	141.6	8.4	85.2
NSO032	Near Sex Rocks	Schist	418127	8.3	4519	319	69	565	25.9	0.2	182.0	12.4	122.0

[a]All units are parts per million (ppm).

Table IV. Results of INAA on Heavy Mineral Fraction of Specularite Samples from Sebilong Mine.

Sample	*Region*	*Fe*[a]	*Sc*	*Ti*	*V*	*Cr*	*Mn*	*Co*	*Sb*	*La*	*Sm*	*Th*
SEB033	Sebilong	472002	2.9	1296	522	108	27	0.6	1.5	2.4	0.5	3.7
SEB035	Sebilong	519507	2.4	2017	353	252	42	1.2	4.4	3.7	0.9	4.3
SEB036	Sebilong	520279	2.6	1615	125	165	25	0.6	0.9	7.8	1.2	4.2
SEB037	Sebilong	506981	2.9	2154	125	218	31	1.0	1.1	10.7	1.6	6.0
SEB038	Sebilong	607889	3.2	3005	664	212	44	0.9	2.8	2.7	0.6	3.0
SEB039	Sebilong	525077	3.0	5774	564	751	55	2.4	7.6	4.0	1.5	8.8
SEB040	Sebilong	620645	4.4	5311	573	1399	57	2.5	3.0	6.6	1.6	10.9
SEB041	Sebilong	634535	4.8	7496	652	2124	69	4.0	4.2	7.2	2.1	11.7
SEB042	Sebilong	526704	7.9	17837	891	3182	164	11.4	5.9	14.7	5.2	32.6
SEB043	Sebilong	571693	4.5	7319	736	843	52	1.8	3.6	5.1	1.4	14.6
SEB044	Sebilong	459044	6.9	42924	1139	1525	160	12.6	72.7	20.2	5.8	35.1
SEB045	Sebilong	458501	8.2	30494	958	1529	151	11.5	64.9	24.4	7.2	43.2
SEB046	Sebilong	453499	13.3	38922	1268	9820	557	22.2	17.9	13.4	7.2	61.0
SEB047	Sebilong	488817	15.6	45812	1325	8012	466	21.3	20.8	15.4	8.4	70.5
SEB048	Sebilong	458089	16.8	62564	1284	5717	220	16.0	39.3	14.9	7.2	54.9
SEB051	Sebilong	573207	2.4	1468	346	19	31	0.6	1.5	3.1	0.6	1.4
SEB052	Sebilong	681000	3.3	487	373	18	41	0.6	2.5	2.0	0.2	0.9
SEB053	Sebilong	538922	2.9	102	369	57	39	0.6	2.7	4.4	0.7	3.0
SEB054	Sebilong	348825	2.8	674	341	52	27	1.3	1.6	3.8	0.7	3.5
SEB055	Sebilong	590432	4.7	4127	969	684	45	1.4	3.4	3.3	0.9	7.6
SEB058	Sebilong	551358	5.8	4947	1018	1062	52	2.1	3.8	3.7	1.5	11.6

[a]All units are parts per million (ppm).

Table V. Results of INAA on Heavy Mineral Fraction of Specularite Samples from Dikgatlampi Mine

Sample	*Region*	*Fe*[a]	*Sc*	*Ti*	*V*	*Cr*	*Mn*	*Co*	*Sb*	*La*	*Sm*	*Th*
DIK059	Dikgatlampi	480870	23.2	11622	1037	435	180	3.4	11.5	17.1	6.4	30.9
DIK060	Dikgatlampi	504170	15.1	3065	1091	195	118	1.8	12.3	68.9	7.1	12.7
DIK061	Dikgatlampi	359160	11.0	3828	809	266	129	2.1	10.5	22.8	5.8	17.0
DIK062	Dikgatlampi	392145	15.5	4285	828	248	148	2.9	15.1	762.3	63.2	48.5
DIK063	Dikgatlampi	450938	21.3	3314	1047	252	205	5.7	15.1	698.0	56.6	34.9
DIK064	Dikgatlampi	465508	37.4	5373	1218	264	328	5.5	32.1	114.3	15.5	27.7
DIK065	Dikgatlampi	382762	27.9	4925	979	262	243	4.7	26.7	206.3	23.7	28.7
DIK066	Dikgatlampi	529025	25.0	5227	1009	321	203	106.5	30.8	31.5	7.6	62.8
DIK070	Dikgatlampi	539791	21.6	5415	1228	308	222	4.7	35.2	27.1	6.0	18.2
DIK071	Dikgatlampi	569009	35.3	4168	1356	341	276	5.5	102.3	216.3	27.8	27.6
DIK072	Dikgatlampi	543620	32.1	4304	1337	418	323	15.1	101.0	221.1	26.6	27.6
DIK073	Dikgatlampi	495998	34.3	6084	1178	612	269	632.6	30.9	431.5	68.5	84.1

[a]All units are parts per million (ppm).

However, this problem makes it unlikely that reliable provenance can be established at the local or individual mine scale in all cases. However, the results are not all negative because in most cases the range of possible sources can still be narrowed down and in some cases an individual mine identified, albeit with much less confidence than at the regional scale.

Conclusions

The findings presented above indicate that specularite deposits in Botswana can be fingerprinted successfully at the regional scale but much less reliably at the local scale. The abundances of the transition metals are most useful in fingerprinting sources using discriminant function analysis, although REEs may provide additional separation power in some cases. The results also suggest that specularite from samples of ore, and therefore also archaeological occurrences, may be concentrated by heavy liquid techniques before INAA analysis. This reduces the contribution of lighter minerals that might confuse the geochemical signal from the hematite and other heavy minerals in samples with relatively little hematite.

Now that the feasibility of geochemically fingerprinting and classifying specularite sources using discriminant function analysis has been established, archaeological samples can be analyzed along with geological samples from probable sources in provenance studies and associations made with confidence. The chemical similarities between specularite and ochreous materials, especially red ochre, also imply that this method could be successfully extended to those materials.

Acknowledgements

Funding was provided by grants from the National Science Foundation (#9520982 and #0313826) to the University of Georgia and NSF #0504015 to the University of Missouri. We are grateful to the Office of the President of Botswana for granting permission to do this work and to the National Museum of Botswana for collaboration.

References

1. Burchell, W. J. *Travels in the Interior of Southern Africa*; C. Struik: Cape Town, 1822.
2. Thackeray, A. I.; Thackeray, J. F.; Beaumont, P. B. *South African Archaeological Bulletin* **1983**, *38*, 17–25.

3. Jercher, M.; Pring, A.; Jones, P. G.; Raven, M. D. *Archaeometry* **1998**, *40*, 383–401.
4. Smith, M. A.; Pell, S. *J. Archaeol. Sci.* **1997**, *24*, 773–778.
5. Smith, M. A.; Frankhauser, B. *An Archaeological Perspective on the Geochemistry of Australian Red Ochre Deposits: Prospects for Fingerprinting Major Ore Sources*; Canberra, 1996.
6. Smith, M. A.; Frankhauser, B.; Jercher, M. *Proc. Prehist. Soc.* **1998**, *64*, 275–292.
7. Robbins, L. H.; Murphy, M. L.; Campbell, A. C.; Brook, G. A. *Curr. Anthro.* **1998**, *39*, 144–150.
8. Denbow, J. R.; Wilmsen, E. N. *Science (Washington, D.C.)* **1986**, *234*, 1509–1515.
9. Robbins, L. H.; Murphy, M. L.; Brook, G. A.; Ivester, A. H.; Campbell, A. C.; Klein, R. G.; Milo, R. G.; Stewart, K. M.; Downey, W. S.; Stevens, N. J. *J. Archaeol. Sci.* **2000**, *27*, 1085–1113.
10. Harding, C. J. M.S. thesis, Rand Afrikaans University, Johannesburg, South Africa, 2004.
11. Wernicke, R. S.; Lippolt, H. J *Geophys. Res. Lett.* **1994**, *21*, 345–347.
12. Weigand, P.; Harbottle, G.; Sayre, E. V. In *Exchange Systems in Prehistory*, Earle, T. K.; Ericsion, J. E., Eds.; Academic Press: New York, 1977; p 15.
13. Robbins, L. H.; Murphy, M. L.; Campbell, A. C.; Brook, G. A. *Botswana Notes and Records* **1996**, *28*, 23–45.
14. Campbell, A. C.; Denbow, J. R.; Wilmsen, E. N. In *Contested Images: Diversity in Southern African Rock Art Research*; Dowson, T. A.; Williams, D. L., Eds.; Witwatersrand University Press: Johannesburg, 1994; pp 131–158.
15. Robbins, L. H. *Nat. Geographic Res.* **1990**, *6*, 329–338.
16. Robbins, L. H.; Murphy, M. L.; Stewart, K. M.; Campbell, A. C.; Brook, G. A. *J. Field Archaeol.* **1994**, *21*, 257–264.
17. Jacobson, L. *Cimbebasia* **1977**, *2*, 228–233.
18. Humphreys, A. J. B. *S. African J. Sci.* **1974**, *70*, 48.
19. Beaumont, P. B. *S. African J. Sci.* **1973**, *69*, 140–146.
20. Campbell, A. C.; Main, M. *Guide to Greater Gaborone*; Botswana Society: Gaborone, 2003; p 292.
21. Aldiss, D. T.; Tombale, A. R.; Mapeo, R. M. B.; Chiepe, M. *The Geology of the Kanye Area*; 33; Geological Survey Department, Republic of Botswana: Gaborone, 1989; p 170.
22. Cohen, G. *Botswana Notes and Records* **1977**, *9*, 17–19.
23. Glascock, M. D.; Neff, H.; Vaughn, K. J. *Hyp. Interact.* **2004**, *154*, 95–105.
24. Aitchison, J., *The Statistical Analysis of Compositional Data*; Chapman and Hall: London; New York, 1986; p xv, 416.
25. Glascock, M. D. *Acc. Chem. Res.* **2002**, *35*, 611–617.

Chapter 26

Instrumental Neutron Activation Analysis of Ochre Artifacts from Jiskairumoko, Peru

Rachel S. Popelka-Filcoff[1,2], Nathan Craig[3], Michael D. Glascock[2], J. David Robertson[1,2], Mark Aldenderfer[4], and Robert J. Speakman[2]

[1]Department of Chemistry and [2]Research Reactor Center, University of Missouri, Columbia, MO 65211
[3]Department of Anthropology, University of California, Santa Barbara, CA 93106
[4]Department of Anthropology, University of Arizona, Tucson, AZ 85721

Ochre is very common in the Terminal Archaic-Early Formative archaeological site of Jiskairumoko, (Rio Ilave, Lake Titicaca Basin, southern Peru). Within the site, ochre was found on tools, palettes, and in burials and soil deposits within structures in several contexts, suggesting both symbolic and functional uses of ochre. Variations in the color and contexts imply possibilities for different uses of ochre.. Instrumental neutron activation analysis was used to analyze the ochre samples found in Jiskairumoko. Multivariate analysis of the elemental data by principal components analysis suggests trends in the data related to the compositional variation of ochres on the site. Further analysis of the ochre will lead to conclusions about the variation in composition of the ochres from Jiskairumoko and possible archaeological conclusions about ancient technologies and uses of ochre on the site.

This paper describes the analysis of 65 archaeological ochres from different contexts from the Terminal Archaic-Early Formative site of Jiskairumoko, Peru. Ochre is very common in many archaeological contexts worldwide, and also on the site of Jiskairumoko. Instrumental neutron activation analysis (INAA) was conducted to evaluate the heterogeneity of the elemental compositions of ochre artifacts used by residents of Jiskairumoko. An important correlation between the chemical signature of artifacts and their place of manufacture is the basis for many artifact provenance or "sourcing" studies. Knowledge of the chemical basis and variation of sources and of artifacts allows distinction between ancient sources, artifact geochemistry and an understanding of ancient technologies and provides further insight into ancient cultures and behavior.

Multi-elemental analysis has been applied to ochre and related materials far less often than it has to either ceramics or obsidian and other archaeological materials. Some exceptions include: the Natufian remains from El-Wad Cave, Israel (*1*); modern ochre source sampling in Australia (*2*); Mousterian deposits at Qafzeh Cave, Israel (*3*); and Paleo-Indian contexts at the Powars II site, Wyoming (*4*). While the earlier literature often quantifies elemental values from ochre analysis, the variation in the samples and sources is not often discussed or quantified. Geochemical variance has to be understood and quantified to interpret the variation in a given source or site. In general, the variance in only the major elements such as Ba, Si, Mn and others was used to address differences between sources or artifacts. Popelka-Filcoff (*5*), Erlandson (*6*), and Mrzlack (*7*) have taken initial steps to measure elemental variation, especially trace elemental variation in ochre. Efforts are taking place to expand the database of chemical analysis, especially trace element data for ochres from the United States and around the world to improve ochre and iron oxide characterization for archaeological questions. The ochre analysis from Jiskairumoko provides information on chemical variability in the ochre used at this site and also contributes to a worldwide collection of analytical data on ochre.

Jiskairumoko-a Late-Terminal Archaic Village

Jiskairumoko is located in the Lake Titicaca basin (Figure 1), which is an important region for understanding cultural change. Some of the earliest evidence for cultural complexity in the Andean highlands is found in the northern basin, at the Formative site of Pucara, and in the Tiwanaku Empire that developed in the southern basin during the Middle Horizon (*8–10*). Despite prolonged archaeological research in the region, comparatively little attention has been paid to early prehistoric time periods. Excavations at Jiskairumoko

Figure 1. Geographic location of Jiskairumoko.

(Figure 2) represent the first systematic study ever conducted in the Lake Titicaca Basin of an open air Archaic Period residential site. The site is located in the Rio Ilave drainage in the southwestern basin. Jiskairumoko was excavated in natural layers employing a *décapage* approach (*11–13*), and horizontal exposures were carefully documented with digital photography and geographic information systems for site level photomapping.

Excavation results indicate that a shift from mobile to increasingly sedentary life-ways occurred in the Rio Ilave drainage about 3300 BC, when people appear to have begun living in pithouses(*13*). During this time, a low-level food producing economy, c.f. (*14*), involving use of domesticated plants and animals, developed and lasted until about 1300 BC. X-ray fluorescence (n=63) (*15*) and portable X-ray fluorescence (PXRF) (n=550) (*16)* geochemical sourcing of obsidian from the Rio Ilave, Rio Huenque, Rio Ramis and Rio Azangaro drainages demonstrates a highly redundant reliance on the Chivay source and implies the existence of regular trade networks (*17*). Gold beads were discovered in a burial that consisted of an adult and a child, and radiocarbon dated to about 2100 BC. About 1300 BC, residential architecture at the site shifted to occupation of above ground dwellings, which are associated with the appearance of ceramics. Ochre was found within a wide variety of contexts spanning the approximately two thousand years of the site's intermittent, but recurrent use during the Archaic to Formative transition.

A comprehensive analysis of temporally diagnostic projectile points from survey data indicates that use of obsidian increased significantly during the end of the Late Archaic and during the Terminal Archaic. This is the period of time during which camelids appear to have been domesticated in the region (*18–27).* Together these data imply that the use of regular camelid caravans that transported goods over long distances developed on the altiplano around the end of the Late Archaic (*28,29*)*.* This is the time period during which Jiskairumoko is first occupied. Jiskairumoko currently represents the earliest known example of an early village in the Lake Titicaca Basin and provides data for understanding the origins of plant cultivation and sedentary life. The deposit is a record of this highly significant, yet poorly understood cultural period (*30).* Figure 2 is a map of the site and outlines the blocks where samples of ochre were found.

Archaeological Contexts of Ochre Encountered at Jiskairumoko

To interpret the significance of the presence of ochre, one must first attempt to discern if the use of the material was practical or symbolic. Practical uses of ochre include application as a preservative in curing hides, as an adhesive for hafting stone tools, and as medicine. Even practical objects can have symbolic

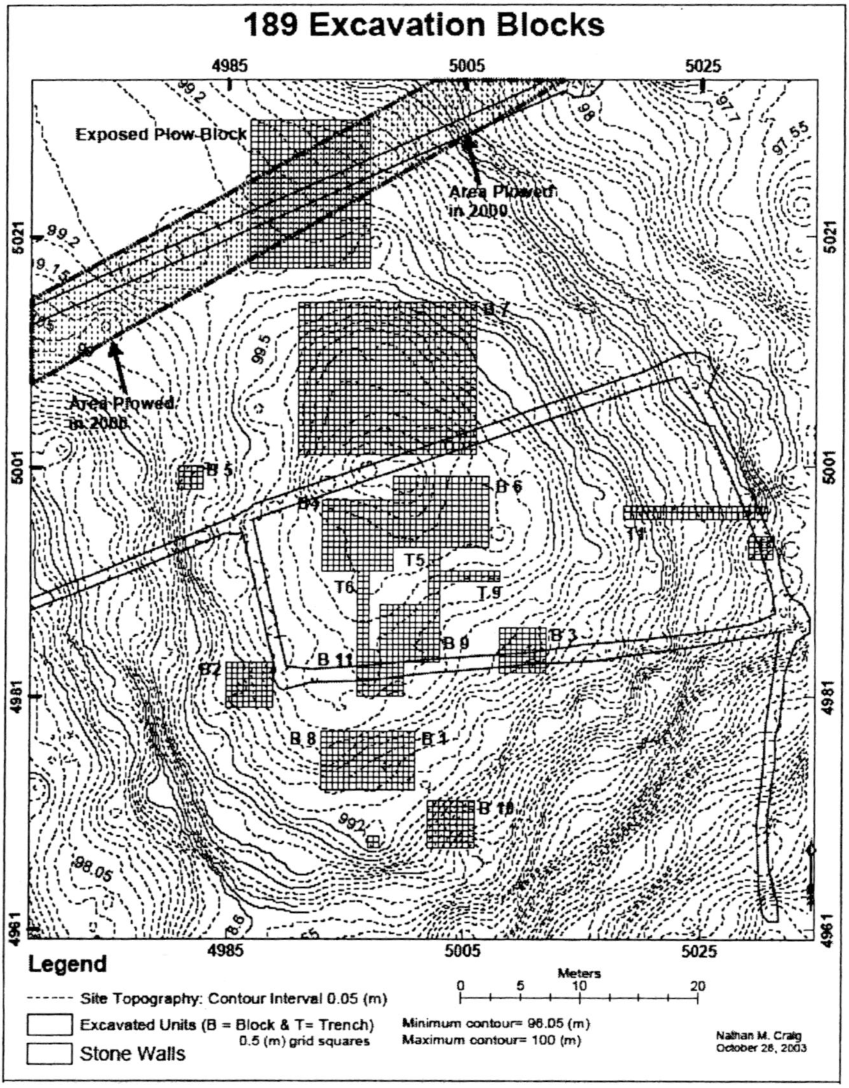

Figure 2. Jiskairumoko site map.

meaning. Most mammalian vision is dichromatic and composed of two cone pigments known as S and M (*31*). The S cone pigment is maximally sensitive to short and the M to medium visible wavelengths.

Some primates, including humans, have a third cone pigment, known as L that is maximally sensitive to long wavelengths. This type of vision is called trichromacy. L cone pigments may have facilitated the detection of red fruit against a green foliage background (*32–34*), or the detection of new red foliage (*35*). Red is one of the three most fundamental color categories present in all known languages (*36*), making it a human universal(*37*). Thus, importance of red appears to have a biological foundation, and ochre's red color makes it a likely symbol in many cultural settings. Symbolically, ochre has many associated connotations including: menstrual blood (*38)*, blood in hunting magic ritual (*4*), renewal (*39)*, potency (*40)*, and life stage transitions (*41)*. Symbolic meanings of the color red are probably multivocal (*42)*.

Key contexts where ochre was encountered at Jiskairumoko are listed in Table I. Ochre was found inside a stone box at the base of a burial cairn at a nearby Terminal Archaic (2578–2302 cal B.C.E.) site of Kaillachuro (*43)*. Ochre was not observed on either the working or hafting edges of any chipped stone tools recovered from Jiskairumoko or Kaillachuro, though ochre has been observed as a hafting adhesive on projectile points in the assemblage from Qillqatani (*18*). None of the contexts listed in Table I suggest purely or even largely practical uses for ochre at Jiskairumoko.(*44*)

Burned ochre fragments were found in a lens of ash on the floor of a Late Archaic pithouse next to the structure's central hearth, which dated to 3385–3078 cal B.C.E. This thin, *in situ*, floor assemblage rests on the surface of the pithouse beneath an uninterrupted stratum of secondary rubbish fill that extended across the structure's extent. A key part of the pithouse's floor assemblage was a perpendicular debris arc that extended from the structure's central hearth. It was composed of ash, lithic debitage, burned camelid bone, fire altered rock, and a large concentration of thermally altered ochre (*44*). Photomapping was applied to piece plot the assemblage. The debris arc was explored using multivariate statistics, and defined using raster based unconstrained clustering.

These analyses revealed a perpendicular arc of debris expected for habitual hearth-side workspaces (*45*). The hearth-associated scatter found on the floor of this Late Archaic pithouse provides clear evidence that ochre was thermally processed to change its color. Further it shows that thermal processing of ochre occurred in a domestic context though it was primarily used in symbolic and ritual contexts. At Jiskairumoko, preparation of ochre for symbolic and ritual purposes appears to have been embedded in the domestic setting. The occupation bears early evidence for prestige goods in burial contexts. Yet, ochre preparation does not appear to have occurred in specialized architecture. Social differentiation was taking place, but preparation of ritual materials took place in domestic contexts.

Table I. Contexts, Dates and Associations for Ochre at Jiskairumoko

Ochre context	*^{14}C Date B.C.E.*	*Associations*
Drop zone from heat treatment in Pithouse 1	3385–3078	Clay lined hearth inside a structure not far from deer bones.
1 palette and 1 abrader	2473–2119 2072–1878	On edge of and next to Pithouse 2
Ground dust at base of Burial 3	1883–1680	Older adult female buried with a lapstone
Ochre stained manos associated with Burial 4 and ochre stained lapstone from just above burial.	No direct dates	Unsexed adult buried with burned and unburned faunal remains from at least two individuals one adult and one juvenile
Ground dust at base of Burial 5	No direct dates	Individual buried with numerous red chert flakes placed at distal end of interment.
External ochre stain outside Rectangular Structures 1 and 2	No direct dates	Rock-soil feature and split rock altar
Ochre stained animal bones	No direct dates	Rock-soil feature and split rock altar
Ochre stained manos, groundstone fragments found in rock pavement.	No direct dates	Rock pavement, perhaps related to cooking

Ochre stained animal bones and a large ochre stain were found: in a rock-soil feature, near a split rock altar, associated with a concentration of obsidian debitage, near early ceramic fragments, all located outside one of Jiskairumoko's later above ground structures. This context for ochre decorated camelid bones implies economic dimensions to ochre's symbolic meaning for Jiskairumoko's Early Formative occupants. Additionally, ochre stained manos and stone pallets were recovered from many burial and refuse contexts at Jiskairumoko. Together these finds indicate that Jiskairumoko's residents used ochre as a pigment, and that it often occurred in symbolically charged contexts like burial assemblages.

Project Goals

This paper intends to accomplish several goals in the analysis of the Jiskairumoko ochre. The chemical variability of ochre artifacts must be determined in order to evaluate its potential role in ancient trade and regional inter-community interaction. Ideally, it would be valuable to know: if ochre was a non-local resource, where it came from, if the source is local or not, and whether multiple sources were used. Primary sources of ochre in the region have not been located, making "sourcing" studies impossible; still, an exploratory study of ochre chemical variability is warranted for several reasons. There is little knowledge about the chemical variability of ochre sources, reporting results will increase this understanding. Chemically variable groups could suggest use of either different sources or different portions of a variable source. Are there chemical differences in ochre remains recovered from different contexts of Jiskairumoko's multi-component occupation? Multiple structure types are located at the site and ochre is found in association with each kind. Are these ochres chemically similar or is there high variability? The site's earliest residential context bears evidence of thermal processing of ochre. There is also ochre in later secondary refuse heaps. Is there large variability in the ochres found in the thermal processing context? Some have suggested that the color of ochre, (perhaps a result of thermal treatment) is independent of geographic location (*46*). Do the ochre samples found in this early residential context appear chemically similar to ochres recovered in secondary refuse contexts? By understanding the chemical composition of ochre in different contexts, variability in use and procurement of the material can be explored.

Ochre Sampling

A total of 65 ochre samples recovered from a range of excavated contexts at Jiskairumoko were analyzed by INAA (Table II). Ochre samples were recovered as both singular lumps and powdered material. Only ochre masses were analyzed to prevent any possible contamination issues. Samples ranged in color from yellow to dark brown, with varying shades of red and red-brown in between. These color variations may indicate differing iron concentrations or particle sizes, and may be due to inherent impurities in the mineral (*47*) or heat treatment. Table II lists the examined samples with Munsell values.

Samples were obtained from pithouse floor assemblages, the perpendicular debris arc found next to the hearth of a Late Archaic pithouse, later expedient hearth features situated in secondary pithouse fill, secondary debris fill in pithouses, from along the edges of later rectangular structures, and from an exterior ritual area. Future studies will include a sample of archaeological ochre recovered from each of the burial contexts where it was encountered.

Table II. Sampled Ochre Characteristics: Block, Provenance, Level, Context and Munsell Color

ID	*Block*	*Provenance*	*Context*	*Munsell*	*Color*
305	3	X35a/7/VIIIa	Pithouse fill	2.5 YR 3/4	Dark red-brown
306	3	Y36c/7/VIIIa	Pithouse fill	2.5 YR 5/6	Red
307	3	Y34a/oa/VII	Matrix	10R 3/6	Dark red
308	3	W34b/5/VII	Pithouse Fill	2.5 YR 4/8	Red
309	3	Y35d7/Viiia	Pithouse External Stain	2.5 YR 6/6	Light red
310	3	Y34d/oa/VII	Matrix	2.5 YR 2.5/4	Dark red-brown
311	4	GG24c/oa/IV	Outside Rectangular Structure 1	2.5 YR 3/3	Dark red-brown
312	4	FF21c/F14/VII	Hardpan-Occupation Interface	2.5 YR 2.5/3	Dark red-brown
313	4	KK24b/oa/IV	Outside Ritual Area	2.5 YR 3/3	Dark red-brown
314	4	JJ21c/F1/VII	Hardpan-Occupation Interface	2.5 YR 2.5/4	Dark red-brown
315	4	II21d/F1/VII	Hardpan-Occupation Interface	2.5YR 3/4	Dark red-brown
316	6	KK-25-C/F3/III	Outside Ritual Area	2.5 YR 3/6	Dark red
317	6	LL-31B/F6/III	Rectangular Structure 2 Edge	10 YR 4/8	Dark yellow brown
318	6	MM32A/Oa/III	Outside Rectangular Structure 2	2.5 YR 3/2	Dark red-brown

319	6	LL28d/HF1/III	Rectangular Structure 2 Edge	2.5 YR 2.5/3	Dark red-brown
320	6	LL28c/ob/IV	Rectangular Structure 2 Edge	2.5 YR 3/4	Dark red-brown
321	6	LL27c/ob/IV	Rectangular Structure 2 Edge	2.5 YR 2.5/3	Dark red-brown
322	6	LL25d/ob/IV	Outside Ritual Area	5 R 3/2	Dark brown
323	8	P22a/10/IV	Pithouse floor	10 R 3/3	Dusky red
324	8	Q23c/10/IV	Pithouse Occupation Surface	2.5 YR 3/4	Dark red-brown
325	8	P23a/10/IV	Pithouse Occupation Surface	10 R 3/2	Dusky red
326	9	KK21D/F1/IV	Late Archaic Pithouse Fill	10 R 5/2	Weak red
327	9	x26c/F8/x	Late Archaic Pithouse Fill	2.5 YR 3/3	Dark red-brown
328	9	X26B/F8/X	Late Archaic Pithouse FIll	7.5 YR 3/2	Dark brown
329	9	X28d/F8/X	Late Archaic Pithouse Lower Fill	7.5 YR 3/2	Dark brown
330	9	Y28a/F8/X	Late Archaic Pithouse Fill	10 R 3/2	Dusky red
331	9	Y28a/F6/IX	Late Archaic Pithouse Fill	10 R 3/2	Dusky red
332	9	Y28b/F6/IX	Late Archaic Pithouse Fill	7.5 YR 3/1	Very dark gray
333	9	Y28c/F6/IX	Late Archaic Pithouse Fill	7.5 YR 3/2	Dark brown
334	9	Y28d/F6/IX	Late Archaic Pithouse Fill	7.5 R 3/4	Dark brown
335	9	Y28d/F6/IX	Late Archaic Pithouse Fill	7.5 R 4/1	Dark gray
336	9	Y27a/F1/ii	Plowzone	7.5 R 4/3	Brown

Continued on next page.

Table II. *Continued.*

ID	*Block*	*Provenance*	*Context*	*Munsell*	*Color*
337	9	X26c/F3/X	Late Archaic Pithouse Fill	2.5 YR 3/3	Dark red-brown
338	9	AA/28c/F2/IX	Hearth in Upper Late Archaic Fill	10 R 3/2	Dusky red
339	9	X28c/F12/IX	Late Archaic Pithouse Fill	10 R 3/3	Dusky red
340	9	Y28d/F8/X	Late Archaic Pithouse Fill	10 R 3/2	Dusky red
341	9	Z28b/F3/IX	Late Archaic Pithouse Fill	10 R 3/2	Dusky red
342	9	Y27/F7/vii	Late Archaic Pithouse Fill	10 R 3/2	Dusky red
343	9	X26d/F9/vii	Late Archaic Pithouse Fill	7.5 R 3/6	Dark brown
344	9	Y27b/F7/vii	Late Archaic Pithouse Fill	7.5 R 3/4	Dark brown
345	9	Y27a/F7/vii	Late Archaic Pithouse Fill	7.5 R 3/2	Dark brown
346	9	BB25d/F1/ii	Plowzone	2.5 YR 3/3	Dark red-brown
347	9	A25c/F1/ii	Plowzone	5 R 3/4	Dark brown
348	9	BB25b/F1/ii	Plowzone	10 R 3/4	Dusky red
349	9	Y25b/F1/ii	Plowzone	7.5 R 3/2	Moderate red brown
350	9	AA25b/F1/ii	Plowzone	7.5 R 3/2	Moderate red brown
351	9	BB24d/F1/ii	Plowzone	2.5 YR 3/3	Dark red-brown
352	9	AA28a/F1/ii	Plowzone	7.5 R 3/2	Moderate red brown
353	9	X26c/F8/viii	Late Archaic Pithouse Fill	7.5 R 3/3	Moderate red

					brown
354	9	Z27b/F7/IX	Late Archaic Pithouse Fill	10 R 3/4	Dusky red
355	9	Y27c/F8/viii	Late Archaic Pithouse Fill Near a Small Hearth	7.5 R 3/4	Dark brown
356	9	BB28a/F7/v	Hearth in Upper Late Archaic Fill	7.5 R 3/6	Dark brown
357	9	Z28c/F4/IX	Hearth in Upper Late Archaic Fill	7.5 R 3/2	Dark brown
358	9	AA26c/F1/ii	Plowzone	7.5 YR 4/4	Dark brown
359	9	AA24c/F2/iii	Plowzone	10 R 4/2	Dusky red
361	9	X24d/oa/v	Late Archaic Pithouse Fill	10 R 3/2	Dusky red
362	9	AA27b/F8/V	Late Archaic Pithouse Fill	10 R 3/2	Dusky red
363	9	Y25d/F9/viii	Outside Stain (Shallow)	7.5 R 2.5/4	Dark brown
364	9	Z27c/F2/IX	Hearth in Upper Late Archaic Fill	10 R 3/4	Dusky red
365	9	BB27d/oa/vii	Matrix	10 R 3/3	Dusky red
366	9	X26B/F9/xiv	Hearth in Base of Late Archaic Pithouse	7.5 R 2.5/4	Dark brown
367	9	Y27bF11/XIII	Hearth in Base of Late Archaic Pithouse	10 R 3/4	Dusky red
368	9	Y27d/F11/XII	Hearth in Base of Late Archaic Pithouse	10 R 3/3	Dusky red
369	9	Y27d/F11/XII	Hearth in Base of Late Archaic Pithouse	10 R 3/4	Dusky red
370	9	AA24c/F2/iii	Plowzone	7.5 YR 4/4	Dark brown

Experimental

The INAA analysis followed standard analysis parameters used in the Archaeometry Lab at the University of Missouri Research Reactor (*48*). Each sample was broken into smaller pieces with a rock hammer. The crumbled sample was then pulverized into a powder using a Brazilian agate mortar and pestle. The powdered ochre material was dried at 100 °C overnight before preparation for INAA. For both the short and long irradiations, approximately 60 mg of sample was used. In the short irradiation, the samples were irradiated in 1.2 mL high-density polyethylene vials for five seconds at a thermal flux of approximately 8.0×10^{13} neutrons cm^{-2} s^{-1}. After a decay of 25 minutes, the samples were counted for 720 seconds on a high resolution HPGe detector. For the mid and long irradiations, the samples were sealed in high purity quartz vials and irradiated for 24 hours at a thermal neutron flux of approximately 5.2×10^{13} neutrons cm^{-2} s^{-1}. The sample was allowed to decay for seven days, and the "mid" count data was acquired for 2,000 seconds on automated sample changers. The samples were allowed to decay an additional three weeks, and the "long" count data were acquired for 10,000 seconds. The comparator standards used in the INAA measurements were NIST SRM 1633a (Fly Ash) and SRM 688 (Basalt), and the quality control standards were NIST SRM 278 (Obsidian Rock) and Ohio Red Clay.(*48*) The elemental data is presented in Table III.

Mathematical and Statistical Treatment of Data

The raw concentration data were subjected to several mathematical and statistical transformations. The ratio of the element of interest to Fe helps offset inherent variation in Fe across the data set. A $\log_{10}$ transform is a standard statistical conversion for elemental data. This transformation reduces the "weighting effect" from very small to very large concentrations in the data (*48*). More details on the calculations and reasoning behind these transforms can be found elsewhere (*5*).

Although INAA can routinely measure approximately 30 elements by the methods described, several elements were below the detection limits, or were otherwise unreliable elements for ochre. For this study, 16 elements were used: As, Ce, Co, Cr, Dy, Eu, La, Mn, Nd, Sb, Sc, Sm, Sr U, V, Yb and Zn. These elements are similar to other ochre studies as those related to the "Fe-oxide signature" and not the surrounding minerals (*5*).

Statistical Operations

An initial study of the data was undertaken with a cluster analysis, to outline possible clusters and groups within the elemental data. This was performed with the Clustan software (ClustanGraphics, Edinburgh, Scotland). The cluster analysis included the use of a hierarchical tree diagram to display the results. Distances were calculated using a squared Euclidean distance. The linkage between groups was calculated by the increased sum of squares. The results of the cluster analysis outlined five distinct groups, henceforth referred to as Groups 1, 2a, 2b, 3a, and 3b.

A principal components (PC) analysis was also performed to evaluate the elements in the data set that contributed to the variance. PC plots graphically indicate a linear combination of original variables, oriented in the direction of greatest variance. PC space also displays the elements with the greatest variation by graphically displaying them with the longest vectors.

In addition, a canonical discriminant analysis (CDA) was performed on the groups defined by the cluster analysis. This statistical procedure was performed to evaluate group differences as defined by the cluster analysis. CDA analysis assumes that the groups are different and calculates the largest difference between the groups (*48*).

Results of Sample Variance

The concentrations of Fe ranged from as low as 4 weight percent to as high as 67 weight percent, with the majority of the samples in range from 35–50 % Fe. The mathematical transform of the elemental ratio to Fe helped to normalize the data for this range of Fe concentration in the samples. After the mathematical transform and PC analysis, several observations were made. Principal components 1 thru 8 described 95% of the total variance for the dataset. Figure 3 is a plot of PC2 vs PC4 as an example of one of several possible permutations. On this plot, the longest vectors represent the most variance in this PC space. Elements with the longest vectors in the various permutations of PC bivariate plots in total included: Co, Mn, Zn, Eu, Sm, Ce and La. Bivariate plots of these elements were explored to ascertain which pairs of elements could be used to visually describe group associations.

The cluster analysis identified five discrete groups based on the trace element chemistry of the selected ochre samples. These five groups were analyzed several ways to discern possible relationships between the variation in the elemental comparison of the artifacts and possible archaeological significance.

Bivariate plots based on archaeological context were examined to investigate potential spatio-temporal contextually specific groupings (Figures 4–6).

Table III. Elemental Data from the Jiskairumoko Ochres

ID	*As*	*Ce*	*Co*	*Cr*	*Dy*	*La*	*Mn*	*Nd*	*Sb*	*Sc*	*Sm*	*Sr*	*U*	*V*	*Yb*	*Zn*
305	449.2	166.0	57.7	12.5	1.1	169.5	1726.5	53.3	15.5	9.1	4.0	1778.4	5.4	311.9	0.6	193.5
306	392.3	128.8	5.7	34.0	2.3	130.5	176.3	40.3	27.4	12.9	4.2	893.5	6.6	240.9	1.7	121.3
307	191.3	273.6	5.2	9.7	60.7	298.9	200.3	56.1	8.0	8.5	14.7	3530.9	18.2	262.2	31.4	60.0
308	32.8	52.0	1.1	12.8	1.4	35.2	581.2	14.2	1.1	18.3	2.1	1021.6	2.6	321.1	1.7	84.1
309	9.2	93.0	6.7	63.8	2.7	53.3	226.1	35.4	0.9	12.6	5.1	652.5	5.1	115.9	1.7	116.0
310	734.7	34.5	9.1	31.7	3.4	33.0	538.9	18.9	3.8	4.2	3.3	0.0	2.7	390.9	1.7	313.0
311	81.3	161.6	1.8	3.0	1.1	221.4	58.8	26.2	4.6	0.9	2.6	842.5	5.9	60.8	0.3	11.3
312	297.1	171.4	14.3	8.9	1.6	127.5	1391.3	34.8	9.0	3.4	5.3	509.7	9.7	327.5	0.7	81.0
313	62.8	54.3	5.6	9.7	0.9	49.5	396.8	0.0	3.2	3.7	2.5	704.9	9.1	69.3	0.4	42.3
314	163.6	113.9	5.9	6.4	0.9	103.2	919.7	47.3	10.4	1.2	3.9	0.0	17.0	126.1	0.0	29.0
315	92.3	5.2	0.6	5.5	0.2	5.5	36.4	0.0	3.5	1.2	0.1	148.6	0.0	74.9	0.0	0.0
316	292.4	172.9	14.0	7.9	1.6	137.8	590.2	22.1	12.6	3.5	3.6	1277.8	6.5	337.6	0.7	62.6
317	106.1	225.8	1.2	164.6	1.2	285.1	65.0	32.4	5.7	6.2	3.8	3699.9	1.3	178.4	0.0	17.4
318	63.6	11.7	0.2	2.3	0.0	19.2	17.2	10.9	5.8	0.3	0.4	776.1	1.2	12.8	0.0	0.0
319	126.5	523.5	0.9	7.4	0.6	773.8	56.5	158.0	1.2	1.1	2.8	3641.9	6.0	42.3	0.7	0.0
320	280.8	567.5	14.6	8.2	2.4	508.8	843.8	96.9	14.6	1.8	9.4	633.4	31.8	317.5	0.0	47.2
321	68.8	30.9	1.5	2.5	0.0	35.8	56.2	0.0	4.4	0.8	0.7	453.5	1.3	67.2	0.0	3.9

322	69.2	172.7	566.9	5.0	0.0	135.3	4773.6	28.3	4.0	1.4	4.3	893.5	9.0	97.3	0.2	51.7
323	79.4	67.3	20.1	7.0	0.4	69.6	1491.6	16.6	4.6	4.2	1.5	1587.3	1.9	94.2	0.3	18.3
324	223.3	202.3	7.8	9.1	1.3	215.0	366.8	21.3	8.8	5.1	4.5	1589.8	14.3	263.6	0.9	57.9
325	200.5	201.5	11.6	8.6	1.1	173.6	838.5	41.6	10.6	2.1	4.6	994.2	14.9	154.2	0.7	27.1
326	108.8	77.6	1.7	53.2	0.0	117.7	71.5	67.5	2.2	3.6	1.3	1186.2	3.1	141.5	0.0	17.5
327	132.3	274.5	7.5	12.0	2.9	232.7	245.4	53.3	5.9	8.2	8.3	1245.1	7.5	166.1	0.9	39.5
328	94.9	190.3	4.3	11.0	2.4	154.9	331.2	52.5	4.4	4.3	6.1	962.5	5.8	121.0	0.7	26.0
329	40.3	70.0	14.2	21.4	0.7	50.4	1212.8	21.0	1.5	8.8	2.5	1010.2	3.6	148.4	0.8	82.8
330	100.2	150.2	1.4	10.3	0.6	168.1	165.1	25.5	3.0	5.2	1.7	2680.3	3.7	185.1	0.0	0.0
331	192.6	326.9	186.4	3.5	0.0	290.4	12144.8	55.5	15.4	1.4	8.8	0.0	18.1	190.4	0.0	22.6
332	102.2	111.2	4.1	11.3	0.8	100.6	266.9	0.0	2.5	8.3	2.3	1755.4	4.8	191.8	0.4	0.0
333	41.9	76.5	1.1	20.9	1.2	68.9	151.4	14.1	2.1	14.9	2.1	1515.6	4.2	162.0	0.7	16.2
334	93.1	210.1	4.0	7.1	2.1	172.6	187.3	53.6	4.9	6.6	6.8	1869.4	10.6	161.4	0.8	27.6
335	92.7	234.8	1.4	8.4	3.6	169.0	93.5	65.1	3.9	4.1	9.7	348.2	16.6	127.4	1.2	8.9
336	226.9	153.0	4.8	16.5	0.7	157.3	817.2	35.9	13.4	5.5	3.4	152.1	6.4	279.0	0.6	24.5
337	171.6	402.2	12.1	16.0	2.8	414.5	344.8	69.2	8.0	6.2	7.4	574.3	9.3	228.2	1.0	132.8
338	108.0	133.7	2.7	11.7	0.6	146.2	178.9	27.9	4.1	6.0	2.4	2416.5	3.6	124.7	0.3	18.0
339	137.3	230.7	3.2	15.7	1.1	235.8	202.7	39.8	4.8	3.5	2.9	2444.2	3.9	138.2	0.2	9.0
340	69.3	93.3	6.8	10.8	1.2	93.2	512.0	21.4	2.7	6.2	2.2	1808.9	4.3	275.0	0.4	60.7

Continued on next page.

Table III. *Continued.*

ID	*As*	*Ce*	*Co*	*Cr*	*Dy*	*La*	*Mn*	*Nd*	*Sb*	*Sc*	*Sm*	*Sr*	*U*	*V*	*Yb*	*Zn*
341	206.7	142.6	2.2	12.4	0.7	168.0	83.7	29.8	7.0	2.8	3.2	458.5	11.1	200.6	0.0	0.0
342	214.6	45.7	5.1	5.5	0.7	22.7	2060.8	0.0	12.0	3.3	2.5	0.0	13.6	140.2	0.0	27.9
343	131.7	524.9	1.9	11.2	4.3	446.2	132.2	124.7	9.3	3.2	14.8	655.0	6.6	175.7	1.4	0.0
344	349.5	95.7	2.9	7.8	1.0	87.6	118.9	28.7	4.5	4.4	3.1	366.3	5.7	99.1	0.5	16.7
345	186.2	65.7	3.8	9.0	0.5	60.2	152.4	21.5	7.9	5.3	2.8	762.5	10.0	121.6	0.2	21.8
346	147.2	240.2	0.9	3.6	2.5	247.0	97.2	57.0	13.9	0.6	6.9	0.0	3.1	171.2	0.6	0.0
347	58.5	13.3	4.2	14.3	0.9	12.7	334.8	0.0	4.4	4.2	1.5	0.0	6.6	179.0	0.0	15.8
348	164.8	42.2	0.5	7.2	0.4	58.2	41.0	0.0	1.8	2.2	1.3	0.0	6.5	49.1	0.0	0.0
349	133.0	11.9	1.9	6.7	1.1	14.4	183.2	0.0	6.6	2.1	1.6	0.0	6.8	147.5	0.0	16.1
350	93.3	61.5	5.4	12.2	2.8	34.0	386.3	20.0	4.0	3.4	3.6	0.0	2.3	441.7	1.8	102.3
351	127.2	201.1	15.8	13.8	1.0	247.6	776.0	40.3	6.0	8.9	5.1	2098.0	19.4	306.9	1.0	109.1
352	69.3	6.4	0.6	7.4	0.0	2.9	63.4	0.0	2.8	2.3	0.4	0.0	1.7	145.4	0.0	0.0
353	160.3	148.9	7.2	14.9	2.7	124.7	351.1	45.9	3.4	12.7	6.4	1077.2	4.5	202.2	1.2	337.5
354	126.6	143.9	10.1	11.7	1.5	124.9	1316.5	39.3	5.5	6.0	8.3	209.4	7.9	201.3	0.4	52.9
355	516.7	57.9	2.6	12.5	0.4	49.8	222.1	6.3	21.1	2.4	1.8	526.8	7.6	225.7	0.0	13.7
356	176.6	180.9	2.7	15.2	1.8	212.0	92.9	36.1	11.5	3.7	5.2	1872.3	3.2	129.8	0.7	31.3
357	66.4	35.2	8.0	11.8	2.3	22.5	335.0	15.4	3.3	32.7	2.9	426.1	2.2	197.9	1.7	106.6
358	11.0	74.1	14.0	49.6	2.9	37.5	765.9	31.2	1.1	11.7	5.7	520.2	2.6	96.2	2.2	148.2

359	132.8	265.4	3.5	8.6	2.3	267.2	218.3	69.5	6.0	5.0	7.8	1008.6	8.6	208.0	0.9	0.0
361	107.6	101.3	20.5	43.1	1.9	72.3	1917.7	18.1	4.6	5.2	4.1	525.5	6.8	143.3	1.0	58.8
362	181.5	167.2	9.5	28.2	2.0	177.7	351.1	37.3	9.3	6.3	4.2	1276.2	6.7	130.4	1.2	52.3
363	110.6	264.7	101.3	8.9	1.8	257.6	855.4	65.6	5.4	3.6	7.1	2103.2	6.1	132.7	0.6	11.9
364	81.6	166.5	5.3	10.5	2.7	142.2	306.4	50.7	3.4	5.1	7.2	813.2	3.6	105.9	1.2	40.5
365	402.5	1707.2	3.6	12.1	16.3	1858.3	262.3	313.6	22.5	1.1	50.4	339.6	93.5	650.1	8.9	0.0
366	61.2	86.4	3.0	25.9	0.8	88.9	233.6	30.7	3.0	4.3	3.2	1286.7	3.4	110.2	0.4	20.8
367	138.3	159.1	14.0	23.0	2.2	120.4	527.1	27.8	6.2	9.0	5.0	337.6	8.1	158.8	1.1	138.4
368	143.5	187.0	3.0	6.2	1.8	165.8	182.3	49.1	6.5	3.2	6.6	343.5	8.6	123.3	0.5	15.4
369	139.4	174.9	8.7	14.9	1.3	136.5	1219.1	37.8	6.3	6.5	4.9	826.4	6.6	183.4	0.6	85.2
370	172.9	46.9	1.4	6.8	0.5	67.8	104.7	7.9	7.8	1.7	1.5	785.0	4.0	64.7	0.4	0.0
371	69.2	14.5	1.5	7.6	0.3	5.5	139.9	11.3	144.3	1.5	1.7	757.4	1.8	0.0	0.4	25.2
372	80.3	57.3	11.7	30.9	2.4	37.1	764.8	31.0	4.9	11.4	4.1	840.3	3.4	110.8	1.5	80.2
373	72.6	0.0	1441.4	9.4	0.3	0.7	291.6	0.0	1.1	2.4	0.3	1111.5	0.0	159.0	0.4	33.2

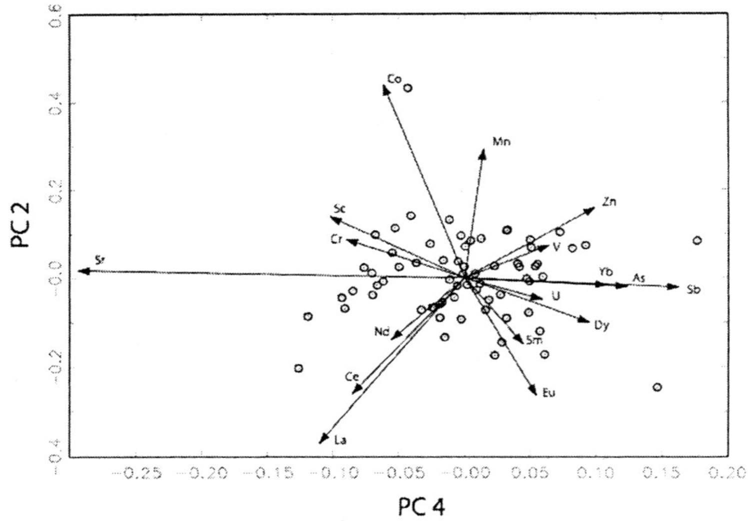

Figure 3. Principal components plot (PC 2 vs. PC 4) demonstrating elements that drive variance in Jiskairumoko ochre

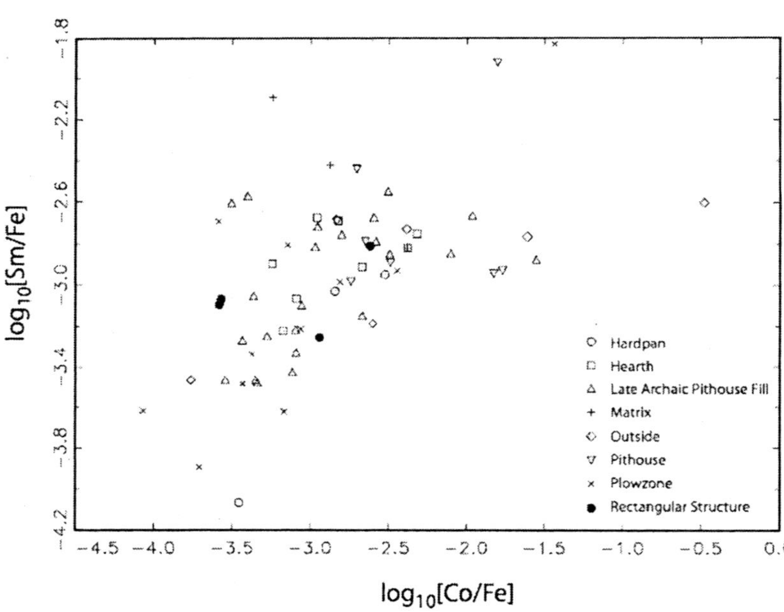

Figure 4. Bivariate plot of $\log_{10}$ [Sm/Fe] vs $\log_{10}$ [Co/Fe]. Samples are plotted by groups as determined by context.

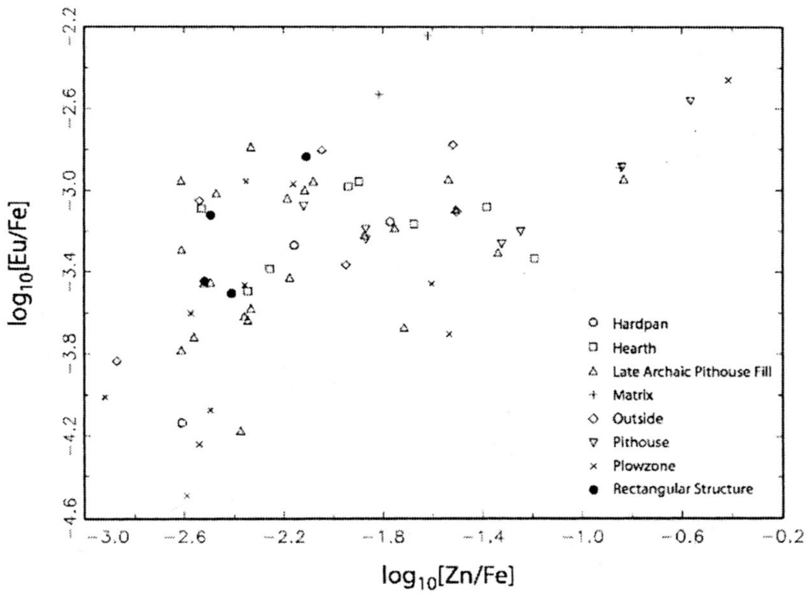

Figure 5. Bivariate plot of $\log_{10}$ [Eu/Fe] vs $\log_{10}$ [Zn/Fe]. Samples are plotted by groups as determined by context.

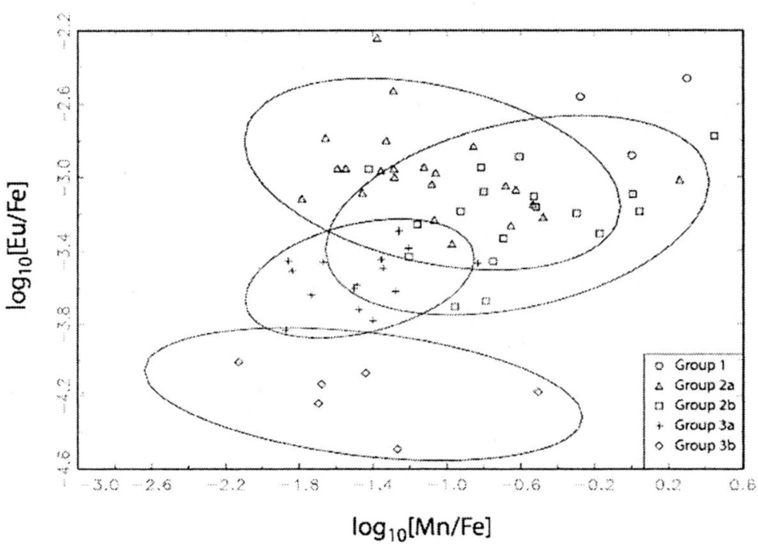

Figure 6. Bivariate plot of $\log_{10}$ [Eu/Fe] vs $\log_{10}$ [Mn/Fe]. Samples are plotted by groups as determined by a cluster analysis. Group 1 does not have enough samples to form an ellipse.

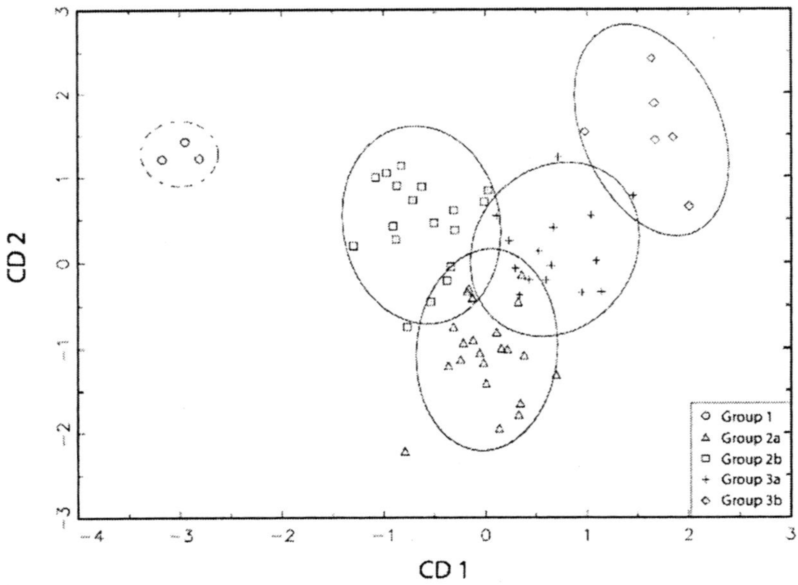

Figure 7. Canonical discriminant analysis plot (CD2 vs CD1) Samples are plotted by groups as determined by a cluster analysis.

Artifacts were divided into several groups based on archaeological context and plotted according to $\log_{10}$[Sm/Fe] vs $\log_{10}$[Co/Fe] (Figure 4), $\log_{10}$[Eu/Fe] vs $\log_{10}$[Zn/Fe] (Figure 5). The third elemental bivariate plot was grouped according to cluster analysis and plotted by $\log_{10}$[Eu/Fe] vs $\log_{10}$[Mn/Fe] (Figure 6). Figure 6 indicates visual grouping of the samples in an elemental plot, with some samples associated in a large group, but other clusters are both visually and statistically distinct.

Canonical discriminant analysis (Figure 7) indicates that, by using CD1 and CD2, groups defined by the hierarchical cluster analysis are well defined and have analytical merit. Group 2a, 2b, and 3a appear to be associated in one central aggregation, while Group 3b is somewhat separated from this central concentration. Group 1 emerges as a distinct cluster that is very different from the others.

The groups found in this project display variation in their chemical composition. Because sources were not analyzed in this study, the variation in the sources cannot be quantified. However, the variation in the groups implies different procurement practices or variation within the procurement site. Depending on the source and the geochemistry of the region, differing elements and element groups can be used to characterize the sources and artifacts. Sub-source variation in the major source may also be present(*5*). It can be seen from these data that the trace element analysis of ochre is helpful for understanding its variability. The following section will discuss archaeological interpretation of the variation in ochre composition at the site of Jiskairumoko.

Archaeological Interpretation

When considering archaeological interpretation of the results from chemical characterization, we return to the five questions described in the project goals.

1) What is the overall variability of ochre? Variability is present, but it is not large. Given the paucity of ochre chemical characterization studies, evaluating the degree of variability is problematic. Hierarchical clustering suggested the presence of five groups, which were corroborated by canonical discriminant analysis. Groups 1 and 3b appear to be the most dissimilar from the majority of samples. Groups 2a, 2b, and 3a may comprise a single heterogeneous cluster.

2) Are there chemical differences in ochre remains recovered from different contexts? Variability is present, but does not appear to be context-specific. The majority of samples are scattered throughout, as seen in Figures 4 and 5. The samples that comprise this aggregation come from all of the depositional contexts that were sampled from Jiskairumoko. Ochre sampled from different archaeological contexts at Jiskairumoko is not chemically distinct.

3) Are ochre artifacts from different structures chemically similar or different? Ochre artifacts from different structures are chemically similar. Samples 368 and 320 are very close together in CD space as well as the $\log_{10}$[Sm/Fe] vs $\log_{10}$[Co/Fe] and $\log_{10}$[Eu/Fe] vs $\log_{10}$[Zn/Fe] bivariate diagrams (Figure 4 and 5). Sample 368 comes from a sealed context at the base of the Late Archaic pithouse while sample 320 was recovered from the edge of a large rectangular structure. Both samples came from clear and unambiguous contexts yet are very similar chemically, indicating that either the same source or the same portion of a heterogeneous source was used throughout the span of Jiskairumoko's occupation.

4) Is there large variability in the ochre from the thermal processing area? Figures 4–6 and the CD plot (Figure 7) indicate that there is considerable chemical variability in the samples taken from the hearth in the base of the Late Archaic pithouse and from the hearth-associated debris arc. This variability suggests that ochre in this activity area was obtained either by 1) various trips to different parts of a single source or 2) various exchanges for ochre from different sources. It is not likely that the thermal treatment of the ochre significantly altered the trace chemical signature of the ochre. Chemical variability of ochre

artifacts in this context suggests either different portions of a single source or that multiple sources are represented. This information serves as an additional line of evidence indicating that the debris arc next to the hearth was not the outcome of a single behavioral episode. Instead, it represents habitual hearthside activity.

5) Is the ochre from this context similar to ochre found in secondary refuse? Elemental and CD space plots all indicate chemical similarities between ochre artifacts recovered from the central hearth and debris arc encountered at the base of the Late Archaic pithouse. The occurrence of chemically similar ochre artifacts in both contexts provides additional evidence that secondary refuse deposits were formed by cleaning hearths and removing debris from inside structures.

Several other observations can be made from Figures 6 and 7. Group 1 exhibits the greatest distance in CD space. This group consists of samples from the plow zone, deep within the fill of Pithouse 3, and a stain outside Pithouse 3. Group 1 may represent a source, or portion of a source, that was not generally preferred by occupants of Jiskairumoko. None of the samples forming Group 1 were recovered from early occupational contexts at Jiskairumoko, though none of them are unambiguously later either. Samples comprising Group 3b are all from plow zone contexts except for samples 315 and 342. These latter two are not from early contexts. Thus, Groups 1 and 3b, which are the most chemically distinct, do not appear to include any samples recovered from early contexts. For example, none of the ochre samples from Late Archaic pithouse contexts are members of either Group 1 or 3b. These two groups may represent two portions of a single heterogeneous source or two additional sources that were not preferred by residents of Jiskairumoko.

Conclusions

While the artifacts were not compared to the ochre sources from the region, some tentative conclusions about the variability of ochre and its use at Jiskairumoko can be drawn. Statistically distinct groups were found in ochre from Jiskairumoko. Major elements, such as Fe and Mn, are as important as trace elements, such as the rare earth elements, in studying ochre variability. Viewed in a number of different dimensions and transformations, the majority of ochre samples appear to form a single rather heterogeneous congregation that comprises all of the depositional contexts that were sampled. Ochre use within any given depositional context is composed of members of more than one statistically defined group. These statistical clusters also consist of samples from multiple depositional contexts. Without locating and geochemically characterizing a range of ochre sources it is impossible to determine if these reflect multiple sources or a single heterogeneous source. Later in time, either two additional portions of a source or two additional sources may have come into use. Additional work locating and characterizing Andean ochre sources is needed. However, it appears that ochre followed a procurement trajectory different from obsidian, which was obtained from a single non-local source in the Colca Canyon near Arequipa(*49*).

Acknowledgments

For the analytical work in this study, the authors wish to acknowledge the Archaeometry Laboratory Staff for preparation of samples, as well as the National Science Foundation Graduate Fellowship (Rachel Popelka-Filcoff) and National Science Foundation grant #0504015 (Principal Investigator: Michael D. Glascock) for funding. Fieldwork was supported by: National Science Foundation grant #9816313 for fieldwork in 1998, and a supplement for fieldwork in 2002; National Science Foundation Equipment grant #9978006 in 1999, and John H. Heinz III Charitable Trust in 1997 and 2002. Mark Aldenderfer was Principal Investigator on all grants for fieldwork.

References

1. Weinstein-Evron, M.; Ilani, S. *J. Archaeol. Sci.* **1994**, *21*, 461–467.
2. David, B.; Clayton, E.; Watchman, A. L. *Austral. Archaeol.* **1993**, *36*, 56–57.
3. Hovers, E.; Ilani, S.; Bar-Yosef, O.; Vandermeersch, B. *Curr. Anthro.* **2003**, *44*, 491–522.
4. Stafford, M. D.; Frison, G. C.; Stanford, D.; Zeimans, G. *Geoarchaeol.* **2003**, *18*, 71–90.
5. Popelka-Filcoff, R. S.; Robertson, J. D.; Glascock, M. D.; Descantes, C. *J. Radioanal. Nucl. Chem.*; in press.
6. Erlandson, J. M.; Robertson, J. D.; Descantes, C. *Am. Antiq.* **1999**, *64*, 517–526.
7. Mrzlack, H. M.A. thesis, University of Colorado-Denver, Denver, CO 2003.
8. Moseley, M. *The Incas and Their Ancestors*. Thames & Hudson: New York, NY, 2001.
9. Stanish, C. *Annu. Rev. Anthro.* **2001**, *30*, 41–64.
10. Stanish, C. In *Ancient Titicaca: The Evolution of Complex Society in Southern Peru and Northern Bolivia*; University of California Press: Berkeley, CA, 2003; p 354.
11. Lavallée, D.; Julien, M.; Wheeler, J.; Karlin, C. In *Telarmachay: Chasseurs et Pasteurs Préhistoriques des Andes*; Institut Français D'Études Andines: Paris, 1985; Vol. I.

12. Aldenderfer, M. S. In *Montane Foragers Asana and the South-Central Andean Archaic*; University of Iowa Press: Iowa City, IA, 1998; p 327.
13. Craig, N. Ph.D. thesis, University of California at Santa Barbara, Santa Barbara, CA, 2005.
14. Smith, B. D. *J. Archaeol. Res.* **2001**, *9*, 1–43.
15. Shackley, S. M.; Eklund, E.; Ogasawara, C. Report. *Source Provenance of Obsidian Artifacts Jiskairumoko (189), Peru*; University of California Berkeley Archaeological XRF Laboratory: Department of Anthropology, Berkeley, CA May 2004; p 10.
16. Speakman, R. J.; Popelka, R. S.; Glascock, M. D.; Robertson, J. D. Report. *Analysis of Obsidian Artifacts from Southern Peru Using a Field-Portable X-Ray Fluorescence Spectrometer*; University of Missouri Research Reactor Center: Columbia, MO, 2005.
17. Craig, N.; Aldenderfer, M. In *Advances in Titicaca Basin Archaeology II*; Klarich, E. A.; Stanish, C., Eds.; Cotsen Institute of Archaeology: Los Angeles, CA; in press.
18. Aldenderfer, M. S. *Qillqatani and the Evolution of Pastoral Societies in the Titicaca Basin*; in press.
19. Baid, C. A.; Wheeler, J. *Mountain Research and Development* **1993**, *13*, 145–156.
20. Dransart, P. *World Archaeol.* **1991**, *22*, 304–319.
21. Dransart, P. *Earth, Water, Fleece, and Fabric: An Ethnography and Archaeology of Andean Camelid Herding*. Routledge: New York, NY 2002.
22. Hesse, B. *J. Ethnobiology* **1982**, *2*, 1–15.
23. Kuznar, L. A. *Awatimarka: The Ethnology of an Andean Herding Community*. Harcourt Brace: Orlando, FL 1995.
24. Nuñez, L. *Boletin de Antropologia Americana* **1981**, *2*, 87–120.
25. Nuñez, L. *Chungará* **1982**, *9*, 80–122.
26. Rick, J. *Prehistoric Hunters of the High Andes. Studies in Archaeology*. Academic Press: New York, NY, 1980.
27. Zlatar, V. *Chungará* **1983**, *10*, 21–28.
28. Dillehay, T. D.; Núñez, L. In *Recent Studies in Pre-Columbian Archaeology*; Saunders, N. J.; Montmollin, O. D., Eds.; BAR International Series: Oxford, 1988; Vol. 421, pp 603–634.
29. Browman, D. *Am. Scientist* **1981**, *69*, 408–419.
30. Quilter, J. *J. World Prehistory* **1991**, *5*, 387–415.
31. Rowe, M. H. *News in Physiol. Sci.* **2002**, *17*, 93–98.
32. Mollon, J. D. *J. Experimental Biology* **1989**, *146*, 21–38.
33. Sumner, P.; Mollon, J. D. *J. Experimental Biology* **2000**, *203*, 1963–1986.
34. Sumner, P.; Mollon, J. D. *J. Experimental Biology* **2000**, *203*, 1987–2000.
35. Dominy, N. J.; Lucas, P. W. *Nature* **2000**, *410*, 363–366.

36. Berlin, B.; Kay, P. *Basic Color Terms*; University of California Press: Berkeley, CA, 1969.
37. Brown, D. E. *Human Universals*; McGraw-Hill, Inc.: New York, NY, 1991.
38. Knight, C.; Power, C.; Watts, I. *Cambridge Archaeol. J.* **1995**, *5*, 75–114.
39. Wreschner, E. E. *Curr. Anthro.* **1980**, *21*, 631–644.
40. Bolton, R. *Curr. Anthro.* **1980**, 21, 633–635.
41. Stevenson, P. *Curr. Anthro.* **1980**, *21*, 633–635.
42. Turner, V. W. *The Forest of Symbols: Aspects of Ndembu Ritual*; Cornell University Press: Ithaca, NY, 1967.
43. Aldenderfer, M. S. In *Kaillachuro: A Formative Mortuary Complex from the Southwestern Lake Titicaca Basin*; Institute of Andean Studies, Berkeley, January, 1998; Berkeley, CA, 1998.
44. Craig, N.; Moyes, H.; Aldenderfer, M. *J. Archaeol. Sci.*; in press.
45. Binford, L. In *Pursuit of the Past: Decoding the Archaeological Record*; Thames and Hudson: New York, NY, 1983; pp 1–256.
46. Elias, M.; Chartier, C.; Prévot, G.; Garay, H.; Vignaud, C. *Mater. Sci. Eng. B* **2006**, *127*, 70–80.
47. Tankersley, K. B.; Tankersley, K. O.; Shaffer, N. R.; Hess, M. D.; Benz, J. S.; Turner, F. R.; Stafford, M. D.; Zeimans, G.; Frison, G. C. *Plains Anthropologist* **1995**, *40*, 185–194.
48. Glascock, M. In *Chemical Characterization of Ceramic Pastes in Archaeology*; Neff, H., Ed.; Prehistory Press: Madison, WI, 1992; pp 11–26.
49. Burger, R. L.; Asaro, F.; Salas, G.; Stross, F. *Andean Past* **1998**, *5*, 202–223.

Chapter 27

Feasibility of Field-Portable XRF to Identify Obsidian Sources in Central Petén, Guatemala

Leslie G. Cecil[1], Matthew D. Moriarty[2], Robert J. Speakman[1], and Michael D. Glascock[1]

[1]Research Reactor Center, University of Missouri, Columbia, MO 65211
[2]Department of Anthropology, Tulane University, New Orleans, LA 70118

Recent research concerning the Postclassic (A.D. 1000–1524) period in central Petén has focused on defining changes in architectural features and pottery manufacturing techniques. Obsidian is frequently excavated from these structures and occurs as offerings in cache vessels. Field-portable XRF can be used to identify obsidian sources based on artifacts excavated from Trinidad de Nosotros. These data will be compared to previous analyses from other archaeological sites in the Petén lakes region. In addition to identifying obsidian sources, obsidian trade in the Petén lakes region will be better understood because of the samples analyzed from the trading port (Trinidad de Nosotros) on the north shore of Lake Petén Itzá.

Archaeological research and artifact analysis involves obtaining any number of permits and following regulations as to the extent to which one can export artifacts for analyses that are not feasible in the field. Many times the permits for exportation of artifacts limit the number of artifacts that can be used for a study as well as the extent to which the artifact can be damaged in the analysis before it must be returned. With the development of portable technology, such as the field-portable XRF, that does not damage the artifact, it is possible to analyze artifacts in the field without obtaining the often elusive export permits. Additionally, because the technique is rapid, it is possible to greatly increase the sample size of any analysis allowing the archaeologist to examine more artifacts and to obtain more robust data sets. This is especially needed in the analysis of artifacts from inland trade ports and/or trans-shipment locations given the scarcity of information known about these archaeological contexts.

Background

To test the feasibility of the portable XRF and to create a robust obsidian data set from an inland trade port/trans-shipment locale, we analyzed a number of Postclassic obsidian artifacts from Trinidad de Nosotros in the central Petén lakes region (Figure 1). Obsidian is ubiquitous in the archaeological record, and can be easily traced to its source area because each volcanic eruption that produces volcanic glass has a unique chemical composition. In Mesoamerica, most obsidian artifacts originate from about 40 known sources of which only about ten were heavily used by the Maya (*1*).

Although obsidian sourcing studies have focused on the Preclassic and Classic periods in the Maya region, more recent research is concentrating on the Early Postclassic period coastal sites. Even though this research is advancing our knowledge concerning coastal trade, it rarely examines the inland trade that must have occurred in conjunction with the coastal trade routes. With the excavations at Trinidad de Nosotros and the previous analysis of obsidian at other sites in the region, it is possible to examine the extent to which inland trade routes differed from coastal trade routes during the Early Postclassic period.

Trinidad de Nosotros (or *Sik'u'* in Itzaj Maya) is located on the north shore of Guatemala's Lake Petén Itzá, 2.6 km southeast of the Late Classic site of Motul de San José and approximately 35 km southwest of the major center of Tikal. Two seasons of investigations at Trinidad de Nosotros have defined it as a medium-sized center, covering an area of about 1 km^2 and including approximately 150 structures. Its occupation extends from the Middle Preclassic (ca. 600 B.C.) period to the Historical era, with major peaks in settlement during the Late Preclassic (300 B.C.-A.D. 250) and Late Classic periods (A.D. 600–830) (*2*). The central portion of the site is characterized by elite residences, several small temples, a ballcourt, and a series of five public plazas (Figure 2). Soil chemical analyses in the largest of these plazas have suggested that it may have served as a setting for periodic markets (*3*).

Trinidad de Nosotros's location situates the site at one of the best natural portages on the north shore of the lake. Along most of this shore, steep natural terraces step down to narrow rocky beaches. At Trinidad de Nosotros, however, the discharge from a seasonal drainage has created a wide beach ideal for landing and loading small boats (*2*). Further, a series of artificial and modified natural features provide a small harbor. At the base of the slope running down

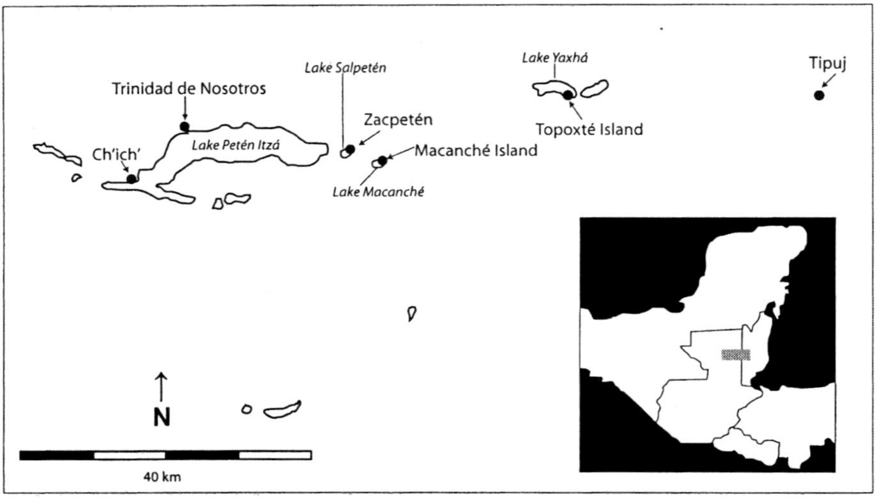

Figure 1. Map of archaeological sites in the central Petén lakes area, El Petén, Guatemala.

from the center of the site, a low platform with a cut stone outer retaining wall extends around the inner circumference of the harbor and provides a level surface for loading and unloading canoes. On the outer edge of the harbor, a small hillock or island was modified and extended to create a 70m long outer harbor wall. Finally, within the harbor, a 7x15 m platform served as a dock. Excavations within these features have delineated a long history of construction and modification beginning in the Preclassic and particularly heavy during the Early Postclassic period (*4*).

The combination of Trinidad de Nosotros's location, the presence of a harbor, and high densities of exotic trade goods recovered during excavations provide multiple lines of evidence for identifying the site as an ancient Maya port (*2*). If this was the case, then Trinidad de Nosotros was well situated to have served several pre-Columbian trade routes. First, the central Petén lakes form a natural transportation route for waterborne travel between the east-flowing rivers of Belize and the northwest flowing rivers of western Petén (*5*). The recent identification of port facilities at Nixtun-Ch'ich' on the west end of Lake Petén Itzá, canalized rivers on the east end, and a series of possible inter-lacustrine canals provide some confirmation for this possibility (*2 ,4, 6*). Second, Trinidad de Nosotros may also have served as a point of trans-shipment between the lake and a northwestern trade route utilizing the Río K'änte't'u'ul, a small river that starts 5 km north of Trinidad de Nosotros and leads to the Río San Pedro Martír, a major pre-Columbian transportation artery. Access from the lake to the river

Figure 2. Map of Trinidad de Nosotros, Petén, Guatemala. The Postclassic portion (and harbor) of the site is indicated with a dashed rectangle.

and other points north was facilitated by a natural path of least resistance through the rolling topography north of the lake (*2*).

One of the many classes of artifacts that may have been traded through Trinidad de Nosotros was obsidian. Because there are no obsidian sources in the Maya lowlands, the material had to be transported from México and highland Guatemala. The three main Guatemalan obsidian sources utilized by the lowland Maya were San Martín Jilotepeque, El Chayal, and Ixtepeque (*1, 2, 7, 8*) (Figure 3). In addition, obsidian from several sources in the Mexican highlands was also imported into the Maya lowlands (*1*).

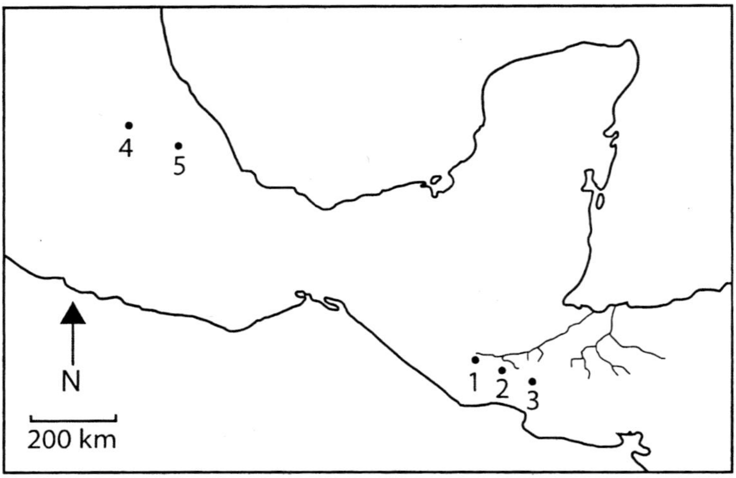

Figure 3. Regional map showing locations of Guatemalan (1=San Martín Jilotepeque; 2=El Chayal; 3=Ixtepeque) and Mexican (4=Pachuca; 5=Zaragoza) obsidian sources used in the study.

San Martín Jilotepeque

This obsidian source zone is located within the department of Chimaltenango, west of the Valley of Guatemala (ca. 275 km from Trinidad de Nosotros, as the crow flies) (*7*). It was the dominant import into the Maya lowlands during the Preclassic period (*8, 9*), however, it continued to be imported to the lowlands throughout the Late Classic and Postclassic periods (*9, 10*). San Martín Jilotepeque obsidian was not under the control of any one socio-political group, as the architecture and lack of dominant workshops suggest an egalitarian society (*1,9*). Braswell (*9*) suggests that this obsidian reached the

Maya lowlands through multiple down-the-line exchanges. During the Classic period, society became more complex and stratified and the archaeological record indicates the presence of an obsidian lithic industry (*9*).

San Martín Jilotepeque obsidian is more commonly excavated from archaeological sites in central Maya lowlands than from sites in Belize and northern Yucatán as well as coastal sites such as Wilde Cane Cay and Moho Cay. This presence suggests overland and riverine routes for trade of obsidian from this source (*11*). During the Late Classic period and with the increasing dominance of Kaminaljuyú in the Guatemala highlands (and the El Chayal obsidian source), San Martín Jilotepeque was traded west to the Pacific coast, north to the Chiapas, and the central Maya lowlands (*12*).

El Chayal

This obsidian source is located in the upland flanks of the Motagua Valley in the highlands of Guatemala. It is approximately 40 km east of the San Martín Jilotepeque source area and 70 km northwest of the Ixtepeque source area (ca. 250 km from Trinidad de Nosotros, as the crow flies). During the Late Classic period, El Chayal was the dominant obsidian source traded throughout the Maya region. However, obsidian from El Chayal was being traded to El Mirador as early as the Late Preclassic period (*10*) and occurs in Postclassic deposits in central Petén (*13*). Kaminaljuyú, one of the larger sites in highland Guatemala during the Late Classic period, controlled the technology of blade production of obsidian from the El Chayal source (20 km from Kaminaljuyú) (*14*) .

Obsidian artifacts sourced to El Chayal have been excavated from archaeological sites around the Usumacinta River basin, in northeastern Petén, in the Belize Valley, and the Toledo District of southern Belize (*12*). Hammond (*15*) suggests that highland Maya transported and traded obsidian to the lowland communities by inland routes through the Usumacinta and Sarstoon River basins or it could have been transported down the Río Pasión to Seibal and then north to Tikal (*11, 16*).

Ixtepeque

The Ixtepeque obsidian source is located 85 km from the El Chayal source zone (ca. 300 km from Trinidad de Nosotros, as the crow flies). It is found at archaeological sites east and north of the source, along the coast of Belize, northeastern Petén, the Belize Valley, and northern Yucatán (*1, 14*). Obsidian artifacts dating to the Late Preclassic, Terminal Classic and Postclassic periods are predominately from this source and it was the main source during the Postclassic period (*11, 15, 17*).

Ixtepeque obsidian most likely reached the lowlands by being transported overland during the Preclassic period (*11*) or down the Río Motagua and north along the Caribbean coast and then inland (*1, 11, 15*) during the Terminal and Postclassic periods. This route is supported by the large quantities of Ixtepeque obsidian at Copán and Quirigua (*12*), as well as the plethora of it at coastal sites that may have served as trans-shipment nodes (*1, 16, 17, 18*).

Pachuca

Green obsidian originates from the Pachuca obsidian source. This source is located between the modern cities of Pachuca and Tulancingo, Hidalgo, México (ca. 900 km from Trinidad de Nosotros, as the crow flies). The nearest prehistoric centers of population were Pachuca, Teotihuacán (50 km from the source) and Tula (70 km from the source). During the Classic period, Pachuca workshops were located at Teotihuacán (*19, 20*). Artifacts made from this source excavated from Classic period archaeological sites in the Maya region may be the result of gifts among elites (*20, 21*). During the Terminal Classic and Early Postclassic periods, Tollan workshops appear to have used the Pachuca resource (*22*). Trade of Pachuca obsidian during the Classic Period appears to have been through Kaminalyujú and followed the El Chayal pattern to the Maya lowlands. On the other hand, during the Postclassic period Pachuca obsidian trade routes may have included previously established overland routes, as well as routes along the coast of the Yucatán Peninsula (*11*) and then inland.

Zaragoza, Puebla, México

The Zaragoza-Oyameles obsidian source is approximately 100 km southeast of the Pachuca obsidian source (ca. 800 km from Trinidad de Nosotros, as the crow flies). It was used from the Preclassic to the Aztec Periods in central México. During the Classic Period, the Veracruz region dominated the Zaragoza prismatic blade obsidian network (*23*). Within this region, Cantona may have had more control of the source (*23*). After the fall of Teotihuacán, the distribution of Zaragoza obsidian was not disrupted thus suggesting that Teotihuacán did not have control of this obsidian source (*23*). The most likely path of distribution into the Maya lowlands from A.D. 850—1050 was along the Gulf Coast around the Yucatán Peninsula and inland through trans-shipment points. This is supported by the presence of blades made of Zaragoza obsidian found at Isla Cerritos (*24*), Cozumel (*25*), and Mayapán (*1*).

Analytical Methods

Two seasons of excavations at Trinidad de Nosotros produced approximately 1800 obsidian artifacts. Of these, approximately one-quarter (450) were recovered from Postclassic contexts within the harbor area and other parts of the site. To examine patterns in obsidian source procurement at Trinidad de Nosotros during the Early Postclassic period, a sample of artifacts from Postclassic contexts were selected for analysis.[1] Samples less than 3 cm in length were excluded from consideration and only artifacts from chronologically secure contexts were analyzed. With this limitation and employing a stratified random sampling technique, a total of 70 obsidian artifacts were selected for analysis. These samples came from Postclassic contexts within the site's harbor area including a residence, the inner harbor platform, the outer harbor wall, and a high-density midden within the harbor itself.

Nondestructive elemental analysis of the obsidian samples was conducted at MURR using an ElvaX desktop energy-dispersive x-ray fluorescence (EDXRF) spectrometer. The instrument consists of an x-ray generator, an x-ray detector, and a multi-channel analyzer (MCA). The detector is a solid state Si-pin-diode with a resolution of 180 eV at 5.9 keV (at 1000 counts per second) with an area of 30 mm^2. The output signal of the detector is formed by a time-variant time processor with pile-up rejector, base line restorer, and automatic adaptation of shaping time to the input count rate. The MCA consists of a fast shaping amplifier (FSA) and a 4K-channel spectrometric analog-to-digital converter (SADC), built as a successive approximations ADC with conversion time of 2 μs, 4096 channels, a 32–bit per channel buffer RAM, "sliding scale" linearization of differential nonlinearity, and dead time correction circuit (*26*). The x-ray tube is air-cooled and low powered with a tungsten anode and 140 μm beryllium end-window.

The analyses were conducted at 30 kV with a tube current of 45 μA and a 400 second live time using a 0.8 mm primary aluminum filter. The following elements were measured: titanium (Ti), manganese (Mn), iron (Fe), zinc (Zn), gallium (Ga), rubidium (Rb), strontium (Sr), yttrium (Y), zirconium (Zr), and niobium (Nb). Concentration values (in parts per million) were determined using *ElvaX Regression*—a program based on the quadratic regression model using data derived from ten reference samples (obsidian from Guatemala and México) of similar composition. The quadratic regression model of the form

$$C_i = \sum_{\substack{j=0 \\ k=0}}^{S} A_{ijk} I_j I_k$$

describes the relationship between the set of analytical intensities and analyte concentrations where C_i is the concentration of analyte i in the sample, $I_0 = 1$, I_j

and I_k are the analytical intensities of analytes j and k respectively, and S is the number of analytes in the product. The regression coefficients $A_{ijk} = A_{ikj}$ are determined from calibration by a set of reference samples. As a rule, as their number exceeds the number of reference samples at hand, not all of the coefficients can be estimated, and only the most significant ones are taken into account (*26*). The number of significant coefficients present in the model for each analyte i is referred to as the number of degrees of freedom of the regression model.

Elemental concentration values of the obsidian artifacts were correlated to known obsidian sources in Guatemala (El Chayal, Ixtepeque, and San Martín Jilotepeque), and México (Pachuca, Ucareo, and Zaragoza) using the ElvaX software comparison spectra feature. To verify matches made with the comparison spectra, elemental concentrations were transformed using base-10 logarithms, plotted, and grouped according to statistical group membership. Use of log concentrations rather than raw data compensates for differences in magnitude between the major and minor elements. Transformation to base-10 logarithms also yields a more normal distribution for many elements.

The interpretation of compositional data obtained from the analysis of archaeological materials is discussed in detail elsewhere (e.g., *27–32*). The main goal of data analysis is to identify distinct homogeneous groups within the analytical database. Based on the provenance postulate, different chemical groups may be assumed to represent geographically restricted sources (*33*). For obsidian, raw material samples are frequently collected from known outcrops or secondary deposits and the compositional data obtained on the samples is used to define the source localities or boundaries. Obsidian sources tend to be more localized and compositionally homogeneous, making artifact-source comparisons straightforward. Groups are characterized by the locations of their centroids and the unique relationships (i.e., correlations) between the elements. Decisions about whether to assign a specimen to a particular compositional group are based on the overall probability that the measured concentrations for the specimen could have been obtained from that group.

All obsidian samples were analyzed as unmodified samples; they were washed in the field. Each sample was placed in the sample chamber with the flattest part of the surface facing the x-ray beam. All samples were at least 3 cm in length with varying widths and thicknesses. The width of the sample did not produce errors when comparing obsidian artifact to potential obsidian source. Accuracy errors result from inaccuracies of the regression model, statistical error of the calibration spectra, inaccuracy of the intensity of the calibration curve and the energy calibration. When the error is taken into account, the relative analytical uncertainty for this project is less than seven percent with this portable XRF unit (*26*).[2]

Results and Discussion

Source determination of all 70 Postclassic obsidian samples was possible using the field-portable XRF instrument. Although the majority of the samples (56%, n=39) represent the Ixtepeque source, 29% (n=20) of the sample comes from El Chayal, and 11% (n=8) can be sourced to San Martín Jilotepeque. In addition to these Guatemalan sources, two different Mexican sources were identified in the Postclassic sample: Pachuca (3%, n=2) and Zaragoza (1%, n=1) (Figure 4).

The high percentage of Ixtepeque samples is consistent with the change in source dominance for the Postclassic period as noted elsewhere in the Maya lowlands (*1, 16, 17, 34, 35, 36*). This is different from the frequency of obsidian during the Late Classic period (primarily El Chayal with smaller amounts from the other highland Guatemala sources) (Figure 5). Although there is a marked change, it is not the caliber of change documented at the coastal sites. The coastal trans-shipment points (such as Wild Cane Cay, Moho Cay, and San Gervasio) had a much higher frequency (approximately 84%) of Ixtepeque obsidian and much lower frequencies (under 10% each) of El Chayal and San Martín Jilotepeque obsidian (*17, 18, 36*). On the other hand, the percentage and sources of Mexican obsidian at Trinidad de Nosotros and coastal trading sites is consistent with other inland Postclassic sites.

When examining the frequency and sources of obsidian found at Trinidad de Nosotros with those at other archaeological sites in the Petén lakes region, similarities exist suggesting that Trinidad de Nosotros may have served as a trans-shipment port for the region (Table I). As a trans-shipment port, Trinidad de Nosotros most likely participated in overland and coastal/riverine trade routes.

During the Early Postclassic period, Ixtepeque was transported down the Motagua River to places like Wild Cane Cay in the Bay of Honduras and around the Yucatan Peninsula (*1, 8, 15, 17, 18, 36*). In all likelihood, Ixtepeque obsidian was transported to central Petén via a number of rivers (Mopan, Mojo, Belize, and New) and overland routes. In addition to this route, El Chayal and San Martín Jilotepeque traders may have followed the Río Pasión to Seibal and then used natural overland routes similar to those used in the Preclassic and Classic periods (*1, 11, 15*). This route easily could have included Trinidad de Nosotros as it is in direct line on the route to Tikal further north. Because of the higher frequencies of these two obsidian sources at the sites in the Petén lakes region, it appears that the trade route did not cease with the end of the Classic period. Mexican obsidian came from a third trade route that originated in central México and Veracruz and was transported east along the Gulf of Tuantepec and the Yucatán Peninsula. There were many possible entry points of obsidian into central Petén that included Isla Cerritos, San Gervasio, Moho Caye, and Cerros (*1, 11, 36, 37*).

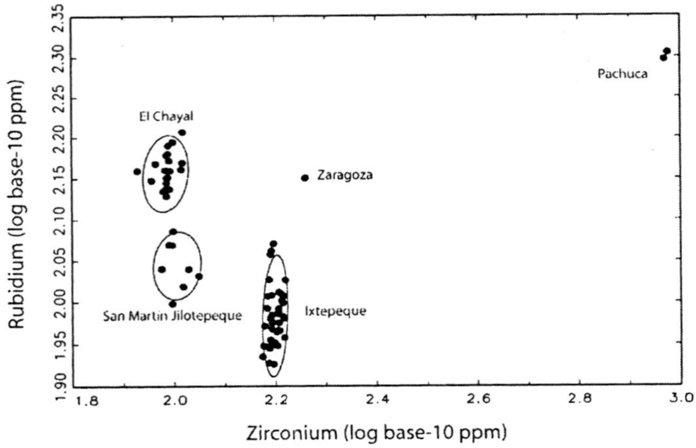

Figure 4. Plot of zirconium and rubidium base-10 logged concentrations showing the separation of the Early Postclassic period obsidian excavated from Trinidad de Nosotros. Ellipses represent 90% confidence interval for group membership.

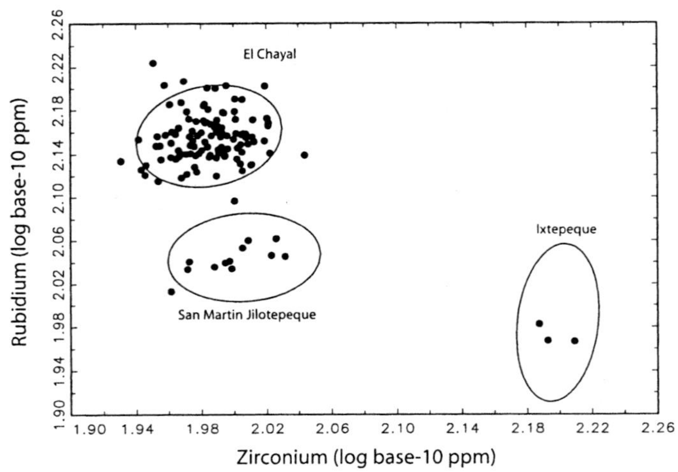

Figure 5. Plot of zirconium and rubidium base-10 logged concentrations showing the separation of the Classic period obsidian excavated from Trinidad de Nosotros. Ellipses represent 90% confidence interval for group membership.

Table I. Frequency of Postclassic Period Guatemalan and Mexican Obsidian Artifacts from Selected Archaeological Sites [3]

Archaeological Site	*San Martín Jilotepeque*	*El Chayal*	*Ixtepeque*	*Mexican Sources*
Trinidad de Nosotros	8 (11%)	20 (29%)	39 (56%)	3 (4%)
Lake Macanché (*38*)	2 (8%)	10 (38%)	12 (46%)	2 (8%)
Lake Salpetén (*38*)	7 (20%)	7 (20%)	19 (58%)	1 (2%)
Lake Quexil (*38*)	5 (29%)	5 (29%)	7 (42%)	0 (0%)
Topoxté (39)	8 (17%)	18 (38%)	21(45%)	0 (0%)
Tipuj (*34*)	15 (9%)	19 (11%)	135 (79%)	2 (1%)
Wild Cane Cay (*17*)	1 (2%)	6 (8%)	63 (85%)	4 (5%)
Colha (*16, 35*)	0 (0%)	0 (0%)	10 (100%)	0 (0%)
Isla de Ceritos (*24*)	0 (0%)	6 (34%)	0 (0%)	12 (66%)
San Gervasio (*1*)	1(3%)	0 (0%)	28 (91%)	2 (6%)
Laguna de On *(*40*)	21(3%)	178(27%)	441 (67%)	0 (0%)

* Twenty samples (3%) could not be sourced.

The obsidian obtained and analyzed from Lakes Macanché, Yaxhá, Salpetén, and Quexil as part of the Central Petén Historical Ecology Project (CPHEP) reflects a similar distribution of sources as seen at Trinidad de Nosotros (*13, 38*). Archaeological sites in central Petén with Postclassic occupation (including those tested by the CPHEP) reflect a dynamic socio-political milieu. The Early Postclassic period is typified by an introduction of new architectural and pottery styles that may indicate the influx of new socio-political groups in the region (*41*). On the other hand, some ceramic types seem to mimic Late Classic gloss wares suggesting the presence of an extant local population (*42*). The material culture distribution during the Early Postclassic period is similar throughout the region (from Nixtun Ch'ich' to Tipuj) suggesting ease of movement in the area. During the Middle and Late Postclassic periods a distinctive east-west dichotomy of cultural patterns is observable, reflecting historical alliance shifts, changing dominance relations, and repeated migrations to and from the northern Yucatán peninsula was established by the Itzá (western lake sites) and the Kowoj (eastern lake sites) (*41, 42*). These distinctive characteristics resulted from the presence of various socio-political groups that established and maintained principal provinces headed by a leader at a provincial capital. Each controlled different subprovinces and each had a distinct origin and migration myths. As a result, east-west boundaries were created and maintained thus restricting movement between Itzá (west) and Kowoj (east) territories/archaeological sites. If Trinidad de Nosotros was the central Petén lakes trans-shipment point of obsidian (and presumably other goods) during the Early Postclassic period, obsidian frequencies at Trinidad de Nosotros should be similar to those at other central Petén archaeological sites. As a trans-shipment port, Trinidad de Nosotros would have received obsidian from a number of trade routes that may have been very similar to those used during the Classic Period. After receiving the obsidian, its inhabitants may have spread equivalent amounts to other archaeological sites in the region. This may be the case since similar quantities and kinds of obsidian are found throughout the central Petén lakes region (with the exception of Tipuj, see below) during the Early Postclassic period, further suggesting that the socio-political distinctions (and boundary maintenance) that occurred in the Late Postclassic period were not yet strong (*41, 42*).

Unlike the other central Petén Postclassic archaeological sites, Tipuj appears to be a participant in another or an additional trade alliance that brought them almost twice the amount of Ixtepeque obsidian. This may be the result of Tipuj being the end point of the Belize River transportation network that was part of a coast trans-shipment route involving Moho Cay. However, obsidian from Tipuj presents a different pattern of distribution. Tipuj, the eastern most archaeological site that is considered part of the central Petén lakes region, has frequencies of obsidian that are more similar to those of the coastal sites than those of the Petén lakes region (*34*). Tipuj may have functioned as the eastern outpost site for the central Petén Maya during the Early Postclassic period as the Maya migrated from northern Yucatán to the central Petén occupation via the east coast of the Yucatán peninsula, down the New and Mopan Rivers to Tipuj. Additionally, it is from here that Fuensalida, Orbita, and Avendaño began their efforts to conquer and convert the Petén Itza in the 17th century (*43*). Therefore, there is a history of Maya using coastal and inland river routes to reach Tipuj suggesting multiple paths by which central Petén Postclassic Maya acquired obsidian.

Conclusion

The obsidian excavated from Trinidad de Nosotros demonstrates that obsidian trade during the Postclassic period was as complex as other periods. And, inland trade, as seen through evidence of a trans-shipment port, was different than coastal trade. In addition to the cultural aspect of this study, we have demonstrated that it is possible to successfully use a field-portable XRF to correlate obsidian artifacts to their sources. Not only were sources identified, but the process was rapid (4–6 minutes per sample), cost-effective, and as accurate and precise as more traditional methods of analysis. Above all, and perhaps most importantly, when working with artifacts from museum collections and foreign countries, the process was non-destructive to the artifacts.

Acknowledgements

The authors wish to acknowledge the Archaeometry Laboratory at the University of Missouri Research Reactor (funded by NSF Grant No. 0504015 and US Department of Energy Office of Nuclear Energy, Science and Technology Award No. DE-FG07–03ID14531) for the XRF analyses. Fieldwork at Trinidad de Nosotros was funded by grants from FAMSI, the Middle American Research Institute, the Amherst College Fellowship Program, and Williams College. We would also like to thank IDEAH for the permission to export the samples from Guatemala. All errors and omissions are our own.

Notes

[1] Dating of the Terminal Classic and Early/Middle Postclassic periods in the central Petén Lakes region is difficult because there is no abrupt and absolute distinct change in pottery (Late Classic to Postclassic) or obsidian frequencies. Some of the pottery types have characteristics of Terminal Classic forms and paste constituents while using a different clay base. For example, the most prominent ceramic grouping in the area during the Postclassic period, Paxcamán, is characterized by an early version that contains a snail inclusion clay (Postclassic) with ash temper (Late/Terminal Classic) whereas Late Postclassic Paxcamán pottery lacks an ash temper. The pottery from Postclassic contexts at Trinidad de Nosotros is most closely associated with these earlier Postclassic ceramic types, suggesting that the contexts reflect Postclassic occupation and not Terminal Classic occupation.

[2] Given the error rate, two samples from each obsidian compositional group determined with XRF were analyzed with INAA to determine the veracity of the XRF groups. There were no differences between INAA and XRF data with regard to source identification.

[3] For the most complete counts of all known sites with obsidian in the Postclassic period see reference (*1*).

References

1. Braswell, G. E. In *The Postclassic Mesoamerican World*; Smith, M.E.; Berdan, F. F., Eds; University of Utah Press: Salt Lake City, UT., 2003; pp 131–158.
2. Moriarty, M. D.; Lawton, C.; Spensley, E. Paper Presented at the Annual Meeting of the Society for Archaeology, San Juan, 2006.

3. Blair, D.; Terry, R.; Moriarty, M. D. Poster presented at the Annual Meeting of the Society for Archaeology, San Juan, 2006.
4. Spensley, E.; Moriarty, M. D Paper Presented at the Annual Meeting of the Society for American Archaeology, Puerto Rico, 2006.
5. Guderjan, T. H. In *Maya Maritime Trade, Settlment, and Populations on Ambergris Caye, Belize*; Guderjan, T. H.; Garber, J. F., Eds.; Labyrinthos Press: Lancaster, CA, 1995; pp 1–8.
6. Rice, D. In *Arqueología Mesoamericana: Homenaje a William T. Sanders*; Mastache, A. G.; Parsons, J. R.; Santley, R. S.; Sera, M. C., Eds.; INAH: México, 1996; pp 109–122.
7. Stross, F. H.; Sheets, P.; Asaro, F.; Michel, H. V. *Am. Antiq.* **1983**, *48*, 323–346.
8. Ford, A.; Stross, F. H.; Asaro, F.; Michel, H. V. *Ancient Mesoamerica* **1997**, *8*, 101–110.
9. Braswell, G. E. In *Ancient Maya Political Economies*; Masson, M. A.; Freidel, D., Eds.; Altamira Press: Walnut Creek, CA, 2002; pp 285–306.
10. Fowler, Jr., W. R.; Demarest, A. A.; Michel, H. V.; Asaro, F.; Stross, F. *Am. Anthropol.* **1989**, *91*, 158–168.
11. Nelson, F. W. *Scanning Electron Microscopy* **1985**, *II*, 631–649.
12. Brown, D. O.; Dreiss, M. L.; Hughes, R. E. *Lat. Am. Antiq.* **2004**, *15*, 222–240.
13. Rice, P. M. *J. Field Archaeol.* **1984**, *11*, 181–194.
14. Mejía, H. E.; Quezada, H.; Chocón, H. J. In *XI Simposio de Investigaciones Arqueológicas en Guatemala, 1997*; Laporte, J. P.; Escobedo, H., Eds.; Museo Nacional de Arqueología y Etnología: Guatemala City, 1998; pp 171–190.
15. Hammond, N. *Science (Washington, D.C.)* **1972**, *178*, 1092–1093.
16. Dreiss, M. L.; Brown, D. O.; Hester, T. R.; Glascock, M. D.; Neff, H.; Stryker, K. S. *Ancient Mesoamerica* **1993**, *4*, 271–283.
17. McKillop, H. *Ancient Mesoamerica* **1996**, *7*, 49–62.
18. McKillop, H. In *Coastal Maya Trade*; McKillop, H.; Healy, P. F., Eds.; Trent University: Peterborough, Ontario, 1996; pp 1–17.
19. Spence, M. W. *Am. Antiq.* **1981**, *46*, 769–788.
20. Spence, M. W. *Lat. Am. Antiq.* **1996**, *7*, 21–39.
21. Drennan, R. D.; Fitzgibbons, P. T.; Dehn, H. *Res. Econ. Anthropol.* **1990**, *12*, 177–199.
22. Lighthart-Ponomarenko, A. *Geoarch.* **2004**, *19*, 71–91.
23. Stark, B. L.; Heller, L.; Glascock, M. D.; Elam, J. M.; Neff, H. *Lat. Am. Antiq.* **1992**, *3*, 221–239.
24. Cobos, R. In *XI Simposio de Investigaciones Arqueológicas en Guatemala, 1997*; Laporte, J. P.; Escobedo, H., Eds.; Museo Nacional de Arqeiología y Etnología: Guatemala City, 1998; pp 791–801.

25. Nelson, F. W.; Phillips, Jr., D. A.; Barrera Rubio, A. In *Investigations at Edzná Campeche, México, Volume 1, Part 1: The Hydraulic System*; Matheny, R. T.; Gurr, D. L.; Forsyth, D. W.; Hauck, F. R., Eds; New World Archaeological Foundation: Provo, UT, 1983; pp 204–219.
26. *ElvaX 2.4*; *Elvatech*, Ltd: Kiev, Ukraine, 2004.
27. Baxter, M. J.; Buck, C. E. In *Modern Analyitcal Methods in Art and Archaeology*; Ciliberto, E.; Spoto, G., Eds.; John Wiley and Sons: New York, 2000; pp 681–746.
28. Bieber, Jr., A. M.; Brooks, D. W.; Harbottle, G.; Sayre E. V. *Archaeometry* **1976**, *18*, 59–74.
29. Bishop, R. L.; Neff, H. In *Archaeological Chemistry IV*; Allen, R. O., Ed.; Advances in Chemistry Series No. 220; American Chemical Society: Washington, DC, 1989; pp 576–586.
30. Glascock, M. D. In *Chemical Characterization of Ceramic Pastes in Archaeology*; Neff, H., Ed.; Prehistory Press: Madison, WI, 1992; pp 11–26.
31. Harbottle, G. *Radiochemistry* **1976**, *3*, 33–72.
32. Neff, H. In *Modern Analyitcal Methods in Art and Archaeology*; Ciliberto, E.; Spoto, G., Eds.; John Wiley and Sons: New York, 2000; pp 81–134.
33. Weigand, P. C.; Harbottle, G.; Sayre, E. V. In *Exchange Systems in Prehistory*; Earle, T. K.; Ericson, J. E., Eds.; Academic Press: New York, 1977; pp 15–34.
34. Baxter, K. H. M. A. thesis, State University of New York, Albany, NY, 1983.
35. Hester, T. R.; Dreiss, M.; Michel, H. V. *Obsidian at Colha and Ancient Maya Trade*; Unpublished Manuscript; University of Texas Center for Archaeological Research: San Antonio, TX, 1983.
36. McKillop, H.; Healy, P. F. *Coastal Maya Trade*; Trent University: Peterborough, Ontario, 1996.
37. Scholes, F. V.; Roys, R. L. *The Maya Chontal Indians of Acalan-Tixchel: A Contribution to the History and Ethnography of the Yucatan Peninsula*; University of Oklahoma Press: Norman, OK, 1968.
38. Rice, P. M.; Michel, H. V.; Asaro, F.; Stross, F. *Am. Antiq.* **1985**, *50*, 591–604.
39. Braswell, G. E. In *El sitio maya de Topoxté: Investigaciones en un isla del lago Yaxhá, Petén, Guatemala*; Wurster, W. W. Ed.; Verlag Philip von Zabern: Mainz am Rhein, 2000; pp 208–221.
40. Mazeau, D. M. A. thesis, State University of New York at Buffalo, Buffalo, NY, 2000.
41. Rice, D. S.; Rice, P. M.; Sánchez-Polo, R.; Jones, G. D. Report submitted to the Instituto de Anthropología e Historía de Guatemala (IDEAH), 1996.
42. Cecil, L. G. *Archaeometry* **2004**, *46*, 385–404.
43. Bowditch, C. P.; Rivera, G. *Relation of Two Trips to Peten*; Labyrinthos: Culver City, CA, 1987.

Chapter 28

Sources of Archaeological Obsidian in Peru: Descriptions and Geochemistry

Michael D. Glascock[1], Robert J. Speakman[1], and Richard L. Burger[2]

[1]Research Reactor Center, University of Missouri, Columbia, MO 65211
[2]Department of Anthropology, Yale University, New Haven, CT 06520

Use of obsidian in South America has been documented from about 13,000 years B.P. through the beginning of the Spanish Conquest. Several of the most archaeologically important sources of obsidian in South America are located in the Andes mountain region of southern Peru. Due to the difficult terrain and volatile political environment, the locations for many of these sources were unknown to archaeologists until quite recently. Neutron activation analysis and X-ray fluorescence have been used to measure the compositional patterns for individual sources. The comprehensive source database established by this work enables definitive and cost-effective assignments of provenance to obsidian artifacts from archaeological sites in Peru and northern Bolivia.

Introduction

Like other regions of the world, the prehistoric inhabitants of South America showed a preference for obsidian as a raw material for the production of stone tools. Evidence of obsidian consumption exists at both highland and coastal sites (*1–3*) in southern Peru dating to at least 13,000 years B.P. The appeal for obsidian was likely due to its visual attractiveness and physical properties that yield sharp-edged, conchoidal-shaped fractures. Obsidian makes sharper edges than most other types of tool-making stones. This property made obsidian a desirable resource for cutting tools such as knives, arrowheads, and scrapers. However, despite the preference of prehistoric people for fine chipping materials, the sources of obsidian were scarce and were often located in difficult terrain.

The mid-1960s discovery that the chemical compositions of obsidian artifacts and sources could be used to link artifacts to their sources increased the level of interest in obsidian research around the world (*4*). The degree of confidence with which an obsidian artifact could be assigned to a specific source was almost unparalleled in archaeology. Archaeologists working in the Mediterranean, Mesoamerica, and other areas conducted hundreds of investigations of obsidian artifacts and sources to study prehistoric human behavior, interaction, and exchange (*5–7*). However, progress on obsidian source characterization research in South America was delayed by factors such as accessibility (i.e., difficult terrain and volatile political environment). Consequently, most information about the locations of obsidian sources and prehistoric long-distance exchange networks in South America has been obtained only recently (*8–16*).

One of the authors [MDG] of this work has been a leading proponent of a systematic approach to obsidian source characterization (*17*). By adhering to this approach, the maximum amount of new information possible will be obtained from each provenance investigation. The purpose of this work is to synthesize the current geological and compositional information for obsidian sources from Peru stored in the University of Missouri Research Reactor (MURR) Archaeometry Lab database. This work is important because Peru was one of the most significant archaeological regions of South America during prehistoric times.

A Systematic Approach to Source Characterization

Clark (*18*) identified several properties of obsidian sources and artifacts made from obsidian that give obsidian a unique role in archaeology:

- The number of available sources was limited
- Obsidian artifacts were widely distributed
- Vast amounts of obsidian debitage occur near the ancient quarries
- Obsidian artifacts retain most of their physical properties from fabrication
- The fragile nature of obsidian caused a high replacement rate

- Obsidian artifacts are nearly indestructible in most archaeological environments
- Different obsidian sources are usually compositionally distinct
- A freshly-made obsidian surface absorbs water over time

Studies of obsidian artifacts and sources can be used to examine resource procurement patterns, to identify long-distance exchange networks, to study manufacturing processes, and to establish site chronologies. In addition, obsidian artifacts may be used to extract dating information through hydration dating.

Undeniably, one of the most widely recognized roles for obsidian involves provenance research (i.e., sourcing). Provenance research connects artifacts to their sources such that interpretations about the movement of obsidian can be made with a high degree of confidence. The process of obsidian artifact sourcing relies on making comparisons of one or more characteristics of obsidian artifacts with those same characteristics for all possible sources. In order to have a successful outcome, one must demonstrate that the characteristics of the artifact and the source are the same to the exclusion of all other possible sources. Usually, the greater the number of characteristics used in the comparison, the more reliable the result.

Archaeologists have long sought methods for obsidian characterization that were rapid, reliable, non-destructive, and low-cost. Among the various methods investigated were visual techniques (*19*), density measurements (*20*), magnetic properties (*21*), thermoluminescence (*22*), fission-track analysis (*23*), Mossbauer spectroscopy (*24*), and natural radioactivity (*25*). Although some of the methods occasionally identified differences between sources, the overlap between sources was such that their overall reliability was unsatisfactory. The most successful method of characterization for obsidian provenance research has been compositional analysis (*26*).

The success of an obsidian provenance study employing compositional analysis is dependent upon the availability of a database of precise and accurate trace element profiles for the sources. Although most previous obsidian characterization studies were successful, a few of the earliest obsidian studies reported conclusions (*27–29*) that were later refuted (*30–31*). These failed studies suffered from a variety of problems such as assignment of artifacts while failing to locate all possible sources, reliance on second-hand information about sources, analyzing too few specimens from each source, measuring too few trace elements, and failure to identify the most discriminating elements. In order to avoid these problems, a more systematic approach to sample collection, chemical analysis, and statistical evaluation of the compositional data from obsidian sources has been recommended before studying significant numbers of artifacts (*17*). By following this approach, the reliability of obsidian provenance studies will be enhanced, and the information obtained from each archaeological investigation of obsidian will have a mucher greater intrinsic value.

In accord with the systematic approach, obsidian researchers are encouraged to locate all sources of obsidian within their region of interest. After locating a source, the source area should be surveyed thoroughly to locate all primary and secondary deposits; and the source area also should be sampled intensively such that the amount of internal variation within the source can be determined. As large a number of source specimens as possible should be analyzed (e.g., a dozen or more from each type instead of two or three). Source samples should be analyzed completely by the available techniques such that the correlations among and the variations between the measured elements within sources can be examined. If compositional differences are discovered within a source, the compositional subgroups should be associated with specific geographic zones.

With regard to the analysis of artifacts, the compositional data for all sources should be carefully examined to discover the most discriminating elements which might lead to identification of lower-cost, abbreviated procedures for use on artifacts. Finally, before implementing an abbreviated procedure for the analysis of artifacts, the procedure should be thoroughly tested to verify its reliability and to understand its limitations.

Obsidian Geochemistry and Methods for Characterization

Before describing the Archaeometry Lab at MURR's involvement in research on obsidian sources and artifacts from South America, some background information on obsidian geochemistry is helpful. In addition, a description of advantages and disadvantages of various analytical methods employed to characterize obsidian is presented.

Obsidian Geochemistry

Obsidian forms when viscous volcanic lava with high silica and alumina content cools rapidly, usually at the margins of a lava flow, such that the process of mineral crystallization is precluded and, instead, a glassy matrix is formed. The glass is generally black or gray in color, but other colors are possible, depending upon the composition and the circumstances of formation. The glass sometimes appears banded or streaky. Due to an atomic structure that is entirely disordered, glass is physically amorphous and isotropic; this is the main reason why obsidian makes such effective tools, since flakes can be obtained by striking an obsidian core from almost any direction. Some low-quality obsidian may contain a significant proportion of phenocrysts (i.e., crystals) made of minerals similar in composition to the obsidian, and due to their poor fracturing properties they produce inferior tools.

Most obsidian is rhyolitic and the composition ranges from about 70–75% SiO_2, 10–15% Al_2O_3, 4–5% Na_2O, 3–4% K_2O, and 1–5% total Fe_2O_3+FeO. In addition, the intrinsic water content of obsidian ranges from 0.1 to 2.0%. The remaining elements in obsidian occur in concentrations well below 1%, and these elements are generally referred to as trace elements.

Because glass is unstable at ambient temperatures, obsidian gradually undergoes hydration through the diffusion of water into the outer surface. As the water content increases to about 5 wt %, concentric "onionskin" cracking patterns occur gradually converting the glass into perlite over hundreds of thousands of years. Thus, obsidian has a relatively short life-time by geological standards; but the life-span for obsidian is still quite long in human terms. In spite of hydration, obsidian artifacts are very durable in most environments, and their compositions are unchanged relative to their sources. The hydration process gives rise to the line of research known as obsidian hydration dating.

Most obsidian sources are chemically homogeneous, with variations in composition on the order of a few percent or less. However, the individual sources have different trace-element compositions as a reflection of the compositions of parent rocks and changes taking place in the magma chamber prior to eruption. The major elements are restricted to a relatively narrow range of composition; but the abundances of trace elements can differ by orders of magnitude between obsidian sources. If the variations within sources are smaller than the differences between sources, then the provenance of obsidian artifacts can be successfully established.

The silicic volcanism from which obsidian is formed is a complex phenomenon, having its origins in partial melting and melt segregation in the upper mantle, usually occurring tens of kilometers beneath the surface and near the tectonic margins. Magma forms at depth with high temperatures (> 1000 °C) by the melting of rock. The melt is in equilibrium with solid materials and the trace elements present are distributed between the liquids and solids. As the magma evolves, fractional crystallization takes place. During the process of crystallization, several trace elements such as Co, Cr, Ga, and Ni are strongly absorbed by the solid phases. These elements are typically known as the *compatible* elements because they are compatible with the crystallization of solids and are readily removed from the magma. Other trace elements are *incompatible* with the solid phase and they tend to become more concentrated in the liquid.

The incompatibility of certain trace elements with the solid phase results from two factors. First, the large-ion lithophile elements (LILE) such as Ba, Cs, Rb, and Sr have large ionic radii. The LILE are too large for the available ionic sites in the solid and they tend to remain in the liquid phase. A second cause of incompatibility is that elements such as Hf, Nb, Ta, and Zr, commonly referred to as the high-field strength elements (HFSE), and elements such as La, Ce, Sm, Eu, etc., usually referred to as the rare earth elements (REE), have large ionic valences that inhibit their inclusion within the crystalline structure of the solid phase. The concentrations of incompatible elements present in the obsidian are affected by the initial amounts present in the melted rock, the temperature and pressure of the magma, and additional melting that takes place as the magma rises to the surface and cools rapidly. As a result, the mixture of incompatible elements within each obsidian source is unique; and, the practice of measuring trace element fingerprints of artifacts to compare and contrast with sources for the purpose of assigning provenance to the artifacts is highly successful.

The sources of obsidian exploited by prehistoric peoples were almost entirely restricted to more recent volcanic eruptions due to the hydration process. Few obsidian sources are greater than 10 million years old, and many are less than 100,000 years of age. In addition to the Andes Mountains of South America, obsidian is found in the Mediterranean, Turkey, Africa, central Europe, central Mexico, western United States, Alaska, Japan, and the islands of the South Pacific.

Methods for Chemical Characterization of Obsidian

Since the mid-1960s, a variety of analytical chemistry techniques have been used to characterize obsidian sources and artifacts for provenance research (*4, 32–36*). The most common of these methods include optical emission spectroscopy (OES), atomic absorption spectroscopy (AAS), particle-induced X-ray emission spectroscopy (PIXE), inductively coupled plasma-mass spectrometry (ICP-MS), laser ablation-inductively coupled plasma mass spectrometry (LA-ICP-MS), X-ray fluorescence spectroscopy (XRF), and neutron activation analysis (NAA). When selecting a method of analysis for obsidian, one must consider accuracy, precision, cost, promptness of results, existence of comparative data, and availability. Most of the above-mentioned techniques are capable of determining a number of elements, but some of the methods are more labor-intensive, more destructive, and less precise than others. The two methods with the longest and most successful history of success for obsidian provenance research are XRF and NAA.

X-ray fluorescence (XRF)

With XRF, the sample is exposed to a beam of X-rays and some of the X-rays are absorbed by atoms on or near the surface of the sample. The absorbed

X-rays cause electrons to be ejected from innermost atomic energy levels, usually from the K level (*37*). The presence of a vacant energy level creates an unstable condition for the atom. In order to return to stability, electrons from the outer (higher energy) levels of the atom repopulate the vacant inner (lower energy) levels. During this process, a secondary (or fluorescent) X-ray is emitted with an energy equal to the difference between the two energy levels. Because the energy levels are unique for each atom, measurement of the emitted X-rays enables identification of the element and quantification of the number of element atoms. The amount of x-ray fluorescence occurring is highly sample dependent and quantitative analysis requires calibration of the spectrometer with standards that are similar in composition to the unknown sample matrix.

X-ray fluorescence is a rapid and low-cost method that can be performed on solid samples. However, the depth of penetration of X-rays in most solid samples is relatively shallow. High-precision XRF on geological samples such as obsidian requires preparation of homogeneous, powdered samples pressed into pellet form. If some loss of precision and accuracy due to irregular size, shape, and thickness of samples is acceptable, obsidian specimens can be analyzed non-destructively. Samples smaller than 1 cm in diameter or with element concentrations less than 5 ppm are generally not suitable for XRF. XRF can determine about 10–15 elements in obsidian (K, Ti, Mn, Fe, Zn, Ga, Rb, Sr, Y, Zr, Nb, Pb, and Th). Fortunately, many of the measurable elements are the incompatible elements which provide discrimination between sources.

Neutron activation analysis (NAA)

The principles of NAA are slightly different from XRF in that samples are irradiated by thermal neutrons from a nuclear reactor (*38–39*). During irradiation, neutrons are captured by the nuclei of atoms in the sample. This process, called activation, causes some of the nuclei to become unstable. During and after irradiation, the unstable nuclei emit gamma rays as transitions between different excited states in the nucleus. The gamma-ray energies are unique for each radioactive nucleus according to the differences in energy between the excited states, and the emission rates are distinct according to the half-lives of the radioactive isotope. By measuring the intensities of the characteristic gamma rays with a spectrometer, a compositional profile for the elements present in the sample can be established.

A variety of irradiation, decay, and counting schemes can be employed by NAA in order to enhance the sensitivities of different subsets of elements. For example, short irradiation times followed short decay times are used to measure the short-lived elements, and long irradiation times followed by long decay times are used to measure the long-lived elements. For obsidian, the short-lived elements measurable are Al, Ba, Cl, Dy, K, Mn, and Na. The long-lived elements measurable in obsidian are Ba, Ce, Co, Cs, Eu, Fe, Hf, Rb, Sb, Sm, Ta, Tb, Th, Zn and Zr. Although some elements are possible using either procedure, data for the element Ba is usually best after long irradiation. Fortunately, a large number of the elements measurable by NAA are the *incompatible* elements.

By comparison, NAA is more expensive and less rapid than XRF. However, NAA can measure more elements, has greater sensitivity, and has superior precision and accuracy for most elements compared to XRF. Although NAA requires sample destruction during sample preparation, it can be used to analyze obsidian samples on the order of 5–10 mg which are usually too small for XRF.

History of Obsidian Research in Peru

The first investigation of obsidian from the Andean region was conducted during the early 1970s, nearly a decade after the earliest obsidian characterization studies were conducted in Southwest Asia (*4–5*), Mesoamerica (*6*), and North America (*7*). A small collection of obsidian artifacts from sites in Cuzco and Puno, Peru were submitted by Karen Mohr Chavez to Adon Gordus at the University of Michigan. The artifacts were analyzed by NAA to investigate patterns of Early Horizon obsidian utilization (*40*). Although the locations of sources were unknown, the artifacts divided into two distinct chemical types that Chavez referred to as the Cuzco and Puno Types.

In 1973, a much larger program of obsidian analyses was initiated at the Lawrence Berkeley National Laboratory (LBNL) by Richard Burger, Frank Asaro, and Helen Michel using NAA and XRF (*41–42*). The LBNL analyzed a total of 846 obsidian artifacts (i.e., 812 of the samples were analyzed by XRF and 141 were analyzed by NAA). The artifacts came from 98 archaeological sites located in Peru and northern Bolivia. Based on the frequencies of artifacts with similar compositions, sixteen different geochemical types were recognized. Eight of the highest-frequency obsidian types were found to be responsible for 98% of the artifacts. The remaining eight low-frequency types (i.e., with only one or two specimens each) were responsible for only 2% of the artifacts. The eight highest-frequency chemical types were assigned names according to the regions where the artifacts occurred archaeologically, with the single exception of Quispisisa Type obsidian. Unfortunately, geographic locations for the sources of obsidian were unknown at that time.

Due to political violence in the Andean highlands, obsidian research in Peru continued only sporadically throughout the 1980s. However, Katharina Schreiber and Paul Trawick successfully located and collected geological samples from the Jampatilla and Alca obsidian sources (see below). Unfortunately, little more was accomplished prior to 1990.

Between 1990 through 2005, Richard Burger and a number of graduate student archaeologists submitted 161 source samples and 850 artifacts from Peru and northern Bolivia to MURR for analysis by NAA. The most active individuals were Sarah Brooks, Kirk Frye, Martin Giesso, Justin Jennings, and Nico Tripcevich. As artifacts and source samples were being analyzed, the results were used to verify the discovery of archaeologically important sources. When artifact data or source data suggested possible locations for sources, this information was relayed back to the archaeologists. A comprehensive NAA database for Peruvian obsidian was gradually acquired. Portions of the source samples were retained for reference and future analysis whenever possible.

In 2006, a table-top energy-dispersive XRF (ED-XRF) spectrometer was acquired by the Archaeometry Lab to facilitate non-destructive analysis of obsidian and other types of artifacts. One of the first projects performed on the new XRF spectrometer was the re-analysis of the geological samples from sources in Peru. As a result, it is now possible for the Archaeometry Lab to use either XRF or NAA to successfully determine the provenance of obsidian artifacts from Peru. Due to its light weight, the spectrometer also has the potential to be transported from the laboratory to museums and to archaeological sites for *in situ* analysis.

Experimental Procedures

Geological samples were collected from primary sources and secondary deposits of obsidian. In most cases, the geographic coordinates of source samples were also recorded. The geological samples were shipped to MURR for sample preparation and analysis.

Sample Preparation

Preparation of obsidian samples for NAA generally involves washing the samples in tap water and cleaning with a toothbrush to remove dirt and other loose debris from the surface. When necessary, acetone and ethyl alcohol were used to remove identification markings made with ink, paint, or fingernail polish. The cleaned samples were cut with a diamond-edged saw blade and minimally reduced to smaller fragments weighing about 50–100 mg using a clean ceramic mortar and pestle. Individual fragments were inspected under a magnifying glass to remove samples with inclusions, crush fractures, or metallic streaks which might contain undesirable contamination. The largest left-over geological samples after sawing were retained for subsequent analysis by XRF.

All geological samples were prepared for the short- and long-irradiation NAA procedures employed at MURR. However, artifacts were only prepared for the long irradiation if the results from short irradiation were inconclusive. The fragments created from sawing and crushing were weighed into clean high-density polyethylene vials for short irradiation and clean high-purity quartz vials for long irradiations. For short irradiations, a 100 mg aliquot of fragments was used, and for long irradiations, a 250 mg aliquot was used. For obsidian artifacts, much smaller samples are possible. In both instances, sample weights were recorded to the nearest 0.01 mg. Along with the source samples, reference standards were similarly prepared from SRM-278 Obsidian Rock and SRM-1633a Fly Ash.

NAA Procedure

The short-irradiation NAA procedure involves sequential irradiations of samples and standards in a neutron flux of 8 x 10^{13} n cm^{-2} s^{-1} for five seconds followed by a 25–minute decay and 12–minute count with a high-purity germanium (HPGe) detector. By measuring the emitted gamma rays and comparing the counting rates of samples to standards, the concentrations of seven short-lived elements are usually measured (i.e., Al, Ba, Cl, Dy, K, Mn, and Na). Use of this short-irradiation NAA procedure has been described previously as a satisfactory method for sourcing a large proportion of the obsidian artifacts in several geographic regions (*43*).

The long-irradiation procedure involves irradiation of a bundle of quartz vials containing 30 unknown samples and six standards. The sample bundle is irradiated in a neutron flux of 5 x 10^{13} n cm^{-2} s^{-1} for 70 hours and followed by two measurements with an HPGE detector coupled to an automatic sample changer. The first measurement occurs one week after the end of irradiation and uses a counting time of 1800 seconds for each sample; and the second measurement takes place about four weeks later with counting times of three hours each. The long-irradiation procedure enables measurement of seven elements during the first count: Ba, La, Lu, Nd, Sm, U, and Yb; and 15 additional elements during the second count: Ce, Co, Cs, Eu, Fe, Hf, Rb, Sb, Sc, Sr, Ta, Tb, Th, Zn, and Zr.

XRF Procedure

The light-weight Elva-X energy dispersive XRF spectrometer employed for this study has an air-cooled rhodium target anode X-ray tube with 140 micron Be window and a thermoelectrically cooled Si-PIN diode detector. The detector resolution is 180 eV for the 5.9 keV X-ray from iron. The beam dimensions are 3 x 4 mm. The minimum allowable sample size for the XRF is about 1 cm diameter for obsidian fragments.

To analyze the geological samples in this study, the X-ray tube was operated at 35 kV using a tube current of 45 μA and measurement times of 400 seconds on each sample. The elements measured by XRF were K, Ti, Mn, Fe, Zn, Ga, Rb, Sr, Y, Zr, and Nb. Peak deconvolution and calculations of elemental concentrations were accomplished using the ElvaX analysis package. The instrument was calibrated using data from a series of previously characterized source samples in the MURR obsidian reference collection, including eleven well known Mesoamerican sources (El Chayal, Ixtepeque, San Martin Jilotepeque, Guadalupe Victoria, Pico de Orizaba, Otumba, Paredon, Sierra de Pachuca, Ucareo, Zaragoza, and Zacualtipan) previously analyzed by NAA (*44*).

Descriptions of Obsidian Sources in Peru

Locations for the geological sources mentioned in this study are shown in Figure 1. Table I lists the geographic coordinates. Descriptions of the sources are presented here.

The Alca Source

The Alca source was discovered by Paul Trawick in 1984 about 3 km southwest of the village of Alca in north central Arequipa. Alca is located in the Cotahuasi Province, Department of Arequipa. The village of Alca is about 16 km northeast of the provincial capital of Cotahuasi, 190 km west of Arequipa, and 195 km southwest of Cuzco. The primary obsidian deposit was found on Cerro Santa Rosa at an altitude of 4600 masl and overlooks the Cotahuasi River Valley, the world's deepest canyon from rim to bottom.

Analysis of the Alca source materials at LBNL (*10*) using NAA found a match with a geochemical type discovered earlier by LBNL (*41–42*) and at the University of Michigan (*40*) known as the Cuzco Type. The Cuzco obsidian compositional profile was present in 14% of the artifacts in the original study at LBNL (*41–42*) and was predominantly linked to artifacts from archaeological sites in southern and central Peru.

In the mid 1990s, a more detailed survey of the Alca source was conducted by Justin Jennings, graduate student from UC Santa Barbara. Jennings collected geological samples over a much wider area in order to understand possible intra-source variability. The samples were submitted to MURR for NAA and the resulting chemical analyses determined that three different chemical fingerprints

Table I. Locations of obsidian sources in Peru.

Map Location	*Locality*	*Latitude °S*	*Longitude °W*
1	Yanarangra	13.24	75.17
2	Puzolana	13.20	74.23
3	Quispisisa	14.07	72.69
4	Jampatilla	14.21	74.22
5	Potreropampa	14.36	73.32
6	Lisahuacho	14.37	73.36
7	Cerro Ticllago	14.62	73.00
8	Alca	15.12	72.69
9	Caylloma	15.19	71.68
10	Uyo Uyo	15.62	71.76
11	Chivay	15.64	71.54
12	Aconcahua	16.84	69.86

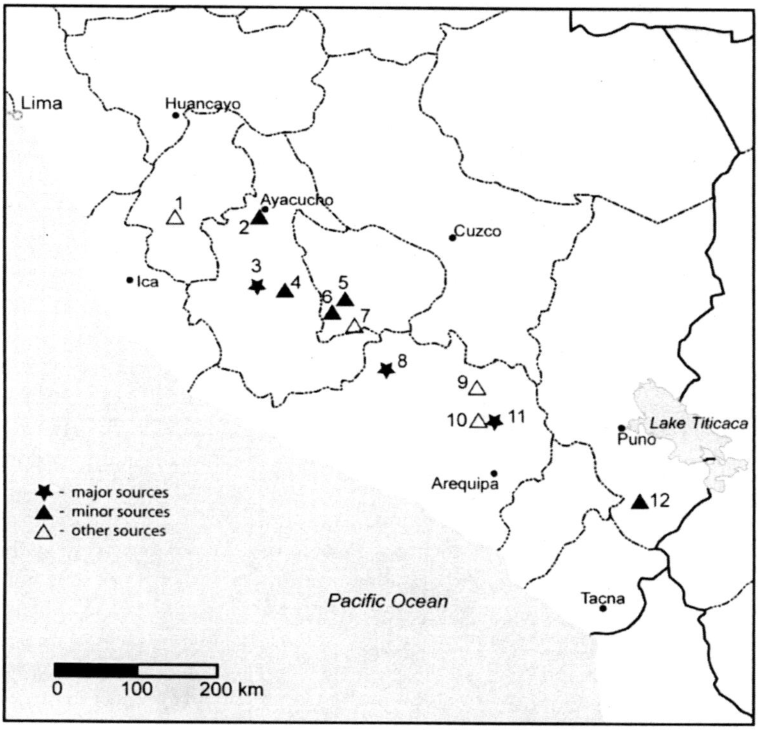

Figure 1. A map of southern Peru showing the locations of known obsidian sources. The source names are listed in Table I.

were present (*45*). Subsequent analyses of artifacts has verified that obsidian with all three of the Alca compositional fingerprints was used prehistorically.

Obsidian from the Alca source is of excellent quality. The glassy matrix is very uniform and easy to use for tool manufacture. Large nodules measuring up to 20–30 cm length are widely available at the main deposit. Smaller nodules are common in nearby streams. Alca obsidian comes in a range of colors from black, chocolate-brown, to a rare aqua blue (*45*).

The Chivay Source

The most common variety of obsidian found on archaeological sites in the southern provinces of Peru was assigned the name Titicaca Basin Type by Burger and Asaro (*41–42*). More than 26% of the artifacts analyzed at LBNL were identified as belonging to the Titicaca Basin Type. The highest frequencies of occurrence for this variety of obsidian were in the Lake Titicaca Basin and northern Bolivia regardless of time period. The distribution of Titicaca Basin Type obsidian also appeared to be similar to that for the artifacts from Puno reported earlier (*40*).

In 1991, Sarah Brooks who was a graduate student at Wisconsin found a number of tools and waste flakes, ranging up to 2–4 cm long, at archaeological sites in the Colca Valley. The existence of large unused flakes suggested a source might be nearby. A small collection of obsidian fragments were sent to Burger who forwarded the fragments to MURR for measurement by NAA. The analyses indicated that the samples from Colca Valley were an excellent match for the Titicaca Basin Type obsidian.

Additional surveys of the area by Sarah Brooks were successful in locating a large deposit of obsidian 5 km east of the the town of Chivay. An independent survey by Burger and Peruvian geologist Guido Salas in 1995 also located the source, which they dubbed the Chivay source; their publications of the results along with the analysis at LBNL provided the first detailed description of the source location and the trace element chemistry (*11*). The source was referred to as Cotallalli by Brooks (*46*); but most of the literature now refers to the source as Chivay (*47–49*).

The main obsidian deposit contains large blocks of obsidian ranging up to 30 cm on a side. The deposit is located at an altitude of 4900 masl near the foot of the slope of Cerro Ancachita. Smaller nodules of obsidian can be found in streams near the town of Chivay due to moraine deposits caused by glacial activity. Raw materials from the Chivay source generally have a heavily weathered cortex. When fractured to produce fresh fragments, the glass has a purple hue and is of high quality.

The Quispisisa Source

The chemical profile known as Quispisisa obsidian was observed in 45% of the artifacts analyzed by LBNL and in more than 90% of the artifacts from archaeological sites located in central and northern Peru (*41–42*). The northernmost extent of Quispisisa obsidian artifacts was in the Department of Junin. Based on incorrect information, the source of this obsidian was originally presumed to be near San Genaro in the Province of Castrovirreyna, Department of Huancavelica (*50*). The assumption that the Quisipisia mine was located near San Genaro was widely accepted for a number of years, before the error was discovered during the course of this work.

Explorations of the San Genaro area in the 1990s by Burger were unable to locate the source of Quispisisa obsidian, but an obsidian type known as Yanarangra was located about 8 km south of San Genaro. After several years of searching for the Quispisisa source and uncovering many incorrect leads, Burger learned of a source of geologic obsidian in the Huanca Sancos area of the Department of Ayacucho. With the assistance of local residents, Burger eventually discovered an obsidian outcrop called Queshqa, which in Quechua dialect of Ayacucho, means glass (*13*). A more complete description of the Quispisisa geological source and its archaeological impact has been reported more recently (*48*). Analysis of geological samples from this location proved to be an excellent match for the Quispisisa artifacts.

The main deposit of Quispisisa obsidian is located 15 km south of the town of Sacsamarca, Department of Ayacucho. The visible part of the obsidian deposit is located at an altitude of 3750 masl in a seam along an exposed bluff overlooking the Rio Urubamba, a tributary of the Rio Caracha. The obsidian seam has a thickness of at least 30 m and displays horizontal banding. The entire area is covered with worked and unworked artifact-grade obsidian nodules ranging up to 30 cm on a size. Cobbles from other stones used as hammerstones also are present. This was obviously an extremely important source of obsidian.

The obsidian from Quispisisa is of excellent quality and comes in various hues. Colors such as a dark red or "wine red" are common. The glass contains very few phenocrysts and little or no natural cracking.

The Puzolana Source

From their LBNL work Burger and Asaro (*41–42*) observed a geochemical type of obsidian known as the Ayacucho Type in 1% of the artifacts. Artifacts of this type were limited to the Ayacucho Basin. This obsidian type represented only 12% of the artifacts from the Ayacucho-Huanta Archeological-Botanical Project which ran from 1977–1979.

In April 1999, Burger and Jose Pinalla, a Peruvian archaeologist, traveled to the city of Ayacucho to follow up on leads concerning the possible location of the Quispisisa source. A local volcanic tuff referred to as *puzolana* was found to have varied amounts of obsidian nodules within the stratum. The greatest density of obsidian nodules was found near the hamlet of Chupas in a geological layer exposed by the construction of the Cachi Canal.

During exploration of the source area, nodules ranging from 1 to 4 cm on a side were found, but nothing larger. Most of the nodules from Puzolana exhibit a black streaking, but a few samples had reddish streaks which are quite uncommon. Analyses of the nodules at MURR found a match between the Puzolana obsidian and Ayacucho Type (*14*). Judging from the rarity of Puzolana obsidian at archaeological sites outside of Ayacucho, it is unlikely that Puzolana was ever involved in a long-distance exchange network, perhaps due to the small nodule sizes.

The Jampatilla Source

Katharina Schreiber located a primary deposit of obsidian known as Jampatilla in 1981, during a survey in nearby Huaycahuacho. Huaycahuacho is located in the Carhuarazo Valley and Province of Lucanas, Department of Ayacucho. The obsidian deposit was discovered 3–4 km east-northeast of the town of Huaycahuacho and 55 km north of the larger town of Puquio. Obsidian workshop areas and associated archaeological sites were reported (*51*). Nodules ranging up to 10–12 cm in diameter were found at the source, but smaller eroded samples were more common in secondary deposits.

Analysis of the Jampatilla samples collected by Schreiber were conducted at MURR. The compositional profile was found to match closely with that of the Pampas Type (*12*) obsidian of Burger and Asaro (*41–42*). They found that this variety of obsidian occurred with greatest frequency in the Carhuarazo Valley. The low frequency of Jampatilla obsidian elsewhere, suggested that Jampatilla obsidian was not part of a long-distance exchange network.

The Aconcahua Source

Geological specimens of obsidian from the Aconcahua source were collected by Kirk Frye in 1996. Cerro Aconcahua is located approximately 15 km southwest of Mazo Cruz in the Department of Puno. The geological samples from Aconcahua matched two artifacts from the nearby archaeological site of Quelcatani according to analyses by NAA at MURR. Exploitation of obsidian from the Aconcahua source appears to have been limited to the local area.

The Cerro Ticllago, Lisahuacho, and Potreropampa Sources

Analyses conducted at LBNL (*41–42*) on artifacts from on a multi-component site known as Waywaka in the town of Andahuaylas in the Department of Apurimac exhibited two chemical fingerprints rarely detected outside the Andahuaylas area. The chemical types were assigned the names Andahuaylas A and Andahuaylas B by Burger and Asaro who were strongly convinced that the sources were located in this region.

During July 2003, Burger and archaeologist Fidel Fajardo Rios traveled to Abancay, capital of the Department of Apurimac, to follow up on several leads regarding obsidian deposits. With the help of a local civil engineer, Carlos Valdes, they traveled to a small village to the south by the name of Iscahuaca. In Iscahuaca, they met a local alpaca herder who directed them to an obsidian deposit 48 km southeast known as Cerro Ticllago or "stained mountain". Cerro Ticllago is located at an altitude of 4890 masl, in the Province of Aymaraes, along the southern border that separates the Department of Apurimac from the Department of Ayacucho. Small unworked nodules no larger than 3 cm were found at the source, but most of the obsidian was friable and tended to have linear rather than conchoidal fracture patterns. A few source samples from Cerro Ticllago and artifacts from the local Yanamachay rock shelter were collected for analysis.

Analyses of the Cerro Ticllago obsidian and Yanamachay artifacts at MURR compositionally matched the artifacts to the source. The data also suggested some similarities between Cerro Ticllago and the Andahuaylas A and B obsidian types, but Cerro Ticllago source was definitely not the source of either type. The evidence suggested, however, that the Andahuaylas A and B sources might be this general area. Use of obsidian from the Cerro Ticllago source was probably limited to the local area.

Based on these results, Burger and Fajardo returned to the Aymaraes region in July 2004. Discussions with a farmer from near Chalhuanca, the provincial capital of Aymaraes, provided a lead to obsidian outcrops along a west-to-southwest winding road in the Huayllaripa drainage. The first outcrop was found at a site named Potreropampa or "plain of pasture land" about 32 km from Chalhuanca at an altitude of 4128 masl. Obsidian nodules were embedded in volcanic tuff and the nodules were mostly 3–6 cm in diameter but a few were up to 10–15 cm. A fair quantity of lithic debris was observed.

The second outcrop was discovered along the same road a distance of 42 km southwest of Chalhuanca at an altitude of 3978 masl. Again, a fair quantity of lithic debris covered the area. Nodules were mostly 4–7 cm in diameter with some ranging as large as 9–12 cm.

Chemical analysis of geological specimens from the three obsidian outcrops proved that the Andahuaylas A type matched the Potreropampa source, the

Andahuaylas B type matched the Lisahuacho source, and Cerro Ticllago was a new chemical type (*52*). The low-quality Cerro Ticllago obsidian was probably never used to make artifacts.

The Yanarangra Source

The Yanarangra obsidian source was discovered by Burger about 8 km south of San Genaro in the Castrovirreyna Province, Department of Huancavelica during the search for the source of Quispisisa obsidian. Nodules from the Yanarangra source were submitted to MURR for NAA. Obsidian from the Yanarangra source was low quality, heavily cracked, with many mineral inclusions in all of the samples collected. Analysis of samples at MURR found Yanarangra did not to match the composition of Quispisisa (*48*) or artifacts with any other compositional profile. There is no evidence that the Yanarangra obsidian source was used during prehistoric times.

The Uyo Uyo and Caylloma Sources

During archaeological surveys in the Department of Arequipa, Sarah Brooks located two additional obsidian sources and collected geological samples from each for NAA. A secondary source deposit was found 5 km west of Chivay near the pre-Inca archaeological site of Uyo Uyo. The second obsidian source named Caylloma was found about 14 km northwest of the town of Caylloma in a northern province of Arequipa by that name. Compositional analysis of both sources failed to match the data for any artifacts. Thus, it is believed that neither source was used prehistorically.

Results

Geological Samples

The geological samples arriving at MURR between 1990 and 2005 were analyzed using the short- and long-NAA procedures described above. In 2006, remaining portions of the geological samples were analyzed by XRF. Examination of the chemical data for the entire collection of 161 samples established the presence of 14 distinct compositional profiles for the twelve sources. Three compositional profiles are required to describe subgroups for the geological samples from the Alca source.

Through intercomparisons of the geological source compositional data with compositional data for artifacts also analyzed at MURR and with the mid-1970s LBNL analyses, the source compositional profiles were matched with the original chemical types. In order make a comparison between the MURR data and the mid-1970s LBNL data, intercalibration factors were determined for the SRM-278 Obsidian Rock and SRM-1633a Fly Ash used at MURR against the Perlman Standard Pottery in use at LBNL (*53*). The MURR obsidian data was found to compare favorably with the mid-1970s NAA data from LBNL. The MURR NAA results for the geological sources found excellent agreement with seven of the highest frequency compositional types identified by Burger and Asaro (*41–42*).

The geological source data was subdivided into three categories according to the relative archaeological importance of the sources (major, minor, and non-archaeological). Table II lists the compositional data means and standard deviations for the elements measured by short-NAA, long-NAA, and XRF in the five compositional groups describing the three major sources and their subgroups: Chivay, Quispisisa, and Alca (i.e., Alca-1, Alca-2, and Alca-3). Table III lists the compositional data means and standard deviations for the five minor obsidian sources rarely found in areas far from the sources: Aconcahua, Jampatilla, Lisahuacho, Potreropampa, and Puzolana. Table IV lists the compositional data means and standard deviations for the four geological sources of obsidian not observed archaeologically: Caylloma, Cerro Ticllago, Uyo Uyo, and Yanarangra.

Artifacts

Between 1990 and 2005, 850 artifacts from Peru and northern Bolivia were submitted to MURR for analysis by NAA. During the first couple of years, samples were analyzed using both short- and long-NAA procedures. Sources for the artifacts were determined through multi-element comparisons with the geological sources. As confidence in the predictability of compositional patterns from the short-NAA data began to emerge, the long-NAA procedure became less necessary.

A total of 830 of the 850 artifacts analyzed at MURR were assigned to the ten archaeologically-important sources described in this study. The remaining 20 artifacts were unable to be linked to a known source. Table V summarizes the artifact results from MURR and LBNL. There is considerable agreement between the MURR and LBNL findings. The MURR data show that the major sources at Quispisisa (31.4%), Chivay (36.5%), and Alca (26.2%) account for

Table II. Element concentration means and standard deviations for obsidian sources from Peru of major archaeological importance.

Element	*Alca-1 (n=36)*	*Alca-2 (n=2)*	*Alca-3 (n=2)*	*Chivay (n=21)*	*Quispisisa (n=16)*
Short-lived NAA data					
Al (%)	6.91 ± 0.26	7.34 ± 0.03	7.28 ± 0.07	6.81 ± 0.29	6.70 ± 0.22
Ba	995 ± 25	952 ± 10	1018 ± 3	163 ± 15	718 ± 13
Cl	678 ± 97	653 ± 96	706 ± 40	396 ± 106	363 ± 42
Dy	1.93 ± 0.39	1.80 ± 1.01	2.00 ± 0.24	2.66 ± 0.31	1.55 ± 0.31
K (%)	3.69 ± 0.18	3.60 ± 0.05	3.42 ± 0.05	3.69 ± 0.19	3.91 ± 0.24
Mn	476 ± 5	459 ± 5	564 ± 5	710 ± 12	366 ± 3
Na (%)	3.16 ± 0.06	3.34 ± 0.03	3.31 ± 0.02	3.06 ± 0.05	2.84 ± 0.03
Long-lived NAA data					
La	28.8 ± 0.4	43.5 ± 0.4	36.2 ± 0.1	19.0 ± 0.3	26.0 ± 0.3
Lu	0.19 ± 0.01	0.21 ± 0.01	0.22 ± 0.02	0.32 ± 0.02	0.18 ± 0.04
Nd	21.8 ± 4.7	31.0 ± 0.9	19.1 ± 0.3	21.3 ± 4.2	17.3 ± 2.5
Sm	3.50 ± 0.13	4.20 ± 0.07	3.81 ± 0.34	4.15 ± 0.10	3.28 ± 0.05
U	3.39 ± 0.27	3.80 ± 0.66	2.76 ± 0.60	7.86 ± 0.70	10.1 ± 0.3
Yb	1.02 ± 0.04	1.07 ± 0.02	1.29 ± 0.04	1.61 ± 0.10	1.10 ± 0.04
Ce	57.6 ± 1.5	80.0 ± 0.2	69.2 ± 0.1	41.0 ± 0.4	48.1 ± 0.7
Co	0.23 ± 0.01	0.44 ± 0.03	0.46 ± 0.01	0.34 ± 0.05	0.47 ± 0.01
Cs	2.79 ± 0.03	2.71 ± 0.10	3.44 ± 0.03	9.91 ± 0.11	11.0 ± 0.1
Eu	0.49 ± 0.01	0.65 ± 0.01	0.68 ± 0.01	0.28 ± 0.01	0.42 ± 0.01
Fe	5422 ± 101	7580 ± 182	8050 ± 59	4880 ± 83	5626 ± 56
Hf	3.60 ± 0.10	4.76 ± 0.12	4.58 ± 0.03	3.73 ± 0.07	3.25 ± 0.04
Rb	136 ± 2	141 ± 1	125 ± 1	244 ± 4	175 ± 2
Sb	0.17 ± 0.02	0.12 ± 0.02	0.27 ± 0.01	0.89 ± 0.05	1.29 ± 0.03
Sc	1.77 ± 0.03	1.94 ± 0.02	1.85 ± 0.01	3.08 ± 0.05	1.36 ± 0.01
Sr	113 ± 21	242 ± 13	320 ± 7	49 ± 10	160 ± 19
Ta	0.94 ± 0.01	0.95 ± 0.01	0.90 ± 0.01	1.75 ± 0.02	1.17 ± 0.01
Tb	0.34 ± 0.03	0.33 ± 0.02	0.39 ± 0.02	0.47 ± 0.06	0.28 ± 0.02
Th	13.7 ± 0.2	16.0 ± 0.1	13.5 ± 0.1	23.2 ± 0.3	19.5 ± 0.2
Zn	43 ± 5	50 ± 1	48 ± 1	35 ± 6	32 ± 3
Zr	116 ± 8	166 ± 3	160 ± 9	132 ± 9	154 ± 8
X-ray fluorescence data					
K (%)	3.68 ± 0.03	3.24 ± 0.16	3.15 ± 0.20	3.74 ± 0.04	3.73 ± 0.03
Ti	786 ± 17	1062 ± 55	1269 ± 94	547 ± 17	836 ± 17
Mn	459 ± 24	487 ± 27	559 ± 69	691 ± 22	332 ± 12
Fe	5537 ± 124	6847 ± 260	7856 ± 407	5021 ± 35	5623 ± 39
Zn	45 ± 1	50 ± 3	60 ± 5	32 ± 2	35 ± 1
Ga	16 ± 1	13 ± 1	11 ± 1	18 ± 1	16 ± 1
Rb	137 ± 3	144 ± 5	127 ± 4	259 ± 4	181 ± 1
Sr	76 ± 10	129 ± 8	194 ± 21	46 ± 2	121 ± 5
Y	15 ± 1	15 ± 2	15 ± 3	27 ± 1	19 ± 2
Zr	103 ± 6	135 ± 10	141 ± 17	78 ± 2	104± 3
Nb	14 ± 1	8 ± 2	9 ± 2	20 ± 1	11 ± 1

All concentrations are in parts per million (ppm) unless percent (%) is indicated.

Table III. Element concentration means and standard deviations for obsidian sources from Peru of minor archaeological importance.

Element	Puzolana (n=16)	Jampatilla (n=6)	Potrero-pampa (n=6)	Lisahuacho (n=6)	Aconcahua (n=11)
Short-lived NAA data					
Al (%)	6.96 ± 0.35	8.65 ± 0.15	7.11 ± 0.21	7.39 ± 0.16	7.15 ± 0.25
Ba	220 ± 35	764 ± 16	378 ± 6	1054 ± 19	944 ± 13
Cl	718 ± 55	618 ± 25	551 ± 36	749 ± 36	1063 ± 61
Dy	1.43 ± 0.24	4.62 ± 0.26	2.26 ± 0.15	2.23 ± 0.38	1.98 ± 0.30
K (%)	4.12 ± 0.49	3.83 ± 0.22	4.18 ± 0.49	3.98 ± 0.15	4.12 ± 0.30
Mn	506 ± 23	637 ± 9	500 ± 5	444 ± 4	479 ± 8
Na (%)	3.13 ± 0.22	3.57 ± 0.04	2.79 ± 0.25	3.04 ± 0.03	2.89 ± 0.05
Long-lived NAA data					
La	19.6 ± 0.7	40.5 ± 0.3	19.9 ± 0.7	56.8 ± 1.1	44.4 ± 0.4
Lu	0.14 ± 0.04	0.44 ± 0.01	0.19 ± 0.01	0.14 ± 0.02	0.25 ± 0.01
Nd	12.7 ± 0.9	27.6 ± 0.8	18.2 ± 3.8	42.6 ± 9.5	28.8 ± 1.5
Sm	2.61 ± 0.09	5.39 ± 0.06	4.64 ± 0.51	3.89 ± 0.08	3.94 ± 0.04
U	5.48 ± 0.20	8.16 ± 0.20	5.41 ± 0.70	3.48 ± 1.10	9.68 ± 1.59
Yb	0.84 ± 0.05	2.68 ± 0.18	1.57 ± 0.5	1.19 ± 0.35	1.11 ± 0.04
Ce	38.2 ± 1.5	75.8 ± 0.6	42.4 ± 0.4	104.2 ± 2.5	78.7 ± 1.2
Co	0.14 ± 0.02	0.56 ± 0.01	0.10 ± 0.01	0.59 ± 0.02	0.61 ± 0.08
Cs	3.74 ± 0.08	12.4 ± 0.1	5.40 ± 0.04	4.19 ± 0.10	9.50 ± 0.13
Eu	0.33 ± 0.02	1.00 ± 0.01	0.56 ± 0.01	0.94 ± 0.02	0.64 ± 0.01
Fe	5009 ± 143	8817 ± 80	4647 ± 37	8673 ± 195	7377 ± 360
Hf	3.79 ± 0.09	4.95 ± 0.17	3.28 ± 0.03	5.40 ± 0.10	4.57 ± 0.16
Rb	120 ± 7	153 ± 1	166 ± 8	150 ± 3	191 ± 14
Sb	0.24 ± 0.01	1.72 ± 0.02	0.31 ± 0.01	0.22 ± 0.01	0.99 ± 0.02
Sc	1.74 ± 0.05	2.35 ± 0.02	1.68 ± 0.01	1.81± 0.04	1.90 ± 0.05
Sr	60 ± 10	371 ± 28	104 ± 29	362 ± 16	158 ± 17
Ta	2.04 ± 0.04	1.62 ± 0.02	1.26 ± 0.01	1.10 ± 0.03	1.14 ± 0.02
Tb	0.26 ± 0.01	0.67 ± 0.04	0.43 ± 0.01	0.41 ± 0.02	0.30 ± 0.03
Th	14.7 ± 0.3	12.5 ± 0.1	17.1 ± 0.1	18.7 ± 0.4	27.9 ± 0.4
Zn	39 ± 6	78 ±8	41 ± 1	59 ± 2	45 ± 2
Zr	121 ± 10	204 ± 9	112 ± 4	194 ± 9	197 ± 10
X-ray fluorescence data					
K (%)	3.84 ± 0.28	3.68 ± 0.10	3.96 ± 0.26	4.03 ± 0.08	3.97 ± 0.13
Ti	377 ± 40	1164 ± 63	534 ± 63	1389 ± 59	1096 ± 83
Mn	554 ± 38	610 ± 43	547 ± 43	472 ± 17	487 ± 26
Fe	4969 ± 139	8692 ± 301	4593 ± 167	8547 ± 135	6862 ± 245
Zn	42 ± 3	110 ± 7	42 ± 3	85 ± 8	72 ± 4
Ga	20 ± 4	11 ± 1	21 ± 3	18 ± 1	16 ± 1
Rb	125 ± 4	158 ± 3	170 ± 6	150 ± 2	186 ± 5
Sr	63 ± 6	252 ± 16	88 ± 7	310 ± 8	186 ± 16
Y	15 ± 2	27 ± 2	16 ± 3	14 ± 3	16 ± 1
Zr	107 ± 10	169 ± 10	91 ± 9	198 ± 6	149 ± 8
Nb	15 ± 3	27 ± 3	13 ± 3	15 ± 4	17 ± 2

All concentrations are in parts per million (ppm) unless percent (%) is indicated.

Table IV. Element concentration means and standard deviations for obsidian sources from Peru of no archaeological importance.

Element	*Cerro Ticllago (n=16)*	*Yanarangra (n = 6)*	*Uyo Uyo (n=11)*	*Caylloma (n=6)*
Short-lived NAA data				
Al (%)	7.03 ± 0.24	6.77 ± 0.25	7.09 ± 0.28	6.63 ± 0.23
Ba	352 ± 10	586 ± 10	821 ± 13	649 ± 12
Cl	581 ± 79	665 ± 41	706 ± 51	729 ± 59
Dy	1.58 ± 0.30	2.05 ± 0.30	2.00 ± 0.13	2.00 ± 0.14
K (%)	3.95 ± 0.18	4.43 ± 0.28	3.66 ± 0.20	3.94 ± 0.37
Mn	456 ± 5	433 ± 7	458 ± 9	486 ± 2
Na (%)	3.05 ± 0.03	2.13 ± 0.06	2.82 ± 0.03	2.70 ± 0.17
Long-lived NAA data				
La	30.8 ± 0.2	27.6 ± 1.6	34.8 ± 0.5	30.2 ± 1.2
Lu	0.20 ± 0.01	0.30 ± 0.01	0.23 ± 0.01	0.26 ± 0.01
Nd	16.3 ± 0.7	16.6 ± 1.1	23.1 ± 3.6	29.6 ± 1.2
Sm	2.98 ± 0.02	3.51 ± 0.06	3.55 ± 0.05	3.49 ± 0.05
U	6.49 ± 0.20	8.27 ± 0.14	5.82 ± 0.61	9.03 ± 0.15
Yb	1.32 ± 0.03	1.30 ± 0.05	1.19 ± 0.04	1.14 ± 0.05
Ce	55.2 ± 0.4	50.7 ± 3.3	63.8 ± 0.9	57.5 ± 2.3
Co	0.11 ± 0.01	0.29 ± 0.03	0.24 ± 0.03	0.15 ± 0.01
Cs	6.44 ± 0.06	12.7 ± 0.3	5.60 ± 0.07	9.03 ± 0.06
Eu	0.37 ± 0.01	0.43 ± 0.01	0.54 ± 0.01	0.44 ± 0.01
Fe	4953 ± 28	5287 ± 281	5519 ± 192	4666 ± 99
Hf	3.99 ± 0.06	3.33 ± 0.12	3.15 ± 0.06	2.94 ± 0.05
Rb	192 ± 2	208 ± 6	162 ± 2	191 ± 13
Sb	0.21 ± 0.01	0.66 ± 0.06	0.43 ± 0.05	1.39 ± 0.82
Sc	1.66 ± 0.01	2.03 ± 0.03	1.87 ± 0.03	1.85 ± 0.02
Sr	74 ± 7	91 ± 22	182 ± 32	66 ± 7
Ta	1.56 ± 0.01	1.25 ± 0.02	1.06 ± 0.01	1.38 ± 0.01
Tb	0.27 ± 0.01	0.34 ± 0.02	0.32 ± 0.03	0.27 ± 0.02
Th	21.7 ± 0.4	22.4 ± 0.3	16.8 ± 0.2	21.6 ± 0.5
Zn	39 ± 1	33 ± 5	40 ± 7	47 ± 1
Zr	149 ± 4	162 ± 6	118 ± 8	137 ± 6
X-ray fluorescence data				
K (%)	3.83 ± 0.09	3.97 ± 0.13	3.91 ± 0.11	4.48 ± 0.24
Ti	582 ± 33	730 ± 88	795 ± 57	823 ± 102
Mn	539 ± 47	480 ± 21	481 ± 25	715 ± 91
Fe	4919 ± 104	5333 ± 435	5369 ± 168	4660 ± 514
Zn	38 ± 1	39 ± 2	52 ± 5	52 ± 8
Ga	18 ± 1	20 ± 2	18 ± 2	25 ± 4
Rb	196 ± 3	209 ± 9	172 ± 4	199 ± 4
Sr	64 ± 3	83 ± 5	155 ± 18	97 ± 12
Y	15 ± 3	17 ± 3	17 ± 4	13 ± 4
Zr	123 ± 7	103 ± 8	96 ± 5	97 ± 8
Nb	9 ± 2	10 ± 1	13 ± 1	22 ± 9

All concentrations are in parts per million (ppm) unless percent (%) is indicated.

more than 94% of the artifacts; and the sources at Potreropampa (2.3%), Lisahuacho (0.2%), Puzolana (0.7%), Aconcahua (0.2%), and Cerro Ticllago (0.3%) account for an additional 3.7%. Thus, sources for 98% of the artifacts were successfully determined. The remaining 2.0% of artifacts analyzed at MURR could not be assigned to a known source. These percentages also compare favorably to the mid-1970s LBNL study where 98% of the artifacts were linked to the eight highest frequency compositional groups (*41–42*). Locations for seven of the eight source groups have been identified.

Discriminating Elements

Source discrimination was accomplished by examining a series of two- and three-dimensional plots of the obsidian source data. Discovery of graphical representations which show the clearest picture of inter-source versus intra-source variation, makes possible source discrimination with a high degree of confidence. The greater the number of elements that one can use to reinforce the observed discrimination the smaller becomes the chances for misassignment of artifacts when compared to the obsidian source database.

Elements from short-NAA

Examination of the short-NAA data finds that Ba and Mn are the only elements with satisfactory discriminating ability. Figure 2 shows the geological source groups for the ten obsidian sources with major and minor archaeological importance. The shaded ellipses represent the major sources and the hollow ellipses represent the minor sources at the 95% confidence (i.e., two standard deviations). Only in the case of the Alca subsources and the Aconcahua is there a significant overlap for which concern is necessary. Thus, artifacts from either the Quispisisa or Chivay sources are easily provenanced without the need for long irradiations which are expensive and time consuming. When artifacts with compositions similar to Alca-1, Alca-2, and Aconcahua or artifacts that lie outside the confidence ellipse for other source are measured, long-NAA is performed.

Elements from long-NAA

The long-irradiation NAA data measures a number of high-precision elements simultaneously and allows discrimination between sources. Examples some of the most powerful elements are Cs, Fe, Hf, Rb, and Th. Figures 3 and 4

Table V. Summary of provenance assignments for obsidian artifacts from analyzed by MURR and LBNL.

Names of obsidian compositional groups	*Number of artifacts at MURR by NAA*	*Number of artifacts at LBNL by NAA and XRF*
Quispisisa = Quispisisa type	267	366
Chivay = Titicaca Basin type	310	213
Alca-1 = Cuzco type	205	116
Alca-2	3	-
Alca-3	15	-
Jampatilla = Pampas type	0	54
Potreropampa = Andahuaylas A type	20	41
Lisahuacho = Andahuaylas B type	2	14
Puzolana = Ayacucho type	6	8
Aconcahua	2	-
Cerro Ticllago	3	0
Acari type	0	10
Others	17	24
TOTAL	850	846

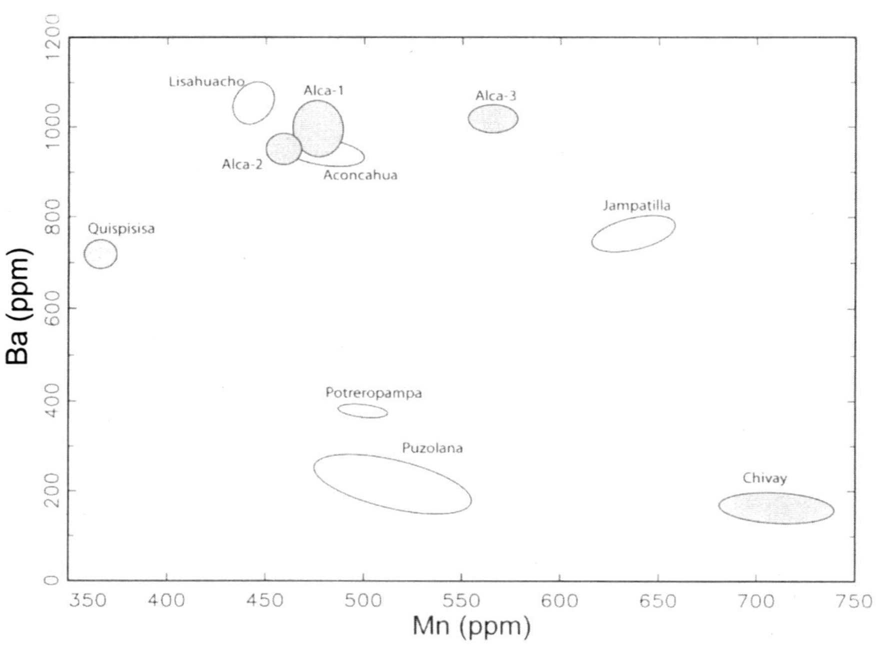

Figure 2. Plot Ba versus Mn from short-NAA for archaeologically-important obsidian sources in southern Peru. Confidence ellipses at the 95% level are shown for the major (shaded) and minor (not shaded) sources.

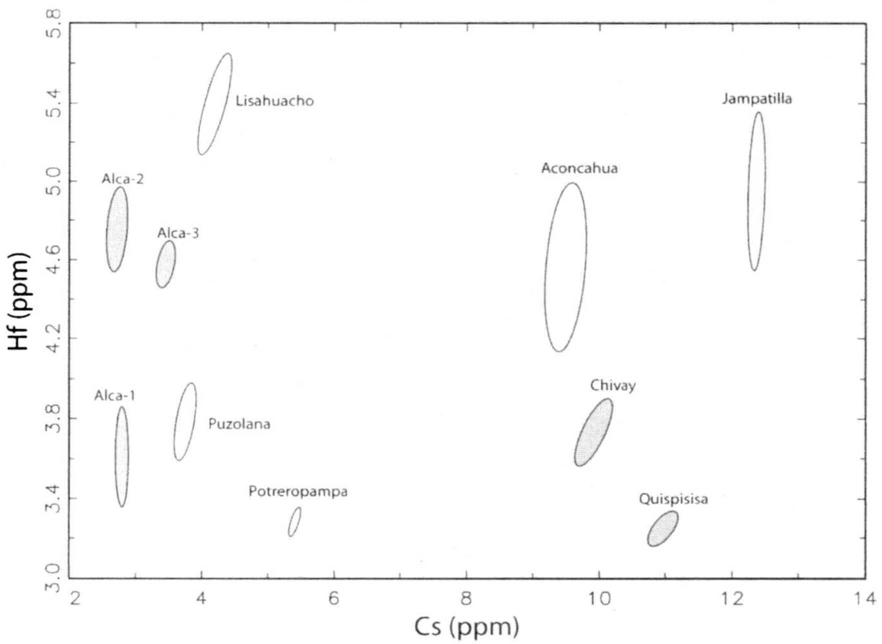

Figure 3. Plot Hf versus Cs from long-NAA for archaeologically-important obsidian sources in southern Peru. Confidence ellipses at the 95% level are shown for the major (shaded) and minor (not shaded) sources.

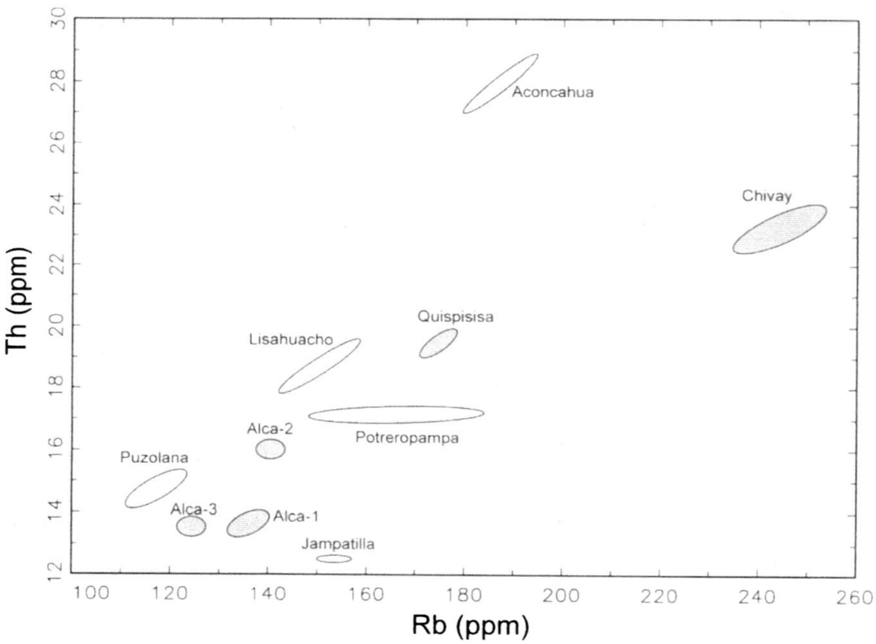

Figure 4. Plot Th versus Rb from long-NAA for archaeologically-important obsidian sources in southern Peru. Confidence ellipses at the 95% level are shown for the major (shaded) and minor (not shaded) sources.

show plots of Hf versus Cs and Th versus Rb, respectively. Both plots show no overlap between 95% confidence ellipses for the ten sources from southern Peru. Sourcing obsidian from these sources can be accomplished with very high degree of certainty.

Elements from XRF

Since MURR acquired the Elva-X ED-XRF spectrometer, investigations of its capability with respect to Andean obsidian have been ongoing. Although the measurement precision for XRF is less than NAA, when a rapid, non-destructive analysis is required XRF can be used quite successfully. As shown in Figures 5 and 6, the elements Rb, Sr, and Fe indicate no overlap at the 95% confidence level. Thus, these elements are then the most powerful discriminating elements available by XRF for the sources in this region.

Discussion

During the past 15 years, archaeologists have identified a dozen geological sources of obsidian in southern Peru which were previously unknown. Fourteen compositional profiles have been recognized for these twelve sources. It appears that only ten of the sources were exploited prehistorically and that three of the sources, Chivay, Quispisisa, and Alca account for more than 94% of the Prehispanic obsidian used in Peru.

The ability to link geochemical profiles from LBNL and MURR to known geologic outcrops has significant implications for understanding Prehispanic obsidian use. At the most basic level, the knowledge of where an obsidian source is located, geographically, enables one to reconstruct trade routes and develop and test hypotheses regarding access to sources, power, and the social dynamics within and between sites and groups of culturally distinct people—all of which is possible because this byproduct of human behavior, the artifact, can be analyzed and its provenance determined with a high degree of certainty. Such knowledge has already made important contributions to our understanding of southern Peruvian archaeology (*3, 47–9*) and will benefit countless future researchers who include an obsidian provenance component in their studies.

Acknowledgements

The authors wish to acknowledge a number of individuals their efforts in collecting the samples for this study including Sarah Brooks, Fidel Fajardo Rios,

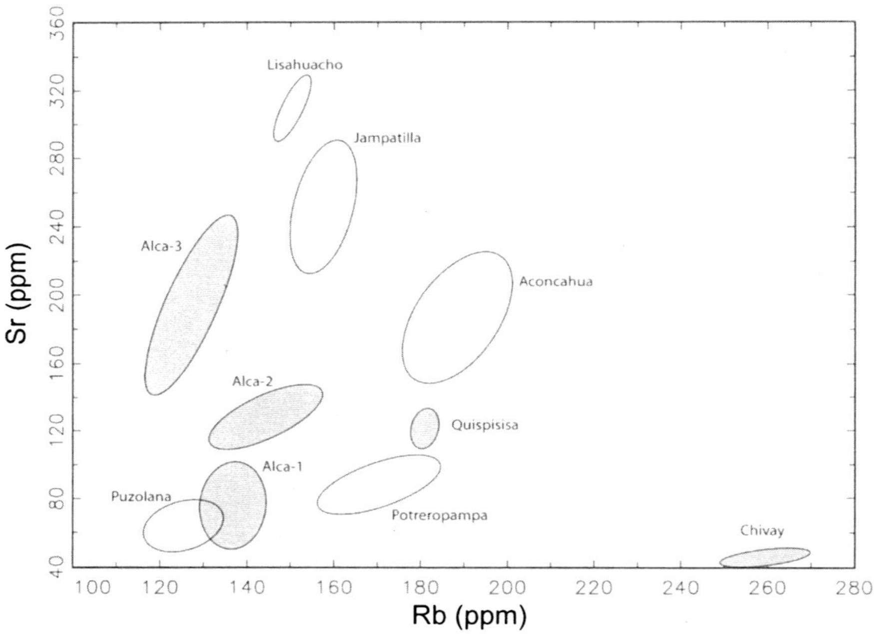

Figure 5. Plot Sr versus Rb from XRF for archaeologically-important obsidian sources in southern Peru. Confidence ellipses at the 95% level are shown for the major (shaded) and minor (not shaded) sources.

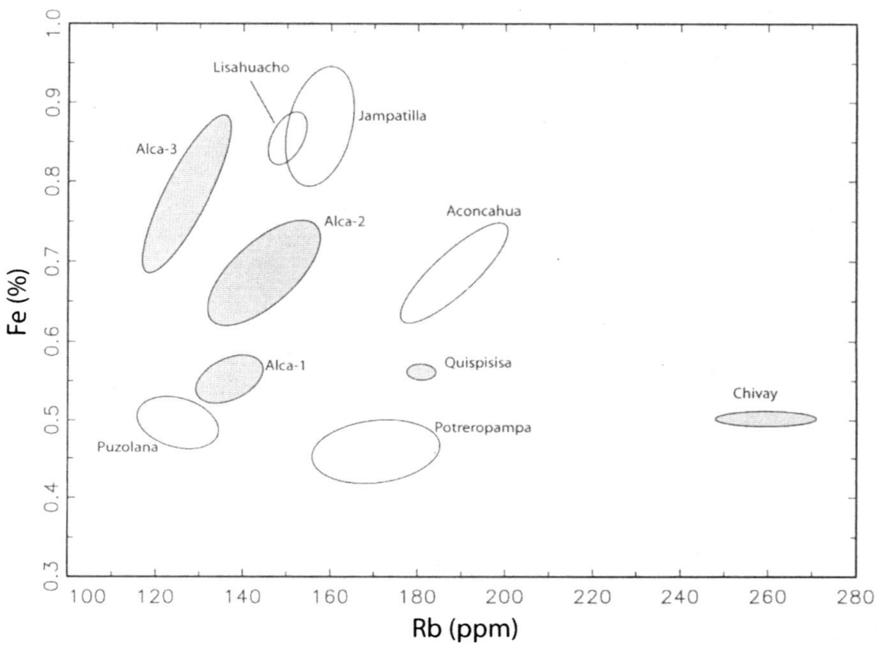

Figure 6. Plot Fe versus Rb from XRF for archaeologically-important obsidian sources in southern Peru. Confidence ellipses at the 95% level are shown for the major (shaded) and minor (not shaded) sources.

Martin Giesso, Justin Jennings, Kirk Frye, Katharina Schreiber, Paul Trawick, and Nico Tripcevich. The analytical work was supported in part by grants from the National Science Foundation including the most recent grant # 0504015.

References

1. Burger, R. L.; Asaro, F. *Contrib. Univ. Calif. Archaeol. Res. Fac.* **1978**, *36*, 61–83.
2. MacNiesh, R. S.; Garcia Cook, A.; Lumbreras, L. G.; Vierra, R. K.; Nelken-Terner, A. *Prehistory of the Ayacucho Basin, Peru. Volume II, Excavations and Chronology*; University of Michigan Press: Ann Arbor; 1971.
3. Sandweiss, D. H.; McInnis, H.; Burger, R. L.; Cano, A.; Ojeda, B.; Paredes, R.; Sandweiss, M.; Glascock, M. D. *Science (Washington, DC)* **1998**, *291*, 1830–1832.

4. Cann, J. R.; Renfrew, C. *Proc. Prehist. Soc.* **1964**, *30*, 111–133.
5. Renfrew, C.; Dixon, J. E.; Cann, J. R. *Proc. Prehist. Soc.* **1966**, *32*, 30–72.
6. Heizer, F.; Williams, H.; Graham, J. A. *Contrib. Univ. Calif. Archaeol. Res. Fac.* **1965**, *1*, 94–103.
7. Griffin, J. B.; Gordus, A. A.; Wright, G. A. *Am. Antiq.* **1969**, *34*, 1–14.
8. Burger, R. L. In *Prehistoric Hunters of the High Andes*; Rick, J. W., Ed.; Academic Press: New York, 1980; pp 257–261.
9. Burger, R. L. *Boletin del Museo Nacional de Antropologia y Arqueología (Lima)* **1982**, 7, 9–10.
10. Burger, R. L.; Asaro, F.; Trawick, P.; Stross, F. *Andean Past* **1998**, *5*, 185–202.
11. Burger, R. L.; Asaro, F.; Salas, G.; Stross, F. *Andean Past* **1998**, *5*, 203–223.
12. Burger, R. L.; Schreiber, K. J.; Glascock, M. D.; Ccencho, J. *Andean Past* **1998**, *5*, 225–239.
13. Burger, R. L.; Glascock, M. D. *Lat. Am. Antiq.* **2000**, *11*, 258–268.
14. Burger, R. L.; Glascock, M. D. *Andean Past* **2000**, *6*, 289–307.
15. Seelenfreund, A.; Rees, C. Bird, R.; Bailey, G.; Barcena, R.; Duran, V. *Lat. Am Antiq.* **1996**, *7*, 7–20.
16. Yacobaccio, H. D.; Escola, P. S. Pereyra, F. X.; Lazzari, M.; Glascock, M. D. *J Archaeol. Sci.* **2004**, *31*, 193–204.
17. Glascock, M. D.; Braswell, G. E.; Cobean, R. H. In *Archaeological Obsidian Studies*; Shackley, M. S., Ed.; Plenum Press: New York, 1998; pp 15–65.
18. Clark, J. E. Paper presented at Society for American Archaeology Meeting, San Diego, CA; 1981.
19. Bettinger, R. L.; Delacorte, M. G.; Jackson, R. J. In *Obsidian Studies in the Great Basin*; Hughes, R. E., Ed.; Contrib. Univ. California Archaeol. Res. Fac., No. 45; 1984.
20. Reeves, R. D.; Armitage, G. C. *New Zealand J. Sci.* **1973**, *16*, 561–572.
21. McDougall, J. M.; Tarling, D. H.; Warren, S. E. *J. Archaeol. Sci.* **1983**, *10*, 441–452.
22. Huntley, D. J.; Bailey, D. C. *Archaeometry* **1978**, *20*, 159–170.
23. Duranni, S. A.; Khan, M. T.; Renfrew, C. *Nature (London)* **1971**, *233*, 242–245.
24. Longworth, G.; Warren, S. E. *J. Archaeol. Sci.* **1979**, *6*, 179–193.
25. Leach, B. F.; Warren, S. E.; Fankhauser, B. *New Zealand J. Sci.* **1978**, *21*, 123–128.
26. Williams-Thorpe, O. *Archaeometry* **1995**, *37*, 217–248.
27. Cobean, R. H.; Coe, M. D.; Perry, E. A., Jr.; Turekian, K. K.; Kharhar, D. P. *Science (Washington, DC)* **1971**, *174*, 141–146.
28. Pires-Ferriera, J.W. *Formative Mesoamerican Exchange Networks with Special Reference to the Valley of Oaxaca*; Memoirs of the Museum of Anthropology, No. 7; Ann Arbor: University of Michigan, 1973.

29. Sappington, R. L. In *Prehistoric Quarries and Lithic Production*; Ericson, J. E.; Purdy, B. A., Eds.; Cambridge University Press: Cambridge; 1984, pp 23–34.
30. Elam, J. M. Ph.D. thesis, University of Missouri-Columbia, Columbia, MO, 1993.
31. Hughes, R. E. *Diachronic Variability in Obsidian Procurement in Northeastern California and South Central Oregon*; University of California Press: Berkeley; 1986.
32. Bird, J. R.; Russell, L. H.; Scott, M. D.; Ambrose, W. R. *Anal. Chem.* **1978**, *50*, 2082–2084.
33. Michels, J. W. *J. Archaeol. Sci.* **1982**, *9*, 113–123.
34. Stross, F. H., Sheets, P.; Asaro, F.; Michel, H. V. *Am. Antiq.* **1983**, *48*, 323–346.
35. Tykot, R. H. *J. Archaeol. Sci.* **1997**, *24*, 467–479.
36. Gratuze, B. *J. Archaeol. Sci.* **1999**, *26*, 869–882.
37. Giauque, R. D.; Asaro, F.; Stross, F. H.; Hester, T. R. *X-ray Spectrom.* **1993**, *22*, 44–53.
38. Glascock, M. D. In *Instrumental Multi-Element Chemical Analysis*; Alfassi, Z. B., Ed.; Kluwer Academic: Dordrecht, The Netherlands; 1998, pp 93–150.
39. Glascock, M. D. *Acc. Chem. Res.* **2002**, *35*, 611–617.
40. Mohr Chavez, K. L. Ph.D. thesis, University of Pennsylvania, Philadelphia, PA, 1977.
41. Burger, R. L.; Asaro, F. *Trace Element Analysis of Obsidian Artifacts from the Andes: New Perspectives on Pre-Hispanic Economic Interaction*; Lawrence Berkeley Laboratory Report 6343; Berkeley, CA; 1977; pp 1–88.
42. Burger, R. L.; Asaro, F. *Revista del Museo Nacional (Lima)* **1977**, *43*, 281–325.
43. Glascock, M. D.; Neff, H.; Stryker, K. S.; Johnson, T. N. *J. Radioanal. Nucl. Chem.* **1994**, *180*, 29–35.
44. Cobean, R. H.; Vogt, J. R.; Glascock, M. D.; Stocker, T. L. *Lat. Am. Antiq.* **1991**, *2*, 69–91.
45. Jennings, J.; Glascock, M. D. *Lat. Am. Antiq.* **2002**, *13*, 107–118.
46. Brooks, S. O.; Glascock, M. D.; Giesso, M. *Nature (London)* **1997**, *386*, 449–450.
47. Burger, R. L.; Mohr Chavez, K. L.; Chavez, S. J. *J. World Prehistory* **2000**, *14*, 267–362.
48. Burger, R. L.; Glascock, M. D. In *Andean Archaeology I: Variations in Sociopolitical Organization*; Isbell, W. H.; Silverman, H., Eds; Kluwer Academic/Plenum Publishers: New York, 2002; pp 341–368.
49. Burger, R. L., Lau, G. F.; Ponte, V. M.; Glascock, M. D. In *La Complejidad Social en la Sierra de Ancash;* Herrera, A.; Orsini, C.; Lane, K., Eds;

Civiche Raccolte d'Arte Applicata del Castello Sforzesco – Raccolte Extraeurope: Milano, 2006; pp 103–120.

50. Ravines, R. *Revista del Museo Nacional (Lima)* **1971**, *37*, p 27.
51. Schreiber, K. J. *Wari Imperialism in Middle Horizon Peru*; Anthropological Papers 87; University of Michigan: Ann Arbor; 1992.
52. Burger, R. L.; Glascock, M. D. *Nawpa Pacha* **2006**, *26*, in press.
53. Perlman, I.; Asaro, F. *Archaeometry* **1969**, *11*, 21–52.

Indexes

Author Index

Adderley, W. Paul, 194
Aldenderfer, Mark, 480
Ariel, Donald T., 258
Armitage, Ruth Ann, 152, 364
Baldia, Christel M., 15, 44
Balsanek, William, 297
Barrett, Raymond, 194
Beckett, Jessica F., 114
Ben-Shlomo, David, 399
Benvenuto, Mark A., 231, 246
Blackwell, Bonnie A. B., 1
Blickstein, Joel I. B., 1
Boats, Jeffe, 231, 246
Brettell, Rhea, 137
Brook, George A., 460
Burger, Richard L., 522
Buxeda i Garrigós, Jaume, 376
Campbell, Alec C., 460
Cecil, Leslie G., 506
Craig, Nathan, 480
Dake, Jonathan Z., 460
Descantes, Christophe, 275
Doolin, Melissa, 364
Dudgeon, John V., 297
Dussubieux, Laure, 336, 349
Eerkens, Jelmer W., 167
Floris, Rosalba, 114
Fonzo, Ornella, 114
Garcia Iñañez, Javier, 376
Glascock, Michael D., 275, 376, 423, 447, 460, 480, 506, 522
Goddard, Ethan, 114
Golitko, Mark, 349
Gregg, Michael W., 137
Hasan, Maysun M., 1
Hawken, James R., 210
Herbert, Gregory S., 167
Herman, Danny, 258
Hill, David V., 423
Hollander, David, 114
Hurry, Silas, 364
Jakes, Kathryn A., 15, 44
Kemp, Brian M., 78
Kiehn, Adam V., 460
Knudson, Kelly J., 99
Lai, Luca, 114
Lienhop, Kyra M., 275
Little, Nicole C., 447
Manunza, Maria Rosaria, 114
Merritt, Christopher W., 447
Minc, Leah, 364
Misner, Jessica, 231
Monroe, Cara, 78
Moriarty, Matthew D., 506
Mouyianis, Meghann, 246
Murphy, Michael L., 460
Neff, Hector, 297, 423
Notis, Michael, 258
Novotny, Claire, 210
Perumplavil, Reshmi, 152
Popelka-Filcoff, Rachel S., 480
Robbins, Lawrence H., 460
Robertson, J. David, 480
Rosenthal, Jeffrey S., 167
Saint, Andrew "Flynn", 297
Scarlett, Timothy J., 447
Shiraki, Ryoji, 167
Shugar, Aaron, 258
Simpson, Ian A., 194
Skaggs, Sheldon, 311

Skinner, Anne R., 1
Skinner, Craig E., 275
Smith, David Glenn, 78
Sosa Suárez, Elena, 376
Speakman, Robert J., 275, 349, 376, 423, 447, 480, 506, 522
Spero, Howard J., 167
Stern, Benjamin, 137
Thatcher, Jennifer J., 275
Thompson, Amanda J., 44
Tung, Tiffiny A., 99
Tykot, Robert H., 114, 275
Usai, Elena, 114
Wells, E. Christian, 210
Wess, Timothy J., 194
Williams, Patrick Ryan, 349

Subject Index

A

Aconcahua source, Peruvian obsidian, 536
aDNA extraction and analysis from archaeological specimens, 78–98
  ancient coastal migration detection from skeletal remains, 80
  authentication and assessment, results, 92–93
  burial context, biological relationships, 80–81
  economic stratification, prehistoric populations, 81–82
  molecular markers, 83
  problems and properties, 82–85
  sex determination techniques, 79–80
  tuberculosis, mummy at Chiribaya Alta site, 82
Africa. *See* Botswana prehistoric mines; Curse tablets from Roman Carthage
Alaska, ancient coastal migration detection by mtDNA from skeletal remains, 80
Alca source, Peruvian obsidian, 532, 534
Ali Kosh, Neolithic site in Iran, 138, 139*f*
Alkaline-based glazes, firing properties, 423–424
American trade monopoly in 16th century, Castilian Kingdom, 397–398
Ancient DNA. *See* aDNA
Andes, geographic origins, trophy heads, 99–113
Apatite values, skeletal remains, five Sardinian sites, 124, 126*f*
Archaeological obsidian sources, Peru, 522–552
Archaeological textiles from eastern North America
  colorant classification, 15–43
  colorant testing protocol, 29–39
  coloration, literature review, 16–17
  fiber and particulate residues, infrared examination, 44–77
Archaeological thin sections, microfocus X-ray analysis methods, 197–206
Artifact exportation limits, 506–507
Augustus. *See* Octavian
Australia, Katipiri Formation, ESR dating of teeth, 3, 5, 6*t*
Ayacucho, strontium isotope ratios, modern guinea pigs, 104–106
Aztec human sacrifices, sex determination by aDNA techniques, 79–80

B

Baltimore Museum of Art, 343
Basra Petrofabric, 436
Bison, demographic modeling by aDNA data, Bayesian statistical techniques, 81
Bitumen traces in ceramic vessels from Neolithic Iran, 137–151
  pottery sherds, sampling and characterization, 141–142
Black willow, charred, infrared spectra, 63, 64*f*
Blue-colored ceramic glazes from Mesopotamia, LA-ICP-MS analyses, 425–426

Bone and enamel, archaeological, strontium isotope analysis, 102–104
Bone chemistry, principles, 116–117
Bone materials in archaeological soils and sediments, 198, 200–204
Botswana prehistoric mines, specular hermatite source fingerprinting, 460–479
Brick Chapel, St. Mary's City, building materials, 364–375
Brigham City Co-Operative Pottery, 448–449
British Columbia, Keatley Creek, prehistoric population economic stratification, aDNA determination, 81–82
Building materials, brick Chapel at St. Mary's City, 364–375
Bulk isotopic analysis ($\delta^{13}C$ and δD), asphaltene in bitumen samples, 141–149
Burial context, biological relationships determination by aDNA variations, 80–81

C

California
  glass beads, lead isotope analysis by LA-TOF-ICP-MS, 305–307*f*
  marine shell beads, *Olivella* source zones, 169–170
  population movement studies, Fish Slough Cave, Owen's Valley, 80
Canary Islands, majolica pottery characterization, 376–398
Canonical discriminant analysis, ochre artifacts from Jiskairumoko, INAA data, 493, 500
Carbon stable isotope ratios in *Olivella biplicata* shell beads sourcing, 180–185*f*
Carthage. *See* Punic Carthage; Roman Carthage
Carylloma and Uyo Uyo obsidian sources, Peru, 538
Casa de Contratación, 377–378, 397–398
Castilian Kingdom
  American trade monopoly in 16th century, 397–398
  Canary Islands, history, 377–378
Central Petén, field-portable XRF identification of obsidian sources, 506–521
Ceramic glaze components, 423
Ceramic glazes from Mesopotamia
  LA-ICP-MS compositional analysis, 424–427
  lead isotope analysis, 427–429
  technology, 422–446
Ceramic pastes, bulk analysis, 282, 284, 285*f*
Ceramic vessels from Neolithic Iran, bitumen traces, 137–151
Cerro Ticilago, Lisahuacho, and Potreropampa obsidian sources, Peru, 537–538
Chagha Sefid, Neolithic site in Iran, 138, 139*f*
Chapel bricks and clays, petrographic analyses, 369–370*f*
Charred fibers, 61–67
Charring fibers, 49–50
Che Tsung, coins, elemental compositions, 233, 240*f*–241*f*
Chen Tsung coins, elemental compositions, 233–236*f*
Chicamuxen Church Formation clays, possible Chapel brick source, 371–372*f*
China
  copper-based coins from Song dynasty, chemical composition by EDXRF, 231–245
  Dragon Jar glazes, characterization by LA-ICP-MS and INAA, 284–287, 288*f*

Chinese copper-based coins from Song dynasty, chemical composition by EDXRF, 231–245
Chinese emperors' coins, elemental compositions, 233–243*t*
Chiribaya Alta site, tuberculosis determination from aDNA studies, 82
Chivay source, obsidian, Peru, 534
Church of Jesus Christ of Latter-day Saints. *See* Latter-day Saint
Clay deposits, St. Mary's County, 365*f*, 368
Clay sampling for brick and tile studies from St. Mary's County, 368–369
Climatic trends
  δ^{15} N and δ^{13}C variations correlation with δ^{18}O variations, 126–131
  δ^{18}O in paleoclimatic reconstruction, 127–131
Cluster analysis
  ochre artifacts from Jiskairumoko, INAA data, 493, 498*f*–499*f*, 500
  Philistia pottery grouping by chemical analyses, 407–411
Coins, copper
  Herodian prutah, physical measurement and elemental composition, 247–249
  Song dynasty, chemical composition by EDXRF, 231–245
Coins, silver, Isfiya and Qumran coin hoards, chemical composition, 258–274
Collagen values for skeletal remains from Sardinian sites, 123–125*f*
Colorado, Mesa Verde pottery sherds, Black-on-white paints, characterization, 287–292
Colorant testing protocol for archaeological textiles, 29–39
Colorants in eastern North American archaeological textiles, 15–43
Comparative Plant Fiber Collection (CPFC), 45, 47
  distinctions in archaeological materials, 55, 59*f*–60
Conchopata ceramics, 350–351
Conchopata trophy heads, geographic origins, 100–109
  strontium isotope results, 106–108
Cone collection. *See* Baltimore Museum of Art
Contamination control, aDNA studies, 84–85
Cooking and eating in ancient Mesoamerican plazas, 210–230
Copper alloy artifacts, applications of laser ablation ICP-MS, 336–348
Copper-based coins from Song dynasty, EDXRF chemical composition determination, 231–245
Copper coins, Herodian prutah, EDXRF elemental compositions, 246–257
Copper in ceramic glazes, pottery from Latter-day Saint Utah, 456
Copper salts in fiber mineralization, 51
Coprolite human mtDNA extraction, Hinds Cave, Texas, dietary reconstruction, 81
Coprolites and population movement, discovery using human mtDNA techniques, 80
Cotallalli source, obsidian. *See* Chivay source, obsidian
CPFC. *See* Comparative Plant Fiber Collection
Cremation ceremonies in fiber charring, 49–50
Crocodylian teeth, ESR dating techniques, 5. 10, 11*f*
^{65}Cu, internal standard for LA-ICP-MS analysis, copper alloy artifacts, 339

Curse tablets from Roman Carthage, lead isotope analysis, 311–335

D

Data preparation, specular hematite source fingerprinting by INAA, 467–468
Data quantification in LA-ICP-MS, 277–278
Data treatment, tin and lead concentrations in majolica pottery production, 383–384
Defixiones. See Curse tablets
Deh Luran Plain, ceramic glaze samples for compositional analysis, 424–427, 434, 436–437, 440
Depth profiling, coating samples from Little Lost River Cave, 162–163
Detection limits in LA-ICP-MS protocol testing
  copper alloy analysis, 341
  Wari ceramics elemental analysis, 353–354*f*
Devni-Khadri, India, ESR crocodilian teeth and elephantid tooth dating, 5, 7–10, 11*f*
Diagenesis in strontium isotope analysis, 101–102
Didrachms. *See* Tyrian shekels and half-shekels
Diet and stable isotopes, western Mediterranean prehistory, 118–120
Dietary reconstruction from coprolites, human mtDNA extraction, Hinds Cave, Texas, 81
Dietary research through stable isotopes, principles and interpretation, 115–117
Dikgatlampi workings, Botswana, specularite sourcing, 465
Discriminant function analysis, INAA geochemical data, 466, 469–477*t*
Disease, aDNA studies, tuberculosis in mummy from Chiribaya Alta site, 82
DNA, ancient. *See* aDNA
Dragon Jar glaze characterization by LA-ICP-MS and INAA, 284–287, 288*f*
Dyeing fiber technology, 51
Dyes and pigments, general characteristics, 18–19
Dyes, infrared identification, 71, 73*f*, 74
Dyestuffs with plant sources, 22*t*–24

E

Eastern North American archaeological textiles
  colorant classification, 15–43
  colorant testing protocol, 29–39
  coloration, literature review, 16–17
  fiber and particulate residues, infrared examination, 44–77
Eating and cooking in ancient Mesoamerican plazas, 210–230
Economic stratification, prehistoric populations aDNA determinations, Keatley Creek, British Columbia, 81–82
EDS. *See* Energy dispersive spectrometry
EDXRF. *See* Energy dispersive X-ray fluorescence
Ekron, Iron IA Philistine Monochrome pottery, 418
El Chayal, Guatemala, obsidian source, 511
Electron microprobe analysis (EMPA)
  Roman Carthage curse tablet analysis, 315, 317–325*t*
  theory and application, 261, 263
Electron spin resonance (ESR) dating, teeth, 1–14

Elemental analysis by XRF, silver Tyrian coins, 263–272*f*
Elemental composition, EDXRF
  copper-based coins from Song dynasty, 231–245
  copper coins, Herodian prutah, 246–257
Elemental concentrations and standard deviations, obsidian sources, Peru, 539, 540*t*–542*t*
Elemental contour maps, sherd surfaces, by LA-ICP-MS, 289–292
ElvaX Regression program, 513
EMPA. *See* Electron microprobe analysis
Enamel and bone, strontium isotope analysis, 102–104
Energy dispersive spectrometry (EDS), scanning electron microscopy, Seip textiles, 35
Energy dispersive X-ray fluorescence (EDXRF), elemental analyses
  copper-based coins, 231–245
  copper coins, Herodian prutah, 246–257
  obsidian samples, 513–514, 516*f*
"English brick," 366
ESR. *See* Electron spin resonance dating
Ethnoarchaeological studies, cooking and eating in Mesoamerica, 212–215
Etowah Mound, Ohio, colorants in archaeological textiles, 44–77
Euclidean distance for chemical grouping by cluster analysis, 407
Eygin Gol Necropolis, Northern Mongolia, biological relations from burial context, determination, 80–81

F

Fiber and particulate residues, archaeological textiles by infrared techniques, 44–77
Fiber charring, 49–50
Fiber dyeing, 51
Fiber mineralization, 51
Field analysis by XRF, obsidian sources in Central Petén, Guatemala, 506–521
Fish Slough Cave, Owen's Valley, California, population movement studies, 80
Food and soil, chemical links, 211–212
Forensic photography, prehistoric textiles, 27, 29

G

Glass beads from central California, lead isotope analysis by LA-TOF-ICP-MS, 305–307*f*
Glaze and paint characterization by LA-ICP-MS, 284, 286–288*f*
  Utah, historic, pottery glazes, chemical analysis, 452–457
  See also Pottery
Guatemala
  Central Petén, field-portable XRF identification of obsidian sources, 506–521
  Las Pozas, Petén, ethnoarchaeological studies, cooking and eating, 215–218*f*
Guinea pigs from Ayacucho, strontium isotope ratios, 104–106

H

Hafnium and neodymium concentrations in Wari ceramics, 361–362
Heavy metal sand metals, trace-element data, St. Mary's County, 371–372*f*
Heavy mineral separation, specular hematite source fingerprinting, 465–466
Herod Agrippa I, copper coins, elemental composition by EDXRF, 246–257
Herod's relationship with Octavian, 269, 271, 273
Herodian prutah, elemental compositions, 246–257
Historical Ecology Project, Central Petén, Guatemala, 517
Honduras, Palmarejo, ethnological studies, cooking and eating, 217–228
Honduras, Petoa, Santa Barbara, ethnological studies, cooking and eating, 215
Hopewell Mound Group, Ohio, archeological textile source. *See* Seip Mound
Huamanga Basin, Ayacucho Valley, food supply for Wari heartland, 104
Human activities, determination by elemental soil data, 221–228
Human bone and teeth, characterization by LA-ICP-MS, 292–294*f*
Human burials at Conchopata, strontium isotope ratios, 105*t*, 106
Human dental enamel, strontium isotope analysis by LA-TOF-ICP-MS, 306–308

I

Iberian Peninsula, production centers, majolica pottery found on Canary Islands, 384, 385–398
Icelandic Norse-trading site, sulfur materials, simultaneous co-incident x-ray micro-fluorescence and micro-diffraction analyses, 204–205
ICP-MS. *See* Inductively coupled plasma-mass spectrometry
ICP-OES. *See* Inductively coupled plasma-optical emission spectroscopy.
Idaho, Little Lost River Cave, black coating on pictographs, surface analysis, 152–166
INAA. *See* Instrumental neutron activation analysis
India, ESR crocodilian teeth and elephantid tooth dating from Devni-Khadri, 5, 7–10, 11*f*
Indian hemp [N.A. plant] fibers, charred and uncharred, 61–64*f*
Inductively coupled plasma-mass spectrometry (ICP-MS)
  and ICP-OES chemical analyses, Philistine pottery, 402–411
  correlation to LA-ICP-MS elemental analysis, Matisse bronze sculptures, 343–346
  Olivella biplicata shell compositional chemistry, geographic patterning, 170–180
  prehistoric textiles, 29, 35
  Roman Carthage curse tablets, 319, 332, 333*t*
  with laser-ablation sample, archaeological research introduction, 275–296
Inductively coupled plasma-optical emission spectroscopy (ICP-OES), soil chemicals, cooking and eating in Mesoamerican plazas, 210–230

Infrared distinctions between fibers in Comparative Plant Fiber Collection, 52, 55–59*f*
Infrared photography, 25
Infrared spectra examination
  comparative modern plant and animal fibers, 52–54*t*
  fiber and particulate residues from archaeological textiles, 44–77
Inorganic mineral pigments, textile fibers, 19–21*t*
Instrumental neutron activation analysis (INAA)
  building materials from St. Mary's City Chapel, 364–375
  majolican pottery sherds. Gran Canaria Island, chemical analyses, 381, 383
  majolican pottery sherds, origin, 392–397
  Mesopotamian glazed ceramics, 430–433, 435*t*, 437–440
  obsidian sourcing, 278, 279–281
  See also Neutron activation analysis
Intermountain Region, North America, historic pottery, 447–459
Iran, Neolithic, bitumen traces in ceramic vessels, 137–151
Iran, Neolithic sites
  Ali Kosh, 138, 139*f*
  Chagha Sefid, 138, 139*f*
Iraq, Mesopotamian ceramic glaze technology, 422–446
Iron Age Philistine pottery from Israel, intra-regional sourcing, 399–421
Iron log ratio transformation, XRF results, majolica pottery origin, 384–385
Iron oxide. *See* Ochre
Isfiya and Qumran coin hoards, chemical composition, 258–274
Isotope ratios for archaeology, 298–309
Isotopic characterization $\delta^{13}C$ and δD values, asphaltene in bitumen samples, 146, 148*f*, 149*f*
Israel. *See* Coins, silver; Philistine Iron Age pottery
Italy. *See* Roman Carthage; Sardinian prehistory
Ixtepeque, Guatemala, obsidian source, 511–512

J

Jampatilla, Peru, obsidian source, 536
Jen Tsung, coins, elemental composition, 233, 237*f*–239*f*
Jiskairumoko, Peru, ochre artifacts
  instrumental neutron activation analysis, 480–505
  mathematical and statistical data treatment, 492–501

K

Katipiri Formation, Australia, ESR dating of teeth, 3, 5, 6*t*
Keatley Creek, British Columbia, prehistoric population economic stratification, aDNA determination, 81–82
King James Bible, Mk 12:41–44, 247
Kriging modeling, ethnological studies, Palmarejo, 221–228

L

La Cueva Pintada, Gran Canaria Island, majolica pottery archaeological site, 378, 379*f*
LA-ICP-MS (Laser ablation-inductively coupled plasma-mass spectrometry)

archaeological research, 275–296
ceramic glazes, 424–427
copper alloy artifacts, 336–348
correlation to ICP-MS results, elemental analysis, Matisse bronze sculptures, 343–346
glaze recipe analysis, techniques, 449–452
historic Latter-day Saint pottery glazes, 447–459
Peruvian Wari ceramics, characterization, 349–363

LA-TOF-ICP-MS (Laser ablation-time of flight-inductively coupled plasma-mass spectrometry)
laser-induced fractionation, 300–301*t*
real world precision, analytical parameters, 299–302

Lake Eyre Basin, Australia, marsupial teeth dating, 3, 5, 6*t*

Lake Titicaca basin, ochre artifacts, open air residential site, 481–483, 484*f*

Lanthanum, log ratio transformation, INAA results, majolica pottery origin, 392

Las Pozas, Peten, Guatemala, ethnoarchaeological studies, 215

Laser ablation-inductively coupled plasma-mass spectrometry. *See* LA-ICP-MS

Laser ablation-time of flight-inductively coupled plasma-mass spectrometry. *See* LA-TOF-ICP-MS

Latter-day Saint pottery glazes, analysis by LA-ICP-MS, 447–459

Lawrence Berkeley National Laboratory (LBNL), obsidian sourcing research, 529, 532, 534, 539, 544*t*

Lead-based glazes, firing properties, 423–424

Lead-based glazes from historic Utah, chemical analysis by LA-ICP-MS, 452–457

Lead-based glazes from Mesopotamia, 422–446

Lead concentrations by XRF, majolican pottery sherds, 383, 384*f*

Lead for glazes, scarcity, 448–449

Lead isotope analysis
archaeological method, 313
ceramic glazes from Mesopotamia, 427–429
LA-TOF-ICP-MS for archaeology, 298, 302–303*f*, 305–306*f*, 307*f*
ore origination studies, Roman Carthage curse tablets, 311–335

Lead isotope ratios, Roman Carthage curse tablets, 326, 328*f*–332*t*

Lighting and sample selection, archaeological textiles colorant testing, 37

Lignins in plant fibers, 49–50*f*

Ligurian origin for majolica pottery from Canary Islands by INAA determination, 392

Limitations
colorant testing protocol, archaeological textiles, 38–39
LA-ICP-MS analyses, 337

Linearity in protocol testing, LA-ICP-MS copper alloy analysis, 340

Literature review, textile colorants, 17–27

Little Lost River Cave, Idaho, black coating on pictographs, surface analysis, 152–166

M

Majolica pottery from Canary Islands, 376–398
origin by XRF, 385–391*f*

Majolica pottery sites, Gran Canaria Island
La Cueva Pintada, 379–381*t*
San Francisco convent, Las Palmas de Gran Canaria, 379–380, 382*t*
Mancos Black-on-white paints from Mesa Verde pottery
characterization by LA-ICP-MS, 287–288*f*
elemental surface maps of pottery sherds, 289–290*f*
Manganese in ceramic glazes, pottery from Latter-day Saint Utah, 456
Manganese variations in transect samples, St. Mary's Hill, 370–371*f*
Manises, production center, majolica pottery, INAA determinations, 392–397
Marine shell beads, sourcing, 167–193
Marsupial teeth, dating by ESR techniques, 3, 5, 6*t*
Maryland, St. Mary's City, Brick Chapel, building materials, determination, 364–375
Maryland's history, 365–366
Maryland's geology, St. Mary's County, 365*f*, 368
Mass bias correction with LA-TOF-ICP-MS, 300*f*, 301–302
Mass spectrometry for stable isotopic analyses, principle, 116
Matisse bronze sculptures, applications of laser ablation ICP-MS (LA-ICP-MS), 343–347
Mesa Verde Black-on-white paints
characterization by LA-ICP-MS, 287–288*f*
elemental surface maps of pottery sherds, 289, 290*f*–292
Mesa Verde painted pottery, black paints, characterization, 287–288*f*
Mesoamerican plazas, cooking and eating evidence from soil chemical data, 210–230
Mesopotamian ceramic glaze technology, 422–446
Mesopotamian glazed ceramics, compositional analysis variations, 430–440
Mesopotamian historical periods, ceramic technology, 443*t*
Metallic inclusions in Roman Carthage curse tablets, electron microprobe analysis, 315–317, 319–325*t*
Method sequencing in colorant testing protocol, archaeological textiles, 38
Mexico
ethnological studies, cooking and eating, 212–214, 215–218*f*
Pachuca, obsidian source, 512
Microfocus x-ray analysis methods, archaeological thin sections, 197–206
Microscale features, archaeological soils and sediments, analysis goals, 195
Middle Woodland period site. *See* Seip Mound
Milkweed, common, charred, infrared spectrum, 63, 65*f*
Mineralized fibers, 51
spectra compared to archaeological textiles, 63, 67–69*f*
Minimal composition variation, discriminating obsidian sources, 281–282, 283*f*
Mississippian period site. *See* Etowah Mound
Molecular sex and human behavior determinations, aDNA applications. *See* aDNA
Mongolia, Northern, Eygin Gol Necropolis, biological relations from burial context, 80–81
Monti Arci, Sardinia, obsidian artifacts, sourcing, 279–281, 283*f*
Muxuccuxcab, Yucatan, Mexico, ethnoarchaeological studies, 214

Mycenaean pottery production by Aegean or Cyprus immigrants, 418

N

NAA. *See* Neutron activation analysis
National Museum of Carthage, 314
Native Americans, coastal migrations, 80
Neodymium and hafnium concentrations in Wari ceramics, 361–362
Neolithic domesticated animals and crops, western Mediterranean, 118–120
Neolithic Iran, bitumen use, 137–151
Neutron activation analysis (NAA)
  obsidian analysis procedure, overview, 531
  obsidian chemical characterization, 528–529, 543–547
  See also Instrumental neutron activation analysis
New Wave UP213 laser, 338, 351
North America, eastern, archaeological textiles
  colorants, 15–43
  fiber and particulate residues, 44–77
North America, intermountain region, historic pottery, 447–459
Norway, bone materials from archaeological sites, 198, 200–204

O

Obsidian
  chemical characterization, experimental procedures, 527–529, 530–532
  geochemistry, 525–527
  procurement at Trinidad de Nosotros, 515–518
  research history in Peru, 529–530
Obsidian provenance determinations, general, 278
  systematic approach, 523–525
Obsidian sources
  Central Petén, Guatemala, determination by field-portable XRF, 506–521
  Oregon, Sycan Marsh and Silver Lake source domes, 281–283*f*
  Pachuca, Mexico, 512
  Peru, 522–552
  western Mediterranean artifacts, 279–281
Ochre artifacts at Jiskairumoko, Peru, INAA, 480–505
  archaeological contexts, 483, 485–486
  characteristics, 487–491*t*
  variability, 501–502
Ochre pigments, 19, 67, 72*f*–73*f*
Octavian, relationship with Herod, 269, 271, 273
Ohio, archaeological textile colorants, 15–43, 44–77
Ohio Historical Society, 32
Olivella biplicata shell beads, geographic sourcing methods, 167–193
On Your Knees Cave, Alaska, ancient coastal migration, mtDNA from skeletal remains, 80
Optical microscopy, 25, 29, 34
Ore origination studies, Roman Carthage curse tablets, 311–335
Oregon, Sycan Marsh and Silver Lake, discriminating obsidian sources, 281–282, 283*f*
Organic constituents, detection, 26–27
Osumacinta Viejo, Chiapas, Mexico, ethnoarchaeological studies, 214
Oxalate inclusions accompanying plant fibers, 48

Oxygen stable isotope ratios, sourcing *Olivella biplicata* shell beads, 180–188

P

Pachuca, Mexico, obsidian source, 512
Painted fabric preparation for comparison to prehistoric textiles, 27–28*t*
Paints and glazes, characterization by LA-ICP-MS, 284, 286–288*f*
Palmarejo site, Honduras, ethnoarchaeological studies, 217–228
Paloma, Peru site, human bone and teeth, characterization by LA-ICP-MS, 292–294*f*
PCA. *See* Principal component analysis
PCO. *See* Plasma-chemical oxidation
PCR aDNA amplification for screening single nucleotide polymorphisms and length polymorphisms, 87–88, 90*t*–91*t*
PCR amplification inhibitors and repeat silica extraction, 88, 92
PCR amplification inhibitors in aDNA extractions, 85
Peru
  archeological obsidian sources, 522–552
  Chiribaya Alta site, tuberculosis determination by DNA studies, 82
  Conchopata, Wari site, trophy heads, geographic origins, 99–113
  discriminating elements in obsidian sources, 543–549
  ochre artifacts, instrumental neutron activation analysis, 480–505
  Paloma site, human bone and teeth characterization by LA-ICP-MS, 292–294*f*
  Wari ceramics, LA-ICP-MS analysis for characterization, 349–363
Petoa, Santa Barbara, Honduras, ethnoarchaeological studies, 215
Petrographic analysis
  Mesopotamian glazed ceramics, 434, 436–437
  sourcing Iron Age Philistine pottery, 412–417
Philistia and southern Israel, geography, 399–400*f*, 412
Philistia pottery, grouping by principal component analysis, 404–411
Philistine Bichrome pottery production, 419
Philistine Iron Age pottery from Israel, 399–421
Philistine Monochrome pottery production, 418
Phosphorus in soils, food preparation and consumption evidence, 215–217, 218*f*
Photography using non-visible regions of the electromagnetic spectrum, 24–25
Pictographs in Little Lost River Cave, black coating, surface analysis, 152–166
Pigments and dyes, general characteristics, 18–19
Pigments and infrared spectra, iron oxide, 67, 72*f*–73*f*
Plant and animal fibers, comparisons, 52–54*t*
Plant sources for dyestuff, 24
Plasma-chemical oxidation (PCO), radiocarbon dating, black coating, 153, 164
Plaza spaces, prehistoric cooking and eating practices, 217, 218–228
Polymerase chain reaction. *See* PCR
Portuguese origin, majolica pottery from Canary Islands by INAA determination, 392

Pottery
majolica, from Canary Islands, characterization, 376–399
Philistine, from Israel, sourcing, 399–421
See also Glaze and paint characterization
Prehistoric populations, economic stratification, aDNA techniques, 81–82
Principal component analysis (PCA), 369
ochre artifacts from Jiskairumoko, INAA data, 493, 498*f*
Protein fibers, infrared spectra, 60, 63, 65*f*
Protocol testing accuracy
LA-ICP-MS compared to INAA, elemental analyses, Wari ceramics, 355–356*f*
LA-ICP-MS copper alloy analysis, 341, 342*f*–343
Provenance research. *See* Source characterization; Sourcing
Punic Carthage, history, 312
Puzolana source, obsidian, Peru, 535–536

Q

Quadratic regression model for elemental concentrations in obsidian artifacts, 513–514
Quispisisa source, obsidian in Peru, 535
Qumran and Isfiya coin hoards, chemical composition, 258–274

R

Repeat silica extraction and PCR amplification inhibitors, 88, 92
Repeatability, protocol testing, LA-ICP-MS elemental analyses, Wari ceramics, 353, 354–355*f*
Reproducibility, protocol testing
LA-ICP-MS compared to INAA, elemental analyses, Wari ceramics, 354, 355
LA-ICP-MS copper alloy analysis, 341–342*f*
Roman Carthage curse tablets, lead isotope analysis, 311–335
Rubidium and zirconium concentrations in obsidian from Trinidad de Nosotros, 516*f*
Ruby Valley Pony Express Station, source of salt-glazed ceramic sherds, 455–456

S

Salt-glazed pottery from historic Latter-day Saint potteries, 455–456
Sample preparation
obsidian samples, 530–531
skeletal remains for Sardin
Sample selection, aDNA studies, 82–83
Sample selection and acquisition, Seip textiles, 32–34
Sampling criteria, sample preparation, skeletal remains for Sardinian prehistory diet studies, 120–123
Sampling strategy, strontium isotope analysis, 102
San Francisco convent at Las Palmas de Gran Canaria, majolica pottery site, 378–379
San Martin Jilotepeque, Guatemala, obsidian source, 510–511
San Vicente Xiloxochitla, Tiaxcala, Mexico, ethnoarchaeological studies, 213
Sand fraction, trace metal association in St Mary's County Chapel bricks, 369–370*f*
Sardinian prehistory, interpreting stable isotopic analyses, 120–132

SAXS. *See* Small-angle x-ray scattering
Scanning electron microscopy (SEM), 25
  prehistoric textiles, 29
  with energy dispersive spectrometry, Seip textiles, 35
Sebilong mine, Botswana, specularite sourcing, 464
Sediments, archaeological, microfocus small-angle x-ray scattering, 197–200
Seip Mound, Ohio, archaeological textile colorants, 15–43, 44–77
Seip textiles, 32–35
SEM. *See* Scanning electron microscopy
Seville, production center, majolica pottery found on Canary Islands, 385–397
Shark teeth dating by ESR techniques, 10–13*f*
Shell bead sourcing, 167–193
Sherd surfaces, elemental contour maps by LA-ICP-MS, 289–292
Sik'u'. *See* Trinidad de Nosotros
Silver coins, Isfiya and Qumran coin hoards, chemical composition, 258–274
Silver content in Tyrian shekels and half-shekels, political condition effects, 269, 271, 273
Silver Lake and Sycan Marsh, Oregon, discriminating obsidian sources, 281–282, 283*f*
Siwalik Group, India, ESR crocodilian teeth and elephantid tooth dating, 5, 7–10, 11*f*
Small-angle x-ray scattering (SAXS), microfocus, archaeological sediments, 197–200
  See also XRD
Social relations from soil chemistry, 215–217, 218*f*
Soil chemistry, cooking and eating in Mesoamerican plazas, 210–230
Soil memory, 211–212
Soils and sediments, thin-section microfocus synchrotron x-ray scattering, diffraction and fluorescence analyses, 194–209
Song dynasty copper-based coins, chemical composition, EDXRF determination, 231–245
Source characterization, obsidian, systematic approach, overview, 523–525
Sourcing obsidian artifacts in western Mediterranean, 279–281, 283*f*
Sourcing specular hematite, discriminant function analysis, INAA data, 466, 469–477*t*
Southern Africa mines. *See* Botswana prehistoric mines
Spain. *See* Canary Islands; Castilian Kingdom; Iberian Peninsula
Specular hematite from mines in Botswana, 460–479
Specularite. *See* Specular hematite
Splash glazes in Mesopotamian pottery, 432, 440–441*f*
St. Barbara's field clays as source for Chapel bricks, 372–373
St. Mary's City, brick Chapel building materials, characterization, 364–375
Stable isotopes, principles and interpretation, 115–117
Standardization in protocol testing, LA-ICP-MS elemental analyses, Wari ceramics, 352–353
Statistical grouping, chemical analyses, Philistia pottery, 404–411
Strontium isotope analysis
  archaeological enamel and bone, 102–104
  by LA-TOF-ICP-MS, 298, 302, 304, 306–308
  investigation, geographic origins of trophy heads, 99–113

overview for archaeology, 100–102
Strontium isotope ratios in the Andes, 104–108
Sulfur materials from Icelandic Norse-trading site, simultaneous co-incident x-ray micro-fluorescence and micro-diffraction analyses, 204–205
Surfer, version 8.01, software, 225
Sycan Marsh and Silver Lake, Oregon, discriminating obsidian sources, 281–282, 283*f*
Synchronous data acquisition, TOF-ICP-MS, 299

T

Tepe Tula, Neolithic site in Iran, 138, 139*f*
Ternary alloy, copper, tin, and lead, coins, 250–255
Terpane distribution patterns, chromatograms from modern bitumen seep, archaeological bitumen and pottery samples, 142–146, 147*t*, 147*f*
Tetradrachms. *See* Tyrian shekels and half-shekels
Texas, Hinds Cave, coprolites, human mtDNA extraction in dietary reconstruction, 81
Textiles, prehistoric, from eastern North America, colorant classification, 15–43
Thermal ionization mass spectrometry (TIMS), Roman Carthage curse table analysis, 315, 318 , 326–332*t*
Thin-section micromorphology for archaeological soil and sediment analysis, overview, 195–197
Tigris-Euphrates drainage network, 436
Time and flexibility, archaeological textile colorant testing protocol, 38
Time-of-flight inductively coupled plasma-mass spectrometry. *See* TOF-ICP-MS
TIMS. *See* Thermal ionization mass spectrometry
Tin-lead glaze coating, opaque-white, on Majolica pottery, 377
Tipuj, Guatemala, obsidian frequencies, 518
Titicaca Basin Type obsidian, 534
TOF-ICP-MS, lead isotope analysis, ceramic glazes, 427–429
Tools from obsidian, properties, 522–523
Trace element abundances in curse tables, 332, 333*t*
Trace metal association with sand fraction in St Mary's County Chapel bricks, 369–370*f*
Trans coniferyl alcohol, 49–50*f*
Trans sinapyl alcohol, 49–50*f*
Transition metal model, discriminant function analysis, INAA data, specular hematite sourcing, 468–478
Trapped charge dating methods, 2–3
Trinidad de Nosotros in Central Petén lakes region, obsidian artifact location, 507–510
Trophy heads in central Andes, geographic origin, archaeological chemistry, 99–113
Tsodilo Hills mines, Botswana, specularite sourcing, 461, 462–464
Tuberculosis in Peruvian mummy, Chiribaya Alta site, by DNA studies, 82
Tunis, Tunisia. *See* Roman Carthage
Tunisian lead ore isotope database, 326–327*t*
Tyre, Roman period history, 269, 271, 273
Tyrian shekels and half-shekels, chemical compositions, 258–274

U

Utah, historic, pottery glazes, chemical analysis by LA-ICP-MS, 452–457
UV fluorescence photography, 24
UV reflected photography, 24
Uyo Uyo and Carylloma obsidian sources, Peru, 538

V

Variability, ochre artifacts from Jiskairumoko, 501—502
Varian ICP-MS, quadrupole mass spectrometer, 337–338, 351
Visual examination in colorant testing protocol, archaeological textiles, 36

W

Wari ceramics, analytical chemical groupings by LA-ICP-MS and INAA, 356–362
Wari heartland sites, food supply, 104
Wari realm, history, 350–351
Wari site, Conchopata, 99–113, 350–351
Western Mediterranean prehistory, stable isotopes and diet, 118–120
Western Mediterranean sourcing obsidian artifacts, 279–281, 283*f*
"Widow's mites" series, copper coins, elemental compositions, 246–257

X

X-ray diffraction (XRD)
  majolican pottery sherds from Gran Canaria Island, 383
  microfocus synchrotron x-ray analysis method, 198–199*f*
  Roman Carthage curse table analysis, 319, 332
  See also Small-angle x-ray scattering
X-ray fluorescence (XRF)
  field-portable equipment for obsidian sourcing, 506–521
  majolican pottery sherds, 380–391*f*
  microfocus synchrotron x-ray analysis method, 198–199*f*
  obsidian analysis from Peru, discriminating elements, 547, 548*f*–549*f*
  obsidian analysis procedure, 531–532
  obsidian chemical characterization, overview, 527–528
  obsidian sourcing, Silver Lake and Sycan Marsh, 281–282, 283*f*
  theory and application, 260–262*t*
X-ray photoelectron spectroscopy (XPS), surface characterization, black deposit from Little Lost River Cave, 153–164
Xaaga, Oaxaca, Mexico, ethnoarchaeological studies, 214
XPS. *See* X-ray photoelectron spectroscopy
XRD. *See* X-ray diffraction
XRF. *See* X-ray fluorescence

Y

Yanarangra obsidian source, Peru, 538

Z

Zaragoza, Mexico, obsidian source, 512
Zinc-tin binary diagram, metal concentrations, Matisse bronze sculptures, 346
Zirconium and rubidium concentrations, Trinidad de Nosotros obsidian, 516*f*